Encyclopaedia of Mathematical Sciences
Volume 49

Number Theory I

Yuri Ivanovic Manin
Alexei A. Panchishkin

Introduction to Modern Number Theory

Fundamental Problems, Ideas and Theories

Second Edition

Springer

Authors

Yuri Ivanovic Manin
Max-Planck-Institut für Mathematik
Vivatsgasse 7
53111 Bonn, Germany
e-mail: manin@mpim-bonn.mpg.de

Alexei A. Panchishkin
Université Joseph Fourier UMR 5582
Institut Fourier
38402 Saint Martin d'Hères, France
e-mail: alexei.pantchichkine@ujf-grenoble.fr

Founding editor of the Encyclopaedia of Mathematical Sciences:
R. V. Gamkrelidze

Original Russian version of the first edition
was published by VINITI, Moscow in 1990

The first edition of this book was published as Number Theory I,
Yu. I. Manin, A. A. Panchishkin (Authors), A. N. Parshin, I. R. Shafarevich (Eds.),
Vol. 49 of the Encyclopaedia of Mathematical Sciences

Mathematics Subject Classification (2000):
11-XX (11A, 11B, 11D, 11E, 11F, 11G, 11R, 11S, 11U, 11Y), 14-XX, 20-XX, 37-XX, 03-XX

ISSN 0938-0396
ISBN-10 3-540-20364-8 Springer Berlin Heidelberg New York
ISBN-13 978-3-540-20364-3 Springer Berlin Heidelberg New York

Springer is a part of Springer Science+Business Media GmbH
springeronline.com

Printed in The Netherlands

Typesetting: by the authors using a Springer LaTeX macro package
Cover Design: E. Kirchner, Heidelberg, Germany
Printed on acid-free paper 46/3142 sz 5 4 3 2 1 0

Preface

The present book is a new revised and updated version of "Number Theory I. Introduction to Number Theory" by Yu.I.Manin and A.A.Panchishkin, appeared in 1989 in Moscow (VINITI Publishers) [Ma-PaM], and in English translation [Ma-Pa] of 1995 (Springer Verlag).

The original book had been conceived as a part of a vast project, "Encyclopaedia of Mathematical Sciences". Accordingly, our task was to provide a series of introductory essays to various chapters of number theory, leading the reader from illuminating examples of number theoretic objects and problems, through general notions and theories, developed gradually by many researchers, to some of the highlights of modern mathematics and great, sometimes nebulous designs for future generations.

In preparing this new edition, we tried to keep this initial vision intact. We present many precise definitions, but practically no complete proofs. We try to show the logic of number-theoretic thought and the wide context in which various constructions are made, but for detailed study of the relevant materials the reader will have to turn to original papers or to other monographs. Because of lack of competence and/or space, we had to - reluctantly - omit many fascinating developments.

The new sections written for this edition, include a sketch of Wiles' proof of Fermat's Last Theorem, and relevant techniques coming from a synthesis of various theories of Part II; the whole Part III dedicated to arithmetical cohomology and noncommutative geometry; a report on point counts on varieties with many rational points; the recent polynomial time algorithm for primality testing, and some others subjects.

For more detailed description of the content and suggestions for further reading, see Introduction.

We are very pleased to express our deep gratitude to Prof. M.Marcolli for her essential help in preparing the last part of the new edition. We are very grateful to Prof. H.Cohen for his assistance in updating the book, especially Chapter 2. Many thanks to Prof. Yu.Tschinkel for very useful suggestions, remarks, and updates; he kindly rewrote §5.2 for this edition. We thank Dr.R.Hill and Dr.A.Gewirtz for editing some new sections of this edition, and St.Kühnlein (Universität des Saarlandes) for sending us a detailed list of remarks to the first edition.

Bonn, July 2004

Yu.I.Manin
A.A.Panchishkin

Contents

Part I Problems and Tricks

Part II Ideas and Theories

Part III Analogies and Visions

Introduction

Among the various branches of mathematics, number theory is characterized to a lesser degree by its primary subject ("integers") than by a psychological attitude. Actually, number theory also deals with rational, algebraic, and transcendental numbers, with some very specific analytic functions (such as *Dirichlet series* and *modular forms*), and with some geometric objects (such as *lattices* and *schemes over* $\mathbb{Z}$). The question whether a given article belongs to number theory is answered by its author's system of values. If arithmetic is not there, the paper will hardly be considered as number–theoretical, even if it deals exclusively with integers and congruences. On the other hand, any mathematical tool, say, homotopy theory or dynamical systems may become an important source of number–theoretical inspiration. For this reason, combinatorics and the theory of recursive functions are not usually associated with number theory, whereas modular functions are.

In this book we interpret number theory broadly. There are compelling reasons to adopt this viewpoint.

First of all, the integers constitute (together with geometric images) one of the primary subjects of mathematics in general. Because of this, the history of elementary number theory is as long as the history of all mathematics, and the history of modern mathematic began when "numbers" and "figures" were united by the concept of coordinates (which in the opinion of I.R.Shafarevich also forms the basic idea of algebra, see [Sha87]).

Moreover, integers constitute the basic universe of discrete symbols and therefore a universe of all logical constructions conceived as symbolic games. Of course, as an act of individual creativity, mathematics does not reduce to logic. Nevertheless, in the collective consciousness of our epoch there does exist an image of mathematics as a potentially complete, immense and precise logical construction. While the unrealistic rigidity of this image is well understood, there is still a strong tendency to keep it alive. The last but not the least reason for this is the computer reality of our time, with its very strict demands on the logical structure of a particular kind of mathematical production: software.

It was a discovery of our century, due to Hilbert and Gödel above all, that the properties of integers are general properties of discrete systems and therefore properties of the world of mathematical reasoning. We understand now that this idea can be stated as a theorem that provability in an arbitrary finitistic formal system is equivalent to a statement about decidability of a system of Diophantine equations (cf. below). This paradoxical fact shows that number theory, being a small part of mathematical knowledge, potentially embraces all this knowledge. If Gauss' famous motto on arithmetic *) needs justification, this theorem can be considered as such.

We had no intention of presenting in this report the whole of number theory. That would be impossible anyway. Therefore, we had to consider the usual choice and organization problems. Following some fairly traditional classification principles, we could have divided the bulk of this book into the following parts:

1. Elementary number theory.
2. Arithmetic of algebraic numbers.
3. Number-theoretical structure of the continuum (approximation theory, transcendental numbers, geometry of numbers Minkowski style, metric number theory etc.).
4. Analytic number theory (circle method, exponential sums, Dirichlet series and explicit formulae, modular forms).
5. Algebraic-geometric methods in the theory of Diophantine equations.
6. Miscellany ("wastebasket").

We preferred, however, a different system, and decided to organize our subject into three large subheadings which shall be described below. Because of our incompetence and/or lack of space we then had to omit many important themes that were initially included into our plan. We shall nevertheless briefly explain its concepts in order to present in a due perspective both this book and subsequent number-theoretical issues of this series.

Part I. *Problems and Tricks*

The choice of the material for this part was guided by the following principles.

In number theory, like in no other branch of mathematics, a bright young person with a minimal mathematical education can sometimes work wonders using inventive tricks. There are a lot of unsolved elementary problems waiting

"... Mathematik ist die Königin von Wissenschaften und Arithmetik die Königin von Mathematik. ...in allen Relationen sie wird zum ersten Rank erlaubt." -Gauss. ..., cf. e.g. `http://www.geocities.com/RainForest/Vines/2977/gauss/deutsch/quotes.html` ("Mathematics is the queen of sciences and arithmetic the queen of mathematics. She often condescends to render service to astronomy and other natural sciences, but in all relations she is entitled to the first rank." -Gauss. Sartorius von Walterhausen: Gauss zum Gedächtniss. (Leipzig, 1856), p.79.)

for fresh approaches. Of course, good taste is still necessary, and this comes with long training. Also, nobody can tell a priori that, say, the ancient problem on the pairs of "friendly numbers" is a bad one, while the Fermat conjecture is a beauty but it cannot be approached without seriously developed technique.

Elementary number theory consists of many problems, posed, solved and developed into theorems in the classical literature (Chapter 1), and also of many tricks which subsequently grew into large theories. The list of such tricks is still growing, as Apéry's proof of the irrationality of $\zeta(3)$ shows. Any professional mathematician can gain by knowing some of these stratagems.

In order not to restrict ourselves to very well known results we emphasize algorithmic problems and such modern applications of number theory as public key cryptography (Chapter 2). In general, the number-theoretical methods of information processing, oriented towards computer science (e.g. the fast Fourier transform) have revitalized the classical elementary number theory.

Part II. *Ideas and Theories*

In this part we intended to explain the next stage of the number-theoretical conceptions, in which special methods for solving special problems are systematized and axiomatized, and become the subject-matter of monographs and advanced courses.

From this vantage point, the elementary number theory becomes an imaginary collection of all theorems which can be deduced from the Peano axioms, of which the strongest tool is the induction axiom. It appears in such a role in meta-mathematical investigations and has for several decades been developed as a part of mathematical logic, namely the theory of recursive functions. Finally, since the remarkable proof of *Matiyasevich's theorem*, a further accomplished number-theoretical fragment has detached itself from this theory – the theory of Diophantine sets.

A Diophantine set is any subset of natural numbers that can be defined as a projection of the solution set of a system of polynomial equations with integral coefficients. The Matiyasevich theorem says that any set generated by an algorithm (technically speaking, enumerable or listable) is actually Diophantine. In particular, to this class belongs the set of all numbers of provable statements of an arbitrary finitely generated formal system, say, of axiomatized set–theoretical mathematics (Chapter 3).

The next large chapter of modern arithmetic (Chapter 4) is connected with the extension of the domain of integers to the domain of algebraic integers. The latter is not finitely generated as a ring, and only its finitely generated subrings consisting of all integers of a finite extension of $\mathbb{Q}$ preserve essential similarity to classical arithmetic. Historically such extensions were motivated by problems stated for $\mathbb{Z}$, (e.g. the Fermat conjecture, which leads to the divisibility properties of cyclotomic integers). Gradually however an essentially new object began to dominate the picture – the fundamental symmetry group of number theory $\mathrm{Gal}(\overline{\mathbb{Q}}/\mathbb{Q})$. It was probably Gauss who first understood this clearly. His earliest work on the construction of regular polygons by ruler-and-compass methods already shows that this problem is governed not by the visible symmetry of the figure but by the well–hidden Galois symmetry. His subsequent concentration on the quadratic reciprocity law (for which he suggested seven or eight proofs!) is striking evidence that he foresaw its place in modern class–field theory. Unfortunately, in most modern texts devoted to elementary number theory one cannot find any hint of explanation as to why quadratic reciprocity is anything more than just a curiosity. The point is that primes, the traditional subject matter of arithmetic, have another avatar as Frobenius elements in the Galois group. Acting as such upon algebraic numbers, they encode in this disguise of symmetries much more number-theoretical information than in their more standard appearance as elements of $\mathbb{Z}$.

The next two chapters of this part of our report are devoted to algebraic-geometric methods, zeta–functions of schemes over $\mathbb{Z}$, and modular forms. These subjects are closely interconnected and furnish the most important technical tools for the investigation of Diophantine equations.

For a geometer, an algebraic variety is the set of all solutions of a system of polynomial equations defined, say, over the complex numbers. Such a variety has a series of invariants. One starts with topological invariants like dimension and (co)homology groups; one then takes into account the analytic invariants such as the cohomology of the powers of the canonical sheaf, moduli etc. The fundamental idea is that these invariants should define the qualitative features of the initial Diophantine problem, for example the possible existence of an infinity of solutions, the behaviour of the quantity of solutions of bounded size etc. (see Chapter 5). This is only a guiding principle, but its concrete realizations belong to the most important achievements of twentieth century number theory, namely A.Weil's programme and its realization by A.Grothendieck and P.Deligne, as well as G.Faltings' proof of the *Mordell conjecture.*

Zeta–functions (see Chapter 6) furnish an analytical technique for refining qualitative statements to quantitative ones. The central place here belongs to the so called *"explicit formulae"*. These can be traced back to Riemann who in his famous memoir discovered the third avatar of primes – zeroes of Riemann's zeta function. Generally, arithmetical functions and zeroes of various zetas are related by a subtle duality. Proved or conjectured properties of the zeroes are translated back to arithmetic by means of the explicit formulae. This duality lies in the heart of modern number theory.

Modular forms have been known since the times of Euler and Jacobi. They have been used to obtain many beautiful and mysterious number-theoretical results. Simply by comparing the Fourier coefficients of a theta-series with its decomposition as a linear combination of Eisenstein series and cusp forms, one obtains a number of remarkable identities. The last decades made us aware that modular forms, via Mellin's transform, also provide key information about the analytic properties of various zeta–functions.

The material that deserved to be included into this central part of our report is immense and we have had to pass in silence over many important developments. We have also omitted some classical tools like *the Hardy–Littlewood circle method* and the *Vinogradov method of exponential sums.* These were described elsewhere (see [Vau81-97], [Kar75], ...). We have said only a few words on Diophantine approximation and transcendental numbers, in particular, the Gelfond–Baker and the Gelfond–Schneider methods (see [FelNes98], [Bak86], [BDGP96], [Wald2000], [Ch-L01], [Bo90]...).

The Langlands program strives to understand the structure of the *Galois group* of all algebraic numbers and relates in a series of deep conjectures the representation theory of this group to zeta–functions and modular forms.

Finally, at the end of Part II we try to present a comprehensive exposition of Wiles' marvelous proof of Fermat's Last Theorem and the Shimura–Taniyama–Weil conjecture using a synthese of several highly developed theories such as algebraic number theory, ring theory, algebraic geometry, the theory of elliptic curves, representation theory, Iwasawa theory, and deformation theory of Galois representations. Wiles used various sophisticated techniques and ideas due to himself and a number of other mathematicians (K.Ribet, G.Frey, Y.Hellegouarch, J.–M.Fontaine, B.Mazur, H.Hida, J.–P.Serre, J.Tunnell, ...). This genuinely historic event concludes a whole epoque in number theory, and opens at the same time a new period which could be closely involved with implementing the general Langlands program. Indeed, the Taniyama–Weil conjecture may be regarded as a special case of Langlands' conjectural correspondence between arithmetical algebraic varieties (motives), Galois representations and automorphic forms.

Part III. *Analogies and Visions*

This part was conceived as an illustration of some basic intuitive ideas that underlie modern number–theoretical thinking. One subject could have been called *Analogies between numbers and functions.* We have included under this heading an introduction to Non–commutative geometry, Arakelov geometry, Deninger program, Connes' ideas on Trace formula in noncommutative Geometry and the zeros of the Riemann zeta function ... Note also the excellent book [Huls94] which intends to give an overview of conjectures that dominate arithmetic algebraic geometry. These conjectures include the Beilinson conjectures, the Birch-Swinnerton-Dyer conjecture, the Shimura-Taniyama-Weil and the Tate conjectures, ... Note also works [Ta84], [Yos03], [Man02],[Man02a] on promising developments on Stark's conjectures.

In Arakelov theory a completion of an arithmetic surface is achieved by enlarging the group of divisors by formal linear combinations of the "closed fibers at infinity". The dual graph of any such closed fiber can be described in terms of an infinite tangle of bounded geodesics in a hyperbolic handlebody endowed with a Schottky uniformization. In the last Chapter 8, largerly based on a recent work of Caterina Consani and Matilde Marcolli, we consider arithmetic surfaces over the ring of integers in a number field, with fibers of genus $g \geq 2$. One can use Connes' theory to relate the hyperbolic geometry to Deninger's Archimedean cohomology and the cohomology of the cone of the local monodromy N at arithmetic infinity.

We use the standard system of cross–referencing in this book.

Suggestions for further reading

A number of interesting talks on Number Theory can be found in the proceedings of the International Congresses of Mathematicians in Beijing, 2002, in Berlin, 1998 and in Zürich, 1994 (see [ICM02], [ICM98], [ICM94]).

A quite complete impression on development of number-theoretic subjects can be obtained from Bourbaki talks : [Des90], [Bert92], [Fon92], [Oe92], [Clo93], [Se94], [Bo95], [Se95], [Oe95], [Goo96], [Kon96] [Loe96], [Wald96], [Abb97], [Fal98], [Mich98], [Colm2000], [Breu99], [Ma99], [Edx2000], [Ku2000], [Car02], [Hen01], [Pey02], [Pey04], [Coa01], [Colm01], [Colm03], [Bi02].

For a more detailed exposition of the theory of algebraic numbers, of Diophantine geometry and of the theory of Transcendental numbers we refer the reader to the volumes Number Theory II, III, and IV of Encyclopaedia of Mathematical Sciences see [Koch97], [La91], [FelNes98], the excellent monograph by J.Neukirch [Neuk99] (completed by [NSW2000]). We recommend also Lecture Notes [CR01] on Arithmetic algebraic geometry from Graduate Summer School of the IAS/Park City Mathematics Institute.

Acknowledgement

We are very grateful to the Institut Fourier (UJF, Grenoble-1) and to the Max-Planck-Institut für Mathematik (Bonn) for the permanent excellent working conditions and atmosphere.

Many thanks to Mrs. Ruth Allewelt, to Dr.Catriona M.Byrne, and to Dr. Martin Peters (Springer Verlag) for their stimulation of our work and for a lot of practical help.

Part I

Problems and Tricks

1

Number Theory

1.1 Problems About Primes. Divisibility and Primality

1.1.1 Arithmetical Notation

The usual decimal notation of natural numbers is a special case of *notation to the base m*. An integer n is written to the base m if it is represented in the form

$$n = d_{k-1}m^{k-1} + d_{k-2}m^{k-2} + \cdots + d_0$$

where $0 \leq d_i \leq m-1$. The coefficients d_i are called *m–ary digits* (or simply digits). Actually, this name is often applied not to the numbers d_i but to the special signs chosen to denote these numbers. If we do not want to specify these signs we can write the m–ary expansion as above in the form $n = (d_{k-1}d_{k-2}\dots d_1 d_0)_m$. The number of digits in such a notation is

$$k = [\log_m n] + 1 = [\log n/\log m] + 1$$

where [ ] denotes the integral part. Computers use the binary system; a binary digit (0 or 1) is called a *bit*. The high school prescription for the addition of a k-bit number and an l–bit number requires $\max(k,l)$ bit–operations (one bit–operation here is a Boolean addition and a carry). Similarly, multiplication requires $\leq 2kl$ bit–operations (cf. [Knu81], [Kob94]). The number of bit–operations needed to perform an arithmetical operation furnishes an estimate of the computer working time (if it uses an implementation of the corresponding algorithm). For this reason, fast multiplication schemes were invented, requiring only $O(k \log k \log\log k)$ bit–operations for the multiplication of two $\leq k$–bit numbers, instead of $O(k^2)$, cf. [Knu81]. One can also obtain a lower bound: there exists no algorithm which needs less than sous certaines restrictions naturelles on peut démontrer qu'il n'existe pas d'algorithme de multiplication des nombres à k chiffres avec le temps d'exécution inférieur à $(k \log k/(\log\log k)^2)$ bit–operations for the multiplication of two general $\leq k$–bit numbers.

Notice that in order to translate the binary expansion of a number n into the m–ary expansion one needs $O(k^2)$ bit–operations where $k = \log_2 n$. In fact, this takes $O(k)$ divisions with remainder, each of which, in turn, requires $O(kl)$ bit–operations where $l = \log_2 m$.

We have briefly discussed some classical examples of *algorithms*. These are explicitly and completely described procedures for symbolic manipulation (cf. [Mar54], [GJ79], [Man80], [Ma99]). In our examples, we started with the binary expansions of two integers and obtained the binary expansion of their sum or product, or their m–ary expansions. In general, an algorithm is called *polynomial* if the number of bit–operations it performs on data of binary length L is bounded above by a polynomial in L. The algorithms just mentioned are all polynomial (cf. [Kob94], [Knu81], [Ma99], [Ries85]).

1.1.2 Primes and composite numbers

The following two assertions are basic facts of number theory: a) every natural number $n > 1$ has a unique factorization $n = p_1^{a_1} p_2^{a_2} \dots p_r^{a_r}$ where $p_1 < p_2 \cdots < p_r$ are primes, $a_i > 0$; b) the set of primes is infinite.

Any algorithm finding such a factorization also answers a simpler question: is a given integer prime or composite? Such *primality tests* are important in themselves. The well known *Eratosthenes sieve* is an ancient (3rd century B.C.) algorithm listing all primes $\leq n$. As a by–product, it furnishes the smallest prime dividing n and is therefore a primality test. As such, however, it is quite inefficient since it takes $\geq n$ divisions, and this depends exponentially on the binary length of n. Euclid's proof that the set of primes cannot be finite uses an *ad absurdum* argument: otherwise the product of all the primes augmented by one would have no prime factorization. A more modern proof was given by Euler: the product taken over all primes

$$\prod_p \left(1 - \frac{1}{p}\right)^{-1} = \prod_p \left(1 + \frac{1}{p} + \frac{1}{p^2} + \dots\right) \qquad (1.1.1)$$

would be finite if their set were finite. However, the r.h.s. of (1.1.1) reduces to the divergent harmonic series $\sum_{n=1}^{\infty} n^{-1}$ due to the uniqueness of factorization.

Fibonacci suggested a faster primality test (1202) by noting that the smallest non–trivial divisor of n is $\leq [\sqrt{n}]$ so that it suffices to try only such numbers (cf. [Wag86], [APR83]).

The next breakthrough in primality testing was connected with Fermat's little theorem (discovered in the seventeenth century).

Theorem 1.1 (Fermat's Little Theorem). *If n is prime then for any integer a relatively prime to n*

$$a^{n-1} \equiv 1 (\text{mod } n), \qquad (1.1.2)$$

(It means that n divides $a^{n-1}-1$). The condition (1.1.2) (with a fixed a) is necessary but generally not sufficient for n to be prime. If it fails for n, we can be sure that n is composite, without even knowing a single divisor of it. We call n *pseudoprime w.r.t.* a if $\gcd(a,n) = 1$ and (1.1.2) holds. Certain composite numbers $n = 561 = 3 \cdot 11 \cdot 17$, $1105 = 5 \cdot 13 \cdot 17$, $1729 = 7 \cdot 13 \cdot 19$ are pseudoprime w.r.t. all a (relatively prime to n). Such numbers are called *Carmichael numbers* (cf. [Kob94], [LeH.80]). Their set is infinite (it was proved in [AGP94]). For example, a square-free n is a Carmichael number iff for any prime p dividing n, $p-1$ divides $n-1$.

A remarkable property of (1.1.2) is that it admits a fast testing algorithm. The point is that large powers $a^m \bmod n$ can be readily computed by repeated squaring. More precisely, consider the binary representation of $n-1$:

$$m = n - 1 = d_{k-1}2^{k-1} + d_{k-2} + \cdots + d_0$$

with $d_{k-1} = 1$. Put $r_1 = a \bmod n$ and

$$r_{i+1} \equiv \begin{cases} r_i^2 \bmod n & \text{if } d_{k-1-i} = 0 \\ a r_i^2 \bmod n & \text{if } d_{k-1-i} = 1 \end{cases}$$

Then $a^{n-1} \equiv r_k \bmod n$ because

$$a^{n-1} = (\dots((a^{2+d_{k-2}})^2 a^{d_{k-3}})^2 \dots) a^{d_0}.$$

This algorithm is polynomial since it requires only $\le 3[\log_2 n]$ multiplications mod n to find r_k. It is an important ingredient of modern fast primality tests using the Fermat theorem, its generalizations and (partial) converse statements.

This idea was used in a recent work of M. Agrawal, N. Kayal and N. Saxena: a polynomial version of (1.1.2) led to a fast deterministic algorithm for primality testing (of polynomial time $\mathcal{O}(\log n)^{12+\varepsilon}$), cf. §2.2.4.

Fermat himself discovered his theorem in connection with his studies of the numbers $F_n = 2^{2^n} - 1$. He believed them to be prime although he was able to check this only for $n \le 4$. Later Euler discovered the prime factorization $F_5 = 4294967297 = 641 \cdot 6700417$. No new *prime Fermat numbers* have been found, and some mathematicians now conjecture that there are none.

The history of the search for large primes is also connected with the *Mersenne primes* $M_p = 2^p - 1$ where p is again a prime. To test their primality one can use the following *Lucas criterion*: $M_k (k \ge 2)$ is prime iff it divides L_{k-1} where L_n are defined by recurrence: $L_1 = 4$, $L_{n+1} = L_n^2 - 2$. This requires much less time than testing the primality of a random number of the same order of magnitude by a general method. Mersenne's numbers also arise in various other problems. Euclid discovered that if $2^p - 1$ is prime then $2^{p-1}(2^p - 1)$ is *perfect* i.e. is equal to the sum of its proper divisors (e.g. $6 = 1 + 2 + 3$, $28 = 1 + 2 + 4 + 7 + 14$, $496 = 1 + 2 + 4 + 5 + 16 + 31 + 62 + 124 + 248$), and Euler proved that all even perfect numbers are of this type. It is not known whether there are any odd perfect numbers, and this is one baffling example of a seemingly reasonable question that has not lead to any number-theoretical insights, ideas or tricks worth mentioning here.

Euler also knew the first eight prime *Mersenne numbers* (corresponding to $p = 2,\ 3,\ 5,\ 7,\ 13,\ 19,\ 31$. Recently computer-assisted primality tests have furnished many new Mersenne primes, e.g. the 42nd known Mersenne prime, discovered by Dr. Martin Nowak on February 26 (2005), is $2^{25,964,951}-1$. It has 7,816,230 decimal digits. It is therefore not only the largest known Mersenne prime, but also the largest known prime of any kind.*)

In Chapter 4 we consider some other modern methods of primality testing, in particular using elliptic curves (ECPP by Atkin–Morain).

1.1.3 The Factorization Theorem and the Euclidean Algorithm

For integers a, b we write $a|b$ if a divides b i.e., $b = ad$ for some integer d. If p is a prime and p^α is the highest power of p dividing n we write $p^\alpha \| n$ and $\alpha = \mathrm{ord}_p n$. The factorization theorem can be easily deduced from its special case: if a prime p divides ab then either $p|a$ or $p|b$. Below we shall prove this property using the *Euclidean algorithm.* Knowing the prime factorizations of a and b one readily sees the existence and the explicit form of the *greatest common divisor* $\gcd(a, b)$ and the *least common multiple* $\mathrm{lcm}(a, b)$. Namely, put $m_p = \min(\mathrm{ord}_p(a), \mathrm{ord}_p(b))$, $g_p = \max(\mathrm{ord}_p(a), \mathrm{ord}_p(b))$. Then

$$\gcd(a, b) = \prod_p p^{m_p}, \quad \mathrm{lcm}(a, b) = \prod_p p^{g_p}.$$

Again, the Euclidean algorithm allows us to prove the existence and to find efficiently $\gcd(a, b)$ without even knowing the prime factorizations. Assume that $a \ge b \ge 1$. The algorithm consists of calculating a sequence $x_0,\ x_1,\ x_2, \dots$ where $x_0 = a,\ x_1 = b$ and x_{i+1} is the residue of x_{i-1} modulo x_i. One stops when $x_k = 0$; then $x_{k-1} = \gcd(a, b)$. The number of required divisions is bounded by $5\log_{10}\max(a, b)$ (*Lamé's theorem*) (cf. [Knu81], [Wun85]). The slowest instances for the Euclidean algorithm are the neighbouring *Fibonacci numbers* $a = u_k,\ b = u_{k-1}$ where $u_0 = u_1 = 1$ and $u_{i+1} = u_i + u_{i-1}$. The Euclidean algorithm also furnishes a representation

$$\gcd(a, b) = Aa + Bb \tag{1.1.3}$$

where A, B are integers. In order to find these, we shall consecutively define pairs (A_i, B_i) such that $x_i = A_i x_0 + B_i x_1$. Put $A_0 = B_1 = 1,\ A_1 = B_0 = 0$ and for $i \ge 1$

* See `http://www.mersenne.org` and `http://mathworld.wolfram.com/news/` for updates and for the history, e.g. two previous values are 20996011 and 24036583. Another recent record is the factorization of M_{953} (Bahr F., Franke J. and Kleinjung T. (2002) (footnotes by Yu.Tschinkel and H.Cohen).

$$A_{i+1} = A_{i-1} - tA_i, \quad B_{i+1} = B_{i-1} - tB_i$$

where t is given by $x_{i+1} = x_{i-1} - tx_i$. Since $\gcd(a, b) = x_{k-1}$ we can take $A = A_{k-1}$, $B = B_{k-1}$. Finally, if $p|ab$ for a prime p and p does not divide a then $\gcd(a, p)=1$ so that $Aa + Bp = 1$ for some integers A, B. Hence $Aab + Bpb = b$ and p divides b.

1.1.4 Calculations with Residue Classes

From the algebraic viewpoint, the set of integers $\mathbb{Z}$ is an associative commutative ring with identity. The general divisibility theory in such rings uses the fundamental notion of an *ideal.* An ideal I in a ring R is a subset which is an additive subgroup with the property $RIR \subset I$.

An ideal of the form $I = aR$, $a \in A$ is called a *principal ideal* and is denoted (a). The divisibility relation $a|b$ is equivalent to the inclusion relation

$$(b) \subset (a) \quad \text{or} \quad b \in (a).$$

Any ideal I of $\mathbb{Z}$ must be principal since its elements are all divisible by the smallest positive element of I. The maximal ideals (ordered by inclusion) are precisely those which are generated by primes. The numbers having the same remainder after division by a fixed N, form N classes with pairwise empty intersections

$$\bar{a} = a + N\mathbb{Z}, \quad 0 \le a \le N - 1,$$

the set of which also has a natural commutative associative ring structure with identity

$$\mathbb{Z}/N\mathbb{Z} = \mathbb{Z}/(N) = \{\bar{0}, \bar{1}, \dots, \overline{N-1}\}.$$

We traditionally write $a \equiv b \pmod N$ in place of $\bar{a} = \bar{b}$. Often one succeeds in reducing some calculations in $\mathbb{Z}$ to calculations in an appropriate *residue ring* $\mathbb{Z}/N\mathbb{Z}$. Besides finiteness, one useful property of this ring is the abundance of invertible elements (while in $\mathbb{Z}$ there are only ± 1). Actually, $\bar{a}$ is invertible iff $\gcd(a, N) = 1$ since the equation $ax + Ny = 1$ or, equivalently, $\bar{a}.\bar{x} = \bar{1}$ can be solved exactly in this case with integers x, y. The group of all invertible residue classes is denoted $(\mathbb{Z}/N\mathbb{Z})^\times$. Its order $\varphi(N)$ is called *Euler's function.* Euler introduced it in connection with his generalization of the Fermat theorem:

$$a^{\varphi(N)} \equiv 1 (\bmod\ N) \tag{1.1.4}$$

for any a relatively prime to N, i.e. $\bar{a}^{\varphi(N)} = \bar{1}$ for any invertible element $\bar{a}$ in $\mathbb{Z}/N\mathbb{Z}$. Euler's conceptual proof shows in fact that in a finite Abelian group of order f the order of an arbitrary element a divides f. In fact, the multiplication by a is a permutation of the set of all elements. The product of all elements is multiplied by a^f under this map. Hence $a^f = 1$.

If $N = N_1 N_2 \dots N_k$ and N_i are pairwise coprime we have a canonical isomorphism

$$\mathbb{Z}/N\mathbb{Z} \cong \mathbb{Z}/N_1\mathbb{Z} \oplus \cdots \oplus \mathbb{Z}/N_k\mathbb{Z}. \tag{1.1.5}$$

The main part of this statement is called *the Chinese Remainder Theorem* : for any $a_i \bmod N_i$, $i = 1, \ldots, k$ there exists an a such that $a \equiv a_i \bmod N_i$ for all i. Again, such an a can be efficiently found using the Euclidean algorithm. Put $M_i = N/N_i$. By assumption, M_i and N_i are relatively prime. Find X_i with $X_i M_i \equiv 1 \bmod N_i$ and put

$$a = \sum_{i=1}^{k} a_i X_i M_i. \tag{1.1.6}$$

This is what we sought. From (1.1.5) we deduce the corresponding factorization of the multiplicative group

$$(\mathbb{Z}/N\mathbb{Z})^\times \cong (\mathbb{Z}/N_1\mathbb{Z})^\times \oplus \cdots \oplus (\mathbb{Z}/N_k\mathbb{Z})^\times, \tag{1.1.7}$$

which shows in particular that $\varphi(N) = \varphi(N_1)\ldots\varphi(N_k)$. Since for a prime p we have $\varphi(p^a) = p^{a-1}(p-1)$ this allows us to find $\varphi(N)$ given the prime factorization of N.

In the special case when $N = q$ is prime the ring $\mathbb{Z}/N\mathbb{Z}$ is a field: all its non–zero elements are invertible. For a prime p, the notation $\mathbb{F}_p$ is used for the field $\mathbb{Z}/p\mathbb{Z}$. The group $(\mathbb{Z}/N\mathbb{Z})^\times$ is *cyclic*: it coincides with the set of all powers of an element $t = t_q$ (it is not unique). No efficient (e.g. polynomial) algorithm for finding such a *primitive root* is known.

Recall *Artin's conjecture* (on primitive roots): If $a \in \mathbb{Z}$ is not -1 or a perfect square, then the number $N(x, a)$ of primes $p \le x$ such that a is a primitive root modulo p is asymptotic to $C(a)\pi(x)$, where $C(a)$ is a constant that depends only on a. In particular, there are infinitely many primes p such that a is a primitive root modulo p. (Note that another famous *Artin's conjecture* (on the holomorphy of L series) will be discussed in §6.4.5). Nobody has proved this conjecture (on primitive roots) for even a single choice of a. There are partial results, e.g., that there are infinitely many p such that the order of a is divisible by the largest prime factor of $p-1$. (See, e.g., [Mor93] and [HB86], [BrGo02]). Neither can one efficiently compute the "discrete logarithm", (or *index*) $x = \mathrm{ind}_t(a)$ defined for an invertible $a \bmod q$ by

$$a \equiv t^x \bmod q, \quad 0 \le x \le q-1. \tag{1.1.8}$$

It is an important unanswered question whether such algorithms exist at all. However, there are fast ways for calculating ind_t if all prime divisors of $q-1$ are small (cf. [Kob94]). First of all, one computes for all p dividing $q-1$ the residue classes

$$r_{p,j} = t^{j(q-1)/p}, \qquad j = 0, 1, \ldots, p-1$$

lying in $(\mathbb{Z}/q\mathbb{Z})^\times$. This can be efficiently done by the *iterated squaring method* (cf. 1.1.2). Let $\alpha_p = \mathrm{ord}_p(q-1)$. It suffices to compute all the residues x mod

p^{α_p} and then to apply the Chinese Remainder Theorem (1.1.5). We fix p, $\alpha = \alpha_p > 0$ and try to to find $x \bmod p^\alpha$ in the form

$$x \equiv x_0 + x_1 p + \cdots + x_{\alpha-1} p^{\alpha-1} (\mathrm{mod}\ p^\alpha), \qquad 0 \le x_i \le p-1.$$

Since $a^{q-1} \equiv 1 \bmod q$ the residue $a^{(q-1)/p}$ is a pth root of unity. From $a \equiv t^x \bmod q$ it follows that

$$a^{(q-1)/p} \equiv t^{x(q-1)/p} \equiv t^{x_0(q-1)/p} \equiv r_{p,x_0} (\mathrm{mod}\ q).$$

Therefore we can find the first digit x_0 by computing $a^{(q-1)/p}$ and comparing it with the precomputed list of $r_{p,j}$. In order to find the next digit x_1 we first replace a by $a_1 = a/t^{x_0}$. Then we have

$$\mathrm{ind}_t(a_1) = \mathrm{ind}_t(a) - x_0 \equiv x_1 p + \cdots + x_\alpha p^{\alpha-1} (\mathrm{mod}\ p^\alpha).$$

As a_1 is a pth power we obtain from here $a_1^{(q-1)/p} \equiv 1 \bmod q$ and

$$a_1^{(q-1)/p^2} \equiv t^{(x-x_0)(q-1)/p^2} \equiv t^{(x_1+px_2+\dots)(q-1)/p} \equiv t^{x_1(q-1)/p} \equiv r_{p,x_1}.$$

Therefore, one can discover x_1 by finding $a_1^{(q-1)/p^2}$ among the precomputed list of $r_{p,j}$. One computes the other digits x_i in the same way. The same list can be used for various a's, q and t being fixed. This is the *Silver–Pollig–Hellman algorithm*, cf. [Kob94]. It becomes impractical if $q-1$ is divisible by a large prime because then the table of $r_{p,j}$ becomes too long. The difficulty of computing ind (and the general factorization problem) is utilized in *cryptography* (cf. Chapter 2, §2.1.6, [DH76], [Hel79], [ARS78] , [Odl84] , [Odl87], [Go02]).

1.1.5 The Quadratic Reciprocity Law and Its Use

Let p and q be odd primes. The main part of the quadratic reciprocity law first proved by Gauss, states that if $p \equiv q \equiv 3 \bmod 4$ then the solvability of one of the congruences $x^2 \equiv p \bmod q$ and $x^2 \equiv q \bmod p$ implies the insolvability of the other; in all other cases they are simultaneously solvable or unsolvable. Gauss used this in order to compile large tables of primes.

To this end, he refined the primality test based on Fermat's congruence (1.1.2). Namely, define *the Legendre symbol* $\left(\frac{a}{n}\right)$ for a prime n by

$$\left(\frac{a}{n}\right) = \begin{cases} 0 & \text{if } a \equiv 0 \bmod n, \\ 1 & \text{if } a \equiv b^2 \bmod n, \\ -1 & \text{otherwise.} \end{cases}$$

Then from the cyclicity of $(\mathbb{Z}/n\mathbb{Z})^\times$ it follows that

$$a^{(n-1)/2} \equiv \left(\frac{a}{n}\right) (\mathrm{mod}\ n). \tag{1.1.9}$$

If n is not prime we define the *Jacobi symbol* by multiplicativity: for an odd positive $n = p_1 p_2 \dots p_k$ where p_i are primes, not necessarily distinct, put

$$\left(\frac{a}{n}\right) = \left(\frac{a}{p_1}\right) \dots \left(\frac{a}{p_k}\right) \tag{1.1.10}$$

Now formula (1.1.9), which holds for the Jacobi symbol when n is prime, can be used as a primality test. Actually, the Jacobi symbol can be extended to all values of the "numerator" and "denominator" and computed without knowing the prime factorization of n. This is done with the help of the *extended quadratic reciprocity law*

$$\left(\frac{Q}{P}\right)\left(\frac{P}{Q}\right) = (-1)^{\frac{(P-1)}{2}\frac{(Q-1)}{2}} \tag{1.1.11}$$

(for odd positive P Q) and two complements to this law:

$$\left(\frac{2}{P}\right) = (-1)^{(P^2-1)/8} \qquad \left(\frac{-1}{P}\right) = (-1)^{(P-1)/2} \tag{1.1.12}$$

together with the multiplicativity property with respect to both "numerator" and "denominator". The computation follows the same pattern as the Euclidean algorithm and requires $\leq \log \max(P, Q)$ divisions with remainder. A natural number n is called an *Eulerian pseudoprime w.r.t. a* if $\gcd(a, n) = 1$ and (1.1.9) holds. Using the chinese remainder theorem, one can prove that if n is pseudoprime w.r.t. all $a \in (\mathbb{Z}/n\mathbb{Z})^\times$ then n is prime. Thus, there are no Eulerian analogues of the Carmichael numbers. Moreover, it was argued in [Wag86] that if n is composite then there is an $a \leq 2\log n \log\log n$ such that n is not an Eulerian pseudoprime w.r.t. a.

The congruence (1.1.9) is used in the modern fast primality tests which will be considered in Chapter 2 (cf. [ARS78], [Mil76], [LeH.80], [Vas88]).

The primality tests work much faster than all known methods for factorizing *"random" large integers*, see §2.3.

To conclude this subsection we say a few words about a subject which has traditionally caught the attention of many unselfish amateurs of number theory: that of finding "a formula" for primes. Euler noticed that the polynomial $x^2 + x + 41$ takes many prime values. However, it was long known that the values of an arbitrary polynomial $f(x_1, \dots, x_n) \in \mathbb{Z}[x_1, \dots, x_n]$ at integer points *cannot all be prime*, e.g. because if p, q are two large primes, then the congruence $f(x_1, \dots, x_n) \equiv 0 \bmod pq$ is always solvable. Nevertheless, using methods from the theory of recursive functions, one can construct a polynomial (in fact, many) whose set of *positive values* taken at lattice points coincides with the set of all primes. The following specimen was suggested in [JSWW76]. It depends on 26 variables that can be conveniently denoted by the letters of the English alphabet:

$$\begin{aligned}
&F(a,b,c,d,e,f,g,h,i,j,k,l,m,n,o,p,q,r,s,t,u,v,w,x,y,z) = \\
&(k+2)\{1-[wz+h+j-q]^2-[(gk+2g+k+1)(h+j)+h-z]^2- \\
&[2n+p+q+z-e]^2-[16(k+1)^3(k+2)(n+1)^2+1-f^2]^2- \\
&[e^3(e+2)(a+1)^2+1-o^2]^2-[(a^2-1)y^2+1-x^2]^2- \\
&[16r^2y^4(a^2-1)+1-u^2]^2- \\
&[((a+u^2(u^2-a))^2-1)(n+4dy)^2+1-(x+cu)^2]^2- \\
&(n+l+v-y)^2-[(a^2-1)l^2+1-m^2]^2-(ai+k+1-l-i)^2- \\
&[p+l(a-n-1)+b(2an+2a-n^2-2n-2)-m]^2- \\
&[q+y(a-p-1)+s(2ap+2a-p^2-2p-2)-x]^2- \\
&[z+pl(a-p)+t(2ap-p^2-1)-pm]^2\}.
\end{aligned}$$

We also mention an inductive description of the sequence of all primes that can be derived by combinatorial reasoning (cf. [Gan71]):

$$p_{n+1} = [1 - \log_2 \alpha_n] \tag{1.1.13}$$

where

$$\alpha_n = \sum_{r=1}^{n} \sum_{1\le i_1<\dots<i_r\le n} \frac{(-1)^r}{2^{p_{i_1}\cdots p_{i_r}}-1}.$$

1.1.6 The Distribution of Primes

A first glance at a table of primes leaves an impression of chaos. For several centuries, mathematicians compiled large tables of primes in an attempt to see some order in them. Pell's table (1668) lists all primes not exceeding 10^5. Lehmer D.H. in [Leh56] published his well known tables containing all primes up to 10^7. In [PSW80] one can find all *Fermat pseudoprimes* $n < 25 \cdot 10^9$ verifying the congruence $2^{n-1} \equiv 1 \bmod n$.

Already the first tables allowed the experimental study of the statistical distribution of primes, which seemed to be more accessible at least asymptotically. Put

$$\pi(x) = \mathrm{Card}\{p \mid p \text{ prime} \le x\}.$$

The graph of this step function even up to $x = 100$ looks pretty regular. For $x \le 50000$ where the jumps are hidden by the scale, the regularity is striking (cf. Fig. 1.1 and 1.2).

Computing $x/\pi(x)$ we see that for large x it becomes close to $\log x$. One sees also from Table 1.1 that that when we multiply x by 10, then

$$\frac{10x}{\pi(10x)} \approx \frac{x}{\pi(x)} + \log 10, \text{ and } \log(10x) = \log(x) + \log 10 \approx \log(x) + 2,3.$$

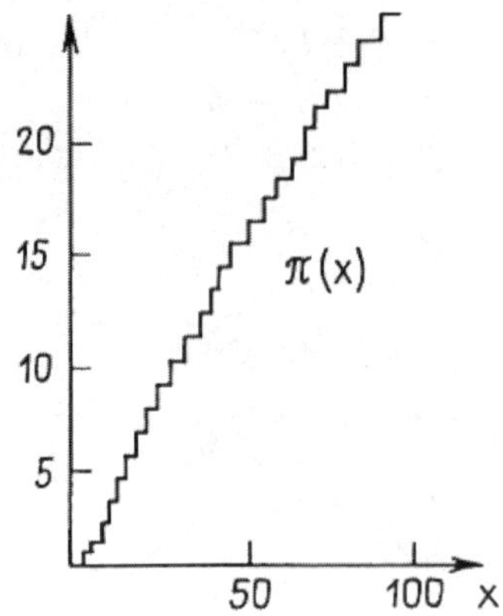

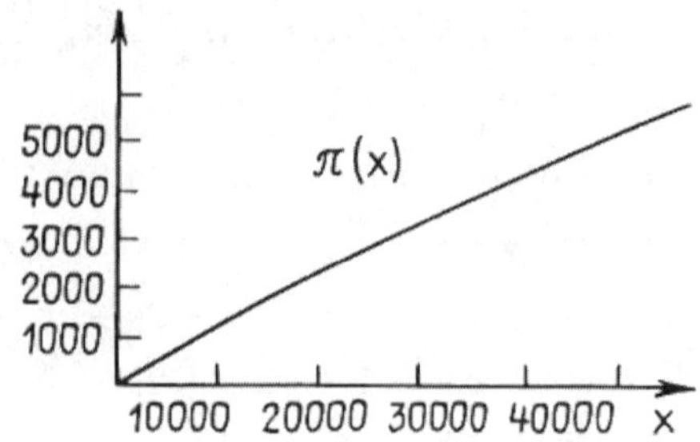

Fig. 1.1.

Fig. 1.2.

Table 1.1. For large x the ratio $x/\pi(x)$ becomes close to $\log x$:

10	4	2,5
100	25	4,0
1 000	168	6,0
10 000	1 229	8,1
100 000	9 592	10,4
1 000 000	78 498	12,7
10 000 000	664 579	15,0
100 000 000	5 761 455	17,4
1 000 000 000	50 847 534	19,7
10 000 000 000	455 052 512	22,0

Actually, the asymptotic law of the distribution of primes (or *prime number theorem*),

$$\pi(x) \sim \frac{x}{\log x} \tag{1.1.14}$$

(meaning that the quotient of the two sides tends to 1 as x tends to infinity) was conjectured by the fifteen year old Gauss on the basis of his studies of the available tables of primes, and proved by analytical methods only in 1896 by Hadamard and de la Vallée-Poussin [Pra57], [Kar75]). Before that, in 1850, P.L.Chebyshev (cf. [Cheby55]) found a very ingenious elementary proof of the inequality

$$0,89\frac{x}{\log x} < \pi(x) < 1,11\frac{x}{\log x}.$$

For this he used only the divisibility properties of the binomial coefficients. The asymptotic law itself was finally proved in an elementary way in 1949 by Selberg and Erdös (cf. [Sel51]).

Gauss also suggested a much better approximation to $\pi(x)$. Computing his tables of primes he noticed that if one counts primes in sufficiently large intervals around a large x their density tends to be close to $1/\log x$. For this reason he decided that a better approximation to $\pi(x)$ would be the integral logarithm

$$\mathrm{Li}(x) = \int_2^x \frac{dt}{\log t}.$$

This observation was refined by Riemann, cf. [Rie1858]. Investigating the zeta-function he came to an heuristic conclusion that $\mathrm{Li}(x)$ should be a very good approximation to the function counting all *powers of primes* $\leq x$ with the weight equal to the power, that is

$$\pi(x) + \frac{1}{2}\pi(\sqrt{x}) + \frac{1}{3}\pi(\sqrt[3]{x}) + \cdots \approx \mathrm{Li}(x). \tag{1.1.15}$$

If one wants to express $\pi(x)$ via $\mathrm{Li}(x)$ from here one should use *the Möbius function*

$$\mu(n) = \begin{cases} 1 & \text{if } n = 1, \\ 0 & \text{n is divisible by a square of a prime,} \\ (-1)^k & \text{otherwise,} \end{cases} \tag{1.1.16}$$

where k is the number of primes dividing n. Let us consider the function

$$F(x) = \sum_{n=1}^{\infty} \frac{1}{n}\pi(x^{1/n}). \tag{1.1.17}$$

Then

$$\pi(x) = \sum_{n=1}^{\infty} \frac{\mu(n)}{n} F(x^{1/n}), \tag{1.1.18}$$

and

$$\pi(x) \approx \sum_{n=1}^{\infty} \frac{\mu(n)}{n} \mathrm{Li}(x^{1/n}). \tag{1.1.19}$$

The special case (1.1.18) of a general inversion formula easily follows from the main property of the Möbius function:

$$\sum_{d|n} \mu(d) = \begin{cases} 1, & \text{if } n = 1 \\ 0, & \text{if } n > 1. \end{cases} \tag{1.1.20}$$

In fact, if $n = \prod_{i=1}^{s} p_i^{\alpha_i}$, $\alpha_i > 0$ then for $s \geq 1$ we have

$$\sum_{d|n} \mu(n) = \sum_{k=0}^{s} (-1)^s \binom{s}{k} = (1-1)^s = 0.$$

The right hand side of (1.1.19) is denoted $R(x)$. Table 1.2 (cf. [Ries85], [RG70], [Zag77]) shows how well it approximates $\pi(x)$.

Table 1.2.

x	$R(x)$	$\pi(x)$
100000000	5761455	5761552
200000000	11078937	11079090
300000000	16252325	16252355
400000000	21336326	21336185
500000000	26355867	26355517
600000000	31324703	31324622
700000000	36252931	36252719
800000000	41146179	41146248
900000000	46009215	46009949
1000000000	50847534	50847455

It is useful to slightly renormalize $\mathrm{Li}(x)$ taking instead the complex integral

$$\mathrm{li}(e^{u+iv}) = \int_{-\infty+iv}^{u+iv} \frac{e^z}{z} dz \qquad (v \neq 0). \tag{1.1.21}$$

For $x > 2$, $\mathrm{li}(x)$ differs from $\mathrm{Li}(x)$ by the constant $\mathrm{li}(2) \approx 1,045$. *The Riemann function*

$$R(x) = \sum_{n=1}^{\infty} \frac{\mu(n)}{n} \mathrm{li}(x^{1/n})$$

is an entire function of $\log x$. It can be expanded into a rapidly convergent power series

$$R(x) = 1 + \sum_{m=1}^{\infty} \frac{t^m}{m! m \zeta(m+1)}, \tag{1.1.22}$$

where $x = e^t$, and

$$\zeta(s) = \sum_{n=1}^{\infty} n^{-s} = \prod_{p \text{ prime}} (1 - p^{-s})^{-1}. \tag{1.1.23}$$

Of course, this *Riemann zeta function* is the main hero of the story. Its properties, established or conjectural, govern the behaviour of $\pi(x)$. Riemann showed how to extend $\zeta(s)$ meromorphically to the whole complex plane (notice that (1.1.23) converges only for $\mathrm{Re}(s) > 1$) and he deduced the astonishing *explicit formula* for $\pi(x)$. This looks as follows:

$$F_0(x) = \mathrm{li}(x) - \sum_{\rho} \mathrm{li}(x^{\rho}) + \int_0^{\infty} \frac{du}{(u^2-1)u\log u} - \log 2, \qquad (1.1.24)$$

where the sum is taken over all zeros ρ of $\zeta(s)$, and

$$F_0(x) = \lim_{\varepsilon \to 0} \frac{F(x+\varepsilon) + F(x-\varepsilon)}{2}.$$

The formula (1.1.24) was published by Riemann in 1859 and proved by Mangoldt in 1895. The series in (1.1.24) is only conditionally convergent. If one excludes the *"trivial zeroes"* $\rho = -2, -4, -6, \dots$ whose contribution is insignificant the remaining summation should be made in the order of increasing $|\rho|$. The set of non–trivial zeros is symmetric with respect to complex conjugation and lies in the critical strip $0 \le \mathrm{Re}(s) \le 1$. The first five roots with positive imaginary part, up to eight decimal digits, are (cf. [Zag77], [Ries85], [RG70])

$$\begin{aligned}
\rho_1 &= \frac{1}{2} + 14,134735i,\\
\rho_2 &= \frac{1}{2} + 21,022040i,\\
\rho_3 &= \frac{1}{2} + 25,010856i,\\
\rho_4 &= \frac{1}{2} + 30,424878i,\\
\rho_5 &= \frac{1}{2} + 32,935057i.
\end{aligned}$$

Let us consider the number $\theta = \sup \mathrm{Re}(\rho)$. From (1.1.24) it follows that

$$\pi(x) - \mathrm{li}(x) = O(x^{\theta}\log x). \qquad (1.1.25)$$

This estimate would be non–trivial if we knew that $\theta < 1$. Unfortunately, it is only known that there are no roots on $\mathrm{Re}(s) = 1$ and in a small neighbourhood of this line whose width tends to zero as $|s|$ grows (cf. [Pra57]). The famous *Riemann hypothesis*, that all non–trivial roots lie on the line $\mathrm{Re}(s) = \frac{1}{2}$, is still unproved. A corollary of this would be

$$\pi(x) = \mathrm{li}(x) + O(x^{1/2}\log x).$$

These questions, however, lie far outside elementary number theory.

We shall return to the *Riemann–Mangoldt* type *explicit formulae* below, cf. Part II, Chapter 6, §6.2.

1.2 Diophantine Equations of Degree One and Two

1.2.1 The Equation $ax + by = c$

In this section, all coefficients and indeterminates in various equations are assumed to be integers unless otherwise stated. Consider first a linear equation with two indeterminates. The set

$$I(a,b) = \{c \mid ax + by = c \quad \text{is solvable}\}$$

coincides with the ideal generated by a and b that is, with $d\mathbb{Z}$ where $d = \gcd(a,b)$. It follows that the equation

$$ax + by = c \tag{1.2.1}$$

is solvable iff d divides c. A special solution can be found with the help of the Euclidean algorithm: first compute X, Y with $aX + bY = d$ and then put $x_0 = eX, y_0 = eY$ where $e = c/d$. One easily sees that the general solution is given by the formula

$$x = x_0 + (b/d)t, \quad y = y_0 - (a/d)t,$$

where t is an arbitrary integer.

Equation (1.2.1) is the simplest example of the general Diophantine problem of investigating systems of polynomial equations

$$F_1(x_1, \dots, x_n) = 0, \quad \cdots, \quad F_m(x_1, \dots, x_n) = 0 \tag{1.2.2}$$

with integral coefficients. We see that all the main questions can be effectively answered for (1.2.1): the existence of solutions, computation a single solution, description of the set of all solutions, counting solutions in a box etc. We shall consider more complicated instances of (1.2.2) and attempt to extend these results.

1.2.2 Linear Diophantine Systems

The Euclidean algorithm allows us to investigate in the same way a general linear Diophantine system

$$Ax = b, \tag{1.2.3}$$

where

$$A = \begin{pmatrix} a_{11} & a_{12} & \cdots & a_{1n} \\ \cdots & \cdots & \ddots & \cdots \\ a_{m1} & a_{m2} & \cdots & a_{mn} \end{pmatrix} \in M_{m,n}(\mathbb{Z}), \quad x = \begin{pmatrix} x_1 \\ \cdots \\ x_n \end{pmatrix}, \quad b = \begin{pmatrix} b_1 \\ \cdots \\ b_m \end{pmatrix}.$$

This can be done with the help of the *elementary divisor theorem.* Recall that an elementary operation on the rows of a matrix over $\mathbb{Z}$ adds to one row an integral multiple of another. One defines an elementary column operation similarly. An elementary operation is equivalent to multiplication of the initial matrix on the left (resp. on the right) by a matrix of the form $E_{ij} = E + \lambda e_{ij}$ belonging to $\mathrm{SL}_m(\mathbb{Z})$ (resp. $\mathrm{SL}_n(\mathbb{Z})$). By repeated application of elementary operations we replace A by UAV where U and V are unimodular matrices with integral entries. On the other hand, the system

$$UAVy = Ub \tag{1.2.4}$$

is equivalent to (1.2.3) since their solutions are in one-to-one correspondence: $x = Vy$. We can use this if we manage to replace A by a simpler matrix $A' = UAV$. In fact, using the Euclidean algorithm and a version of the Gaussian elimination procedure avoiding divisions, one can find a matrix A' of the form

$$D = \begin{pmatrix} d_1 & 0 & 0 & \dots & 0 \\ 0 & d_2 & 0 & \dots & 0 \\ \dots & \dots & \ddots & \dots & 0 \\ 0 & 0 & \dots & d_r & \dots \\ \dots & \dots & \dots & \dots & \dots \end{pmatrix} = UAV. \tag{1.2.5}$$

Hence we either see that our system has no solutions even in $\mathbb{Q}$, or we obtain the set of all rational solutions from the very simple system $d_i y_i = c_i, \quad c = Ub$ for $i \leq r$, $y_i = 0$ for the other i. The set of integral solutions is non-empty iff d_i divides c_i for $i \leq r$, and can then be parametrized in an obvious way. The product $d_1 \cdot \dots \cdot d_i$ coincides with gcds of all minors of A of order i and $d_i | d_{i+1}$. They are called the *elementary divisors* of A. It follows that (1.2.3) is solvable iff the elementary divisors of A of orders $\leq m$ coincide with those of the extended matrix (with the column b added). In turn, this is equivalent to the simultaneous solvability of the congruences

$$Ax \equiv b (\mathrm{mod}\ N)$$

where N is an arbitrary integer. Such a condition can be readily extended to a completely general system of Diophantine equations. Clearly, it is necessary for the existence of a solution. The above argument shows that for (1.2.3) it is also sufficient. When this is true for a class of equations one says that *the Minkowski–Hasse principle* is valid for this class. The question of the validity of the *Minkowski–Hasse principle* is a central problem in this theory. We shall discuss it below in §1.2.4 and in Part II, §4.5, §5.3.

More difficult problems arise if one wants to find "the smallest solution" to (1.2.3) with respect to some norm. These questions are considered in the *geometry of numbers.* Siegel (cf. [Sie29], [Fel82]) has shown that the system of linear equations

$$a_{i1}x_1 + \dots + a_{in}x_n = 0 \qquad (i = 1, \dots, m)$$

with $n > m$ in which the integers a_{ij} are bounded by B has a non–trivial integral solution with coordinates bounded by $1 + (nB)^{m/(n-m)}$. If the rows of $A = (a_{ij})$ are linearly independent and d denotes the gcd of the minors of order m of A, one can obtain the more precise upper bound $(d^{-1}\sqrt{\det(A^tA)})^{1/(n-m)}$. This estimate and its generalization to algebraic number fields was proved by Bombieri and Vaaler (cf. [BV83]) using fairly subtle results from geometric number theory (Minkowski's theory of the successive minima of quadratic forms [Cas59a]).

For applications, it is essential to develop efficient methods for finding *solutions of a linear Diophantine system with non–negative coordinates.* This is the central problem of *integral linear programming.* It belongs to the class of intractable problems i.e. those for which polynomial algorithms are not known. The intractability of the *knapsack problem* has been used in *cryptography* (see Ch.2). It consists of finding a solution of the equation $a_1x_1 + \cdots + a_nx_n = b$ with $x_i \in \{0, 1\}$ where a_i, b are given integers (see [Kob94], [LeH.84]).

1.2.3 Equations of Degree Two

Consider the following Diophantine equation with integral coefficients

$$f(x_1, x_2, \ldots, x_n) = \sum_{i,j}^{n} a_{ij}x_ix_j + \sum_{i=1}^{n} b_ix_i + c = 0. \tag{1.2.6}$$

Here we shall begin by finding the set of all rational solutions, which is easier than finding the integral solutions but far from trivial.

A classical example is furnished by the *rational parametrization of the circle* $x^2 + y^2 = 1$:

$$x = \frac{2t}{1+t^2}\ y = \frac{1-t^2}{1+t^2} \quad (x = \cos\varphi,\ y = \sin\varphi,\ t = \tan\left(\frac{\varphi}{2}\right)). \tag{1.2.7}$$

This parametrization allows us in turn to describe all primitive Pythagorean triples (X, Y, Z), that is, natural solutions of $X^2+Y^2 = Z^2$ with gcd(X, Y, Z) $= 1$. The answer is: $X = 2uv$, $Y = u^2 - v^2$, $Z = u^2 + v^2$, where $u > v > 0$ are relatively prime integers. To prove this it suffices to put $t = u/v$ in (1.2.7).

Similarly, finding rational solutions to (1.2.6) is equivalent to finding integral solutions to the homogeneous equation

$$\begin{aligned} F(X_0, X_1, \cdots, X_n) &= \sum_{i,j=0}^{n} f_{ij}X_iX_j \\ &= \sum_{i,j=1}^{n} f_{ij}X_iX_j + 2\sum_{i,j=1}^{n} f_{i0}X_iX_0 + f_{00}X_0^2 \end{aligned} \tag{1.2.8}$$

where $f_{ij} = f_{ji} = a_{ij}/2$ for $1 \le i < j \le n$ and $f_{0i} = f_{i0} = b_i/2$ for $i = 1, 2, \ldots, n$, $f_{00} = c$. The *non–homogeneous coordinates* $x_1, \ldots, x_n$ are related to the *homogeneous coordinates* $X_0, \ldots, X_n$ by $X_i = x_i X_0 \quad (i = 1, 2, \ldots, n)$. The quadratic form $F(X)$ can be conveniently written as

$$F(X) = X^t A_F X, \quad X^t = (X_0, X_1, \ldots, X_n),$$

where $A_F = (f_{ij})$ is the matrix of coefficients. If there exists a non–trivial integral solution to $F(X) = 0$ we say that F *represents zero over* $\mathbb{Z}$. This equation defines a *quadric* Q_F. Its points are all complex solutions (except the trivial one) considered as points in the *complex projective space* $\mathbb{CP}^n$:

$$Q_F = \{(z_0 : z_1 : \cdots : z_n) \in \mathbb{CP}^n \quad | \quad F(z_0, z_1, \ldots, z_n) = 0\}.$$

Any non–trivial rational solution of $F(X) = 0$ gives a point on this quadric. If we know one solution X_0 then we can find all the others by considering intersections of Q_F with the (projective) lines defined over $\mathbb{Q}$ and containing X_0. Algebraically, a line passing through X^0 and Y^0 consists of all points $uX^0 + vY^0$. The equation $F(uX^0 + vY^0) = 0$ reduces to

$$uv \sum_{i=1}^{n} \frac{\partial F}{\partial X_i}(X^0) Y_i^0 + v^2 F(Y^0) = 0.$$

In general, not all the partial derivatives $\frac{\partial F}{\partial X_i}$ vanish at X^0. If this is the case, then for any Y^0 we can find an intersection point of Q_F with our line:

$$v = -u \sum_{i=1}^{n} \frac{\partial F}{\partial X_i}(X^0) Y_i^0 / F(Y^0). \tag{1.2.9}$$

(If by chance $F(Y^0) = 0$ then Y^0 is already on Q_F). Again, this point will in general be unique. Limiting cases can be well understood in geometric terms: if all partial derivatives vanish at X^0 then our quadric is a cone with vertex X^0, and the problem is reduced to that of finding rational points on the base of the cone, this base being a quadric of lower dimension; if a line happens to lie entirely on Q_F then all its rational points should be taken into account etc.

This *stereographic projection method*, applied to $x^2 + y^2 = 1$ and the point (0,-1) gives precisely (1.2.7) if one denotes by t a coefficient of the equation of the line passing through (0,-1) and (x, y) : $\quad y + 1 = tx$.

Considering the equation

$$F(X_0, X_1, \ldots, X_n) = 0 \tag{1.2.10}$$

(with F as in (1.2.8)) over the rationals, we could alternatively begin by diagonalizing F by a non–degenerate linear substitution $X = CY$ where $C \in \mathrm{M}_{n+1}(\mathbb{Q})$. The matrix C can be found effectively by Lagrange's method of successively completing the squares. The previous geometric analysis then becomes quite transparent.

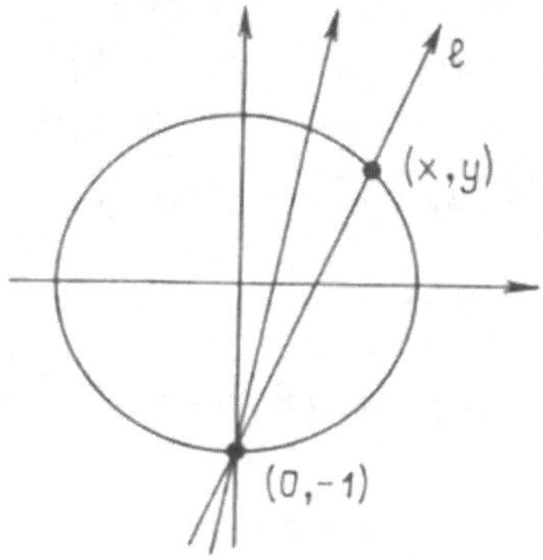

Fig. 1.3.

For homogeneous equations such as (1.2.10) the problems of finding solutions in $\mathbb{Q}$ and in $\mathbb{Z}$ are essentially equivalent. Since we can find all solutions starting from one of them, the key question is that of deciding whether there is one. An answer is given by the following result.

1.2.4 The Minkowski–Hasse Principle for Quadratic Forms

Theorem 1.2. *A quadratic form $F(x_1, x_2, \ldots, x_n)$ of rank n with integral coefficients represents zero over the rationals iff for any N, the congruence $F(x_1, \ldots, x_n) \equiv 0 \pmod N$ has a primitive solution and in addition F represents zero over the reals, i.e. it is indefinite.*

For a general proof see [BS85], [Cas78]. Of course, the necessity of this condition is obvious.

We reproduce here the beautiful proof of sufficiency in the case $n = 3$ due to Legendre ([BS85], [Ire82]). Let

$$F = a_1x_1^2 + a_2x_2^2 + a_3x_3^2 \qquad (a_1a_2a_3 \neq 0).$$

Since F is indefinite we may assume that the first two coefficients are positive while the third one is negative. Furthermore, we can and will assume that they are square-free and relatively prime: this may be achieved by obvious changes of variables and by dividing the form by the gcd of its coefficients. Denote the form with such properties by

$$ax^2 + by^2 - cz^2. \tag{1.2.11}$$

Consider a prime p dividing c. Since $F \equiv 0 (\mathrm{mod}\ p)$ has a primitive solution, we can find a non–trivial solution (x_0, y_0) to the congruence $ax^2 + by^2 \equiv 0 (\mathrm{mod}\ p)$. Therefore

$$ax^2 + by^2 \equiv ay_0^{-2}(xy_0 + yx_0)(xy_0 - yx_0) \pmod p.$$

For $p = 2$ we clearly have

$$ax^2 + by^2 - cz^2 \equiv (ax + by - cz)^2 \pmod 2.$$

Hence for all $p|2abc$ we can find linear forms $L^{(p)}$, $M^{(p)}$ of x, y, z with integral coefficients such that $F \equiv L^{(p)}M^{(p)} (\mathrm{mod}\ p)$. Using the Chinese Remainder Theorem, we find L (resp. M) with integral coefficients congruent to those of $L^{(p)}$ (resp. $M^{(p)}$) (mod p) for all $p|abc$. We then have

$$ax^2 + by^2 + cz^2 \equiv L(x, y, z)M(x, y, z) \pmod{abc}. \tag{1.2.12}$$

Consider now the integral points in the box

$$0 \le x < \sqrt{bc}, \quad 0 \le y < \sqrt{ac}, \quad 0 \le z < \sqrt{ab}. \tag{1.2.13}$$

If we exclude the trivial case $a = b = c = 1$, not all square roots are integers so that the total number of points will exceed the volume of this box which is abc. Hence there are two different points where L takes the same value mod abc. Taking their difference, we find

$$L(x_0, y_0, z_0) \equiv 0 \pmod{abc} \tag{1.2.14}$$

for some $|x_0| \le \sqrt{bc}$, $|y_0| \le \sqrt{ac}$, $|z_0| \le \sqrt{ab}$. Hence

$$ax_0^2 + by_0^2 - cz_0^2 \equiv 0 \pmod{abc} \tag{1.2.15}$$

and

$$-abc < ax_0^2 + by_0^2 - cz_0^2 < 2abc.$$

It follows that either

$$ax_0^2 + by_0^2 - cz_0^2 = 0 \tag{1.2.16}$$

or

$$ax_0^2 + by_0^2 - cz_0^2 = abc. \tag{1.2.17}$$

In the first case the theorem is proved. In the second case we obtain the following non–trivial solution

$$a(x_0 z_0 + by_0)^2 + b(y_0 z_0 - ax_0)^2 - c(z_0^2 + ab)^2 = 0.$$

Legendre's original statement is that $ax^2 + by^2 - cz^2 = 0$ is solvable iff all the residue classes $bc \pmod a$, $ac \pmod b$, $-ab \pmod c$ are squares.

One can prove that an indefinite quadratic form of rank ≥ 5 always represents zero over the rationals. For smaller rank, the Minkowski-Hasse principle can be combined with an a priori minimization of the moduli to be tested to give an effective way of establishing the existence of a solution. Below we shall reformulate this approach using the more convenient language of p-adic numbers (cf. Part II Chap. 4 §4.2.5, 4.3.1, Chap. 5 §5.3.6).

1.2.5 Pell's Equation

For non–homogeneous problems, the difference between rational and integral solutions becomes essential. For example, consider *Pell's equation*

$$x^2 - dy^2 = 1, \tag{1.2.18}$$

where d is a positive integer (and not a square). Since we know one trivial rational solution $(1,0)$ the others can be easily found by the method described above. However, to obtain only integral solutions we must act in a totally different way.

First of all, assume that the set of non–trivial integral solutions is non–empty (in fact, this can be proved by various methods). It is sufficient to consider only solutions with positive coordinates. We shall call such a solution (x_1, y_1) *minimal* if the linear form $x + \sqrt{d}y$ takes its minimal value on it. This solution is unique since $\sqrt{d}$ is irrational. The central result of the theory of Pell's equation states that all solutions are of the form $(\pm x_n, \pm y_n)$ where $x_n + \sqrt{d}y_n = (x_1 + \sqrt{d}y_1)^n$, n being an arbitrary non–negative integer.

The most natural proof, which admits a far-reaching generalization, is based on studying the *quadratic field* $K = \mathbb{Q}(\sqrt{d}) = \{a + b\sqrt{d} \mid a, b \in \mathbb{Q}\}$. The set $A = \mathbb{Z} + \mathbb{Z}\sqrt{d}$ is a *subring* in K. The *norm* of $\alpha = a + b\sqrt{d}$ is by definition

$$N(\alpha) = N_{K/\mathbb{Q}}(\alpha) = a^2 - db^2.$$

Clearly,

$$N(\alpha\beta) = N(\alpha)N(\beta) \tag{1.2.19}$$

for all $\alpha, \beta \in K$. Solutions of Pell's equation are numbers in α with norm 1. From (1.2.19) it follows that they form a *group* (with multiplication as the group law), in which the positive elements form the cyclic subgroup generated by $x_1 + y_1\sqrt{d}$.

In classical papers several methods were suggested for finding the minimal solution, or at least some solution. One of these algorithms is based on approximation theory (cf. §4 below). Dirichlet in 1837 published explicit formulae giving some solutions of Pell's equation expressed through trigonometric functions. For example, for $d = 13$ his general formulae show that $x_1 + y_1\sqrt{13} = \eta^2$ where

$$\eta = \frac{\sin\frac{2\pi}{13}\sin\frac{5\pi}{13}\sin\frac{6\pi}{13}}{\sin\frac{\pi}{13}\sin\frac{3\pi}{13}\sin\frac{4\pi}{13}} \in \mathbb{Q}(\sqrt{13})$$

(cf. [Dir68], [BS85], [Maz83]). In 1863 Kronecker published an expression for $x_1 + y_1\sqrt{d}$ via special values of *elliptic functions* (cf. [Kr1863], [Sie65], [Wei76]).

Finally, it is worth mentioning that a general quadratic Diophantine equation in two variables over the integers may be reduced by linear substitutions to a Pell-like equation if one solution is known.

A solution of Pell's equation using continued fractions is described in §1.4.5.

1.2.6 Representation of Integers and Quadratic Forms by Quadratic Forms

Consider two quadratic forms with integral coefficients

$$f(x) = f(x_1, \ldots, x_n) = \sum_{i,j=1}^{n} a_{ij} x_i x_j = A[x] = x^t A x,$$

$$g(y) = g(y_1, \ldots, y_m) = \sum_{i,j=1}^{m} b_{ij} y_i y_j = B[y] = y^t B y,$$

where A and B are symmetric matrices. We shall say that f *represents* g *over* $\mathbb{Z}$ if for some $C \in M_{n,m}(\mathbb{Z})$ we have

$$f(Cy) = g(y), \quad \text{or, equivalently} \quad A[C] = B. \tag{1.2.20}$$

In particular, for $m = 1$ and $g(y) = by^2$, f represents g iff $f(c_1, \ldots, c_n) = b$ for some integers $c_1, \ldots, c_n$.

Pell's equation considered above is a special case of the much more difficult general problem of representing integers and quadratic forms by quadratic forms. We shall sketch below some results and approaches to this vast domain.

Lagrange proved that every positive integer is a sum of four squares. A more difficult result due to Gauss states that $b > 0$ is a sum of three integer squares iff it is not of the form $4^k(8l - 1)$, $k, l \in \mathbb{Z}$. Lagrange's theorem can be easily deduced from this fact (cf. [Se70], [Cas78]).

Put

$$r_k(n) = \mathrm{Card}\{(n_1, \ldots, n_k) \in \mathbb{Z}^k \mid n_1^2 + \cdots + n_k^2 = n\}. \tag{1.2.21}$$

For example, $r_2(5) = 8$, as one may convince oneself by listing all solutions. There exist many formulae for this arithmetical function (cf. a vast bibliography in [Kog71]). Most of them are descendants of the classical formula of Jacobi ([Mum83], [Se70], [And76]):

$$r_4(n) = \begin{cases} 8 \sum_{d|n} d, & \text{if } n \text{ is odd,} \\ 24 \sum_{\substack{d|n \\ d \equiv 1(2)}} d, & \text{if } n \text{ is even.} \end{cases} \tag{1.2.22}$$

The proof is based on a study of the generating function for the sequence $r_k(n)$, that is, the series

$$\sum_{n=0}^{\infty} r_k(n) q^n = \sum_{(n_1, \ldots, n_k) \in \mathbb{Z}^k} q^{n_1^2 + n_2^2 + \ldots n_k^2} = \theta(\tau)^k$$

where

$$\theta(\tau) = \sum_{n \in \mathbb{Z}} q^{n^2}, \qquad q = e^{2\pi i \tau}. \tag{1.2.23}$$

This *theta–function* is a holomorphic function on the complex upper half–plane $H = \{\tau \in \mathbb{C} \mid \mathrm{Im}(\tau) > 0\}$. It has many remarkable analytic properties. They can be summarized by saying that $\theta^4(\tau)$ is a *modular form of weight* 2 with respect to the group $\Gamma_0(4)$ where

$$\Gamma_0(N) = \left\{ \begin{pmatrix} a & b \\ c & d \end{pmatrix} \in SL(2, \mathbb{Z}) \;\middle|\; N|c \right\}. \tag{1.2.24}$$

This means that the holomorphic differential $\theta^4(\tau)d\tau$ is invariant with respect to the substitutions $\tau \mapsto (a\tau + b)(c\tau + d)^{-1}$ for every matrix $\begin{pmatrix} a & b \\ c & d \end{pmatrix}$ in $\Gamma_0(4)$. Modular functions will be considered more systematically in Part II, Ch. 6, §6.3. The space of all such differentials is two-dimensional, and one can construct a basis of this space with the help of *Eisenstein series* whose Fourier coefficients are more or less by construction certain divisor sums. Examining the first two coefficients of the series one finds an expression for $\theta^4(\tau)$ as a linear combination of the Eisenstein series. On comparing coefficients one obtains (1.2.22). This method is very general. When the number of squares grows one has to take into account not only the Eisenstein series but also *cusp forms* whose Fourier coefficients have a more complicated arithmetical nature but in many cases allow a non–trivial direct interpretation. If one manages to construct an explicit basis of the relevant modular forms, one can then express the *theta-series* of a quadratic form $f(x_1, \ldots, x_k) = A[x]$ with respect to this basis

$$\theta(\tau; f) = \sum_{x \in \mathbb{Z}^k} e(f(x)\tau) = \sum_{n=0}^{\infty} r(f; n) q^n \tag{1.2.25}$$

where

$$e(\tau) = \exp(2\pi i \tau) = q,$$
$$r(f; n) = \mathrm{Card}\{x \in \mathbb{Z}^k \mid f(x) = n\}.$$

This theta–series is a modular form of the weight $k/2$ with respect to a congruence subgroup of the modular group.

For a recent progress by G.Shimura on the representation of integers as sums of squares, we refer to [Shi02], [Shi04].

We quote as an example a formula proved by A.N.Andrianov ([An65], [Fom77]). Let $f = x^2 + y^2 + 9(z^2 + t^2)$. The theta–series of this form is a modular form of weight 2 w.r.t. $\Gamma_0(36)$. For any prime $p \neq 2, 3$ we have

$$r(f; p) = \frac{4}{3}(p + 1) - \frac{8}{3} \sum_{x=0}^{p-1} \left(\frac{x^3 + 1}{p} \right) \tag{1.2.26}$$

where the sum in the right hand side contains the Legendre symbols, cf. §1.1.4.

Generating functions are traditionally used in combinatorics and the theory of partitions. The simple partitions of n into sums of non–increasing natural summands are counted by the *partition function* $p(n)$:

$$\begin{aligned} &p(1) = 1 &:& \quad 1 = 1; \\ &p(2) = 2 &:& \quad 2 = 2, \quad 1+1; \\ &p(3) = 3 &:& \quad 3 = 3, \quad 2+1, \quad 1+1+1; \\ &p(4) = 5; &p(5)& = 7. \end{aligned}$$

Its generating function satisfies the *Euler identity* (cf. [Cha70], [And76]): for $|q| < 1$ one has

$$1 + \sum_{n=1}^{\infty} p(n) q^n = \prod_{m=1}^{\infty} (1 - q^m)^{-1}. \tag{1.2.27}$$

To prove this, it suffices to represent the r.h.s. as the product of the power series and to notice that $p(n)$ is the number of solutions of a linear Diophantine equation with an infinite set of non–negative indeterminates

$$a_1 + 2a_2 + 3a_3 + \cdots = n.$$

Remarkably, the theta-series of certain quadratic forms are also connected with certain infinite products similar to (1.2.27). For example, if $|q| < 1, \ z \neq 0$ we have (cf. [And76])

$$\sum_{n=-\infty}^{\infty} z^n q^{n^2} = \prod_{m=0}^{\infty} (1 - q^{2m+2})(1 + zq^{2m+1})(1 + z^{-1} q^{2m+1}) \qquad \text{(Jacobi)},$$

$$\sum_{n=0}^{\infty} q^{n(n+1)/2} = \prod_{m=1}^{\infty} (1 - q^{2m})(1 - q^{2m-1})^{-1} \qquad \text{(Gauss)}.$$

These identities follow from a more general result of Cauchy, valid for $|q| < 1, \ |t| < 1, \ a \in \mathbb{C}$:

$$1 + \sum_{n=1}^{\infty} \frac{(1-a)(1-qa)\dots(1-aq^{n-1})t^n}{(1-q)(1-q^2)\dots(1-q^n)} = \prod_{m=0}^{\infty} \frac{(1-atq^m)}{(1-tq^m)}. \tag{1.2.28}$$

Recently this list of such identities has been greatly enlarged, thanks to the discovery that they are connected with the representation theory of the simple *Lie algebras*, *root systems* and *finite simple groups* (cf. [Mac80]).

An impressive example of the use of generating functions is given by Borcherds [Borch92] in his proof of the Conway and Norton conjectures concerning connections between the monster simple group, M (and also other finite sporadic simple groups), and modular functions. This group is the largest sporadic finite simple group, its order is

$$8080, 17424, 79451, 28758, 86459, 90496, 17107, 57005, 75436, 80000, 00000 =$$
$$2^{46} \cdot 3^{20} \cdot 5^9 \cdot 7^6 \cdot 11^2 \cdot 13^3 \cdot 17 \cdot 19 \cdot 23 \cdot 29 \cdot 31 \cdot 41 \cdot 47 \cdot 59 \cdot 71.$$

The degree of the smallest nontrivial irreducible complex representation of M is 196883, which is 1 less than the first nontrivial q coefficient of the famous $j(q)$ or elliptic modular function. In fact

$$j(q) = q^{-1} + 196884q + 21493760q^2 + \dots$$

and the other coefficients of j turn out to be simple combinations of the degrees (traces on the identity) of representations of M.

Conway and Norton conjectured in [CoNo79] that the functions $j_g(q)$ obtained by replacing the traces on the identity by the traces on other elements g of M are "genus zero" modular functions. In other words if G_g is the subgroup of $\mathrm{SL}_2(\mathbb{R})$ which fixes $j_g(q)$, then the quotient of the upper half of the complex plane by G_g is a sphere with a finite number of points removed corresponding to the cusps of G_g, cf. §6.3.

The proof is just as remarkable as the original moonshine conjectures and involves the theory of vertex operator algebras and generalized Kac-Moody algebras, cf. [Kon96].

It turns out, that some questions of the quantum field theory are related with modularity properties of such q–expansions, cf. [DGM90]. For example, this property is an open question for the q–expansions:

$$\sum_{\substack{X \in \mathbb{Z}^n \\ x_i \geq 0}} \frac{q^{\frac{1}{2}X^t AX + BX + c}}{(q)_{x_1} \cdot \ldots \cdot (q)_{x_n}}$$

where $X = (x_1, \cdots, x_n) \in \mathbb{Z}^n$, $n \geq 1$, $(q)_m = (1-q)(1-q^2) \cdot \ldots \cdot (1-q^m)$, $A \in \mathrm{M}_n(\mathbb{Q})$, $B \in \mathbb{Q}^n$, $c \in \mathbb{Q}$ (private communication by H.Cohen).

Symmetry properties of generating functions were used in Wiles' proof of Fermat's Last Theorem and of the Shimura–Taniyama–Weil Conjecture (see [Wi], [Ta-Wi], [DDT97] and Chapter 7). In this truly marvelous proof, a traditional argument of *reductio ad absurdum* is presented in the following form: if $a^p + b^p = c^p, abc \neq 0$, for a prime $p \geq 5$ (a primitive solution (a, b, c)), then one can associate to (a, b, c) a certain generating function $f = f_{a,b,c}$: $\mathbb{H} \to \mathbb{C}$ on the Poincaré upper half–plane $\mathbb{H}$, defined by a Fourier series with the first coefficient equal to 1, as explained in §7.1. It turns out that this function has too many symmetries, expressing the fact that it is a modular cusp form of weight 2 and level 2, and forcing $f \equiv 0$ by §6.3, a contradiction with the construction of f.

1.2.7 Analytic Methods

Generating functions are also used to obtain various asymptotic formulae for functions like $r(f;n)$ and $p(n)$ as $n \to \infty$. In particular, many results have been derived using the *Hardy-Littlewood circle method* , its variants and generalizations (cf. [Vin52], [Vin71], [VK], [Mal62], [HW81], [Vau81-97], [Des90]).

The application of this method to a generating function

$$F(\tau) = \sum_{n=0}^{\infty} a(n)q^n \qquad (q = e(\tau) = \exp(2\pi i\tau))$$

starts with Cauchy's formula:

$$a(n) = \frac{1}{2\pi i}\int_{|q|=r<1} F(\tau)q^{-n-1}dq. \tag{1.2.29}$$

The following discussion can be efficiently applied to many situations when the unit circle is the natural boundary for the function $F(\tau)$ and roots of unity on this boundary behave as "the worst essential singularities" (to get some feeling for this, look at the r.h.s. of (1.2.27)). The idea is to break the integration domain into two parts: I_1 (the contribution of roots of unity of comparatively small degree) and I_2 (everything else) and to attempt to prove that I_2 is much smaller than I_1. To understand the asymptotic behaviour of I_1 and to majorize I_2 one often uses exact or approximate functional equations for $F(\tau)$, Poisson summation etc.

For example, to estimate $p(n)$, Hardy, Littlewood and Ramanujan put $r = e^{-2\pi/n^2}$. In terms of τ, they integrated over the segment $L_n = \{\tau = x+iy \mid 0 \le x \le 1,\ y = 1/n^2\}$, which they divided up as follows: I_1 is the union of the pairwise disjoint segments $\beta_{p,q} = \{x \mid |x-p/q| < 1/2qn^\delta\ (\delta \ge 1)\}$ where p/q runs through the rational numbers between 0 and 1 with denominator $\le n$; I_2 is the complement of I_1.

For (1.2.27) this furnishes the Hardy-Ramanujan asymptotic formula

$$p(n) = \frac{e^{K\lambda_n}}{4\sqrt{3}\lambda_n^2} + O(e^{K\lambda_n}/\lambda_n^3),$$

where

$$\lambda_n = \sqrt{n-1/24}, \quad K = \pi\sqrt{2/3}$$

(cf. [Cha70]).

Later this method was perfected by K.Rademacher who gave an exact formula for $p(n)$ as an infinite sum whose summands correspond to (all) roots of unity.

In one of the applications of the circle method to the theory of quadratic forms, A.V.Malyshev proved in [Mal62] the following general result. Let $k \geq 4$, $f(x_1, \ldots, x_k)$ a positive quadratic form with integral coefficients and determinant d. Then as $n \to \infty$ we have

$$r(f;n) = \frac{\pi^{k/2}}{d^{1/2}\Gamma(\frac{k}{2})} n^{\frac{k}{2}-1} H(f;n) + O(d^{(k+12)/8} n^{(k-1)/4+\epsilon}).$$

Here the constant in O depends only on k and $\epsilon > 0$ and $H(f;n)$ is the so called *singular series.* This series is obtained in the process of computing of the contribution of I_1 as an infinite product over all primes including the "infinite prime":

$$H(f;n) = r_\infty(f;n) \prod_p r_p(f;n),$$

where

$$r_p(f;n) = \lim_{m\to\infty} p^{-m(k-1)} \mathrm{Card}\{x \in (\mathbb{Z}/p^m\mathbb{Z})^k \mid f(x) \equiv 0 \bmod p^m\}$$

and $r_\infty(f;n)$ is a certain "real density" of the solutions of $f(x) = n$.

It follows that if n is sufficiently large and is representable by f modulo all prime powers, then it is representable by f. This method however does not work for 2 or 3 variables, where more subtle approaches are needed (cf. [Lin79], [GF77], [Fom77], [Lom78]).

The circle method was considerably modified and perfected by I.M.Vinogradov (cf. [Vin52], [Vin71], [VK]), who suggested replacing generating functions by *exponential sums*, which are essentially their partial sums restricted to the unit circle, e.g.

$$\theta_N(\tau; f) = \sum_{\substack{(x_1,\ldots,x_k)\in\mathbb{Z}^k \\ |x_i|<N}} e(f(x)\tau). \tag{1.2.30}$$

As a function of the real variable τ this sum oscillates vigorously and has local maxima (of its modulus, real, and imaginary parts) at rational numbers with small denominators. This behaviour reflects the singular behaviour of the generating function in the vicinity of its natural boundary but is much less wild and more easily controllable. This is one of the reasons for the success of Vinogradov's method.

Figures 1.4 and 1.5 show the (scaled) graphs of the two simplest exponential sums featuring this behaviour. Instead of Cauchy's formula (1.2.29), one uses in Vinogradov's method the integral formula

$$\int_0^1 \theta_N(\tau; f) e(-n\tau) d\tau = \mathrm{Card}\{x \in \mathbb{Z}^k \mid f(x) = n, \ |x_i| < N\} \tag{1.2.31}$$

which follows directly from the orthogonality of the basic exponential functions.

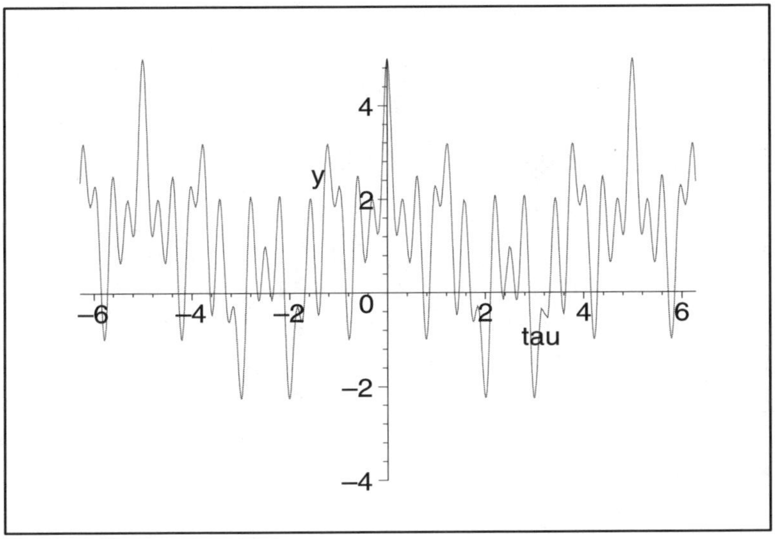

Fig. 1.4. $y(\tau) = \sum_{k=0}^{4} \cos(2\pi \frac{k^2}{5} \tau)$

Vinogradov's version of the circle method enabled him to prove that every large odd integer is a sum of three primes (Goldbach conjectured in 1742 that every even integer is a sum of two primes) and to considerably diminish the number of summands in *Waring's problem* (1770) on the representation of large integers as sums of k-th powers. An improvement on Vinogradov's bound due to Karacuba, [Kar85] is $k(2\log k + 2\log\log k + 12)$. Interesting results for $G(k)$ asymptotic to $k \log k$ has been obtained by R.C.Vaughan and T.D.Wooley (cf. [VaWo91], [VaWoIV]). Further details of analytic methods are outside the scope of this report and we refer the interested readers to the monographs [Vin71], [VK], [Pos71], [AKCh87], [HW81], [Cha70], [Vau81-97] and others. We should mention only the wide applicability of formulae of the type (1.2.31) counting various numbers of solutions and the important role of exponential sums like (1.2.30) in arithmetical problems (*Gauss sums, Jacobi sums, Kloostermann sums etc.*, cf. Ch.2, §2.2).

More generally, harmonic analysis is now used in number theory in its non–commutative and multi–dimensional versions. For example, the construction of the *Hecke basis* in the space of modular forms which is orthonormal with respect to the *Petersson inner product (scalar product)* can be considered as a two–dimensional analogue of the orthogonality relations for the exponentials mentioned above (Part II, Ch. 6, §6.3).

1.2.8 Equivalence of Binary Quadratic Forms

Two quadratic forms over the integers f, g are called *equivalent* (over $\mathbb{Z}$) if they represent each other (cf. §1.2.6). We shall denote a binary quadratic form

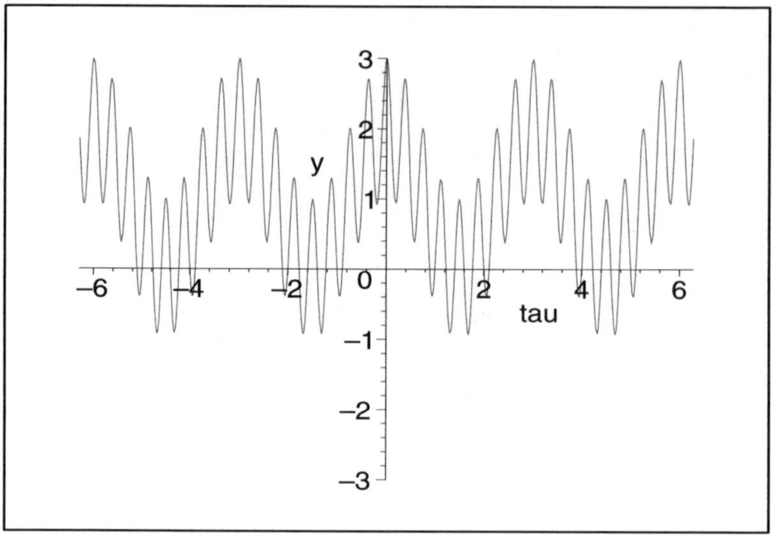

Fig. 1.5. $y(\tau) = \sum_{k=0}^{2} \cos(2\pi \frac{k^3}{3} \tau)$

$f(x, y) = Ax^2 + Bxy + Cz^2$ also (A, B, C). Such a form is called *primitive* if A, B and C have no common factor. Its discriminant is denoted $\Delta = B^2 - 4AC$. Two forms f and g are called *properly equivalent* if we have

$$f(x, y) = g(mx + ny, kx + ly)$$

for an appropriate matrix

$$\begin{pmatrix} m & n \\ k & l \end{pmatrix} \in \mathrm{SL}_2(\mathbb{Z}).$$

Gauss founded the equivalence theory of binary quadratic forms. He proved that if Δ is not a square, then the set $Cl(\Delta)$ of proper equivalence classes of forms with discriminant Δ can be made into a finite Abelian group with respect to a natural *composition law.* (Actually, this was one of the first abstract Abelian groups discovered in number theory). Very recently M.Bhargava (a PhD student of A.Wiles, cf. [Bha04]) found higher composition laws, giving a new view on Gauss composition.

In order to define this composition law in modern terms, consider the quadratic number field $K = \mathbb{Q}(\sqrt{\Delta}) = \mathbb{Q}(\sqrt{d}) = \{x + y\sqrt{d} \mid x, y \in \mathbb{Q}\}$ where d is a square-free integer. We have $\Delta = Dc^2$ where D is the *discriminant* of the quadratic field K, $D = d$ if $d \equiv 1 \bmod 4$ and $D = 4d$ otherwise. An element $\alpha = x + y\sqrt{d} \in K$ is called an *integer* if its *trace* $2x$ and its *norm* $x^2 - dy^2$ are integers. The set of all integers in K forms a ring

$$\mathcal{O} = \langle 1, \omega \rangle = \{m + n\omega \mid m, n \in \mathbb{Z}\},$$

where $\omega = \sqrt{d}$ if $d \equiv 2,3 \bmod 4$ and $\omega = (1+\sqrt{d})/2$ if $d \equiv 1 \bmod 4$. For any integer c we can define a subring $\mathcal{O}_c = \mathbb{Z} + c\mathcal{O} =< 1, c\omega >$. A *fractional ideal* M in $\mathcal{O}_c$ is a free additive subgroup with two generators which is stable with respect to multiplication by elements of $\mathcal{O}_c$. The product of two fractional ideals is, by definition, the subgroup generated by the products of elements from one ideal with elements of the other. The fractional ideals form an Abelian group with identity $\mathcal{O}_c$. To each such ideal M corresponds a quadratic form with discriminant $D = dc^2$ which can be constructed as follows. Define the *norm* of M by $N(M) = \mathrm{Card}(\mathcal{O}_c/M)$. Choose a basis $\{\alpha, \beta\}$ for M in such a way that $\gamma = -\beta/\alpha = x + y\sqrt{d}$ satisfies the condition $y > 0$. Then the quadratic form in question is

$$f(x,y) = Ax^2 + Bxy + Cy^2 = \frac{N(\alpha x + \beta y)}{N(M)}.$$

One can check that this is a primitive integral form.

Two fractional ideals M and M_1 are called *equivalent* in the narrow sense if $M = \gamma M_1$ for some $\gamma \in K$ with positive norm. The equivalence classes of fractional ideals correspond bijectively to the proper equivalence classes of primitive binary forms of discriminant Dc^2. Multiplication of the fractional ideals induces a group structure on this set. The identity of this group is represented by the quadratic form $(1, 0, -\Delta/4)$ (resp. $(1, 1, (1-\Delta)/4)$ if Δ is even (resp. odd). In computations it is convenient to work with the *reduced* forms (A, B, C) for which $A > 0$, $-A \le B < A$, $\gcd(A, B, C) = 1$. If $\Delta < 0$ then the group $Cl(\Delta)$ is trivial exactly for the following values: $-\Delta = 4, 8, 3, 7, 11, 19, 43, 67, 163$ ($c = 1$); $16, 12, 28$ ($c = 2$); 27 ($c = 3$) (cf. Part II, Ch. 5, §5.4.1).

1.3 Cubic Diophantine Equations

1.3.1 The Problem of the Existence of a Solution

For *cubic forms* $F(X,Y,Z)$ in three variables with integral coefficients, nobody has succeeded in devising a general algorithm which provably decides whether the equation $F = 0$ has a non–trivial integral solution. Large classes of such equations have been studied both theoretically and numerically; see for example the early influential papers by E.S.Selmer (cf. [Selm51] and [Selm54]) devoted to the equations

$$aX^3 + bY^3 + cZ^3 = 0.$$

Even some of the simplest equations like $3X^3 + 4Y^3 + 5Z^3 = 0$ fail to satisfy the Minkowski–Hasse principle: they have no non–trivial integral solutions although they do have both real solutions and primitive integral solutions modulo any $N > 1$. The degree of such failure can be measured quantitatively by the *Shafarevich–Tate group*: cf. §5.3.

D.R.Heath-Brown has shown (cf. [HB84]) that any non–singular cubic form in ten variables represents zero non–trivially, and C.Hooley in [H88] has established the Minkowski–Hasse principle for non–singular nonary cubic forms (a form is called non–singular if it and all its first partial derivatives have no common non–trivial complex zeroes). Previously Davenport and Birch had shown that there exist non–singular cubic forms in nine variables which do not represent zero modulo a power of every prime.

Birch in [Bir61] established that forms of any odd degree d represent zero if the number of variables is sufficiently large (with the bound depending only on d). These results have since been generalized, extended and made more precise by several authors. They are proved by the circle method, cf. [Vau81-97], [Des90].

1.3.2 Addition of Points on a Cubic Curve

Any ternary cubic form $F(X,Y,Z)$ defines a cubic curve $\mathcal{C}$ in the complex projective plane $\mathbb{P}^2$:

$$\mathcal{C} = \{(X : Y : Z) \mid F(X,Y,Z) = 0\}. \tag{1.3.1}$$

If $\mathcal{C}$ (that is, F) is non–singular, and if $F = 0$ has at least one rational solution, then one can find a non–degenerate change of projective coordinates with rational coefficients which reduces F to a *Weierstrass normal form*

$$Y^2Z - X^3 - aXZ^2 - bZ^3 \qquad (a, b \in \mathbb{Q}). \tag{1.3.2}$$

One may also assume that the initial solution becomes the obvious solution $(0 : 1 : 0)$ of (1.3.2). The non–singularity condition for (1.3.2) is equivalent to the non–vanishing of the discriminant $4a^3 + 27b^2$. Non-singular cubic curves are also called *elliptic*. Passing to non–homogeneous coordinates $x = X/Z, y = Y/Z$ we reduce $F = 0$ to the form

$$y^2 = x^3 + ax + b, \tag{1.3.3}$$

where the cubic polynomial in the r.h.s. has no multiple roots. In this affine form, our initial solution becomes the infinite point O. There is a beautiful geometric description of a composition law on the set of rational points of $\mathcal{C}$ making it an Abelian group with O as identity (or zero). This is called *the secant–tangent method* (cf. [Sha88], [Cas66], [Kob84]). Namely, for a given pair of points $P, Q \in \mathcal{C}(\mathbb{Q})$, we first draw a line containing them both. This line also intersects $\mathcal{C}$ at a well-defined third rational point P'. Now we again draw a line through P' and O. Its third intersection point with $\mathcal{C}$ is, by definition, *the sum* $P + Q$. If $P = Q$, the first line to be drawn should of course touch $\mathcal{C}$ at P.

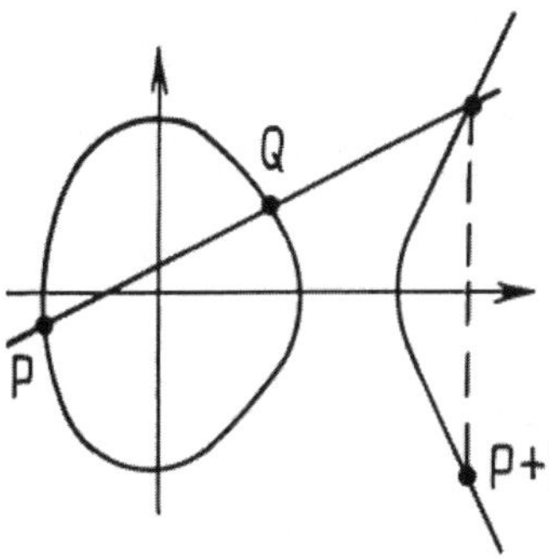

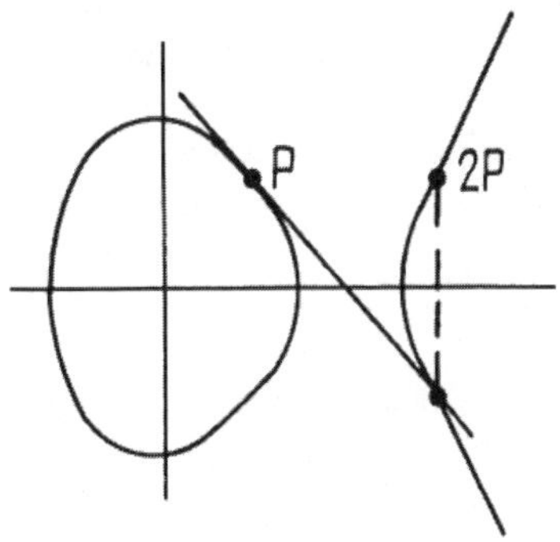

Fig. 1.6.

Fig. 1.7.

Calculating in non–homogeneous coordinates $P = (x_1, y_1),\ Q = (x_2, y_2)$ one finds $P + Q = (x_3, y_3)$ where

$$x_3 = -x_1 - x_2 + \left(\frac{y_1 - y_2}{x_1 - x_2}\right)^2,$$

$$y_3 = \frac{y_1 - y_2}{x_1 - x_2}(x_1 - x_3) - y_1. \tag{1.3.4}$$

In the limit case $P = Q$ we have instead

$$x_3 = -2x_1 + \left(\frac{3x_1^2 + a}{2y_1}\right)^2, \ y_3 = \frac{3x_1^2 + a}{2y_1}(x_1 - x_3) - y_1. \tag{1.3.5}$$

If $x_1 = x_2$ and $y_1 = -y_2$ then $P + Q = O$, the infinite point which is zero for the group law.

This method allows us to construct new rational points starting with some known ones. They will be the elements of the group generated by the initial points, e.g. mP, $m \in \mathbb{Z}$, if just one point P (except O) was found initially.

For *singular cubic curves* this construction fails. For example, consider the curve

$$\mathcal{C} : \; y^2 = x^2 + x^3, \tag{1.3.6}$$

which is drawn in Fig. 1.8. Any line passing through $(0,0)$ has only one more common point with $\mathcal{C}$: on $y = tx$ it is defined by the equation $x^2(t^2 - x - 1) = 0$. Besides the trivial solution $x = 0$, we obtain $x = t^2 - 1$ and $y = t(t^2 - 1)$ so that we have found all points on $\mathcal{C}$ with the help of a rational parameterization. In the non–singular case no such parameterization exists. On the other hand, in our example we could have still defined the group law on the set of non–singular points as above. However, this becomes simply multiplication (for a suitably chosen rational parameterization).

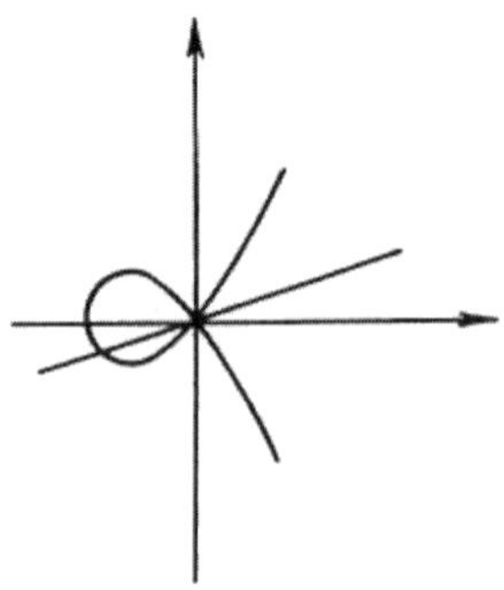

Fig. 1.8.

A curve admitting a rational parameterization is called rational. How one can establish that such a parameterization exists or otherwise, and how its existence influences the problem of describing all rational points, is answered by algebraic-geometric methods.

1.3.3 The Structure of the Group of Rational Points of a Non–Singular Cubic Curve

The most remarkable qualitative feature of the secant–tangent method is that it allows one to construct all rational solutions of a non–singular cubic equation (1.3.3) starting with only a finite number of them. In group-theoretical language, the following result is true.

Theorem 1.3 (Mordell's Theorem). *The Abelian group $\mathcal{C}(\mathbb{Q})$ is finitely generated.*

(cf. ([Mor22], [Cas66], [Mor69], [La83], [Se97] and Appendix by Yu.Manin to [Mum74]). From the structure theorem for finitely generated Abelian groups, it follows that

$$\mathcal{C}(\mathbb{Q}) \cong \Delta \times \mathbb{Z}^r$$

where Δ is a finite subgroup consisting of all torsion points, and $\mathbb{Z}^r$ is a product of r copies of an infinite cyclic group. The number r is called *the rank* of $\mathcal{C}$ over $\mathbb{Q}$.

The group Δ can be found effectively. For example, Nagell and Lutz (cf. [Lu37]) proved that torsion points on a curve $y^2 = x^3 + ax + b$ for which a and b are integers, have integral coordinates. Furthermore, the y–coordinate of a torsion point either vanishes or divides $D = -4a^3 - 27b^2$.

B.Mazur proved in 1976 that the torsion subgroup Δ over $\mathbb{Q}$ can only be isomorphic to one of the following fifteen groups:

$$\mathbb{Z}/m\mathbb{Z}\ (m \leq 10, m = 12),\ \mathbb{Z}/2\mathbb{Z} \times \mathbb{Z}/2n\mathbb{Z}\ (n \leq 4), \tag{1.3.7}$$

and all these groups occur, cf. [Maz77].

It is still an open question whether r can be arbitrarily large. Mestre (cf. [Me82]) constructed examples of curves whose ranks are at least 14. *)

A comparatively simple example of a curve of rank ≥ 9 is also given there: $y^2 + 9767y = x^3 + 3576x^2 + 425x - 2412$. One can conjecture that rank is unbounded. B. Mazur (cf. [Maz86]) connects this conjecture with *Silverman's conjecture* (cf. [Silv86]) that for any natural k there exists a cube-free integer which can be expressed as a sum of two cubes in more than k ways.

Examples. 1) Let $\mathcal{C}$ be given by the equation

$$y^2 + y = x^3 - x$$

whose integer solutions list all cases when a product of two consecutive integers equals a product of three consecutive integers. Here Δ is trivial while the free part of $\mathcal{C}(\mathbb{Q})$ is cyclic, with a generator $P = (0,0)$. Points mP (labeled by m) are shown in Figure 9.

The following Table 1.3, reproduced here from [Maz86] with Mazur's kind permission, shows the absolute values of the X–coordinates of points mP, for even m between 8 and 58.

* Martin–Mcmillen (2000) found an elliptic curve of rank ≥ 24:

$$\begin{aligned} y^2 + xy + y = x^3 &- 120039822036992245303534619191166796374x \\ &+ 504224992484910670010801799168082726759443756222911415116 \end{aligned}$$

(see `http://www.math.hr/~duje/tors/rankhist.html` for more examples). (footnote by Yu.Tschinkel).

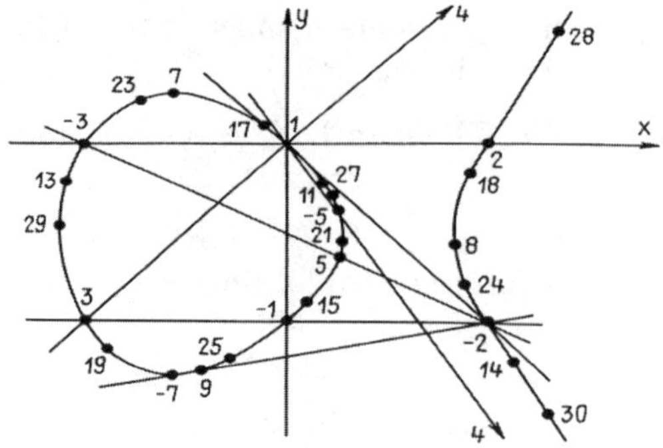

Fig. 1.9.

Table 1.3.

20
116
3741
8385
239785
59997896
18490337896
270896443865
16683000076735
2786836257692691
3148929681285740316
342115756927607927420
280251129922563291422645
804287518035141565236193151
743043134297049053529252783151
3239336802390544740129153150480400
2613390252458014344369424012613679600
12518737094671239826683031943583152550351
596929565407758846078157850477988229836340351
2385858586329829631608077553938139264431352010155
56186054018434753527022752382280291882048809582857380
2389750519110914018630990937660635435269956452770356625916
65008789078766455275600750711306493793995920750429546912218291
8633815035886806713921361263456572740784038065917674315913775417535
43276783438948886312588030404441444313405755534366254416432880924019065
5930760454696426589489567617397943244827292346871145123187277732855876671389

One sees that the last figures lie approximately on a parabola. This is not an accident, but a reflection of the *quadratic nature of heights on elliptic curves* (cf. below).

2) Table 1.4 was kindly calculated for this edition by H.Cohen, using PARI computing system, [BBBCO]. This table lists ranks r and generators for curves $X^3 + Y^3 = AZ^3$ with natural cube-free $A \leq 500$; it corrects and completes the Tables of Selmer (cf. [Selm51], [Selm54]) which were reproduced in the first edition [Ma-Pa]. Note the 3 missing values $A = 346, 382, 445$ for which H.Cohen proved that $r = 1$, but the method of Heegner points for computing

generators (see §6.4.4) takes too much time. However, this computation was completed by Ch. Delaunay, see Table 1.5.

Table 1.4. Number of generators r and basic solutions of $X^3 + Y^3 = AZ^3$ with A cube–free, $A \leq 500$.

A	r	(X, Y, Z)	A	r	(X, Y, Z)
6	1	(37, 17, 21)	94	1	(15642626656646177, -15616184186396177, 590736058375050)
7	1	(2, -1, 1)	97	1	(14, -5, 3)
9	1	(2, 1, 1)	98	1	(5, -3, 1)
12	1	(89, 19, 39)	103	1	(592, -349, 117)
13	1	(7, 2, 3)	105	1	(4033, 3527, 1014)
15	1	(683, 397, 294)	106	1	(165889, -140131, 25767)
17	1	(18, -1, 7)	107	1	(90, 17, 19)
19	2	(36, -17, 13),(109, -90, 31)	110	2	(181, -71, 37),(629, 251, 134)
20	1	(19, 1, 7)	114	1	(9109, -901, 1878)
22	1	(25469, 17299, 9954)	115	1	(5266097, -2741617, 1029364)
26	1	(3, -1, 1)	117	1	(5, -2, 1)
28	1	(3, 1, 1)	123	1	(184223499139, 10183412861, 37045412880)
30	2	(163, 107, 57),(289, -19, 93)	124	2	(5, -1, 1),(479, -443, 57)
31	1	(137, -65, 42)	126	2	(5, 1, 1),(71, -23, 14)
33	1	(1853, 523, 582)	127	2	(7, -6, 1),(121, -120, 7)
34	1	(631, -359, 182)	130	1	(52954777, 33728183, 11285694)
35	1	(3, 2, 1)	132	2	(2089, -901, 399),(39007, -29503, 6342)
37	2	(4, -3, 1),(10, -1, 3)	133	1	(5, 2, 1)
42	1	(449, -71, 129)	134	1	(9, 7, 2)
43	1	(7, 1, 2)	139	1	(16, -7, 3)
49	1	(11, -2, 3)	140	1	(27397, 6623, 5301)
50	1	(23417, -11267, 6111)	141	1	(53579249, -52310249, 4230030)
51	1	(730511, 62641, 197028)	142	1	(2454839, 1858411, 530595)
53	1	(1872, -1819, 217)	143	1	(73, 15, 14)
58	1	(28747, -14653, 7083)	151	1	(338, -95, 63)
61	1	(5, -4, 1)	153	2	(70, -19, 13),(107, -56, 19)
62	1	(11, 7, 3)	156	1	(2627, -1223, 471)
63	1	(4, -1, 1)	157	1	(19964887, -19767319, 1142148)
65	2	(4, 1, 1),(191, -146, 39)	159	1	(103750849, 2269079, 19151118)
67	1	(5353, 1208, 1323)	161	1	(39, -16, 7)
68	1	(2538163, -472663, 620505)	163	2	(11, -3, 2),(17, -8, 3)
69	1	(15409, -10441, 3318)	164	1	(311155001, -236283589, 46913867)
70	1	(53, 17, 13)	166	1	(1374582733040071, -1295038816428439, 136834628063958)
71	1	(197, -126, 43)	169	1	(8, -7, 1)
75	1	(17351, -11951, 3606)	170	1	(26353, 14957, 5031)
78	1	(5563, 53, 1302)			
79	1	(13, -4, 3)			
84	1	(433, 323, 111)			
85	1	(2570129, -2404889, 330498)			
86	2	(13, 5, 3),(10067, -10049, 399)			
87	1	(1176498611, -907929611, 216266610)			
89	1	(53, 36, 13)			
90	1	(1241, -431, 273)			
91	2	(4, 3, 1),(6, -5, 1)			
92	1	(25903, -3547, 5733)			

Table 1.4. (continued)

A	r	(X, Y, Z)
171	1	(37, 20, 7)
172	1	(139, -103, 21)
177	1	(2419913540753, 1587207867247, 468227201520)
178	1	(110623913, 8065063, 19668222)
179	1	(2184480, -1305053, 357833)
180	1	(901, 719, 183)
182	2	(11, 5, 2),(17, 1, 3)
183	2	(14, 13, 3),(295579, -190171, 46956)
186	1	(56182393, 15590357, 9911895)
187	1	(336491, -149491, 57070)
193	1	(135477799, -116157598, 16825599)
195	1	(68561, -54521, 9366)
197	1	(2339, -2142, 247)
198	1	(1801, -19, 309)
201	2	(16, 11, 3),(3251, 124, 555)
202	1	(2884067, 257437, 491652)
203	2	(229, 32, 39),(2426, -2165, 273)
205	1	(8191, -6551, 1094)
206	1	(5211, -4961, 455)
209	2	(52, -41, 7),(125, -26, 21)
210	2	(1387, 503, 237),(3961, -2071, 633)
211	1	(74167, 66458, 14925)
212	1	(337705939853, -315091652237, 32429956428)
213	1	(64313150142602539-525717, 46732739212871-851099283, 12000095230-802028099750)
214	1	(307277703127, -244344663377, 40697090945)
215	1	(6, -1, 1)
217	2	(6, 1, 1),(9, -8, 1)
218	2	(7, -5, 1),(279469, -61469, 46270)
219	2	(17, 10, 3),(168704, -36053, 27897)
222	1	(5884597, 858653, 972855)
223	1	(509, 67, 84)
228	1	(46323521, -27319949, 7024059)
229	1	(745, -673, 78)
231	1	(818567, -369503, 129186)
233	1	(124253, -124020, 3589)
236	1	(248957, 209827, 47106)
238	1	(53927, 3907, 8703)
241	1	(292, -283, 21)
244	1	(99, -67, 14)
246	2	(571049, -511271, 59787), (2043883, -1767133, 230685)
247	1	(20, -11, 3)
249	1	(2756573072910450752O3-684958997, -275522784-968298556737485593813, 4974480998065387679-603368524)
251	1	(4284, -4033, 373)
254	2	(238013, -206263, 26465), (238393, -222137, 21676)
258	1	(2195839, -2047231, 198156)
259	1	(13, -5, 2)
265	1	(36326686731109813, 9746422253537867, 5691757727610864)
267	1	(861409, -342361, 130914)
269	1	(800059950, -786434293, 45728263)
271	2	(10, -9, 1),(487, -216, 73)
273	2	(19, 8, 3),(190, -163, 21)
274	1	(111035496427236122887, -43257922194314055637, 16751541717010945845)
275	1	(424560439, -309086839, 55494828)
277	1	(209, -145, 28)
278	1	(13, 3, 2)
279	1	(7, -4, 1)
282	2	(117217, -96913, 13542), (2814607, 1571057, 452772)
283	1	(20824888493, -8780429621, 3090590958)
284	1	(7722630462000896449-941136589, -12938136226219393-03367981, 1174877194362780234-594343698)
285	1	(18989, 1531, 2886)
286	1	(323, -37, 49)
287	1	(248, 121, 39)
289	1	(199, 90, 31)
294	1	(124559, -103391, 14118)
295	1	(34901, -16021, 5068)

Table 1.4. (continued)

A	r	(X, Y, Z)
301	1	(382, 5, 57)
303	1	(2659949, 67051, 396030)
305	1	(86, -81, 7)
306	1	(6697, -3943, 921)
308	1	(199, 109, 31)
309	2	(20, 7, 3),(272540932, -142217089, 38305371)
310	1	(5011613, -190493, 740484)
313	1	(22, -13, 3)
314	1	(241, -223, 21)
316	1	(7, -3, 1)
319	1	(6462443919765751305499, -6182025219694143438499, 472407353310304561590)
321	1	(13755277819, 8670272669, 2164318002)
322	1	(1873, 703, 278)
323	1	(252, 71, 37)
325	1	(128, 97, 21)
330	1	(1621, 1349, 273)
331	1	(11, -10, 1)
333	1	(397, -286, 49)
335	2	(7, -2, 1),(390997, 260243, 61362)
337	1	(53750671, -53706454, 1043511)
339	1	(1392097139, -345604139, 198626610)
341	1	(6, 5, 1)
342	2	(7, -1, 1),(1253, -1205, 86)
345	2	(16543, 8297, 2454), (389699, -190979, 53292)
346	1	
348	2	(40283, -15227, 5622), (2706139, 425861, 385230)
349	1	(23, -14, 3)
355	1	(2903959, 2617001, 492516)
356	1	(15026630492061476041947013, -4709632110011335573393177, 2098221141580681446554589)
357	1	(19207, 6497, 2742)
358	1	(7951661, 2922589, 1138095)
359	1	(77517180, 50972869, 11855651)
363	1	(1909159356457, -1746345039913, 165073101648)
366	1	(2087027, -1675277, 228885)
367	1	(42349, 526, 5915)
370	2	(7, 3, 1),(70523, 19387, 9891)
372	1	(2717893, 630107, 379470)
373	1	(1604, -1595, 57)
377	1	(469, -237, 62)
379	2	(15, -7, 2),(917, -908, 39)
380	1	(1009, -629, 127)
382	1	
385	1	(20521, -17441, 2054)
386	1	(9, -7, 1)
387	1	(8, -5, 1)
388	1	(4659, -3287, 553)
390	2	(3043, 467, 417),(4373, -863, 597)
391	1	(590456252061289, -171359229789289, 80084103077160)
393	1	(4045451855513988711059, 2369372172284459347309, 5870469694135369683336)
394	1	(1439245403, -573627403, 192088390)
395	1	(7891, -7851, 266)
396	1	(46789273, -37009657, 5074314)
397	2	(12, -11, 1),(360, 37, 49)
399	2	(22, 5, 3),(401, 328, 63)
402	1	(585699417548405371, 102798361240815491, 79502362839530631)
403	1	(53, -22, 7)
407	2	(7, 4, 1),(33733, -33634, 939)
409	1	(22015523, 21425758, 3687411)
411	1	(186871897, 49864103, 25292280)
413	1	(2575, -2103, 266)
414	1	(68073157, 32528843, 9454410)
418	1	(76267, 25307, 10323)

Table 1.4. (continued)

A	r	(X, Y, Z)
420	2	(2213, 1567, 327),(10459, -6679, 1263)
421	1	(19690, 4699, 2639)
422	1	(15, 1, 2)
425	1	(2393, 1007, 326)
427	1	(25, -16, 3)
428	1	(1294057, -1190053, 104013)
429	1	(16739, 14149, 2598)
430	1	(5989967, 3449393, 841204)
431	1	(701, -270, 91)
433	2	(37, 35, 6),(252, 181, 37)
435	2	(32779, -1459, 4326), (3784049, 2981071, 570276)
436	2	(19, 17, 3), (1667465, 307927, 220362)
438	1	(12636764083, 11127850973, 1979215602)
439	1	(571, -563, 26)
441	1	(13, 11, 2)
444	1	(4174254535499, -726500109131, 546201297768)
445	1	
446	2	(23, -5, 3), (4286417, -4285265, 52212)
447	1	(4405301, -382301, 576030)
449	1	(323, 126, 43)
450	1	(21079, 11321, 2886)
452	1	(851498679025552429, 224535817897760071, 111626729681785675)
453	2	(23, 4, 3),(50167097, 39331207, 7447188)
454	1	(753389202595029867- 852290245746241110629, -204264638826527324- 892641927694862943879, 97368775947767167139- 892682703702288385)
457	1	(41, 31, 6)
458	1	(953039, -761375, 97482)
460	1	(248768189, -234795689, 17466345)
462	2	(3779, 379, 489),(11969, -7811, 1389)
463	1	(403, -394, 21)
465	1	(1212356942047, -1197072217207, 52307828958)
466	1	(464540708319337302841, 88798763256715446551, 60057801943830995598)
467	1	(1170, -703, 139)
468	2	(7, 5, 1),(859, -763, 74)
469	2	(13, -12, 1),(26, -17, 3)
474	1	(568871, -453689, 57627)
477	2	(89, 70, 13), (12040, -11881, 523)
481	1	(43, 29, 6)
483	1	(2401741, 1945259, 352830)
484	1	(236521, -176021, 25235)
485	1	(8, -3, 1)
490	1	(193229, -74159, 24039)
493	1	(8432715268961, -1057596310369, 1066758076384)
494	1	(59, -33, 7)
495	1	(342361, -57241, 43212)
497	2	(55, 16, 7), (7411, -6772, 579)
498	2	(611137, -490123, 60543), (15811001, -15250751, 933765)
499	1	(80968219, 17501213, 10242414)

Table 1.5. Basic solutions of $X^3 + Y^3 = AZ^3$ with $A = 346,\ 382,\ 445$.

A	r	(X, Y, Z)
346	1	(47189035813499932580169103856786964321592777067, 42979005685698193708286233727941595382526544683, 8108695117451325702581978056293186703694064735)
382	1	(58477534119926126376218390196344577607972745895728749, 16753262295125845463811427438340702778576158801481539, 8122054393485793893167719500929060093151854013194574)
445	1	(362650186970550612016862044970863425187, -58928948142525345898087903372951745227, 47432800292536072666333861784516450106)

These computations have now been extended up to $A \leq 70000$ (Stephens). Don Zagier noticed that in this range there are about 38.3% of curves with $r = 0$; 48.9% with $r = 1$; 11.7% with even $r \geq 2$ and 1.1% with odd $r \geq 3$, and these values vary only slightly within large intervals of the tables. We refer to [Si01] for a survey of open questions in arithmetic algebraic geometry.

3) Let $\mathcal{C}$ be given by the equation

$$y^2 + y = x^3 - 7x + 6.$$

Then $\mathcal{C}(\mathbb{Q}) \cong \mathbb{Z}^3$, and the points (1,0), (6,0), (0, 2) form a basis of this group.

4) For $y(y+1) = x(x-1)(x+2)$ we have $r = 2$; for $y(y+1) = x(x-1)(x+4)$, $r = 2$ (compare this with example 1).

5) Consider the curve $y^2 = x^3 + px$, $p = 877$. A generator modulo torsion of the group of rational points of this curve has x–coordinate

$$x = \frac{375494528127162193105504069942092792346201}{6215987776871505425463220780697238044100}.$$

This shows that naive methods of seeking points quickly become inefficient (cf. [Cas66], [CW77], [Coa84] for an educated approach).

1.3.4 Cubic Congruences Modulo a Prime

Let p be a prime and $F(X_0, X_1, X_2)$ a cubic form with integral coefficients. Reducing F modulo p, we obtain a cubic form over the prime finite field $\mathbb{F}_p$. This reduction is called non–singular if it has no common zeroes with its first partial derivatives in any extension of $\mathbb{F}_p$. We can also apply elementary algebraic-geometric ideas to a field K of finite characteristic. The normal forms are then slightly more complicated. By making a change of projective coordinates and passing to the non–homogeneous equation, we can always reduce the equation $F = 0$ to the form

$$y^2 + a_1xy + a_3y = x^3 + a_2x^2 + a_4x + a_6,$$

where $a_1, a_2, a_3, a_4, a_6 \in K$ and

$$\Delta = -b_2^2b_8 - 8b_4^3 - 27b_6^2 + 9b_2b_4b_6 \neq 0,$$

where

$$b_2 = a_1^2 + 4a_2, \quad b_4 = 2a_4 + a_1a_3, \quad b_6 = a_3^2 + 4a_6.$$

The notation $j = \dfrac{c_4^3}{\Delta}$ is used, where

$$c_4 = b_2^2 - 24b_4, c_6 = -b_2^3 + 36b_2b_4 - 216b_6.$$

Then this equation can be further simplified using the transformation $x \mapsto u^2x' + r$, $y \mapsto u^3y' + su^2x'r + t$ in order to obtain the following (cf. [Ta73], [Kob87] :

1) For $p \neq 2, 3$:

$$y^2 = x^3 + a_4 x + a_6 \text{ with } \Delta = -16(4a_4^3 + 27a_6^2 \neq 0). \tag{1.3.8}$$

2) For $p = 2$ we have that the condition $j = 0$ is equivalent to $a_1 = 0$, and the equation transforms as follows: if $a_1 \neq 0$ (i.e. $j \neq 0$), then choosing suitably r, s, t we can achieve $a_1 = 1$, $a_3 = 0$, $a_4 = 0$, and the equation takes the form

$$y^2 + xy = x^3 + a_2 x^2 + a_6, \tag{1.3.9}$$

with the condition of smoothness given by $\Delta \neq 0$. Suppose next that $a_1 = 0$ (i.e. $j = 0$), then the equation transforms to

$$y^2 + a_3 y = x^3 + a_4 x + a_6, \tag{1.3.10}$$

and the condition of smoothness in this case is $a_3 \neq 0$.

3) For $p = 3$:

$$y^2 = x^3 + a_2 x^2 + a_4 x + a_6, \tag{1.3.11}$$

(here multiple roots are again disallowed).

The projective curve defined by the respective homogeneous equation always has a rational point $O = (0 : 1 : 0)$.

How many points over $\mathbb{F}_p$, that is, solutions of the congruence $F \equiv 0 \bmod p$, should we expect? Clearly, the total number (counting O) cannot exceed $2p + 1$, since every finite x gives no more than two values of y. On the other hand, of all the non–zero residue classes, only half of them are squares (for odd p). Hence we might expect that $x^3 + ax + b$ is a square only for about a half of the x's.

More precisely, let $\chi(x) = \left(\dfrac{x}{p}\right)$ be the Legendre symbol (cf. §1.1.5). Then, by definition, the number of solutions of $y^2 = u$ in $\mathbb{F}_p$ is $1 + \chi(u)$. Therefore,

$$\begin{aligned}\text{Card } C(\mathbb{F}_p) &= 1 + \sum_{x \in \mathbb{F}_p} (1 + \chi(x^3 + ax + b)) \\ &= p + 1 + \sum_{x \in \mathbb{F}_p} \chi(x^3 + ax + b).\end{aligned}$$

N.Koblitz in [Kob94] compares the last sum with the result of a random walk on a line. After p steps one might expect to be at distance roughly $\sqrt{p}$ from zero. Actually, one can prove the following remarkable theorem (cf. [Ha37]):

Theorem 1.4 (Hasse's Theorem). *Let* $N_p = \text{Card } C(\mathbb{F}_p)$. *Then*

$$|N_p - (p + 1)| \leq 2\sqrt{p}.$$

An elementary proof of this theorem was given in 1956 (cf. [Man56]). Since then, both the algebraic-geometric and the elementary proofs have been greatly extended. For a review of the elementary methods, cf. [Step74], [Step84], [Step94].

We refer to [LaTr76] for the problem of the distribution of Frobenius automorphisms for varying p, and of the difference $N_p-(p+1)$, which is related to the *Sato–Tate Conjecture* (cf. Chapter I in [Se68a] and §6.5.1).

The Abelian group structure on the group of points $E(\mathbb{F}_p)$ on an elliptic curve is used in many arithmetical questions. In particular, the case when this group is cyclic of large size leads to ECDLP ("Elliptic curve discrete logarithm problem") which is very important for applications in public-key cryptography, see [Kob87].

1.4 The Structure of the Continuum. Approximations and Continued Fractions

1.4.1 Best Approximations to Irrational Numbers

Since $\sqrt{2}$ is irrational, the quadratic form $x^2 - 2y^2$ cannot vanish at integral points $(x, y) \neq (0, 0)$. The smallest values taken by this form at such points are

$$x^2 - 2y^2 = \pm 1. \tag{1.4.1}$$

This is an instance of Pell's equation, which we discussed in §1.2.5; we are now interested in it because its successive solutions give the best approximations to $\sqrt{2}$ by rational numbers.

More precisely, a/b is said to be a best approximation to α if

$$|b\alpha - a| < |d\alpha - c|$$

for all $0 < d \leq b,\ a \neq c$. Every solution to (4.1) can be obtained by setting

$$a + \sqrt{2}b = (1 + \sqrt{2})^n.$$

Table 1.6.

x	y	x/y
1	1	1,0
3	2	1,5
7	5	1,4
17	12	1,416...
41	29	1,4137...
99	70	1,41428...
239	169	1,414201...
577	408	1,414215...
1393	985	1,4142132...
3363	2376	1,4142136...
$\vdots$	$\vdots$	$\vdots$

1.4.2 Farey Series

One way of finding good approximations is connected with a specific procedure for enumerating all rational numbers between 0 and 1. Denote by $\mathcal{F}_n$ the *Farey*

Table 1.7.

$\frac{1}{5}$	$\frac{1}{4}$	$\frac{1}{3}$	$\frac{2}{5}$	$\frac{1}{2}$	$\frac{3}{5}$	$\frac{2}{3}$	$\frac{3}{4}$	$\frac{4}{5}$	$\frac{1}{1}$	$\frac{5}{4}$	$\frac{4}{3}$	$\frac{3}{2}$	$\frac{5}{3}$	$\frac{2}{1}$	$\frac{5}{2}$	$\frac{3}{1}$	$\frac{4}{1}$	$\frac{5}{1}$

series of order n, which consists of all such numbers in increasing order whose denominators are $\leq n$:

$$\mathcal{F}_n = \{a/b \mid 0 \leq a \leq b \leq n,\ (a,b) = 1\}. \tag{1.4.2}$$

Theorem 1.5 ([HaWr]). *For every real number* $\alpha \in [0,1]$ *there exists* $a/b \in \mathcal{F}_n$ *such that*

$$\left|\alpha - \frac{a}{b}\right| \leq \frac{1}{b(n+1)}. \tag{1.4.3}$$

The proof is based on the fact that if a/b, c/d are neighbours in $\mathcal{F}_n$ then $ad - bc = \pm 1$. This in turn can be seen by noting that one can go from $\mathcal{F}_n$ to $\mathcal{F}_{n+1}$ by inserting between a/b and c/d all mediants $(a+c)/(b+d)$ with $c + d = n + 1$.

In this theorem α need not be irrational, so we obtain some information about rational approximations to rational numbers with large denominators.

If α is irrational, this theorem shows that the inequality

$$\left|\alpha - \frac{a}{b}\right| < \frac{1}{b^2} \tag{1.4.4}$$

has infinitely many solutions a/b. If a/b is a best approximation, then (1.4.4) follows from (1.4.3) with $n = b$. An efficient way of finding best approximations is furnished by *continued fractions.* This tool also allows us to show that for irrational β the following stronger inequality has infinitely many solutions

$$\left|\beta - \frac{a}{b}\right| < \frac{1}{\sqrt{5}b^2} \tag{1.4.5}$$

1.4.3 Continued Fractions

(cf. [Khi78], [Dav52], [HaWr]). For an arbitrary real number α, we define a sequence of integers a_i and real numbers α_i by the following rules: $a_0 = [\alpha]$ (the integral part), $\alpha_0 = \alpha$, $\alpha_{i+1} = 1/(\alpha_i - a_i)$, $a_{i+1} = [\alpha_{i+1}]$ $(i \geq 0)$. We obtain a continued fraction

$$\alpha = a_0 + \cfrac{1}{a_1 + \cfrac{1}{a_2 + \cfrac{1}{a_3 + \dots \cfrac{1}{a_m + \cfrac{1}{\alpha_{m+1}}}}}},$$

which can be written in a more compact notation as

$$\alpha = [a_0; a_1, a_2, \dots, a_m, \alpha_{m+1}]. \tag{1.4.6}$$

Deleting α_{m+1} in (1.4.6), we get the finite continued fraction

$$C_m = [a_0; a_1, \dots, a_m]$$

called the mth *convergent* of α. The numerators and denominators of the successive convergents $C_m = A_m/B_m$ can be calculated recursively starting from $A_{-2} = B_{-1} = 0$, $A_{-1} = B_{-2} = 1$ with the help of the following relations:

$$\begin{aligned} A_{k+1} &= a_{k+1}A_k + A_{k-1}, \\ B_{k+1} &= a_{k+1}B_k + B_{k-1} \ (k = -1, 0, 1, \dots). \end{aligned} \tag{1.4.7}$$

If α is irrational, then $\alpha_m \neq 0$ for all natural m. Convergents of even order increase; those of odd order decrease, and both sequences converge to α. The limit is denoted as the (infinite) continued fraction

$$\alpha = [a_0; a_1, a_2, \dots, a_n, \dots].$$

This all follows easily from (1.4.7): first we see that

$$\begin{aligned} B_k A_{k-1} - A_k B_{k-1} &= (-1)^k, \ k \geq 1, \\ B_k A_{k-2} - A_k B_{k-2} &= (-1)^{k-1} a_k, \ k \geq 0, \end{aligned} \tag{1.4.8}$$

and then

$$\begin{aligned} \frac{A_{k-1}}{B_{k-1}} - \frac{A_k}{B_k} &= \frac{(-1)^k}{B_k B_{k-1}}, \\ \frac{A_{k-2}}{B_{k-2}} - \frac{A_k}{B_k} &= \frac{(-1)^{k-1} a_k}{B_k B_{k-2}}. \end{aligned} \tag{1.4.9}$$

From (1.4.9) one also deduces that every best approximation to α is equal to a convergent A_m/B_m, because

$$\frac{1}{B_m(B_m + B_{m+1})} < \left| \alpha - \frac{A_m}{B_m} \right| < \frac{1}{B_m B_{m+1}}. \tag{1.4.10}$$

1.4.4 SL_2–Equivalence

The numbers α_m defined by (4.6) are related to α via fractional linear transformations

$$\alpha = \frac{A_{m-1}\alpha_m + A_{m-2}}{B_{m-1}\alpha_m + B_{m-2}}. \tag{1.4.11}$$

Moreover, the determinants of these transformations are $(-1)^m$ (see (1.4.8)). In general, two numbers related by a fractional linear transformation of determinant 1 are called $\mathrm{SL}_2(\mathbb{Z})$–equivalent. Hence α and $(-1)^m\alpha_m$ are equivalent in this sense. Conversely, α and β are equivalent iff $\alpha_m = \beta_n$ for appropriate m and n (cf. e.g. [Khi78]). In particular, all rational numbers are equivalent to one another.

1.4.5 Periodic Continued Fractions and Pell's Equation

Consider an infinite continued fraction which becomes periodic after a certain place k_0, with a period of length k:

$$\alpha = [a_0; a_1, \dots, a_{k_0-1}, \overline{a_{k_0}, \dots, a_{k_0+k-1}}]. \tag{1.4.12}$$

Then from (1.4.11) it follows that α is a quadratic irrational number.

Example. We have

$$\sqrt{3} = [1; 1, 2, 1, 2, 1, 2, \dots],$$

since, denoting by x the r.h.s. continued fraction, we have

$$x = 1 + \frac{1}{1 + \frac{1}{1+x}}$$

that is,

$$2x + x^2 = 3 + 2x$$

and, finally, $x = \sqrt{3}$.

The following algorithm efficiently calculates α_m for a quadratic irrationality α. Let N be square-free,

$$\alpha = (P_0 + \sqrt{N})/Q_0,$$

$N - P_0^2$ being divisible by Q_0. Find successively

$$P_{i+1} = a_i Q_i - P_i,$$

$$Q_{i+1} = (N - P_{i+1}^2)/Q_i, \; i = 0, 1, 2, \dots$$

Then the P_i and Q_i are all integers; Q_i divides $N - P_{i+1}^2$, and

$$\alpha_{i+1} = \frac{P_{i+1} + \sqrt{N}}{Q_{i+1}}.$$

In general, P_i and Q_i do not grow as rapidly as the numerators and denominators of the successive convergents. For example, if $|P_0| < \sqrt{N}, 0 < Q_0 < \sqrt{N}$, we have for all $i \geq 1$:

$$0 < P_i < \sqrt{N}, 0 < Q_i < 2\sqrt{N},$$

$$A_i^2 - NB_i^2 = (-1)^{i+1} Q_{i+1}$$

(cf. [Ries85], [Knu81]). At the ith stage, the calculations consist of four steps.

1) $P_{i+1} = [\sqrt{N}] - R_i$ $(R_0 = 0)$,
2) $Q_{i+1} = (N - P_{i+1}^2)/Q_i$ $(Q_{-1} = (N - P_0^2)/Q_0)$,
3) $a_{i+1} = [(P_{i+1} + [\sqrt{N}])/Q_{i+1}]$,
4) $R_{i+1} =$ the residue of $P_{i+1} + [\sqrt{N}]$ modulo Q_{i+1}.

This algorithm can be used to calculate efficiently the smallest solution to Pell's equation. In fact, if $a^2 - Nb^2 = 1$, then we have $a^2 \geq 1 + N$, $b^2 \geq 1$, so that

$$\left|\frac{a}{b} - \sqrt{N}\right| = \frac{1}{2b^2\sqrt{N}}.$$

Hence a/b is one of the convergents of $\sqrt{N}$.

Example. The smallest solution to $x^2 - 43y^2 = 1$ is $x = 3482$, $y = 531$. Its calculation by the method described above is protocolled in table 1.8.

Table 1.8.

i	-2	-1	0	1	2	3	4	5	6	7	8	9
a_i			6	1	1	3	1	5	1	3	1	1
P_i			0	6	1	5	4	5	5	4	5	1
Q_i			1	7	6	3	9	2	9	3	6	7
R_i			0	5	1	2	1	1	2	1	5	0
A_i	0	1	6	7	13	46	59	341	400	1541	1941	3482
B_i	1	0	1	1	2	7	9	52	61	235	269	531
$A_i^2 - 43B_i^2$			-7	6	-3	9	-2	9	-3	6	-7	1

1.5 Diophantine Approximation and the Irrationality of $\zeta(3)$

1.5.1 Ideas in the Proof that $\zeta(3)$ is Irrational

One of the amazing mathematical inventions of recent time showing the vast undiscovered power of elementary methods in number theory, was the proof of the irrationality of $\zeta(3) = \sum_{n=1}^{\infty} n^{-3}$ found by the French mathematician Apéry. This proof was first presented in June 1978 in the conference Journée Arithmétique de Marseille–Luminy.

We follow here an informal exposition of the proof due to van der Poorten (cf. [vdP79]), who notes the original mistrust of the proof among other mathematicians, which was at first taken as a collection of mysterious statements.

1) For all integers a_1, a_2, ...

$$\sum_{k=1}^{\infty} \frac{a_1 a_2 \cdots a_{k-1}}{(x+a_1)\cdots(x+a_k)} = \frac{1}{x} \qquad (1.5.1)$$

2)

$$\zeta(3) = \frac{5}{2} \sum_{n=1}^{\infty} \frac{(-1)^{n-1} n^3}{\binom{2n}{n}}. \qquad (1.5.2)$$

3) Consider the recurrence relation: for $n \geq 2$

$$n^3 u_n - (34n^3 - 51n^2 + 27n - 5)u_{n-1} + (n-1)^3 u_{n-2} = 0, \qquad (1.5.3)$$

and let b_n be a sequence defined by the initial conditions $b_0 = 1, b_1 = 5$ and the relation (1.5.3). Let a_n be the sequence defined by (1.5.3) and the initial conditions $a_0 = 0$, $a_1 = 6$. Then the denominators of the rational numbers a_n divide $2[1, 2, \ldots, n]^3$ where $[1, 2, \cdots, n]$ denotes the least common multiple of the numbers $1, 2, \ldots, n$.

4) The sequence a_n/b_n converges to $\zeta(3)$ rapidly enough for one to establish irrationality of $\zeta(3)$. Moreover, for $\varepsilon > 0$ and for all integers $p, q > 0$ with q sufficiently large the inequality holds

$$\left|\zeta(3) - \frac{p}{q}\right| > \frac{1}{q^{\theta+\varepsilon}}, \quad \theta = 13.41782... \qquad (1.5.4)$$

One has the following continued fraction expansion:

$$\zeta(3) = \cfrac{6}{p(0) - \cfrac{1^6}{p(1) - \cfrac{2^6}{p(2) - \cfrac{\cdots}{\cdots \cfrac{\cdots}{p(n-1) - \cfrac{n^6}{p(n) - \cdots}}}}}} \qquad (1.5.5)$$

i.e.

$$\zeta(3) = \frac{6}{5-} \; \frac{1 \mid}{117-} \; \frac{64 \mid}{535-} \; \frac{729 \mid}{1436-} \; \frac{4096 \mid}{3105-} \cdots$$

$$\cdots \frac{\mid \quad n^6}{34n^3 + 51n^2 + 27n + 5} \cdots .$$

1.5.2 The Measure of Irrationality of a Number

In §4 of Chapter 1 we noted a link between the property of a number β being irrational and the existence of infinitely many good rational approximations p/q to β, i.e. such that the equality holds

$$\left|\beta - \frac{p}{q}\right| < \frac{1}{q^2}.$$

Analogously one could state the following criterium for the irrationality of a number: if there exists $\delta > 0$ and a sequence $\{p_n/q_n\}$ of rational numbers $\{p_n/q_n\} \neq \beta$ such that

$$\left|\beta - \frac{p_n}{q_n}\right| < \frac{1}{q^{1+\delta}} \quad (n = 1, 2, \ldots,), \tag{1.5.6}$$

then β is an irrational number. The use of this criterium gives an interesting measure of irrationality: if $|\beta - \frac{p_n}{q_n}| < \frac{1}{q^{1+\delta}}$ and q_n steadily increase in such a way that $q_n < q_{n-1}^{1+\kappa}$ for sufficiently large n and $\kappa > 0$ then for any fixed $\varepsilon > 0$ and for all sufficiently large $p, q > 0$ the following equality holds:

$$\left|\beta - \frac{p_n}{q_n}\right| > \frac{1}{q^{(1+\delta)/(\delta-\kappa)+\varepsilon}}. \tag{1.5.7}$$

In the interesting case when q_n increases geometrically, i.e. $q_n, C, \alpha > 0$ one could take for κ an arbitrarily small positive integer, and the exponent in (1.5.7) becomes $1 + (1/\delta)$ which is called *the irrationality degree* of β.

Surprisingly, the method of Apéry turned out also to be applicable to the number

$$\zeta(2) = \sum_{n=1}^{\infty} n^{-2} = \pi^2/6,$$

whose transcendence is well known. However Apéry's proof implies the inequality

$$\left|\pi^2 - \frac{p}{q}\right| > \frac{1}{q^{\theta'+\varepsilon}} \quad \theta' = 11.85078..., \tag{1.5.8}$$

for all $\varepsilon > 0$ and q sufficiently large. One also knows that the irrationality degrees of π^2 and $\zeta(3)$ are not greater than θ and θ' respectively.

1.5.3 The Thue–Siegel–Roth Theorem, Transcendental Numbers, and Diophantine Equations

(cf. [Roth55], [Dav58], [Spr82], [Fel82], [Shid87], [Maz86]). This famous theorem states that if β is an algebraic number, i.e. a root of a polynomial $f(X) = a_n X^n + a_{n-1} X^{n-1} + \cdots + a_0$ ($a_i \in \mathbb{Z}$), then for an arbitrary fixed $\varepsilon > 0$ and all sufficiently large q the following inequality holds:

$$\left|\beta - \frac{p}{q}\right| > \frac{1}{q^{2+\varepsilon}}. \qquad (1.5.9)$$

In other words, if we take arbitrary positive constants C and ε, then there exist only a finite number of approximations x/y of β satisfying the inequality

$$\left|\beta - \frac{x}{y}\right| < \frac{1}{y^{2+\varepsilon}}. \qquad (1.5.10)$$

In particular, if the inequality (1.5.6) holds for a sequence (p_n/q_n) with a fixed $\delta > 1$, then the number β must be transcendental (i.e. not algebraic). However it turns out that this condition defines only a subset of the transcendental numbers of measure zero.

Note that the theorem of Thue–Siegel–Roth has very important applications in the theory of Diophantine equations, which can be explained by the example of the equation

$$X^3 - 5Y^3 = m \quad (m \neq 0) \qquad (1.5.11)$$

for a fixed integer m. This equation resembles Pell's equation, but its degree is greater then 2. If (x, y) a solution of (1.5.11) then the following equality holds:

$$|\frac{x}{y} - \sqrt[3]{5}| \le \frac{c}{|y|^3} \quad (c = \sqrt[3]{m}). \qquad (1.5.12)$$

However if we take $\varepsilon > 0$ such that $2+\varepsilon < 3$ then Roth's theorem implies that there are only finitely many solutions for the inequality (1.5.12) and hence for the equation (1.5.11).

Using algebraic geometric methods, but resting on essentially the same idea, Siegel established the following result:

Theorem 1.6 (Siegel C.–L. (1929)). *Let $f(X,Y)$ be an irreducible polynomial with integer coefficients. Then the equation*

$$f(X,Y) = 0 \qquad (1.5.13)$$

has only finitely many integral solutions excluding the two special cases:

a) The curve $f(X,Y) = 0$ admits a rational parameterization: substituting to (1.5.13) non-zero rational fractions $X = p(t)/q(t)$, $Y = r(t)/s(t) \in \mathbb{Q}(t)$ this equation becomes an identity of rational functions of t.

b) The projective envelope of the curve (1.5.13) has not more than two points at infinity.

In particular, the *Thue equation* $f(x,y) = m$ where $f(x,y) \in \mathbb{Z}[x,y]$ is an irreducible form of the degree $n \geq 3$, has only a finite number of integral solutions. A. O. Gelfond (cf. [Ge83]) has shown that an effective bound for solutions of the Thue equation can be obtained if one has a good lower bound for the module of linear forms of logarithms of algebraic numbers $\alpha_1, \ldots, \alpha_n$ (with integer coefficients). Such estimates were obtained by [Ba71], making it possible to solve a number of important arithmetic problems. These problems include besides bounds for solutions of Diophantine equations ([Spr82], [Step84], [Schm79], [Bak86], [La60], [La62]), also effective bounds for the class numbers of algebraic number fields and the numbers of equivalence classes of quadratic forms ([Ba71], [St67], [St69]). An effective upper bound (see [Bak86] [Spr82], [ShT86])

$$y^v < x^u < \exp\exp\exp\exp 10^3$$

was obtained by *Baker's method* for solutions of the *Catalan equation*

$$x^v - y^u = 1$$

which provide an example of an *exponential Diophantine equation* systematically studied in [ShT86]. Catalan asked in 1843 whether 8 and 9 are the only consecutive perfect powers. A recent solution of this problem by P. Mihăilescu (who answered the question affirmatively) has become one of the main arithmetical highlights of the past few years, cf. [Mih03], [Bi02].

1.5.4 Proofs of the Identities (1.5.1) and (1.5.2)

First of all the equality

$$\sum_{k=1}^{K} \frac{a_1 a_2 \cdots a_{k-1}}{(x+a_1)\cdots(x+a_k)} = \frac{1}{x} - \frac{a_1 a_2 \cdots a_{K-1}}{(x+a_1)\cdots(x+a_K)} \tag{1.5.14}$$

is easy to check. We may write the right hand side in the form

$$A_0 - A_K, \quad A_k = \frac{a_1 a_2 \cdots a_{k-1}}{(x+a_1)\cdots(x+a_k)}$$

and note that each term in the left hand side is equal to $A_{k-1} - A_k$. The identity (1.5.1) follows immediately from (1.5.14).

Now substituting $x = n^2$ and $a_k = -k^2$ and taking $k \leq K \leq n-1$ we obtain

$$\begin{aligned}\sum_{k=1}^{n-1} \frac{(-1)^{k-1}(k-1)!^2}{(n^2-1^2)\cdots(n^2-k^2)} &= \frac{1}{n^2} - \frac{(-1)^{n-1}(n-1)!^2}{n^2(n^2-1^2)\cdots(n^2-(n-1)^2)}, \\ &= \frac{2(-1)^{n-1}}{n^2\binom{2n}{n}}.\end{aligned}$$

Writing $\varepsilon_{n,k} = \dfrac{1}{2}\dfrac{k!(n-k)!}{k^3(n+k)!}$, we have

$$(-1)^k n(\varepsilon_{n,k} - \varepsilon_{n-1,k}) = \frac{(-1)^{k-1}(k-1)!^2}{(n^2-1^2)\cdots(n^2-k^2)}$$

from which follows the identity

$$\sum_{n=1}^{N}\sum_{k=1}^{n-1}(-1)^k(\varepsilon_{n,k} - \varepsilon_{n-1,k}) = \sum_{n=1}^{N}\frac{1}{n^3} - 2\sum_{n=1}^{N}\frac{(-1)^{n-1}}{n^3\binom{2n}{n}} =$$
$$\sum_{k=1}^{N}(-1)^k(\varepsilon_{N,k} - \varepsilon_{k,k}) = \sum_{k=1}^{N}\frac{(-1)^k}{2k^3\binom{N+k}{k}\binom{N}{k}} + \sum_{k=1}^{N}\frac{(-1)^{k-1}}{k^3\binom{2k}{k}}. \tag{1.5.15}$$

The equality (1.5.2) is implied by (1.5.15) on noting that the sum

$$\sum_{k=1}^{N}\frac{(-1)^k}{2k^3\binom{N+k}{k}\binom{N}{k}}$$

tends to zero as $N \to \infty$.

1.5.5 The Recurrent Sequences a_n and b_n

Write the recurrence relation (4.3) satisfied by a_n and b_n:

$$\begin{aligned} n^3 a_n - P(n-1)a_{n-1} + (n-1)^3 a_{n-2} &= 0,\\ n^3 b_n - P(n-1)b_{n-1} + (n-1)^3 b_{n-2} &= 0, \end{aligned}$$

where $P(n-1) = p(n) = 34n^3 - 51n^2 + 27n - 5$. If we multiply the first equality by b_{n-1} and the second by a_{n-1}, and then subtract second from the first, we get

$$n^3(a_n b_{n-1} - a_{n-1}b_n) = (n-1)^3(a_{n-1}b_{n-2} - a_{n-2}b_{n-1}).$$

Recall that by the initial conditions we have $a_1 b_0 - a_0 b_1 = 6\cdot 1 - 0\cdot 5 = 6$, which implies

$$a_n b_{n-1} - a_{n-1}b_n = \frac{6}{n^3}. \tag{1.5.16}$$

This easily leads to the relation

$$\left|\zeta(3) - \frac{a_n}{b_n}\right| = \sum_{k=n+1}^{\infty}\frac{6}{k^3 b_k b_{k-1}} = O(b_n^{-2}). \tag{1.5.17}$$

This is proved by the induction starting from the equality $\zeta(3) - \frac{a_0}{b_0} = \zeta(3)$. The absolute values of the numbers b_n can be easily estimated using the relation

$$b_n - (34 - 51n^{-1} + 27n^{-2} - 5n^{-3})b_{n-1} + (1 - 3n^{-1} + 3n^{-2} - n^{-3})b_{n-2} = 0.$$

Using the fact that the "linearized characteristic polynomial" of this recurrence relation is $x^2 - 34x + 1$ and has roots $17 \pm 2\sqrt{2} = (1 \pm \sqrt{2})^4$, we obtain the estimate

$$b_n = O(\alpha^n), \quad \alpha = (1 + \sqrt{2})^4.$$

Assume for a moment that the statement in 3) on integrality of the numbers b_n and on the denominators of a_n dividing $2[1, 2, \cdots, n]^3$ is already proved. Then it is easy to complete the proof as follows. Let

$$p_n = 2[1, 2, \cdots, n]^3 a_n, \; q_n = 2[1, 2, \cdots, n]^3 b_n,$$

where $p_n, q_n \in \mathbb{Z}$. The value of $[1, 2, \cdots, n]$ can be estimated using for example a rough form of the prime number theorem: $\sum_{p \le x} 1 \approx x/\log x$. Then

$$[1, 2, \cdots, n] = \prod_{p \le n} p^{[\log n/\log p]} \le \prod_{p \le n} n \approx n^{n/\log n} = e^n.$$

Hence $q_n = O(\alpha^n e^{3n})$ and

$$\left| \zeta(3) - \frac{p_n}{q_n} \right| = O(b_n^{-2}) = O(\alpha^{-2n}) = O(q_n^{-(1+\delta)})$$

with the constant $\delta = (\log \alpha - 3)/(\log \alpha + 3) = 0.080529... > 0$. According to the irrationality criterium in §1.5.2, we obtain the statement (1.5.6) in which the irrationality degree is not greater than $1 + (1/\delta) = \theta$.

The statement on the denominators of the numbers a_n and b_n is one of the most difficult points of the proof. Apéry proved this fact by explicitly constructing the sequences a_n and b_n:

$$c_n = \sum_{k=0}^{n} \binom{n}{k}^2 \binom{n+k}{k}^2, \quad a_n = \sum_{k=0}^{n} \binom{n}{k}^2 \binom{n+k}{k}^2 c_{n,k},$$

where

$$c_{n,k} = \sum_{m=1}^{n} \frac{1}{m^3} + \sum_{m=1}^{k} \frac{(-1)^{m-1}}{2m^3 \binom{n}{m}\binom{n+m}{m}}. \tag{1.5.18}$$

It follows from these formulae that the numbers a_n are integral. The bound on the denominators of b_n is given by the fact that all of the numbers

$$2[1, 2, \cdots, n]^3 c_{n,k} \binom{n+k}{k}$$

are integers. The proof of this uses an estimate for the maximal power with which a prime p can arise in the denominator of each term in the sum (1.5.18) defining $c_{n,k}$.

Further irrationality properies of zeta values at odd positive integers were studied recently in [Riv01], [Zu95], [BaRi01], see also [Zu]. In particular, W.Zudilin proved that at least one of $\zeta(5), \zeta(7), \zeta(9), \zeta(11)$ is irrational.

There are interesting attempts to prove the irrationality of Euler's constant and to understand its arithmetic nature, cf. [Son04]. In [BelBr03] Euler's constant γ is interpreted as an *exponential period.* The ring P of periods is generated by the numbers of the form $\int_\gamma \omega$ where X is a smooth algebraic variety of dimension d defined over $\mathbb{Q}$, $D \subset X$ is a divisor with normal crossings, $\omega \in \Omega^d(X)$ is an algebraic differential form of degree d on $X, \gamma \in H_d(X(\mathbb{C}), D(\mathbb{C}); \mathbb{Q})$, cf. Chapter 5. This ring was introduced by M. Kontsevich and D. B. Zagier in [KoZa01] (see also [Del79]).

1.5.6 Transcendental Numbers and the Seventh Hilbert Problem

It is useful to compare the given elementary proof with the highly developed theory of transcendental numbers, i.e. the numbers, which are not roots of polynomials with rational coefficients. The existence of such numbers was first established by Liouville in 1844; then Hermite proved the transcendence of e (in 1873), and Lindeman in 1883 proved the transcendence of π ([Ba75], [Bak86], [Shid87]). In the framework of a general theory A. O. Gelfond (see in [Ge73]) and Th. Schneider (cf. [Sch57]) obtained a solution to the seventh Hilbert problem (Hilbert D. (cf. [Hil1900]), [Fel82]): to prove that "the power α^β of an algebraic base α to an irrational algebraic exponent β (e.g. $2^{\sqrt{2}}$ or $e^\pi = i^{-2i}$ is always transcendental, or at least irrational"; "... we found it very probable that such a function as $e^{\pi i z}$, which evidently takes algebraic values for all rational values of the argument z, will take, on the other hand, for algebraic irrational values of z, only transcendental values".

1.5.7 Work of Yu.V. Nesterenko on e^π, [Nes99]

One of the most impressive achievments of the last decade in the theory of transcendental numbers was the work of Yu.V. Nesterenko on the algebraic independence of π and e^π, see [Nes99] and [Nes02]. This result is based on the study of the transcendence degree of a field generated by numbers connected with the modular function $j(\tau)$. In [Nes99] the algebraic independence of π, e^π and $\Gamma(\frac{1}{4})$ is also established by this powerful method. For proving this result, the problem is reduced to estimating the measure of algebraic independence for the numbers π and $\Gamma(\frac{1}{4})$.

2

Some Applications of Elementary Number Theory

2.1 Factorization and Public Key Cryptosystems

2.1.1 Factorization is Time-Consuming

In order to multiply two primes $p \leq q$ given their binary expansions, it suffices to perform $C(\log q)^2$ bit-operations (see section 1.1.1). Suppose now that we are given $n = pq$ and are asked to find p and q. If $p \sim q \sim \sqrt{n}$ then the naive repeated trial of all $d \leq \sqrt{n}$ would require more than

$$C\sqrt{n} = C\exp\left(\frac{1}{2}\log n\right)$$

divisions with remainder. This exponential growth of the running time makes the factorization of even rather small numbers unfeasible, at least unless one invents more efficient algorithms. For example, consider the factorization

$$(10^{71} - 1)/9 = 241573142393627673576957439049\times \\ 45994811347886846310221728895223034301839. \tag{2.1.1}$$

With some patience, one can multiply the two numbers on the right hand side in an hour or two on a sheet of paper. However, the factorization of the result by the trial-and-error method would take about 10^{10} years of running time (if one division requires 10^{-9} sec: cf. [Sim79], [Pet85], [Wun85], [Ya02]).

In real life, the factorization (2.1.1) was first found in 1984 with the assistance of a CRAY supercomputer and fairly advanced factorization methods, which made this task feasible if not inexpensive.

2.1.2 One–Way Functions and Public Key Encryption

We may consider the binary expansion of $n = pq$ as a message which can be encoded in many other ways, e.g., by giving expansions of p and q. The rules explaining how to pass from one form to another from the information-theoretical viewpoint can be called *enciphering, encryption* and *deciphering.* Experimentally, one knows that some functions are easy to compute but difficult to invert (*one–way,* or *trap–door* functions). It is then natural to try to use these functions in cryptography. We recall that cryptography studies problems of information handling concerned with keeping and breaking secrecy of messages. One–way functions are used in the so called public key encryption schemes, which were suggested in the seventies and revolutionized this domain.

Before explaining the design of one such scheme, we must stress however that there are no *theoretical* lower bounds on computational complexity justifying our experimental observation that complexity of factorization far exceeds that of multiplication. In principle, we cannot exclude the possibility that a very efficient algorithm for factorization (or for inverting any given trap–door function) might eventually be found. This is one of the basic problems of computational complexity theory (cf. e.g. [GJ79], [DH76] [CoLe84], [ARS78], [Ya02]). If, however, we assume this experimental fact, we can use it in order to generate new encryption schemes with remarkable properties.

We shall now describe the first "public key cryptosystem" suggested by L.Adleman, R.Rivest, and A.Shamir in 1978, cf. [ARS78].

2.1.3 A Public Key Cryptosystem

Imagine a system of users U_1, U_2, $U_3, \ldots$ From time to time any pair of users may need to exchange messages that should remain secret to other users or outsiders.

In a classical cryptosystem, they should first share keys and keep them secret. A public key system avoids this last restriction: secret pairwise communication becomes possible using only information open to everybody. Such a system can be devised as follows.

a) Every user U_i choses two large primes p_i and q_i, and two residue classes $e_i, d_i \bmod n_i$, where $n_i = p_i q_i$, such that $e_i d_i \equiv 1 \bmod \varphi(n_i)$ where $\varphi(n_i) = (p_i - 1)(q_i - 1)$ denotes the Euler function (cf. 1.1.4).
b) The numbers (e_i, n_i) are made public for all users.

   We argue that it is unfeasible to calculate d_i knowing only (e_i, n_i), so that d_i can be considered as a secret known to U_i alone. In fact, we shall show that an efficient algorithm for calculating d_i would also find efficiently the prime factorization of n_i, which we assumed to be difficult. Suppose that we know d_i. We then know that $\varphi(n_i)$ divides $e_i d_i - 1$. If we knew $\varphi(n_i)$ itself then we could easily find p_i and q_i, since $p_i + q_i = n_i + 1 - \varphi(n_i)$ and $p_i - q_i = \sqrt{(p_i + q_i)^2 - 4n_i}$. One can show that even knowing only a multiple of $\varphi(n_i)$ suffices (cf. [Mil76], [Wag86]) to find p_i and q_i.

c) Suppose that a user U_i wishes to transmit to U_j a coded message which is a sequence of bits. He first breaks this sequence up into blocks of length $[\log_2 n_j]$, then considers each block as a residue class $m \bmod n_j$ and finally encodes it as the residue class $m^{e_j} \bmod n_j$. Thus, (n_j, e_j) serves as the encryption key of the jth user (recall that it is common knowledge).
d) Having received the encoded message, U_j decodes any block $b \bmod n_j$ by computing $b^{d_j} \bmod n_j$ (recall that he knows the deciphering key d_j). This is easily checked with the help of Fermat's little theorem (1.1.4).

Clearly, the details of such a scheme can be varied ad infinitum. For example, one can devise an authentification procedure ("electronic signature") which uses a form of a secret message from U_i to U_j allowing U_j to convince a third party (a "judge") that the author of the message is U_i, so that it is not faked by U_j himself. This can be crucial for certain financial transactions.

Denote by E_i the encoding map for messages addressed to U_i and by D_i his deciphering map. Then E_i is public domain while D_i is U_i's property. For an arbitrary plain message M we have $D_i(E_i(M)) = M$ and $E_i(D_i(M)) = M$. The user U_i sending his message M to U_j uses as his signature $S = D_i(M)$ and transmits to U_j its encoded version $E_j(S)$. In his turn, U_j first computes $S = D_j(E_j(S)$ and then $M = E_i(S)$ using the public key E_i. The addressee can convince a judge that M comes from U_i because only by applying E_i can one transform S into a given sensible message M. On the other hand, the addressee cannot fake S since he does not know D_i.

We shall concentrate now on the number–theoretical rather than the information–theoretical aspects of public key cryptosystems. We shall describe how some classical number–theoretical results can be applied to two particular problems in this domain.

Problem 1. How does one produce large primes?

We want to stress that we really need an efficient method for mass production of "sufficiently random" large primes, in order to allow a user to compute (with the assistance of a large computer) his customized pair (p_i, q_i), and to be sure that a different user will get a different pair.

Problem 2. How does one factorize large integers?

This problem is crucial for a third party wanting to break the cryptosystem and, of course, for the designers wanting to secure its infallibility (cf. [DH76], [ARS78], [Kah71]).

According to A.Wiles [Wi2000], one change in number theory over the last twenty years is that it has become an applied subject (Pehaps one should say it has gone back to being an applied subject as it was more than two thousand years ago). Public key cryptography has changed the way we look at secrecy and codes. The RSA system depends on the practical difficulty of factoring a number. The seventeenth century problems of generating large primes, primality testing and factoring now pose new and precise problems. How fast can algorithms for answering these questions be? The question of primality testing is solved (cf. §2.2.4, §2.2.6). The other two are not theoretically solved, although the first of the problems seems much easier in practice than the third.

2.1.4 Statistics and Mass Production of Primes

The asymptotic law of the distribution of primes (or prime number theorem) is $\pi(x) \sim \dfrac{x}{\log x}$ (cf. 1.1.6). We can start then with a naive assumption that if N is not too small with respect to x then between x and $x+N$ there should be about $N/\log x$ primes. For example, if the least prime following x is bounded by $x + (\log x)^M$, then one can just check successively $x, x+1, x+2, \ldots$. The complexity of the prime production would then be of the same order as the complexity of the primality testing algorithm used. If one can take $M = 1$, then to produce a prime of order about 2^{100} one should first produce a random number x of that order, and then test about $(\log 10^{100})/2 \approx 115$ odd integers. If there is a primality test for y, which is polynomial in $\log y$, then this is a feasible task.

We shall discuss in the following subsection efficient probabilistic primality tests.

We should remark however that such absence of large gaps between primes is not proved and probably is not even true. All known results on the gaps give upper bounds which are powers of x (see [HB88], [Hild88], [Zag77]). We quote some of them:

$$\pi(x + x^{7/12}) - \pi(x) = \frac{x^{7/12}}{\log x}\left(1 + O\left(\left(\frac{\log\log x}{\log x}\right)^4\right)\right), \tag{2.1.2}$$

$$\pi(x + x^{\theta}) - \pi(x) \geq C(\theta)\frac{x^{\theta}}{\log x}$$

for $\theta \geq 11/20$ where $C(\theta)$ is a positive function. For almost all x, a stronger result is known:

$$\pi(x + x^{\theta}) - \pi(x) \geq 0.15\frac{x^{\theta}}{\log x},$$

if $\theta \geq 1/12$.

For an interesting discussion of large gaps between primes, see [Ries85], p.84 and [Hild88], [Zag77]).

2.1.5 Probabilistic Primality Tests

Some modern efficient primality tests actually check a weaker property connected with the notion of Eulerian pseudoprimes (cf. §1.1.5). We recall that n is called an Eulerian pseudoprime modulo b if n and b are relatively prime, and

$$b^{(n-1)/2} \equiv \left(\frac{b}{n}\right) \bmod n. \qquad (2.1.3)$$

Primes are pseudoprimes modulo every b: this follows from the fact that $(\mathbb{Z}/n\mathbb{Z})^\times$ is cyclic (cf. section 1.1.5). One readily sees that for composite n, (2.1.3) fails for at least half of the residue classes in $(\mathbb{Z}/n\mathbb{Z})^\times$. A probabilistic primality test based upon this observation consists of checking (2.1.3) for, say, several hundred randomly chosen b. If an n passes such a test it is sometimes called "a commercial prime". Commercial primes are used in public key cryptosystems although, strictly speaking, their primality is not established by the test.

It turns out that such a proof could be given if one assumes the *generalized Riemann conjecture* on the zeroes of the Dirichlet L–functions. Namely, one can deduce from this conjecture that the validity of (2.1.3) for all $b < 2(\log n)^2$ implies that n is prime (cf. [Mil76], [Wag86]). To check this property, it suffices to perform $O((\log n)^{4+\epsilon})$ divisions with remainder, for any $\epsilon > 0$.

This *Solovay-Strassen primality test* admits some interesting variations, e.g., the *Miller–Rabin test* (see [SolSt77]), [Mil76], [Ra80], [Ries85], [Schr84], [Kob94]). It is based on the following notion of *strict pseudoprimality*. Suppose that n is pseudoprime modulo b, so that $b^{n-1} \equiv 1 \bmod n$. We shall now calculate all consecutive square roots of the left hand side, that is, $b^{(n-1)/2^i}$ for $i = 1, \dots, s$ where $t = (n-1)/2^s$ is odd. If n is prime then the first residue class in this sequence distinct from 1 should be -1. We shall call n *strict pseudoprime* if either $b^t \equiv 1 \bmod n$ or for some $0 \le r < s$ we have

$$b^{2^r t} \equiv -1 \bmod n. \qquad (2.1.4)$$

The Miller-Rabin test consists of checking this property for a set of randomly chosen b.

It was noticed by F.Morain [Mor03a] that in practice the algorithm of Miller is quite long, because it needs to compute numerous modular exponents, and a faster ECPP method is discussed in Section 2.2.6.

In Section 2.2.4 we describe an important recent theoretic discovery that primes are recognizable in polynomial time: the work of M. Agrawal, N. Kayal and N. Saxena [AKS] who found that a polynomial version of Fermat's Little Theorem (1.1.2) leeds to a fast deterministic algorithm for primality testing: the time of this algorithm is given by $\widetilde{\mathcal{O}}(\log^{12} N)$, where the notation $\widetilde{\mathcal{O}}(t(N))$ for $\mathcal{O}(t(n) \cdot poly(\log t(N)))$ is used for a function $t(N)$ of N.

2.1.6 The Discrete Logarithm Problem and The Diffie-Hellman Key Exchange Protocol

The Diffie-Hellman key exchange is the first public-key cryptosystem ever published [DH76].

In order to communicate an important information to Bob, Alice wish to use this algorithm as follows: Alice and Bob agree on a prime number p and an integer g that has order $p-1$ modulo p. (So $g^{p-1} \equiv 1\,(\mathrm{mod} p)$, but $g^n \not\equiv 1\,(\mathrm{mod} p)$ for any positive $n < p-1$.) Alice chooses a random number $n < p$, and Bob chooses a random number $m < p$. Alice sends $g^n \mathrm{mod} p$ to Bob, and Bob sends $g^m \mathrm{mod} p$ to Alice. Alice can now compute the secret key:

$$s = g^{mn} = (g^m)^n\,(\mathrm{mod} p).$$

Likewise, Bob computes the secret key:

$$s = g^{mn} = (g^n)^m\,(\mathrm{mod} p).$$

Now Alice uses the secret key s to send Bob an encrypted version of her message. Bob, who also knows s, is able to decode the message.

Non-authorized persons can see both $g^n\,(\mathrm{mod} p)$ and $g^m\,(\mathrm{mod} p)$, but they aren't able to use this information to deduce either m, n, or $g^{mn}\,(\mathrm{mod} p)$ quickly enough.

2.1.7 Computing of the Discrete Logarithm on Elliptic Curves over Finite Fields (ECDLP)

The Abelian group structure on the group of points $E(\mathbb{F}_p)$ on an elliptic curves is used in many arithmetical questions. In particular, the case when this group is cyclic of large size leads to ECDLP ("Elliptic curve discrete logarithm problem") which is extremely important for applications in public-key cryptography, see [Kob87], [Kob98], [Kob01], [Fr01], [Men93], [Kob02]. This idea was independently proposed by Neal Koblitz and Victor Miller in 1985, and since then there has been an enormous amount of research on the topic. The computational problem on which the security depends is the elliptic curve discrete logarithm problem: Given an elliptic curve E over a finite field $\mathbb{F}_q$ and two points $P, Q \in E(\mathbb{F}_q)$, find an integer λ (if it exists) such that $Q = [\lambda]P$. If the field size q is sufficiently large, and if the elliptic curve E avoids various special cases, then this seems to be a difficult computational problem.

Numerous applications of arithmetical algebraic geometry to cryptographic constructions were discussed in [Fr01], [Men93], and in other good sources: [Kob2000], where the problem of computing the orders of elliptic curve groups is discussed in some detail as well. For example, we can learn about the cryptographic significance of old number-theoretic questions such as the existence of infinitely many Sophie Germain primes and Mersenne primes.

2.2 Deterministic Primality Tests

Probablilstic polynomial-time primality tests have been known for many years. There is a well–known almost-polynomial-time ($(\log n)^{\log\log\log n}$) deterministic algorithm due to Adleman, Pomerance and Rumely (1983) cf. [APR83], [LeH.80], [CoLe84], and also a randomized algorithms due to Goldwasser–Kilian, cf. [GK86], [GK99], Atkin–Morain [AtMo93b], and Adleman–Huang [AdHu92] which give certificates for both primality and compositeness in expected polynomial time on all inputs. This method of primality proving using elliptic curves, the ECPP was further developed by F.Morain, [Mor98a].

In August 2002, a deterministic polnomial-time algorithm was found by M. Agrawal, N. Kayal and N. Saxena from the IIT Kanpur. Among other things, we give an exposition of this result in this section.

We describe some *deterministic primality tests*

a) Adleman, Pomerance and Rumely (1983): they have subexponential running time, and the proofs that they work are unconditional (i.e. they do not use any unproved conjectures).
b) A resent discovery that primes are recognizable in polynomial time by M. Agrawal, N. Kayal and N. Saxena who found that a polynomial version of Fermat's Little Theorem (1.1.2) led to a fast deterministic algorithm for primality testing: The time of this algorithme is given by $\widetilde{\mathcal{O}}(\log^{12} n)$, where the notation $\widetilde{\mathcal{O}}(t(n))$ for $\mathcal{O}(t(n)\cdot poly(\log t(n)))$ is used for a function $t(n)$ of n.
c) Elliptic curves and primality proving, the ECPP (Elliptic Curve Primality Proving by F.Morain, see [AtMo93b], [Mor98a], [Mor03].

2.2.1 Adleman–Pomerance–Rumely Primality Test: Basic Ideas

There are two main variants of this algorithm cf. [APR83], [LeH.80], [CoLe84]: a simpler, probabilistic version, and a deterministic one. Its running time is bounded by

$$\log n^{c\log\log\log n},$$

where c is an effective constant. The power in this expression grows so slowly that this bound can be considered "almost polynomial". All previously known deterministic primality tests had exponential running time (e.g., Pollard's test, described in [Pol74], [Ries85], requires about

$$n^{\frac{1}{8}+\epsilon} = \exp\left(\left(\frac{1}{8}+\varepsilon\right)\log n\right)$$

operations).

The algorithm consists of the following steps.

a) One checks a series of conditions generalizing the congruence (2.1.3) for the Jacobi symbol. If n fails to satisfy any of these conditions, then it is composite.
b) If n passes the first stage, the test furnishes a small set of integers containing all divisors r of n not exceeding $\sqrt{n}$. It remains to check whether n is divisible by at least one element of this set.
c) The set of potential divisors r is determined by specifying their residue classes modulo an integer $s > \sqrt{n}$, which in turn is a product of several distinct primes q. In view of the Chinese Remainder Theorem (cf. (1.1.5)), it suffices to specify $r \bmod q$ for all q dividing s.
d) Every q dividing s satisfies the following condition: $q-1$ is a product of several distinct primes taken from a fixed set $\{p_0, \ldots, p_k\}$. These primes are called the *initial primes*, and the q are called the *Euclidean primes*, because they are constructed by the method used in Euclid's proof that the set of primes is infinite:

$$q = 1 + p_0^{\alpha_0} p_1^{\alpha_1} \ldots p_k^{\alpha_k}, \ \alpha_i = 0 \ or \ 1.$$

To estimate the running-time, one has to use a hard theorem from analytic number theory (cf. [Pra57]) which guarantees that even for a small set of initial primes, the product of all Euclidean primes generated by them can be large. More precisely, given n, one can determine a set of initial primes $\{p_0, \ldots, p_k\}$ whose product t is bounded by

$$t = \prod_{i=0}^{k} p_i < \log(n^{c_2 \log\log\log n}) \ (n > e^e), \tag{2.2.1}$$

whereas the product of the corresponding Euclidean primes is bounded from below by

$$s = \prod_{(q-1)|t} q > \sqrt{n}, \tag{2.2.2}$$

where c_2 is a computable positive constant. Notice that in this situation the number of Euclidean primes is bounded by $\pi(t+1) < t+1$. For any $n \le 10^{350}$ one can take $t = 2 \cdot 3 \cdot 5 \cdot 7 \cdot 11 \cdot 13 \cdot 17 \cdot 19$.

e) To determine $r \bmod q$, one actually calculates the discrete logarithms $\mathrm{ind}(r, g, q)$ of all possible r with respect to a fixed generator g of $(\mathbb{Z}/q\mathbb{Z})^\times$. These logarithms are in turn determined by their residue classes $\mathrm{ind}(r, g, q) \bmod p_i$ where p_i runs over all initial primes. Again, this follow from the Chinese Remainder Theorem.

We shall now describe the algorithm in more detail.

2.2.2 Gauss Sums and Their Use in Primality Testing

For an odd q, the Euler criterion $\left(\frac{q}{n}\right) \equiv q^{(n-1)/2} \mod n$ can be rewritten in the form

$$\left(\frac{n}{q}\right) \equiv (-1)^{\frac{n-1}{2}\cdot\frac{q-1}{2}} q^{\frac{n-1}{2}} \mod n, \tag{2.2.3}$$

which gives a formula for calculating the quadratic residue symbol of n modulo q. In the algorithm we discuss here, one uses generalizations of this formula to arbitrary p^{th} power residue symbols for initial primes p. In order to explain these generalizations, we must introduce Gauss sums, which were initially used in one of Gauss' proof of the quadratic reciprocity law (cf. below).

One calculates the number of solutions of a congruence $x^p = a$ in $(\mathbb{Z}/q\mathbb{Z})^\times$ with the help of the *Dirichlet characters* of order p modulo q, that is, the homomorphisms $\chi : (\mathbb{Z}/q\mathbb{Z})^\times \to \mathbb{C}^\times$. Every such character is defined by the image $\exp(k \cdot 2\pi i/p)$ of a generator g of $(\mathbb{Z}/q\mathbb{Z})^\times$. The number of such characters is p if p divides $q-1$, and 1 otherwise. If q is prime, we have

$$\mathrm{Card}\{x \in (\mathbb{Z}/q\mathbb{Z})^\times | x^p = a\} = \sum_{\chi|\chi^p=1} \chi(a). \tag{2.2.4}$$

In particular, for $p = 2$ this is $1 + \left(\frac{a}{q}\right)$. The sum in the right hand side of (2.2.4) vanishes iff $\chi(a) \neq 1$ for some χ. This happens only if $p|(q-1)$ and a is not a pth power modulo q. If p does not divide $q-1$, both sides are equal to 1. Finally, if $p|(q-1)$ and a is a pth power, both sides are equal to p.

One way to understand Gauss sums is to view them as discrete analogues of the *gamma function* $\Gamma(s)$, which for $\mathrm{Re}(s) > 0$ is given by the integral

$$\Gamma(s) = \int_0^\infty e^{-y} y^s \frac{dy}{y}. \tag{2.2.5}$$

Here the integrand is the product of an additive quasicharacter of $\mathbb{R}$ (the homomorphism $y \mapsto e^{-y}$) and a multiplicative quasicharacter $y \mapsto y^s$ of $\mathbb{R}_+^\times$. One integrates this over the positive reals with respect to the multiplicative invariant measure $\frac{dy}{y}$.

In order to get a Gauss sum, one should replace here $\mathbb{R}$ by $\mathbb{Z}/N\mathbb{Z}$ for some $N > 1$; e^{-y} by an additive character $\mathbb{Z}/N\mathbb{Z} \to \mathbb{C}^\times : y \mapsto \zeta_N^y, \zeta_N = \exp\left(\frac{2\pi i}{n}\right)$, and y^s by a multiplicative character $\chi : (\mathbb{Z}/N\mathbb{Z})^\times \to \mathbb{C}^\times$. A Dirichlet character $\chi : \mathbb{Z} \to \mathbb{C}$ corresponding to χ and denoted also χ is defined by $\chi(a) = \chi(a \bmod N)$ for $(a, N) = 1$ and by $\chi(a) = 0$ for $(a, N) > 1$. The *Gauss sum* $G(\chi)$ is, by definition,

$$G(\chi) = \sum_{x=1}^{N-1} \chi(x)\zeta_n^x. \tag{2.2.6}$$

For $a \in \mathbb{Z}$, the following notation is often used:

$$G_a(\chi) = \sum_{x=1}^{N-1} \chi(x)\zeta_n^{ax}.$$

Since the formulae (2.2.5) and (2.2.6) are obviously similar, they define functions with many similar properties.

To state them, we need the important notion of a *primitive* Dirichlet character. A character χ is primitive modulo N if it is not induced by a character modulo M for any proper divisor M of N. Equivalently, the restriction of χ to any subgroup $H_M = ((1+M\mathbb{Z})/(1+N\mathbb{Z}))^\times$ is non–trivial. If χ is primitive, we have

$$G_a(\chi) = \overline{\chi}(a)G(\chi) \quad (a \in \mathbb{Z}), \tag{2.2.7}$$

$$\overline{G(\chi)} = \chi(-1)G(\overline{\chi}), \tag{2.2.8}$$

$$|G(\chi)|^2 = N. \tag{2.2.9}$$

Property (2.2.7) corresponds to the integral formula

$$\int_0^\infty e^{-ay}y^s \frac{dy}{y} = a^{-s}\Gamma(s) \ (\mathrm{Re}(s) > 0),$$

and (2.2.9), rewritten in the form $G(\chi)G(\chi^{-1}) = \chi(-1)N$, corresponds to the functional equation

$$\Gamma(s)\Gamma(1-s) = -\frac{\pi}{s\sin \pi s}.$$

From (2.2.7)–(2.2.9) one readily deduces the *quadratic reciprocity law*. Let us prove, for example, the main formula

$$\left(\frac{l}{q}\right)\left(\frac{q}{l}\right) = (-1)^{\frac{l-1}{2}\cdot\frac{q-1}{2}}, \tag{2.2.10}$$

where l and q are odd primes. Notice first that the quadratic residue symbol $\chi(a) = \left(\frac{a}{q}\right)$ is a primitive Dirichlet character modulo q. The corresponding quadratic Gauss sum $G(\chi)$ is an element of the cyclotomic ring of algebraic integers $R = \mathbb{Z}[\zeta_q]$. In any commutative ring the congruence $(a+b)^l \equiv a^l + b^l \bmod lR$ holds because the binomial coefficients C_l^i are divisible by l. Since $\chi^l(a) = \chi(a) = \pm 1$, we have

$$G(\chi)^l \equiv G_l(\chi^l) \bmod lR, \ G_l(\chi^l) = \overline{\chi(l)}G(\chi),$$

so that

$$G(\chi)^{l-1} \equiv \left(\frac{l}{q}\right) \mod lR. \tag{2.2.11}$$

On the other hand, $\chi = \overline{\chi}$, and from (2.2.9) it follows that

$$G(\chi)^2 = \chi(-1)q = (-1)^{\frac{q-1}{2}} q. \tag{2.2.12}$$

Representing the left hand side of (2.2.11) as $G(\chi)^{2\frac{l-1}{2}}$ we obtain

$$(-1)^{\frac{q-1}{2} \cdot \frac{l-1}{2}} q^{\frac{l-1}{2}} \equiv \left(\frac{l}{q}\right) \mod lR. \tag{2.2.13}$$

Finally, (2.2.13) and Euler's criterion

$$q^{\frac{l-1}{2}} \equiv \left(\frac{l}{q}\right) \mod l$$

give (2.2.10).

For $\mathbb{Z}/N\mathbb{Z}$, there is also an analogue of the *beta-function*

$$B(s,t) = \int_0^1 x^{s-1}(1-x)^{t-1}dx =$$

$$\int_{\mathbb{R}_+^\times} \frac{y^s}{(1+y)^{s+t}} \frac{dy}{y} \qquad (\mathrm{Re}(s), \mathrm{Re}(t) > 0).$$

It is called *the Jacobi sum* depending on two Dirichlet characters $\chi, \psi \bmod N$. By definition,

$$J(\chi,\psi) = \sum_{x \bmod N} \chi(x)\psi(1-x) = \sum_{y \bmod N} \chi(y)\overline{(\chi\psi)}(1+y). \tag{2.2.14}$$

(The equality of these two expressions can be established by the change of variables $y(1-x) \mapsto x, x(1+y) \mapsto y$). If χ, ψ, and $\chi\psi$ are primitive modulo N, we have

$$J(\chi,\psi) = G(\chi)G(\psi)/G(\chi\psi) = J(\psi,\chi), \tag{2.2.15}$$

which corresponds to the classical identity $B(s,t) = \Gamma(s)\Gamma(t)/\Gamma(s+t)$. In fact, let us calculate the product

$$G(\chi)G(\psi) = \sum_{x \bmod N} \chi(x)\zeta_n^x G(\psi) = \sum_{x \bmod N} \chi\psi(x)\zeta_n^x \overline{\psi(x)} G(\psi). \tag{2.2.16}$$

Applying (2.2.7), we get

$$\overline{\psi(x)}G(\psi) = G_x(\psi) = \sum_{y \bmod N} \zeta_N^{xy}\psi(y)$$

so that (2.2.16) becomes

$$\sum_{x,y \bmod N} (\chi\psi)(x)\psi(y)\zeta_N^{x(1+y)} = \sum_{y \bmod N} \psi(y)G_{1+y}(\chi\psi) =$$

$$\sum_{y \bmod N} \psi(y)\overline{(\chi\psi)}(1+y)G(\chi\psi) = J(\psi,\chi)G(\chi\psi).$$

We now establish some congruences useful in primality testing. Let p and q be primes, $p|(q-1)$, χ a Dirichlet character of degree p modulo q. Choose a generator $t = t_q$ of $(\mathbb{Z}/q\mathbb{Z})^\times$ and put $\eta_p = \chi(t_q)$. This is a primitive p^{th} root of unity, and $G(\chi) \in R = \mathbb{Z}[\zeta_p, \zeta_q] = \mathbb{Z}[\zeta_{pq}]$. Now let l be a prime distinct from p and q. From (2.2.7) one deduces that

$$G(\chi)^l \equiv \chi(l)^{-l}G(\chi^l) \bmod lR. \tag{2.2.17}$$

Iterating this $p-1$ times, we obtain

$$G(\chi)^{l^{p-1}} \equiv \chi(l)^{-l^{p-1}}G(\chi^{l^{p-1}}) \bmod lR,$$

so that

$$G(\chi)^{l^{p-1}-1} \equiv \chi(l)^{-1} \bmod lR, \tag{2.2.18}$$

because $l^{p-1} \equiv 1 \bmod p$. Now (2.2.18) can be rewritten in the form

$$(G(\chi)^p)^{(l^{p-1}-1)/p} \equiv \overline{\chi(l)} \bmod lR$$

which generalizes the formula (2.2.13).

It is important that $G(\chi)^p$ belongs to the smaller ring $\mathbb{Z}[\zeta_p]$ (for $p=2$, this is just $\mathbb{Z}$). Moreover, it can be expressed via Jacobi sums: for $p>2$ we have

$$G(\chi)^p = \chi(-1)q\prod_{i=1}^{p-2} J(\chi,\chi^i). \tag{2.2.19}$$

To prove this identity, it suffices to multiply termwise the formulae

$$\frac{G(\chi)G(\chi^i)}{G(\chi^{i+1})} = J(\chi,\chi^i) \ (i = 1,2,\dots,p-2)$$

taking into account (2.2.8) in the form $G(\chi^{p-1})G(\chi) = G(\overline{\chi})G(\chi) = \chi(-1)q$. One uses (2.2.19) in conjunction with a congruence due to Iwasawa ([Iwa75], Theorem 1):

$$J(\chi^a,\chi^b) \equiv -1 \bmod (\lambda)^2,$$

where $(\lambda) = (1-\zeta_p)$ is a *prime ideal* of $\mathbb{Z}[\zeta_p]$. Therefore,

$$G(\chi)^p \equiv -\chi(-1)q \bmod (\lambda)^2, \tag{2.2.20}$$

which becomes an exact equality for $p=2$.

2.2.3 Detailed Description of the Primality Test

a) In the preliminary stage (cf. section 2.2.1 d), one calculates the number $t = \prod_{i=0}^{k} p_i$ which is the product of the initial primes satisfying (2.2.1) and (2.2.2):

$$t < \log n^{c_2 \log\log\log n}, \quad s = \prod_{q, q-1|t} q > \sqrt{n}.$$

As we have already mentioned, for $n \leq 10^{350}$ we can take $t = 2 \cdot 3 \cdot 5 \cdot 7 \cdot 11 \cdot 13 \cdot 17 \cdot 19$. In general, to find t one uses a trial–and–error method, and the primality of the Euclidean primes is tested by the primitive case–by–case check. Since each q is bounded by $t+1$, and the number of q's is bounded by $\pi(t+1) < t+1$, this preliminary stage requires no more than $\log n^{c_3 \log\log\log n}$ operations, with an effective positive constant c_3. At this stage, one should also check that $(n, s) = (n, t) = 1$ (otherwise n is composite, and the algorithm stops).

b) The necessary conditions of primality, essentially of the type (2.2.18), are then checked for every pair p, q with $p|(q-1), q|s$, and every Dirichlet character $\chi \bmod q$ of degree p. It is convenient to fix p and vary q. For each q, one calculates a generator t_q of $(\mathbb{Z}/q\mathbb{Z})^\times$. The Dirichlet characters correspond to primitive roots of unity η_p. The primality condition corresponding to (p, q, χ) is

$$G(\chi)^{n^{p-1}-1} \equiv \eta(\chi) \bmod nR, \tag{2.2.21}$$

where $\eta(\chi)$ is a pth root of unity (for prime n, $\eta(\chi) = \overline{\chi(n)}$, in view of (2.2.18)). To check (2.2.21), one expands the left hand side with respect to the $\mathbb{Z}$–basis of $R = \mathbb{Z}[\zeta_p, \zeta_q]$ and compares it with the right hand side coordinate–wise.

c) If all the congruences (2.2.21) hold true, one calculates a set containing virtual prime divisors r of n not exceeding $\sqrt{n}$. We shall first explain how this is done in the simplest case when $n^{p-1} - 1$ is not divisible by p^2 for any p. Then we have simply

$$r \equiv n^i (\text{mod } s) \text{ for some } i \in \{0, 1, \ldots, t\}.$$

In fact, if $r|n$, put

$$l_p(r) \equiv (r^{p-1} - 1)/(n^{p-1} - 1) \bmod p, l_p(r) \in \mathbb{Z}/p\mathbb{Z}. \tag{2.2.22}$$

Then

$$l_p(rr') = l_p(r) + l_p(r'), \; l_p(n) = 1. \tag{2.2.23}$$

If r is prime, it follows from (2.2.18) that

$$G(\chi)^{r^{p-1}} \equiv \chi(r)^{-1} \bmod\ rR.$$

Let us write $(r^{p-1}-1)/(n^{p-1}-1)$ in the form a/b where $b \equiv 1(\bmod\ p)$, so that $l_p(r) \equiv a(\bmod\ p)$. From (2.2.21) and (2.2.22) it follows that

$$\chi(r) = \chi(r)^b \equiv G(\chi)^{b(r^{p-1}-1)} \equiv G(\chi)^{a(n^{p-1}-1)} \equiv \eta(\chi)^a \bmod\ rR,$$

and finally

$$\chi(r) = \eta(\chi)^{l_p(r)} \qquad (r > 2). \tag{2.2.24}$$

The additivity property (2.2.23) then shows that (2.2.24) holds for all divisors of n, not only the prime ones. In particular, for $r = n$, we find $\eta(\chi) = \chi(n)$, because $l_p(n) = 1$.

Summarizing, we established, that if $n^{p-1}-1$ is not divisible by p^2, then for any triple (p, q, χ) we have

$$\chi(r) = \chi(n)^{l_p(r)},$$

so that $r \equiv n^i \bmod\ q$ where $i \equiv l_p(r) \bmod\ p$ for all p.

d) In general, we have $n^{p-1} - 1 = p^h u$, $h \geq 1$, p does not divide u. The calculations become longer, but the running time is still bounded by $\log n^{c \log\log\log n}$ for a possibly larger constant c. Again, for every triple (p, q, χ) we have the congruence (2.2.21):

$$G(\chi)^{p^h u} \equiv \eta(\chi) \bmod\ nR,\ h = h(p, q, \chi) \geq 1.$$

Let us define $w(\chi)$ as the smallest $i \in \{1, 2, \ldots, h\}$ such that $G(\chi)^{p^i u}$ is congruent to a power of ζ_p modulo nR. If $w(\chi) \geq 2$, the number $G(\chi)^{p^{w(\chi)-1}u} = (G(\chi)^p)^{p^{w(\chi)-2}u}$ belongs to the ring $\mathbb{Z}[\zeta_p]$ with the $\mathbb{Z}-$basis $\{1, \zeta_p, \ldots, \zeta_p^{p-2}\}$. At this stage, one must check the following auxiliary condition:

$$\begin{gathered} for\ every\ j \in \{0, 1, \ldots, p-1\},\ at\ least\ one \\ of\ the\ coefficients\ of \\ G(\chi)^{p^{w(\chi)-1}u} - \zeta_p^j \\ with\ respect\ to\ this\ basis\ is\ relatively\ prime\ to\ n. \end{gathered} \tag{2.2.25}$$

If this assertion is wrong, n is composite, because it has a non–trivial common divisor with one of the coefficients. Otherwise, one can prove, as above, that $r^{p-1} \equiv 1(\bmod\ p^{w(\chi)})$ for all $r|n$, and that for all triples (p, q, χ) with a given q one has

$$\chi(r) = \chi(\nu^j) \text{ for a certain } i \in \{0, 1, \ldots, t\}, \tag{2.2.26}$$

where $\nu \bmod\ q$ is the uniquely defined residue class for which

$$\chi(\nu) = \eta'(\chi),\ \eta'(\chi) \equiv G(\chi)^{p^{w(\chi)u}} \bmod nR. \tag{2.2.27}$$

One can also determine the root of unity $\chi(\nu) \in \mathbb{Z}[\zeta_p]$ using Jacobi sums (cf. [LeH.80]). Choose $a, b \in \mathbb{Z}$ such that

$$p^2 \text{ does not divide } ab(a+b),\ p^2 \text{ does not divide } ((a+b)^p - a^p - b^p)$$

(e.g., $a = b = 1$ for $p < 3.10^9$, $p \neq 1093,\ 3511$). Using (2.19), one can prove then that

$$\nu(\chi) \equiv J(\chi^a, \chi^b) \bmod n\mathbb{Z}[\zeta_p].$$

e) We must now synthesize all the calculations to obtain a residue class ν modulo s such that every potential divisor $r|n, r < \sqrt{n}$, satisfies a congruence $r \equiv \nu^i \bmod s$ for some $0 \le i \le t$. In view of the Chinese Remainder Theorem, it suffices to determine for every $q|s$ a power k such that $\nu \equiv t_q^k \bmod q$. To this end, we choose for every $p|(q-1)$ a character χ with $\chi(t_q) = \zeta_p$. From (2.2.27) it follows that $\chi(t_p^k) = \zeta_p^k = \eta'(\chi)$, which defines $k \bmod p$ and finally $\nu \bmod s$.
f) It remains to check whether one of the numbers r_i defined by

$$r_i \equiv \nu^i \bmod s,\ 0 \le r_i < s,\ 0 \le i \le t,$$

actually divides n.

A number n which passes all these checks is prime. In practice, this algorithm is quite fast (cf. [CoLe84], [Vas88]).

Primality testing can often be speeded up by the following elementary observation. If s is a square-free divisor of $n-1$, and if for every $q_i|s$ there exists such an $a_i \in (\mathbb{Z}/n\mathbb{Z})^\times$ that

$$\gcd(a_i^{(n-1)/q_i} - 1, n) = 1,\ a_i^{n-1} \equiv 1 \bmod n, \tag{2.2.28}$$

then each prime divisor p of n is congruent to 1 modulo s. In fact, from (2.2.28) it follows that the order of $a^{(n-1)/q_i}$ in $(\mathbb{Z}/p\mathbb{Z})^\times$ is equal to q_i. Since $q_i|(p-1)$, we have $s|(p-1)$. In particular, if $s > \sqrt{n}$, then n is prime. Of course, to apply this observation, one must know a sufficiently large divisor s of $n-1$.

A variant of this idea is used in some related primality tests in [LeH.80], [GK86], and in the ECPP, see section 2.2.6. This trick was also used in a proof that R_{1031} is prime [WD86], where $R_n = (10^n - 1)/9$. It is known that for lesser values of n, only R_2, R_{19}, R_{23}, and R_{317} are prime. A very nontrivial prime decomposition of R_{71} was given in the equality (2.1.1).

Since the work of Goldwasser and Kilian, a general primality test was developped by Atkin-Morain which has probably polynomial time (see [AtMo93b], and the end of this section for a discussion of the ECPP (Elliptic Curve Primality Proving). Adleman and Huang in [AdHu92] modified Goldwasser–Kilian algorothm to obtain a randomized polynomial-time algorithm that always produced a certificate for primality.

2.2.4 Primes is in P

`manindra@cse.iitk.ac.in, kayaln@iitk.ac.in,`
`nitinsa@cse.iitk.ac.in`

Let us describe now a resent discovery that primes are recognizable in polynomial time. This is the work of M. Agrawal, N. Kayal and N. Saxena who found that a polynomial version of Fermat's Little Theorem (1.1.2) led to a fast deterministic algorithm for primality testing: the time of this algorithme is given by $\widetilde{\mathcal{O}}(\log^{12} n)$, where the notation $\widetilde{\mathcal{O}}(t(n))$ for $\mathcal{O}(t(n) \cdot poly(\log t(n)))$ is used for a function $t(n)$ of n.

The algorithm is based on the following polynomial version of Fermat's Little Theorem (1.1.2):

Theorem 2.1. *Let p be an integer, and a an integer such that* $\gcd(a,p) = 1$. *Then p is a prime iff* $(x-a)^p \equiv x^p - a (\bmod p\mathbb{Z}[x])$.

Let n be the given number whose primality or compositeness is to be determined. If n is prime then obviously the test

$$\texttt{Test}(a,r): \ (x-a)^n \stackrel{?}{\equiv} x^n - a(\bmod(x^r - 1, n) \tag{2.2.29}$$

will succeed (give the answer "true") for all integers a and r. The result of Agrawal-Kayal-Saxena says that conversely, if $\texttt{Test}(a,r)$ is true for all integers a and r in the range $0 < r \ll \log^6 n$, $0 < a \ll \log^4 n$, and n has no prime factors $\ll \log^4 n$, then n is prime or a power of prime. Here and from now on all constants implied in the sign $\ll$ are absolute).

Since performing $\texttt{Test}(a,r)$ takes time ar most $\mathcal{O}(r^2 \log^3 n)$, or even $\mathcal{O}(r^{1+\varepsilon} \log^{2+\varepsilon} n)$, if FFT is used for multiplication of polynomils and of numbers modulo n, this gives a deterministic polnomial-time algorithm as desired, since obviously checking that n is non-trivial power can be done in polynomial time. Recall that the *Fast Fourier Transform (FFT)* is a fast algorithm which reduces the number of multiplications of coefficients needed for of the multiplication of polynomials of degree r from $\mathcal{O}(r^2)$ to $\mathcal{O}(r \log r)$, and the FFT reduces the time of multiplication of two integers modulo n from $\mathcal{O}(\log n^2)$ to $\mathcal{O}(\log n \log\log n)$. In fact, the result actually proved by them is somewhat stronger: one can find in a deterministic way a *single* number $r \ll \log^6 n$ for which the validity of $\texttt{Test}(a,r)$ for all $a \ll \log^4 n$ suffices to imply the primality of n. This improves the maximal running time of the algorithm from $\mathcal{O}(\log^{18+\varepsilon} n)$ to $\mathcal{O}(\log^{12+\varepsilon} n)$. The actual running time on a well known hypothesis on the density of Sophie Germain primes is in fact only $\mathcal{O}(\log^{6+\varepsilon} n)$ (Sophie Germain primes are odd primes q such that $r = 2q+1$ is also a prime):

Conjecture 2.2 (On the density of the Sophie Germain primes).

$$\#\{\, q \le x \mid q \text{ and } 2q+1 \quad \text{are primes}\} \sim C \cdot \frac{x}{\log_2{}^2 x}$$

Noice that there is an obvious analogy with the asymptotic law of the distribution of primes (1.1.14) (the density of all primes less or equal to x):

$$\#\{\ q \leq x \mid q \text{ is prime}\} \sim K \cdot \frac{x}{\log_2 x}$$

More presisely, the results of Agrawal-Kayal-Saxena, which imply the above statements, are as follows. Here $P(m)$ denotes the largest prime divisor of an integer m ond $o_r(m)$ the order of $m(\bmod r)$, where r is any prime not dividing m.

Proposition 2.3. *For any n, there is a prime $r \ll \log^6 n$ such that $P(o_r(n)) \geq 2\sqrt{r}\log n$*

Proposition 2.4. *Let n be an integer and $r \nmid n$ a prime satisfying*

(a) $P(o_r(n)) \geq l$,
(b) `Test`(a, r) is true for $a = 1, 2, \ldots, l$,
(c) n has no prime factors $< l$,

where $l = 2\sqrt{r}\log n$. Then n is a power of a prime number.

The proof uses a result of Fouvry [Fou85], and Baker-Harman [BaHa96], which says that $P(r-1) > r^{2/3}$ for a positive proportion of all primes r. This result, proved with sieve theory, is difficult but not surprising since it is easy to see that $P(m) > m^{2/3}$ (or even $P(m) > m^c$ for any fixed $c < 1$) for a positive proportion of all integers m. (The number of $m \leq x$ having a prime factor $q > x^c$ for $c \geq \frac{1}{2}$ is $\sum\limits_{\substack{x^c < q \leq x \\ q \text{prime}}} [x/q]$, which is asymptotically equal to $\log(1/c)x$ for x large.) We will show that, for C sufficiently large absolute constant, there exists for every n a prime number r satisfying

$$(2\log n)^6 \leq r \leq (C\log n)^6, o_r(n) > r^{1/3}, P(r-1) > r^{2/3}. \tag{2.2.30}$$

Indeed, by the result just quoted, the number of primes $r < x = (C\log n)^6$ with $P(r-1) > r^{2/3}$ is at least $c\pi(x) \sim c\frac{C^6 \log n^6}{6\log(Cn)}$, where c is an absolute constant. The number of primes $< (2\log n)^6$ is $\ll \frac{(\log n)^6}{\log\log n}$, and the number of $r \leq x$ with $o_r(n) \leq r^{1/3}$ is $\ll \dfrac{C^4 \log n^5}{\log\log n}$, since all these r divide the number $N := \prod_{j=1}^{x^{1/3}} (n^j - 1)$, and

number of prime factors of N is $\ll \displaystyle\sum_{j=1}^{x^{1/3}} \frac{j \log n}{\log(j \log n)} \ll \frac{x^{2/3}\log n}{\log\log n} = \frac{C^4 \log n^5}{\log\log n}.$

It follows that for C sufficiently large there are $\gg \dfrac{\log n^6}{\log\log n}$ primes satisfying (2.2.30). Let r be such a prime and $q = P(r-1)$. Then q is a prime dividing

$r-1$ but not $(r-1)/o_r(n)$ (since $(r-1)/o_r(n) < r^{2/3} < q$), so q divides $o_r(n)$ and $P(o_r(n)) \geq q \geq r^{2/3} \geq 2\sqrt{r}\log n$ □

The idea of the proof of Proposition 2.4 is to consider the integers $n^i p^j$ and $n^k p^l$, and to show that there exist two different couples (i,j) and (k,l) such that $n^i p^j \equiv n^k p^l (\bmod r)$.

Set $q = P(o_r(n))$. Since q is a prime, q must divide $o_r(n)$ for some prime divisor p of n. The field extension $K = \mathbb{F}_p[\zeta]$, where ζ is a non-trivial r^{th} root of unity, has degree $d = o_r(p) \geq q$. Let G be the subgroup of $K^\times$ generated by $\zeta - 1$, $\zeta - 2$, …, $\zeta - l$. Then we have

$$|G| \geq \binom{d+l-1}{l} \tag{2.2.31}$$

because the elements $(\zeta-1)^{d_1}(\zeta-2)^{d_2}\dots(\zeta-l)^{d_1}$ $(d_l \geq 0, \sum d_i < d)$ of G are *distinct*. (The linear functions $x-1$, …, $x-l$ are distinct irreducible polynomials modulo p because of assumption (c), and ζ cannot satisfy a polynomial equation of degree $< d$.) On the other hand, we claim that

$$|G| < n^{2\sqrt{r}} \tag{2.2.32}$$

if n is not a power of p, and this proves the proposition since (2.2.31) and $d \geq q \geq l$ imply

$$|G| \geq \binom{d+l-1}{l} \geq \binom{2l-1}{l} > e^l = n^{2\sqrt{r}}.$$

To prove (2.2.32), we denote by σ_s $(s \in \mathbb{Z})$ the automorphism of K induced by $\zeta \mapsto \zeta^s$. We have $\sigma_p(g) = g^p$ for any $g \in K^\times$ and $\sigma_n(g) = g^n$ for any $g \in G$ by virtue of assumption (b), so $\sigma_s(g) = g^s$ for any s in the form np^j. Let $S = \{np^j \mid 0 \leq i,j \leq \sqrt{r}\}$. If n is not a power of p, then these elements are all distrinct, $|S| > r$ and we can find $s \neq s' \in S$ with $s \equiv s' (\bmod r)$. But then $g^s = \sigma_s(g) = \sigma_{s'} = g^{s'}$. Taking for g a generator of the cyclic group G we deduce from this that $|G| \leq |s-s'| < n^{2\sqrt{r}}$, as desired. □

The algorithm to check primality is therefore as follows. First check that no root $n^{1/k} (2 < k \leq \log_2 n)$ is integral. Then check succesive primes $r > 4\log^2 n$ until one is found for which $r-1$ has a prime factor $q \geq 2\sqrt{r}\log n$ with $n^{(r-1)/q} \not\equiv 1 (\bmod r)$. By Proposition 2.3 the smallest such r is $\ll \log^6 n$. Now check (b) and (c) of Proposition 2.4; n is prime if and only if both hold.

Remark 2.5. It seems that the smallest r satisfying the condition of Proposition 2.3 is not only $\ll \log^6 n$, but is very close to the minimum possible value $r_0 = [4\log^2]$. For instance, a two-line PARI program checks that for $n = 10^{300} + 1$, already 1908707, the second prime $\geq r_0$, works and that for $n = 10^j + 1 (j \leq 300)$ one never needs to try more than 10 primes, or to go further than $r_0 + 186$, before achieving success. (Total computation time is about 2 seconds on a SUN work station).

The given version is due to Dan Bernstein, who slightly improved the original version of August 6, 2002; his contribution is the use of the inequality $\binom{d+l-1}{l} \geq n^{2\sqrt{r}}$, cf. [Mor03], [Ber03], and [Mor03a].

2.2.5 The algorithm of M. Agrawal, N. Kayal and N. Saxena

(see [AKS], p.4), and [Mor03a], p. 4.)

```
Input: integer n > 1

if (n is of the form a a^b, b > 1) output COMPOSITE;
r := 2;
while (r < n) {
        if (r is prime)
           if r divides n output COMPOSITE;
              find the largest prime factor q of r − 1;
              if (q ≥ 4√r log_2 n) and n^((r−1)/q) ≢ 1  mod r ;
                    break;
              r := r + 1;
}
for a = 1 to 2√r log_2 n
          if ((x−a)^n ≡ (x^n−a) mod (x^r−1, n).) output COMPOSITE;
output PRIME;
```

Theorem 2.6. *The algorithm produce PRIME if and only if n is prime.*

Remark 2.7. Practically, one can certainly find r of the size $\mathcal{O}((\log n)^2)$ in order to satisfy the conditions in the algorithm. This leads to the estimate of the complexity $\tilde{\mathcal{O}}((\log n)^6)$ in the best case.

2.2.6 Practical and Theoretical Primality Proving. The ECPP (Elliptic Curve Primality Proving by F.Morain, see [AtMo93b])

The questions of practical primality proving of numbers with thousands of digits and the questions of the mass production of large primes are discussed in [Mor03a].

It is noticed by F.Morain that even the algorithm of Miller is already long, because it needs to compute numerous modular exponents. The quantity $(\log n)^6$ in AKS gives an idea of the order of the degree of polynomials with which one needs to work. In practice, it is almost certain that one can find an $r = c(\log_2 n)^6$ with $c \geq 64$. For example, if $n = 2^{512}$, then in the most optimistic case $r = 64(\log_2 n)^6 = 2^{24} > 16 \cdot 10^6$, leading to manipulate with dense polynomials containing more than 1 Gbytes, which is already rather difficult.

Suppose that we wish to prove the primality of the number $n = 10^9 + 7$ (which is a prime). Using an implementation of AKS by E.Thomé with GMP 4.1 on a PC with 700 MHz, one takes $r = 57287$ which leads to $s = 14340$ (see [Mor03a]). Each intermediate computation takes 44 seconds, giving a total time of more than 7days. If one uses directly the condition (2.2.29), one can take $(r, q, s) = (3623, 1811, 1785)$, and this takes 1.67×1785 seconds or about

49 minutes. The best triplet is $(r, q, s) = (359, 179, 4326)$, leading to a total time of 6 minutes and 9 seconds.

One could compare these algotithms with the algorithm using Jacobi sums (cf. [Coh96] for a presentation of this algorithm which is close to one presented in Section 2.2.1), and with another efficient algorithm, the ECPP (Elliptic Curve Primality Proving) by F.Morain.

The ECCP produces really rapidly a certificate (in $\mathcal{O}((\log n)^4)$) using elliptic curves E over $\mathbb{Z}/n\mathbb{Z}$, and using an elementary observation on congruences (2.2.28) adopted to the groups like $E(\mathbb{Z}/n\mathbb{Z})$. Such a certificate is the program which produces a long list of numbers that constitute the proof of primality for that number. In brief, a decreasing sequence of primes is built, the primality of the successor in the list implying that of the predecessor.

The ECCP can even prove the primality of numbers in 512 bits in 1 seconde, and that of 1024 bits in 1 minute, and that for 10000 bits in a reasonable time (of about one month). According to [Mor03a], it seems that even if one succeed to lower the number r in the algorithm AKS, it will not produce an algorithm, which is practically more efficient than the ECPP, cf. `http://www.lix. polytechnique.fr/Labo/Francois.Morain/Prgms/ ecpp.english.html`.

2.2.7 Primes in Arithmetic Progression

An important recent discovery in [GrTa] by B.J. Green and T.Tao says that the primes contain arbitrary long arithmetic progressions (cf. [Szm75], [Gow01], but also `http://primes.utm.edu/top20/` for interesting numerical examples of long arithmetic progressions of consecutive primes).

It was a well-known classical folklore conjecture that there are arbitrarily long arithmetic progressions of prime numbers. In Dickson's History of the Theory of Numbers [Dic52] it is stated that around 1770 Lagrange and Waring investigated how large the common difference of an arithmetic progression of L primes must be.

It was proved in [GrTa] that there are arbitrarily long arithmetic progressions of primes. There are three major ingredients. The first is Szemerédi's theorem, which asserts that any subset of the integers of positive density contains progressions of arbitrary length. The second is a certain transference principle. This allows one to deduce from Szemerédi's theorem that any subset of a sufficiently pseudorandom set of positive relative density contains progressions of arbitrary length. The third ingredient is a recent result of Goldston and Yildirim, cf. [GoYi03]. Using this, one may place the primes inside a pseudorandom set of "almost primes" with positive relative density.

It was found in 1993 by Moran, Pritchard and Thyssen (cf. [MPTh]) that $11410337850553+4609098694200k$ is prime for $k = 0, 1, ..., 21$. In 2003, Markus Frind found the rather larger example $376859931192959 + 18549279769020k$ of the same length. Main theorem of [GrTa] resolves the above conjecture.

Theorem 2.8 (Theorem 1.1 of [GrTa]). *The prime numbers contain arithmetic progressions of length k for all k.*

A little stronger result was established:

Theorem 2.9 (Theorem 1.2 of [GrTa]). *Let A be any subset of the prime numbers of positive relative upper density, thus*

$$\limsup_{N\to\infty} \pi(N)^{-1}|A \cap [1,N]| > 0,$$

where $\pi(N)$ denotes the number of primes less than or equal to N. Then A contains arithmetic progressions of length k for all k.

2.3 Factorization of Large Integers

2.3.1 Comparative Difficulty of Primality Testing and Factorization

Let $n > 1$ be an integer. The problem of finding integers $a, b > 1$ with $n = ab$ can be divided into two steps: first, to establish their existence (this is solved by any primality test), second, to find them explicitly (factorization). In practice, the primality test described in §2.2.1 does not give a concrete divisor of n. In fact, when n fails such a test, it usually fails already one of the necessary conditions in §2.2.3 b), so that the algorithm stops before we come to the stage of calculating potential divisors. Therefore, this algorithm factorizes only primes and those n which admit small divisors, namely, divisors of the numbers s and t, defined in §2.2.1.

As we mentioned in §2.1, an efficient factorization algorithm could be used for breaking a standard public key cryptosystem. For this reason, factorization has become an applied problem, attracting considerable effort and support ([Pet85], [Sim79], [Kob94]). However, the running times of the best known factorization algorithms do not allow one to factorize a product n of two 150–digit (decimal) primes. The theoretical bound (cf. [Coh2000]) for this running time is of order

$$\exp\left(\sqrt[3]{\frac{64}{9}\log n\cdot(\log\log n)^2}\right), \tag{2.3.1}$$

and for a 300–digit n they may require billions years. This made Odlyzko ask whether we now see the actual level of difficulty of the factorization problem or whether we are just overlooking something essential, cf. [Pet85].

Anyway, the progress in factorization of some concrete large integers ([Wun85], [Wag86]) relied more on the new hardware or parallel computation schemes, than on the discovery of conceptually new algorithms, cf. more recent developments in [Ma99].

2.3.2 Factorization and Quadratic Forms

If $n = x^2 - y^2$, then $x - y$ is in most cases a non–trivial divisor of n. This simple remark leads to the "Fermat factorization algorithm" which generally requires $O(n^{1/2})$ operations but is more efficient if n is a product of two numbers t, s with a small difference. Then $n = x^2 - y^2$ where $x = (t+s)/2$, $y = (t-s)/2$. The algorithm consists of calculating $x^2 - n$ for x starting with $[\sqrt{n}]+1$ until a perfect square is found. Similar considerations can be useful in other problems ([Bril81]). One can also generalize this trick and use other quadratic forms in factorization algorithms ([Kob94], [Ries85]).

Consider an imaginary quadratic field $\mathbb{Q}(\sqrt{-n})$. Let n be square-free. Denote by $Cl(\Delta)$ the *ideal class group* of this field (cf. §1.2.8, §4.2.2). The elements of this group may be identified with the classes under $\mathbb{Z}$–equivalence of the primitive, positive definite quadratic forms $f(x,y) = ax^2 + bxy + cy^2$ with negative discriminant $\Delta = b^2 - 4ac$, where $\Delta = -n$ if $n \equiv 3 \bmod 4$, $\Delta = -4n$ if $n \equiv 1 \bmod 4$. (Here we assume n to be odd). Denote by $\alpha = (a,b,c)$ such a form. We shall call α *ambiguous* if it belongs to one of the types $(a,0,c)$, (a,a,c) or (a,b,a) ([Gau], [Shan71]). The discriminant of an ambiguous form has the explicit factorization: $-\Delta = 4ac$ (resp. $a(4c-a)$, $(2a-b)(2a+b)$) for $\alpha = (a,0,c)$ (resp. $(a,a,c), (a,b,a)$). One easily sees that a converse statement is also true (cf. [BS85]): a factorization of Δ of this type determines an ambiguous form. On the other hand, there are independent methods for constructing ambiguous forms which are based on the following property: they represent elements of order two in the class group $Cl(\Delta)$. In 1971 D.Shanks devised a rather fast algorithm allowing one to factorize n in $O(n^{1/4})$ operations and to determine the structure of the group $Cl(\Delta)$. This method uses the analytic formula due to Dirichlet:

$$L(1,\chi_\Delta) = \frac{\pi h(\Delta)}{\sqrt{|\Delta|}} \qquad (h(\Delta) = |Cl(\Delta)|).$$

Here $\chi_\Delta(m) = \left(\frac{\Delta}{m}\right)$, and $L(1,\chi_\Delta)$ is the value at $s=1$ of the Dirichlet L–function

$$L(s,\chi) = \sum_{m=1}^{\infty} \chi(m) m^{-s} = \prod_p (1 - \chi(p)p^{-s})^{-1}.$$

The approximate formula

$$h(\Delta) \approx \frac{\sqrt{|\Delta|}}{\pi} \prod_{p=2}^{P} \frac{p}{p - \left(\frac{\Delta}{p}\right)}$$

is valid with a relative error $< 0.1\%$ for $P \geq 132000$. The elements of the class group are constructed with the help of small primes p such that $\left(\frac{\Delta}{p}\right) = 1$. They are represented by the forms $F_p = (p, B_p, C_p)$ whose coefficients satisfy the discriminant relation $\Delta = B_p^2 - 4pC_p$ and are found from the condition $\Delta \equiv B_p^2 \bmod p$. Knowing the class number $h(\Delta) = |Cl(\Delta)|$, we can construct the second order elements starting with $x = F_p$, calculating its maximal odd power dividing $h(\Delta)$, and then consecutively squaring until we get 1.

2.3.3 The Probabilistic Algorithm CLASNO

(cf. ([Pom87], [Sey87]). The idea of using $Cl(\Delta)$ in factorization algorithms can be considerably improved. In this algorithm, one bypasses the calculation of $h(\Delta)$, and the running time is estimated by $L = \exp(\sqrt{\log n \cdot \log\log n})$, which grows slower than any positive power of n. Assume first that the prime divisors of $h(\Delta)$ are small, or, rather, that $h(\Delta)$ divides $k!$ for a small k.

Take a random element $x \in Cl(\Delta)$, say, $x = F_p$ for some p with $\left(\frac{\Delta}{p}\right) = 1$ and calculate $B_k = x^{\text{odd power of } k!}$. Then an element of order 2 should be contained in the sequence of consecutive squares of B_k. We need not know the exact value of k; we just hope that some small k will do. If we succeed, we factorize Δ in $O(k)$ operations. If we fail, we can try the same trick for the field $\mathbb{Q}(\sqrt{-an})$ where a is a small square-free number.

In order to justify this procedure in general, one assumes that for variable a, the class number $h(\Delta_a)$ of $\mathbb{Q}(\sqrt{-an})$ behaves like a random number varying in a neighbourhood of $n^{1/2}$ (this estimate follows from the Dirichlet formula). One can then estimate the probability that $h(\Delta_a)$ will be composed of only small primes. To this end, denote by $\Psi(x, y)$ the number of natural numbers $\leq x$ not divisible by any prime $\geq y$ (they can be called "y–smooth"). Put $k = L^\alpha$, $\alpha > 0$. The probability that a random number of order $n^{1/2}$ is L^α–smooth is $\Psi(n^{1/2}, L^\alpha)/n^{1/2}$. We must now understand the behavior of $\Psi(x,y)/y$. Dickman (cf. in [Hild86]) has shown that this depends essentially on the value of $\log x/\log y$. Namely, for every $u > 0$ the limit $\lim_{y\to\infty} \Psi(y^u, y)/y$ exists. This limit is called the Dickman function $\rho(u)$ and is uniquely defined by the following properties:

$$\text{for } 0 \leq u \leq 1,\ \rho(u) = 1,$$

$$\text{for } u > 1, \rho'(u) = -\frac{\rho(u-1)}{u}.$$

At $u = 1$, $\rho(u)$ is continuous. As $u \to \infty$, $\rho(u) = u^{(-1+o(1))u}$. De Bruijn (de Bruijn N.G. (1951)) proved that

$$\Psi(y^u, y) = y^u \rho(u) \left(1 + O_\epsilon\left(\frac{\log(y+1)}{\log y}\right)\right),$$

where $y \geq 2$, $1 \leq u \leq (\log y)^{3/5-\epsilon}$ with a positive ϵ.

In our case, however, L^α grows slower than any positive power of n so that Dickman's theorem is not applicable. The necessary estimate has recently been obtained:

$$\Psi(n^{1/2}, L^\alpha) = n^{1/2}/L^{(1/4\alpha)+o(1)}.$$

For more details, see [Hild86].

Returning to the factorization algorithm under discussion, one sees that its running time for a given $k = [L^\alpha]$ is bounded by L^α and the probability of success is about $L^{-1/4\alpha}$. Hence the total number of attempts should be about $L^{1/4\alpha}$, and the total running time will be bounded by $L^{\alpha+(1/4\alpha)+\epsilon}$, $\epsilon > 0$. This estimate is minimized by choosing $\alpha = 1/2$ (i.e. $k = L^{1/2}$), and the result is then $L^{1+\epsilon}$. Of course, theoretically we may get stuck on an especially bad n, but this is quite improbable.

Let us illustrate the estimate $\dfrac{x}{\Psi(x,y)} \approx u^{-u}$ when u is much smaller than y (for a simple proof of this see [Kob87], p. 137). For example, take $y \approx 10^6$

(so that $\pi(y) \approx 7.10^4$ and $\log y \approx 14$) and $x \approx 10^{48}$. Then the fraction of natural numbers $\leq x$ which are products of primes $\leq y$ is about $1/2^{24}$.

2.3.4 The Continued Fractions Method (CFRAC) and Real Quadratic Fields

(cf. [Kob94], [Wun85], [Ries85], [Wil84]) . Improving the Fermat factorization method, let us try to seek solutions x, y of the congruence $x^2 \equiv y^2 \bmod n$ such that x is not congruent to $\pm y \bmod n$. Then $\gcd(x+y, n)$ or $\gcd(x-y, n)$ is a non–trivial divisor of n because n divides $(x+y)(x-y)$ but neither $x+y$ nor $x-y$. Let us look for x among products of such numbers x_i that the residue $x_i^2 \bmod n$ with the smallest absolute value is a product of small primes. Then y will also be a product of these primes. More precisely, consider a set $B = \{p_1, p_2, \ldots, p_h\}$ all of whose elements are primes, except possibly p_1 which can be -1. Let us call such a set *a factorization basis* for n. We shall refer to any integer b such that the residue of $b^2 \bmod n$ with the smallest absolute value is a product of (powers of) elements of B as a B–number. Let x_i be a family of B–numbers, $a_i = \prod_{j=1}^h p_j^{\alpha_{ij}}$ the respective minimal residues of $x_i^2 \bmod n$. Put

$$\epsilon_i = (\epsilon_{i1}, \epsilon_{i2}, \ldots, \epsilon_{ih}) \in \mathbb{F}_2^h, \text{ where } \epsilon_{ij} \equiv \alpha_{ij} \bmod 2.$$

Suppose that the sum of vectors ϵ_i vanishes mod 2. Put

$$x = \prod_i x_i \bmod n, \qquad y = \prod_{j=1}^h p_j^{\gamma_j},$$

where

$$\gamma_j = \frac{1}{2}\sum_i \alpha_{ij}.$$

Then $x^2 \equiv y^2 \bmod n$.

Example 2.10. ([Kob87], p. 133). Let $n = 4633$, $B = \{-1, 2, 3\}$. Then $x_1 = 67, x_2 = 68, x_3 = 69$ are B–numbers, because

$$67^2 \equiv -144 \bmod 4633, \quad 68^2 \equiv -9 \bmod 4633, \quad 59^2 \equiv 128 \bmod 4633.$$

Moreover, $\epsilon_1 = (1,0,0), \epsilon_2 = (1,0,0), \epsilon_3 = (0,1,0)$, so that we can put $x = x_1 x_2 = 67.68 \equiv -77 \bmod 4633$, $c = 2^{\gamma_2} 3^{\gamma_3} = 2^3 3^2 = 36$. Besides, -77 is not congruent to $\pm 36 \bmod 4633$. Summarizing, we obtain a non–trivial divisor $41 = \gcd(-77+36, 4633)$ of $n = 4633$.

Of course, if we are unlucky, it may happen that $x \equiv \pm y \bmod n$. Then one should choose a new x_i or even a new B. An efficient method for seeking B–numbers utilizes continued fractions of real quadratic irrationalities. Let

$x > 1$ be a real number, $x = [a_0, a_1, \ldots]$ its continued fraction expansion. Put $A_i/B_i = [a_0, a_1, \ldots, a_i]$. These convergents can be calculated from the relations $A_{-2} = B_{-1} = 1, A_{-1} = B_{-2} = 0$ and $A_i = a_i A_{i-1} + A_{i-2},\ B_i = a_i B_{i-1} + B_{i-2}$. From the relation

$$\frac{A_i}{B_i} - \frac{A_{i+1}}{B_{i+1}} = (-1)^{i+1} \frac{1}{B_i B_{i+1}}$$

it follows that

$$|A_i^2 - x^2 B_i^2| < 2x, \tag{2.3.2}$$

because

$$|A_i^2 - x^2 B_i^2| = B_i^2 \left| \frac{A_i}{B_i} - x \right| \cdot \left| \frac{A_i}{B_i} + x \right| < B_i^2 \frac{1}{B_1 B_2} \left(2x + \frac{1}{B_1 B_2} \right).$$

In particular, we can find the continued fraction expansion of $x = \sqrt{n}$ with the help of the algorithm described in §1.4, and a_i form a periodic sequence. Since $A_i^2 \equiv A_i^2 - nB_i^2 \bmod n$, (2.3.2) shows that the absolute value of the smallest residue of $A_i^2 \bmod n$ is bounded by $2\sqrt{n}$ which can help in looking for B–numbers. However, A_i quickly become large even with respect to n, and to facilitate the calculation of $A_i^2 \bmod n$ one can use the congruence

$$A_{i-1}^2 \equiv (-1)^i Q_i \bmod n, \tag{2.3.3}$$

where Q_i is the denominator of $x_i = (\sqrt{n} + P_i)/Q_i$, of $A_i^2 \bmod n$, that is,

$$\sqrt{n} = [a_0, a_1, a_2, \ldots, a_i, x_i].$$

In fact, applying formally the recurrence relations to $\sqrt{n}$ we get

$$\sqrt{n} = x = \frac{A_{i-1} x_i + A_{i-2}}{B_{i-1} x_i + B_{i-2}} = \frac{A_{i-1}\sqrt{n} + P_i A_{i-1} + Q_i A_{i-2}}{B_{i-1}\sqrt{n} + P_i B_{i-1} + Q_i B_{i-2}}.$$

Comparing the coefficients at 1 and $\sqrt{n}$, we obtain

$$Q_i A_{i-2} + P_i A_{i-1} = nB_{i-1},$$

$$Q_i B_{i-2} + P_i B_{i-1} = A_{i-1}.$$

Solving this for Q_i, we see that

$$(A_{i-2}B_{i-1} - A_{i-1}B_{i-2})Q_i = nB_{i-1}^2 - A_{i-1}^2.$$

But the coefficient at Q_i equals to $(-1)^{i-1}$. This proves (2.3.3). Recall also that P_i, Q_i can be calculated using a very efficient algorithm which we restate in a slightly changed form. Let $x_0 = (P_0 + \sqrt{n})/Q_0$ be a quadratic irrationality, with Q_0 dividing $n - P_0^2$. Put $x_i = (P_i + \sqrt{n})/Q_i$. Then

$$P_{i+1} = a_i Q_i - P_i, \qquad a_i = [P_i + \sqrt{n}/Q_i], \tag{2.3.4}$$

$$Q_{i+1} = Q_{i-1} + (P_i - P_{i+1})a_i. \tag{2.3.5}$$

This follows directly from $x_{i+1} = 1/(x_i - a_i)$, or

$$\frac{P_i + \sqrt{n}}{Q_i} = a_i + \frac{Q_{i+1}}{P_{i+1} + \sqrt{n}}.$$

If this method does not provide us with the required amount of B–numbers, we can repeat the calculations with an instead of n where a is a small square-free number. The number of operations required is estimated by

$$L^{\sqrt{3/2}} = \exp\left(\sqrt{\frac{3}{2}\log n \log\log n}\right)$$

(compare with section 2.3.3), and the practical efficiency of this algorithm was demonstrated by its application to the Fermat number $F_7 = 2^{128} - 1$ (cf. [MB75], [Wil84]).

Let us describe also an elegant algorithm SQUFOF due to Shanks which is also based upon the arithmetic of real quadratic fields (cf. [Ries85], [Wil84]). It consists of two stages.

1) Put $x_0 = \sqrt{n}$, that is, $P_0 = 0, Q_0 = 1$ in the formulae (2.3.4), (2.3.5). Calculate x_m until we find an odd integer m such that $Q_{m-1} = t^2$ for some natural t. From (2.3.3) it follows that $A_{m-2}^2 \equiv t^2 \bmod n$. Presumably, one can then find a divisor of n with the help of the Euclidean algorithm as $\gcd(A_{m-2} \pm t, n)$. In practice, however, A_{m-2} is usually too large to be calculated directly, so that one changes tactics.
2) Put $\tilde{P}_0 = P_m, \tilde{Q}_0 = t, \tilde{x}_0 = (\tilde{P}_0 + \sqrt{n})/Q_0$ and calculate the tails of the continued fraction expansion of $\tilde{x}_0$, $\tilde{x}_i = (\tilde{P}_i + \sqrt{n})/\tilde{Q}_i$. We perform this until we find such $\tilde{x}_q$ that $\tilde{P}_q = \tilde{P}_{q+1}$. From (2.3.4) and (2.3.5) it follows, that
$$\tilde{a}_q \tilde{Q}_q = 2\tilde{P}_q, \qquad \tilde{Q}_q \text{ divides } n - \tilde{P}_q^2.$$
Hence either $\tilde{Q}_q$, or $\tilde{Q}_q/2$ divides n. If this divisor is trivial, one should again replace n by an for a small a and repeat the calculations. Using a calculator for factorizing a number $\le 10^9$, it is convenient to write the intermediate results in a table. Table 2.1 illustrates the course of calculations for $n = 11111 = 41 \cdot 271$. In general, q is about $m/2$ (in our example, $m = 7, q = 4$. The algorithm is based on the fact, that the fractional ideal $(1, \tilde{x}_0)$ is of order two in $Cl(4n)$, and on the second stage we calculate the corresponding ambiguous form in disguise. The number of operations is estimated by $n^{1/4}$.

Table 2.1.

0	1	2	3	4	5	6	7
105	2	2	4	5	2	7	
0	105	67	87	97	88	94	81
1	86	77	46	37	91	$25 = 5^2$	
37	3	1	1	3			
81	104	73	25	82	82		
5	59	98	107	41	107		

Since

$$\tilde{x}_4 = \frac{82 + \sqrt{11111}}{41} = 2 + \frac{\sqrt{11111}}{41},$$

the ideal $(1, \tilde{x}_4)$ corresponds to the ambiguous form (41,0,-11111/41), or (41,0,-271), with the discriminant $4n$.

2.3.5 The Use of Elliptic Curves

The general idea of utilizing the calculations in a finite group (such as class group) in order to factorize n found unexpected implementations using groups of different types.

a) *Pollard's* $(p-1)$*–method.* Suppose that n has such a prime factor p that the order of $(\mathbb{Z}/p\mathbb{Z})^\times$ is "smooth", that is, $p-1$ divides $k!$ for a not too large k, say, $k \leq 100000$. Then we can proceed as follows: calculate consecutively $a_i = 2^{i!}-1 \bmod n$ using the recursive relation $a^{i+1} \equiv (a_i+1)^{i+1}-1 \bmod n$ and find $\gcd(a_k, n)$; it will be divisible by p in view of Fermat's little theorem. This will fail if there are no $p|n$ with smooth $p-1$ ([Pol74]). For a change, one can try to use the multiplicative groups of fields $\mathbb{F}_{p^r}$ of order $p^r - 1$. For $r = 2$, we obtain the Williams $p+1$–algorithm ([Wil82]).

b) Much wider perspectives of varying the finite group in the factorization algorithms are opened by *elliptic curves over finite fields.* Their use leads to one of the fastest known factorization algorithms requiring $O(L^{1+\epsilon})$ operations [LeH.87].

   Choose a random elliptic curve Γ and a point P on it. To this end, choose random integers a, x_0, y_0 and put $b = y_0^2 - 4x_0^3 - ax_0, P = (x_0, y_0)$. Then $P = (x_0, y_0)$ is a point on the curve defined by the equation

$$\Gamma : \qquad y^2 = 4x^3 + ax = b$$

   (cf. §1.3.3). It is an elliptic curve over $\mathbb{Q}$, if the right hand side cubic polynomial has no multiple roots. We may also assume that the discriminant of this polynomial is relatively prime to n; otherwise we either get a non–trivial divisor of n, or must change the curve.

In the projective plane, Γ is determined by the homogeneous equation

$$Y^2Z = 4X^3 = aXZ^2 + Z^3 \quad (X = xZ, Y = yZ).$$

Reducing it modulo a prime p, we obtain an elliptic curve over $\mathbb{F}_p = \mathbb{Z}/p\mathbb{Z}$. Its identity is $O_\Gamma = (0:1:0)$, and the order of $\Gamma(\mathbb{F}_p)$ equals $p+1-a_p$ where $|a_p| < 2\sqrt{p}$ (Hasse's theorem, see §1.3.3).

Assuming now that $(p+1-a_p) \mid k!$ for some $p|n$ and small k, we calculate consecutively $P_i = i!P \bmod n$ in the projective plane over $\mathbb{Z}$. The prime p must divide the Z–coordinate of P_k, and the $\gcd(n, Z_k)$. If we are lucky, $O(k)$ operations will provide us with a non–trivial divisor of n. Otherwise one should renew the curve, without wasting too much time on an unsuccessful choice (*"the strategy of early interruption"*). In order to optimize the choice of k for each test curve and the number of tests, let us take $p = n^\beta$. The probability of success with $k = [L^\alpha]$ is approximately $\Psi(n^\beta, L^\alpha)/n^\beta \approx L^{-\beta/2\alpha+o(1)}$ (see 3.3). Hence we shall have to try about $L^{\beta/2\alpha}$ random elliptic curves with a marked point, whereas for each of them the number of operations will be estimated by L^α. The general number of operations $L^{\alpha+\beta/2\alpha}$ is minimal for $\alpha = \sqrt{\beta/2}$. In the worst case, $\alpha = \beta = 1/2$ we get $L^{1+\epsilon}$, $\epsilon > 0$.

Notice that our estimates are based upon the following heuristic conjecture: the orders of the groups $\Gamma(\mathbb{F}_p)$ behave with respect to the smoothness property as the random numbers taken from $(p - 2\sqrt{p}+1, p+2\sqrt{p}+1)$. The belief in this conjecture is strengthened by the study of the set of isomorphism classes of elliptic curves modulo a prime [LeH.87].

We must also notice that some cryptosystems using elliptic curves also were suggested [Kob87].

There exists a probabilistic algorithm with rigorously estimated running time

$$O(L^{\sqrt{5/2}})$$

due to [Dix84], and several probabilistic algorithms using *linear* or *quadratic sieves*, with the expected running time

$$O(L^{\sqrt{2}}) \text{ and } O(L)$$

respectively [Pom82], [Wag86].

Many more interesting algorithms and computer programs can be found in Riesel's book [Ries85]. One can also find there some heuristic arguments in favour of the existence of the algorithms which would be much faster then everything we know now.

More recent information on factoring large integers and new records can be found in [Coh2000] and at the Web page of F.Morain.

Part II

Ideas and Theories

3

Induction and Recursion

3.1 Elementary Number Theory From the Point of View of Logic

3.1.1 Elementary Number Theory

Almost all of part I of this book belongs to elementary number theory (ENT). This notion can be rigorously defined using tools of mathematical logic, but in order to do this one must first introduce a formal language of arithmetic and fix an adopted system of axioms (one or other version of Peano's axioms). In order to avoid such irrelevant details, we restrict ourselves to some intuitive remarks. In ENT there are some initial statements and some axioms, which formalize our intuitive ideas of natural numbers (or integers), as well as certain methods for constructing new statements and methods of proofs. The basic tool for construction is recursion. In the simplest case assume that we want to define some property $P(n)$ of a natural number n. Using the method of recursion we explain how one can decide whether $P(n+1)$ is true if it is already known whether $P(1)$, ..., $P(n)$ are true or not. Say, the property "n is a prime" can be defined as follows: "1 is not a prime; 2 is a prime; $n+1 \geq 3$ is a prime iff none of the primes among 1, 2, ..., n divide $n+1$". Analogously the main tool in the proofs of ENT is induction. In order to prove by induction a statement of type "$\forall n,\ P(n)$ is true" we first prove say $P(1)$ and then the implication "$\forall n$ the property $P(n)$ implies $P(n+1)$".

Even in the earliest research into the axiomatics of number theory (Peano, Frene) it was established that all the notions empirically thought of as belonging to ENT (such as divisibility, primality etc.), functions (the number of divisors, the Euler function $\varphi(n)$, $\pi(x)$) and theorems (Fermat's little theorem, the quadratic reciprocity law etc.) can be respectively constructed by recursion and proved by induction, cf. [Rog67], [Man80].

It happens sometimes that a result admits an elementary formulation, but its elementary proof is not known. For example, the prime number theorem

$\pi(x) \sim \dfrac{x}{\log x}$ can be stated in an elementary way assuming that x runs only through natural numbers, and replacing $\log x$ by the sum $\sum_{i=1}^{x} \frac{1}{i}$; an elementary proof of this theorem was found only in the late 40s by Selberg, cf. [Sel51] while the analytic proof had been known for half a century.

3.1.2 Logic

The study of ENT from the point of view of logic has lead to new concrete number theoretical results which we shall discuss below. However the most important consequence of this study has been that the place of ENT inside mathematics in general has become much clearer. We wish to stress the following three aspects.

a) ENT as a mathematical discipline in principle can not be "self-sufficient". For every choice of axioms there will always be statements which can be formulated in an elementary way, and which are decidable, but which can not be deduced using only elementary methods (cf. the theorem of Gödel [Gö], discussed in [Man80]).
Thus the historical tradition of proving number theoretic facts using analysis (Euler, Jacobi, Dirichlet, Riemann, Hardy, Littlewood, Vinogradov, ...), geometry (Minkowski, Hermit, ...) and generally all possible tools, has deep reasons.
b) ENT can be used by means of formal logic to model any axiomatized mathematical discipline inside elementary number theory (Gödel). In such a modeling we forget the contentive sense of the definitions and theorems of our theory and leave only information concerning their formal structure, and syntactic rules for deducing one statement from others. Enumerating by Gödel's method all syntactically correct statements by natural numbers, we can then write a program or algorithm to list all provable results of our theory (its theorems). Thus a theory is modeled by a function $f : \mathbb{Z}^+ \to \mathbb{Z}^+$ (the first $\mathbb{Z}^+$ is a number generating the theorem, the second is the encoded statement in the theory). Instead of asking whether the theorem with number n is provable we can ask whether the equation $f(x) = n$ is solvable.
Although the equation $f(x) = n$ is defined in terms of ENT, it is far from being a Diophantine equation since the function f is not a polynomial. As was shown by Yu.V.Matiyasevich, it is possible to reduce this problem to a Diophantine one (Hilbert's tenth problem), see [Mat04].
He showed that one can find a polynomial $P_f(x_1, \ldots, x_m; n)$ with integral coefficients such that the solvability of $f(x_1) = n$ is equivalent to the solvability of $P_f(x; n) = 0$ with $x \in (\mathbb{Z}^+)^m$. The calculation of P_f from f is completely effective (as is the construction of f given the system of axioms defining the initial theory). In this sense the problem of provability of any

mathematical result is equivalent to a standard kind of number theoretical problem. (The reader who is used to thinking not in terms of "provability" but of "truthfulness" must at this point take consciously some intellectual precautions. Considering for example the theorem of Gödel - Cohen that the continuum - hypothesis is independent of the standard axioms of set theory, it is clear that "truthfulness", as opposed to "provability", is a rather philosophical notion. It would therefore be unreasonable to expect it to have a precise mathematical definition.)

c) ENT provides a framework for the precise formulation and study of the notions of algorithm and (semi-)computable function. These notions, implemented in the theory of recursive functions, turn out to be much more universal than one could expect *a priory* (the Church thesis, cf. [Man80], [Rog67], [KMP74]). The theory of recursive functions has both a fundamental general mathematical meaning, and an applied meaning. Its methods are used in proving the Matiyasevich theorem mentioned above.

In the next section we formulate some basic facts from the theory of recursive functions, which have independent number theoretical interest. We then give some precise definitions and hints of proofs.

3.2 Diophantine Sets

3.2.1 Enumerability and Diophantine Sets

Definition 3.1. *A subset $E \subset (\mathbb{Z}^+)^m$, $m \geq 1$ is called Diophantine if there exists a polynomial with integral (or, equivalently, with natural) coefficients*

$$P(t_1, \ldots, t_m, x_1, \ldots, x_n),$$

such tha

$$(t_1, \ldots, t_m) \in E \Longleftrightarrow \exists (x_1, \ldots, x_n) \in \mathbb{Z}^n, \ P(t, x) = 0.$$

Every Diophantine set is enumerable in the following informal sense of the word: there is a deterministic algorithm, which produces one–by–one all elements of E (a formal definition will be given in the next section). Indeed, let us check one–by–one all the elements of $\mathbb{Z}^{m+n}$: substitute them into P and, if we get zero, write down the first m coordinates. We thus obtain a growing list of elements of E, which exhausts E when we pass to the limit.

3.2.2 Diophantineness of enumerable sets

Theorem 3.2. *Conversely, every enumerable set is Diophantine. Its defining polynomial can be effectively constructed from the algorithm generating E.*

It seems *a priori* that there are many more enumerable sets than Diophantine sets; it is therefore clear that in proving theorem 3.2, one needs to prove the Diophantineness of some unexpected sets. J. Robinson discovered that this problem can be simplified if one takes for granted the Diophantineness of the set $\{(a, b, c) \mid a = b^c\}$, and Yu. V. Matiyasevich (cf. [Mat72], [Mat04]) established this last step. Below we give some examples and constructions used in the proof, which are purely number–theoretical. We first formulate the following very general property.

3.2.3 First properties of Diophantine sets

Proposition 3.3. *The class of Diophantine sets contains the level sets of polynomials with integral coefficients, and it is closed with respect to the operations of finite direct sum, finite intersection, and projection.*

This follows immediately from the definition. It suffices to note that if $E, F \subset \mathbb{Z}^m$ correspond to polynomials P, Q respectively, then $E \cap F$ corresponds to $P^2 + Q^2$; $E \cup F$ corresponds to PQ, and $E \times F$ corresponds to $P^2 + \tilde{Q}^2$, where $\tilde{Q}$ is obtained from Q by renumbering the first m variables.

Now we give the key arithmetical lemma – the proof of the Diophantineness of a set related to solutions of Pell's equation (it is important that for this set one coordinate grows approximately as the exponent of the other).

Consider Pell's equation $x^2 - dy^2 = 1$ ($d \in \mathbb{Z}^+$ is a square free integer). Its solutions $(x,y) \in \mathbb{Z}^2$ form a cyclic group with respect to the following law of composition: if (x_1, y_1) is a solution with the first coordinate minimal, then any other solution is of the type (x_n, y_n), where $n \in \mathbb{Z}^+$ and

$$x_n + y_n\sqrt{d} = (x_1 + y_1\sqrt{d})^n.$$

The number n is called the solution number (cf. Part I, §1.2.5).

The coordinates x_n, y_n grow exponentially with n, but the set of solutions, and its projections on the $x-$ and $y-$ axes are Diophantine. However this is still not what we need: the main difficulty is to include the solution number into a set of coordinates of a Diophantine set; only then will we be able to use further arguments. This is done below.

It is convenient to use for d the number $d = a^2 - 1$, $a \in \mathbb{Z}^+$, since in this case $(x_1, y_1) = (a, 1)$. The equation $x^2 - (a^2-1)y^2 = 1$ will be called the a–equation. Define two sequences $x_n(a)$, $y_n(a)$ to be the coordinates of its nth solution:

$$x_n(a) + y_n(a)\sqrt{a^2-1} = (a + \sqrt{a^2-1})^n.$$

Formal definitions of $x_n(a)$ and $y_n(a)$ as polynomials in a can easily be given by induction over n. Then $x_n(a)$ and $y_n(a)$ will have sense for all $n \in \mathbb{Z}$ and $a \in \mathbb{C}$. In particular, $x_n(1) = 1$, $y_n(1) = n$; in this extended range all of the formulae given below will be valid.

3.2.4 Diophantineness and Pell's Equation

Proposition 3.4. *The set* $E : y = y_n(a)$, $a > 1$ *is Diophantine in the* (y, n, a)*–space.*

The idea in the Diophantine reconstruction of n from (y, a) is based on the remark, that the congruence

$$y_n \equiv n \bmod (a-1)$$

determines n uniquely for $n < a - 1$. In order to treat the general case, an auxiliary A–equation is introduced, with big A. Its nth solution so that n be used only in Diophantine context.

Besides the main variables y, n, a, one introduces six auxiliary variables: x, x', y'; A; x_1, y_1. Furthermore define the following sets:

$$E_1 : y \geq n,\ a > 1;$$

$$E_2 : x^2 - (a^2-1)y^2 = 1;$$

$$E_3 : y' \equiv 0 \bmod 2x^2y^2;$$

$$E_4 : x'^2 - (a^2-1)y'^2 = 1;$$

$$E_5 : A = a + x'^2(x'^2 - a);$$
$$E_6 : x_1^2 - (A^2 - 1)y_1^2 = 1;$$
$$E_7 : y_1 - y \equiv 0 \mod x'^2;$$
$$E_8 : y_1 \equiv n \mod 2y.$$

The sets E_i are all Diophantine, and $\mathrm{pr}E' = E$, where $E' = \cap_{i=1}^8 E_i$. In order to check this fact we use the following properties:

$$y_k(a) \equiv k \mod (a-1). \tag{3.2.1}$$

$$\text{If } a \equiv b \mod c \text{ then } y_n(a) \equiv y_n(b) \mod c. \tag{3.2.2}$$

$$\text{If } y_i(a) \equiv y_j(b) \mod x_n(a),\ a > 1 \text{ then } i \equiv j \mod 2n \text{ or } i \equiv -j \mod 2n. \tag{3.2.3}$$

$$\text{If } y_i(a)^2 | y_j(a) \text{ then } y_j(a) | j. \tag{3.2.4}$$

Properties (3.2.1) – (3.2.4) are easily deduced from the equalities

$$x_{n\pm m}(a) = x_n(a)x_m(a) \pm (a^2-1)y_n(a)y_m(a),$$
$$y_{n\pm m} = \pm x_n(a)y_m(a) + x_m(a)y_n(a).$$

3.2.5 The Graph of the Exponent is Diophantine

We now prove that the set $E : y = a^n$ in the (y, a, n)–space is *Diophantine.* It suffices to check Diophantineness of $E_0 = E \cap \{a \mid a > 1\}$. For $a > 1$ one easily obtains by induction on n that

$$(2a-1)^n \le y_{n+1}(a) \le (2a)^n$$

in the notation of §3.2.4. From this it follows that

$$a^n = [y_{n+1}(Na)/y_{n+1}(N)]$$

for sufficiently large N. To be more precise, E_0 is the projection of the set E_1:

$$a > 1;\ 0 \le y_{n+1}(N)y - y_{n+1}(y);\ N > 4n(y+1);$$

and Diophantineness of E_1 is then obtained by introducing trivial auxiliary relations $y' = y_{n+1}(N)$ and $y'' = y_{n+1}(Na)$.

3.2.6 Diophantineness and Binomial coefficients

Proposition 3.5. *The set $E : r = \binom{n}{k}$, $n \ge k$ in the (r, k, n)–space is Diophantine.*

3.2.7 Binomial coefficients as remainders

Lemma 3.6. *If $u > n^k$ then $\binom{n}{k}$ is equal to the remainder of the division $[(u+1)^n/u^k]$ by u.*

The proof follows from the binomial formula

$$(u+1)^n/u^k = \sum_{i=k+1}^{n} \binom{n}{i} u^{i-k} + \binom{n}{k} + \sum_{i=0}^{k-1} \binom{n}{i} u^{i-k}.$$

The first sum is divisible by u and the second is less than 1 for $u > n_k$.

The proof of Proposition 3.5 has the same scheme, using the auxiliary variables u and v and the relation

$$E_1 : u > n^k; \quad E_2 : v = [(u+1)^n/u^k];$$

$$E_3 : r \equiv v \bmod u; \; E_4 : r < v; \; E_5 : n \geq k.$$

From the lemma it follows immediately that $E = \mathrm{pr} \cup_{i=1}^{5} E_i$. The Diophantineness of E_1 follows from that of the exponent; the Diophantineness of E_3, E_4 and E_5 is obvious. The Diophantineness of E_2 becomes clear if we represent E_2 in the form

$$(u+1)^n \leq u^k v \; < (u+1)^n + u^k$$

and use again the Diophantineness of the exponent.

3.2.8 Diophantineness of the Factorial

Proposition 3.7. *a) The set $E : m = k!$ is Diophantine.*
b) The set

$$E : \frac{x}{y} = \binom{p/q}{k}, \quad p > qk,$$

is Diophantine in the space (x, y, p, q, k).

The proof is a modification of the arguments in §3.2.6, §3.2.7, using the following lemma.

3.2.9 Factorial and Euclidean Division

Lemma 3.8. *a) If $k > 0$ and $n > (2k)^{k+1}$ then*

$$k! = \left[\frac{n^k}{\binom{n}{k}} \right].$$

b) Let $a > 0$ be an integer such that $a \equiv 0 (\bmod\ (q^k k!))$ and $a > 2^{p-1}p^{k+1}$. Then

$$\binom{p/q}{k} = a^{-1}[a^{2k+1}(1+a^{-2})^{p/q}] - a[a^{2k-1}(1+a^{-2})^{p/q}].$$

The proof of this lemma follows from some elementary computations and Proposition 3.7 is proved using the same methods as above.

3.2.10 Supplementary Results

The Diophantine representations stated above are used in the proof of the general theorem of Matiyasevich. On the other hand they can also be used to find exponential – Diophantine representations for some interesting concrete sets. As an example consider the set of prime numbers. By Wilson's theorem p is a prime $\iff (p-1)!+1$ is divisible by p. The set of prime numbers is therefore a projection of the set of solutions to the following system of equations

$$\begin{cases} p = f + 1 \\ q = f! \\ q - ap = 1. \end{cases}$$

which is Diophantine in view of the Diophantineness of $q = f!$.

Any Diophantine subset $E \subset \mathbb{Z}^+$ coincides with the set of all natural values of a polynomial with integral coefficients (on $\mathbb{Z}^N$). Indeed if E is the projection of $P(t; x_1 \dots x_n) = 0$ then $Q(t; x_1 \dots x_n) = t(1 - P^2)$ is the appropriate polynomial. Thus the set of all primes can be represented as the set of all natural values of a polynomial Q. (It should be noted that Q will take infinitely many other integral values ≤ 0 which is unavoidable).

The Fibonacci numbers form the sequence 1 1 2 3 5 8 $\dots$, $u_{n+2} = u_{n+1} + u_n$. J. Jones found that this sequence can be represented as the set of positive values of a very simple polynomial in two variables (this is not the case for the set of all primes):

$$2a^4b + a^3b^2 - 2a^2d^3 - a^5 - ab^4 + 2a.$$

Although as we have noticed above the question on the provability of any theorem can in principle be reduced to a Diophantine equation some concrete problems admit natural reductions without the use of a formal language. We refer the reader to the very interesting and informative article [DMR74]. In particular this article contains Diophantine forms of the *Riemann Hypothesis* and the *four–colour problem.*

3.3 Partially Recursive Functions and Enumerable Sets

3.3.1 Partial Functions and Computable Functions

In this Section we give a precise definition of a class of partial functions from $\mathbb{Z}^m$ to $\mathbb{Z}^n$. This definition can be considered as an adequate formalization of the class of (semi-)computable functions. Using the definition one is able to define the class of enumerable sets. We shall denote by $D(f)$ the domain of definition of a partial function f, [Rog67], [Man80].

3.3.2 The Simple Functions

$$\begin{aligned} \mathrm{suc} : \mathbb{Z}^+ \to \mathbb{Z}^+, \ \mathrm{suc}(x) = x + 1; \\ 1^{(n)} : \mathbb{Z}^n \to \mathbb{Z}^+, \ 1^{(n)}(x_1, \dots, x_n) = 1 \quad n \geq 0; \\ \mathrm{pr}_i^n : \mathbb{Z}^n \to \mathbb{Z}^+, \ \mathrm{pr}_i^n(x_1, \dots, x_n) = x_i, \quad n \geq 1. \end{aligned}$$

3.3.3 Elementary Operations on Partial functions

(a) *Composition (or substitution).* This operation takes a pair of partial functions $f : \mathbb{Z}^m \to \mathbb{Z}^n$ and $g : \mathbb{Z}^n \to \mathbb{Z}^p$ and gives a partial function $h = g \circ f : \mathbb{Z}^m \to \mathbb{Z}^p$, defined as follows

$$D(g \circ f) = f^{-1}(D(g)) \cap D(f) = \{x \in \mathbb{Z}^m \mid x \in D(f), f(x) \in D(g)\},$$

$$(g \circ f)(x) = g(f(x)) \text{ for } x \in D(g \circ f).$$

(b) *Junction.* This operation takes partial functions f_i from $\mathbb{Z}^m$ to $\mathbb{Z}^{n_i}$, $i = 1, \dots, k$ to the function $(f_1, \dots, f_k)$ from $\mathbb{Z}^m$ to $\mathbb{Z}^{n_1} \times \cdots \times \mathbb{Z}^{n_k}$ defined as follows

$$D((f_1, \dots, f_k)) = D(f_1) \cap \cdots \cap D(f_k),$$

$$(f_1, \dots, f_k)(x_1, \cdots, x_m) = (f_1(x_1, \cdots, x_m), \cdots, f_k(x_1, \cdots, x_m)).$$

(c) *Recursion.* This operation takes a pair of functions f from $\mathbb{Z}^n$ to $\mathbb{Z}^+$ and g from $\mathbb{Z}^{n+2}$ to $\mathbb{Z}^+$, to the function h from $\mathbb{Z}^{n+1}$ to $\mathbb{Z}^+$ defined as follows

$$h(x_1, \dots, x_n, 1) = f(x_1, \dots, x_n) \quad \text{(the initial condition)};$$

$$h(x_1, \dots, x_n, k+1) = g(x_1, \dots, x_n, k, h(x_1, \dots, x_n, k)) \text{ for } k \geq 1$$
$$\text{(the recursive step)}.$$

The domain of definition $D(h)$ is also described recursively:

$$(x_1, \dots, x_n, 1) \in D(h) \iff (x_1, \dots, x_n) \in D(f);$$

$$\begin{aligned} (x_1, \dots, x_n, k+1) \in D(h) \iff (x_1, \dots, x_n) \in D(f) \text{ and} \\ (x_1, \dots, x_n, k, h(x_1, \dots, x_n, k)) \in D(g) \text{ for } k \geq 1. \end{aligned}$$

(d) *The μ–Operation.* This operation takes a partial function f from $\mathbb{Z}^{n+1}$ to $\mathbb{Z}^+$ to the partial function h from $\mathbb{Z}^n$ to $\mathbb{Z}^+$ which is defined as follows:

$$\begin{aligned} D(h) =& \{(x_1, \dots, x_n) \in (\mathbb{Z}^+)^n | \exists x_{n+1} \geq 1 \ f(x_1, \dots, x_n, x_{n+1}) = 1 \\ & \text{and } (x_1, \dots, x_n, k) \in D(f) \text{ for all } k \leq x_{n+1}\}, \end{aligned}$$

$$h(x_1, \dots, x_n) = \min\{x_{n+1} \mid f(x_1, \dots, x_n, x_{n+1}) = 1\}.$$

Generally speaking, the role of μ is to introduce "implicitly defined" functions. The use of μ makes it possible to introduce a one–by–one check of objects in order to find a desired object in an infinite family. The following three features of μ should be stressed immediately. The choice of the minimal y with $f(x_1, \dots, x_n, y) = 1$ is made, of course, in order to ensure that the function h is well defined. Also, the domain of definition of h seems at first sight to be artificially diminished: if, say, $f(x_1, \dots, x_n, 2) = 1$ and $f(x_1, \dots, x_n, 1)$ is not defined, we consider $h(x_1, \dots, x_n)$ to be undefined, rather than being equal to 2. The reason for this is the wish to preserve the property that h is intuitively computable. Finally we remark that all previously defined operations produce everywhere defined functions if applied to everywhere defined functions. This is obviously not the case for the operation μ. Hence this is the only operation responsible for the appearance of partially defined functions.

3.3.4 Partially Recursive Description of a Function

Definition 3.9. *(a) The sequence of functions $f_1, \cdots f_N$ is called a partially recursive (resp. primitively recursive) description of a function $f = f_N$ if f_1 is one of the simple functions; f_i is for all $i \geq 2$ either a simple function or is obtained by applying the elementary operations to some of the functions $f_1, \cdots, f_{i-1}$ (resp. one of the elementary operations apart from μ).*

(b) The function f is called partially recursive (resp. primitively recursive), if it admits a partially recursive (resp. primitively recursive) description.

Polynomials with positive values. We first establish the recursivity of sums and products.

a)
$$\text{sum}_2 : \mathbb{Z}^2 \to \mathbb{Z}^+, \quad (x_1, x_2) \mapsto x_1 + x_2.$$

Use recursion over x_2 starting from the initial condition $x_1 + 1 = \text{suc}(x_1)$, with the recursive step $x_1 + k + 1 = \text{suc}(\sum_2(x_1, k))$.

b)
$$\text{sum}_n : \mathbb{Z}^n \to \mathbb{Z}^+, \quad (x_1, \dots, x_n) \mapsto \sum_{i=1}^{n} x_i, \quad n \geq 3.$$

Assuming that sum_{n-1} is recursive we obtain sum_n using junctions and composition

$$\mathrm{sum}_n = \mathrm{sum}_2 \circ (\mathrm{sum}_{n-1} \circ (\mathrm{pr}_1^n, \cdots, \mathrm{pr}_{n-1}^n), x_n).$$

Another version is the recursion over x_n starting from the initial condition $\mathrm{suc} \circ \mathrm{sum}_{n-1}$ and the recursive step

$$\sum_{i=1}^{n-1} x_i + k + 1 = \mathrm{suc}(\mathrm{sum}_n(x_1, \ldots, x_{n-1}, k)).$$

One finds that the number of recursive descriptions of a function increases step–by–step, even if one only counts the "natural" descriptions.

c)
$$\mathrm{prod}_2 : \mathbb{Z}^2 \to \mathbb{Z}^+, \quad (x_1, x_2) \mapsto x_1 x_2.$$

Use recursion over x_2 starting from the initial condition x_1, with the recursive step $x_1(k+1) = x_1 k + x_1 = \mathrm{sum}_2(x_1 k, x_1)$.

d)
$$\mathrm{prod}_n : \mathbb{Z}^n \to \mathbb{Z}^+, \quad (x_1, \ldots, x_n) \mapsto x_1 \cdot \cdots \cdot x_n, \quad n \geq 3 :$$
$$\mathrm{prod}_n = \mathrm{prod}_2 \circ (\mathrm{prod}_{n-1}(\mathrm{pr}_1^n, \cdots, \mathrm{pr}_{n-1}^n), x_n).$$

e) "*Substraction of one*": $\mathbb{Z}^+ \to \mathbb{Z}^+$:

$$x \mapsto x \dot{-} 1 = \begin{cases} x - 1, & \text{if } x \geq 2; \\ 1, & \text{if } x = 1. \end{cases}$$

We apply recursion to the simple functions

$$f : \mathbb{Z}^+ \to \mathbb{Z}^+, \quad f = 1,$$
$$g = \mathrm{pr}_2^3 \mathbb{Z}^3 \to \mathbb{Z}^+ : (x_1, x_2, x_3) \mapsto x_2,$$

and as a result obtain the function $h(x_1, x_2) = x_2 \dot{-} 1$. Hence $x \dot{-} 1 = h \circ (x, x)$, where $x = \mathrm{pr}_1^1(x)$.

f) "*Truncated difference*"

$$\mathbb{Z}^2 \to \mathbb{Z}^+ :$$

$$(x_1, x_2) \mapsto x_1 \dot{-} x_2 = \begin{cases} x_1 - x_2, & \text{if } x_1 > x_2; \\ 1, & \text{if } x_1 \leq x_2. \end{cases} \tag{3.3.1}$$

This "truncated difference" is constructed by applying recursion to the functions

$$f(x_1) = x_1 \dot{-} 1, \quad g(x_1, x_2, x_3) = x_3 \dot{-} 1.$$

Let $F : \mathbb{Z}^n \to \mathbb{Z}^+$ where F is any polynomial in $x_1, \ldots, x_n$ with integral coefficients taking only values in $\mathbb{Z}^+$. If all of the coefficients of f are non–negative then F is a sum of products of functions $\mathrm{pr}_i^n : (x_1, \ldots, x_n) \mapsto x_i$. Otherwise $F = F^+ - F^-$, where F^+ and F^- have non–negative coefficients, and the values of the untruncated difference coincide with the values of the truncated one $F^+ \dot{-} F^-$ by the assumption on F. In what follows we use the recursivity of the functions $(x_1 - x_2)^2 + 1$ and $h = (f - g)^2 + 1$ where f and g are recursive: this trick makes it possible to identify the "coincidence set" $f = g$ with the "*level set*" $h = 1$ which is easier to tackle.

3.3.5 Other Recursive Functions

"The Step":

$$s_{x_0}^{a,b}(x) = \begin{cases} a, & \text{for } x \le x_0, \\ b & \text{for } x > x_0; \end{cases} \qquad a, b, x \in \mathbb{Z}^+.$$

For $x_0 = 1$ this is obtained using recursion with the initial condition a and the following value b. In the general case

$$s_{x_0}^{a,b}(x) = s_1^{a,b}(x + 1 \dot{-} x_0).$$

$\mathrm{rem}(x, y) =$ *the remainder in* $\{1, \ldots, x\}$ *after dividing* y *by* x (we do not have zero!). We have

$$\mathrm{rem}(x, 1) = 1 :$$

$$\mathrm{rem}(x, y + 1) = \begin{cases} 1, & \text{if } \mathrm{rem}(x, y) = x : \\ \mathrm{suc} \circ \mathrm{rem}(x, y), & \text{if } \mathrm{rem}(x, y) \ne x. \end{cases}$$

We use the following artificial trick. Consider the step $s(x) = 2$ for $x \ge 2$, $s(1) = 1$ and set

$$\varphi(x, y) = s((\mathrm{rem}(x, y) - x)^2 + 1).$$

It is obvious that

$$\mathrm{rem}(x, y) \ne x \Longleftrightarrow \varphi(x, y) = 1,$$
$$\mathrm{rem}(x, y) = x \Longleftrightarrow \varphi(x, y) = 2,$$

hence

$$\mathrm{rem}(x, y + 1) = 2\mathrm{suc}(\mathrm{rem}(x, y)) \dot{-} \varphi(x, y)\mathrm{suc}(\mathrm{rem}(x, y)).$$

This gives us a recursive definition of rem. A generalization of this trick is "conditional recursion":

$$\begin{aligned} h(x_1, \ldots, x_n, 1) &= f(x_1, \ldots, x_n); \\ h(x_1, \ldots, x_n, k + 1) &= g_i(x_1, \ldots, x_n, k, h(x_1, \ldots, x_n, k), \\ &\quad \text{if the condition } C_i(x_1, \ldots, x_n, k; h) \ (i = 1, \ldots, m) \\ &\quad \text{is satisfied.} \end{aligned} \tag{3.3.2}$$

We reduce the mutually exclusive conditions $C_i(x_1, \ldots, x_n, k; h)$ to the form

$$C_i \text{ is satisfied} \Longleftrightarrow \varphi_i(x_1, \ldots, x_n, k; h(x_1, \ldots, x_n, k)) = 1 \tag{3.3.3}$$

(an everywhere defined recursive function taking only values 1 and 2.)

Then the recursive step can be described as follows:

$$\begin{aligned} h(x_1, \ldots, x_n, k + 1) = &2 \sum_{i=1}^{m} g_i(x_1, \ldots, x_n, k, h(x_1, \ldots, x_n, k)) \\ &- \sum_{i=1}^{m} (g_i \varphi_i)(x_1, \ldots, x_n, k, h(x_1, \ldots, x_n, k)). \end{aligned} \tag{3.3.4}$$

This trick makes it possible to establish the primitive recursivity of the following functions which will be used below.

The incomplete quotient:

$$\mathrm{qt}(x,y)=\begin{cases}\text{integral part of } y/x, & \text{if } y/x\geq 1,\\ 1, & \text{if } y/x<1.\end{cases}$$

We have

$$\mathrm{qt}(x,y+1)=\begin{cases}\mathrm{qt}(x,y), & \text{if } \mathrm{rem}(x,y+1)=x,\ y+1\neq x;\\ \mathrm{qt}(x,y)+1, & \text{if } \mathrm{rem}(x,y+1)\neq x,\ y+1\neq x;\\ 1, & y+1=x.\end{cases}$$

One reduces these conditions to the standard form (3.3.3) with the help of the functions

$$\tilde{s}((\mathrm{rem}(x,y+1)-x)^2+1),$$
$$s((\mathrm{rem}(x,y+1)-x)^2+1)\cdot\tilde{s}((x-y-1)^2+1),$$
$$s((x-y-1)^2+1),$$

where

$$s(1)=1,\ s(\geq 2)=2;\ \tilde{s}(1)=2,\tilde{s}(\leq 2)=1.$$

The function $\mathrm{rad}(x)$ – *the integral part of* $\sqrt{x}$. One has

$$\mathrm{rad}(1)=1$$

$$\mathrm{rad}(x+1)=\begin{cases}\mathrm{rad}(x) & \text{if } \mathrm{qt}(\mathrm{rad}(x)+1,x+1)<\mathrm{rad}(x)+1,\\ \mathrm{rad}(x)+1 & \text{if } \mathrm{qt}(\mathrm{rad}(x)+1,x+1)=\mathrm{rad}(x)+1.\end{cases}$$

These conditions can be reduced to the standard form (3.3.3) in a similar way.

The function $\min(x,y)$:

$$\min(x,1)=1;$$

$$\min(x,y+1)=\begin{cases}\min(x,y), & \text{if } x\leq y,\\ \min(x,y)+1, & \text{if } x>y.\end{cases}$$

The function $\max(x,y)$ (similarly).

3.3.6 Further Properties of Recursive Functions

If $f(x_1, \dots, x_n)$ is recursive then

$$Sf = \sum_{k=1}^{x_n} f(x_1, \dots, x_{n-1}, k), \quad Pf = \prod_{k=1}^{x_n} f(x_1, \dots, x_{n-1}, k)$$

are recursive. We can also obtain recursive functions from f in the following ways:

a) by any substitution of the arguments;
b) by introducing any number of extra arguments;
c) by identifying the members of any group of arguments (e.g. $f(x, x)$ instead of $f(x, y)$ etc.)

The map $f : \mathbb{Z}^m \to \mathbb{Z}^n$ is recursive if and only if all of its components $\mathrm{pr}_i^n \circ f$ are recursive.

Definition 3.10. *The set $E \subset \mathbb{Z}^n$ is called enumerable, if there exists a partially recursive function f such that $E = D(f)$ (the domain of definition).*

The discussion of §3.1 and §3.2 shows that enumerability has the following intuitive meaning: there exists a program which recognizes the elements x belonging to E, but which not necessarily recognizes elements which do not belong to E. Below, a different description of the enumerable sets will be given, which will explain the ethimology of the name: they are the sets with the property that all their elements may be obtained (possibly with repetitions and in an unknown order) by a "generating program".

The following simple fact is easily deduced from the properties of partially recursive functions.

3.3.7 Link with Level Sets

Proposition 3.11. *The following three classes coincide: a) the enumerable sets;*
b) the level sets of partially recursive functions;
c) the 1–level sets of partially recursive functions.

A much more difficult statement is the following result and its corollaries.

3.3.8 Link with Projections of Level Sets

Theorem 3.12. *The following two classes coincide:*
a) the enumerable sets;
b) the projections of level sets of primitively recursive functions.

Among the primitively recursive functions are the polynomials with coefficients in $\mathbb{Z}^+$. Recall that *Diophantine sets* are projections of the level sets of such polynomials. The *Matiyasevich theorem* can now be stated precisely as follows:

3.3.9 Matiyasevich's Theorem

Theorem 3.13. *The enumerable sets are Diophantine; hence the two classes coincide.*

We sketch the proof of Theorem 3.12 in this section, and of Theorem 3.13 in the next section.

Let us temporarily call the projections of level–sets of primitively recursive functions the *primitively enumerable sets.* In the first part of the proof of Theorem 3.12 it is established that the primitively enumerable sets are all enumerable; in the second part the opposite inclusion is proved.

We therefore let $f(x_1, \dots, x_n, x_{n+1}, \dots, x_{n+m})$ be a primitively recursive function, and E the projection of its 1–level to the first n coordinates. We shall explicitly construct a partially recursive function g such that $E = D(g)$. This will show that any primitively enumerable set must be enumerable

We divide the proof into three cases depending on the codimension of the projection: $m = 0, 1$ or $m \geq 2$.

Case a): $m = 0$. Then the set E is the 1–level of f and is enumerable by Proposition 3.11.

Case b): $m = 1$. Set

$$g(x_1, \dots, x_n) = \min\{x_{n+1} \mid f(x_1, \dots, x_n, x_{n+1}) = 1\}.$$

It is clear that g is partially recursive and $D(g) = E$.

Case c): $m \geq 2$. We shall reduce this to the previous case using the following lemma, which is interesting in itself (the lack of a notion of "dimension" in "recursive geometry") and plays an important role in various other questions.

3.3.10 The existence of certain bijections

Lemma 3.14. *For all $m \geq 1$ there exists a one–to–one map $t^{(m)} : \mathbb{Z}^+ \to \mathbb{Z}^m$ such that:*

a) the functions $t_i^{(m)} = \mathrm{pr}_i^{(m)} \circ t^{(m)}$ are primitively recursive for all $1 \leq i \leq m$;
b) the inverse function $\tau^{(m)} : \mathbb{Z}^m \to \mathbb{Z}^+$ is primitively recursive

Application of the lemma.

We apply Lemma 3.14 in the case 3.3.9 c) and set for $m \geq 2$

$$g(x_1, \dots, x_n, y) = f(x_1, \dots, x_n, t_1^{(m)}(y), \dots, t_m^{(m)}(y)).$$

It is clear that g, being a composition of primitively recursive functions is itself primitively recursive. It is easy to check that E coincides with the projection of the 1–level set of the function g to the first n coordinates. Since this projection is of codimension 1, we have reduced *Case c* to *Case b*.

Proof of the lemma. The case $m = 1$ is trivial. We shall prove the lemma by induction on m, starting from $m = 2$.

Construction of $t^{(2)}$. We first construct $\tau^{(2)} : \mathbb{Z}^2 \to \mathbb{Z}^+$ by setting

$$\tau^{(2)}(x_1, x_2) = \frac{1}{2}((x_1 + x_2)^2 - x_1 - 3x_2 + 2).$$

It is easy to check that if we index the pairs $(x_1, x_2) \in \mathbb{Z}^2$ in the "Kantor order", and inside each group with given $x_1 + x_2$ in increasing order, then $\tau^{(2)}(x_1, x_2)$ will be exactly the number of the pair (x_1, x_2) in this list. Thus $\tau^{(2)}(x_1, x_2)$ is bijective and primitively recursive (use (3.3.4) and the recursivity of qt from (3.3.5) in order to take into account the $1/2$.

The reconstruction of a pair (x_1, x_2) from its image y is an elementary task and this leads to the following formula for the inverse function $t^{(2)}$:

$$t^{(2)}(y) = y - \frac{1}{2}\left[\sqrt{2y - \frac{7}{4}} - \frac{1}{2}\right]\left(\left[\sqrt{2y - \frac{7}{4}} - \frac{1}{2}\right] + 1\right),$$

$$t_2^{(2)}(y) = \left[\sqrt{2y - \frac{7}{4}} - \frac{1}{2}\right] - t^{(2)}(y) + 2.$$

Here $[z]$ denotes the integral part of z. Using the results and methods of §3.3.5 – §3.3.6, one can verify that these functions are primitively recursive.

Construction of $t^{(m)}$. Assume that $t^{(m-1)}$, $\tau^{(m-1)}$ are already constructed, and their properties are proved. Set first of all

$$\tau^{(m)} = \tau^{(2)}(\tau^{(m-1)}(x_1, \dots, x_{m-1}), x_m).$$

It is clear that $\tau^{(m)}$ is primitively recursive and bijective. Solving the equation

$$\tau^{(2)}(\tau^{(m-1)}(x_1, \dots, x_{m-1}), x_m) = y$$

in two steps, we get the following formulae for the inverse function $t^{(m)}$:

$$t_m^{(m)}(y) = t_2^{(2)}(y), \quad t_i^{(m)}(y) = t_i^{(m-1)}(t_1^{(2)}(y)), \quad 1 \le i \le m - 1.$$

By induction, $t_1^{(m)}$ is primitively recursive.

This finishes the proof of the lemma and the first part of the proof of theorem 3.12.

The second part of the proof. We now prove that every primitively enumerable set is enumerable. We begin with the following easily verified property of the class of primitively enumerable sets.

3.3.11 Operations on primitively enumerable sets

Lemma 3.15. *The class of primitively enumerable sets is closed under the operations of finite direct sum, finite intersection, finite union and projection.*

Now let E be an enumerable set. Using proposition in 3.3.7 we realize it as the 1–level of a partially recursive function $f : \mathbb{Z}^n \to \mathbb{Z}^+$. Note that in order to prove that E is primitively enumerable it suffices to check that the graph $\Gamma_f \subset \mathbb{Z}^n \times \mathbb{Z}^+$ is primitively enumerable. Indeed it is clear that E coincides with the 1–level of the projection onto the first n coordinates of the set $\Gamma_f \cap (\mathbb{Z}^n \times \{1\})$. Also the set $\{1\} \subset \mathbb{Z}^+$ is primitively enumerable in view of the properties listed in §3.3.4, so if we prove that Γ_f is primitively enumerable, then the same would follow for E by lemma 3.15. We have thus reduced our problem to that of proving that the graphs of partially recursive functions f are primitively enumerable.

With this purpose we check that: a) the graphs of simple functions are primitively enumerable; b) if we are given functions whose graphs are primitively enumerable, then any function obtained from them using one of the elementary operations also has a primitively enumerable graph.

Stability under recursion and the μ operation are the most delicate points. In order to prove these, the following nice lemma is used.

3.3.12 Gödel's function

Lemma 3.16. *There exists a primitively recursive function* $\mathrm{Gd}(k,t)$ *(Gödel's function) with the following property: for each* $N \in \mathbb{Z}^+$ *and for any finite sequence* $a_1, \ldots, a_N \in \mathbb{Z}^+$ *of the lenth* N *there exists* $t \in \mathbb{Z}^+$ *such that* $\mathrm{Gd}(k,t) = a_k$ *for all* $1 \le k \le N$ *(In other words,* $\mathrm{Gd}(k,t)$ *may be regarded as a sequence of functions of the argument* k *indexed by the parameter* t *such that any function of* k *on an arbitrarily large interval* $1, \ldots, N$ *can be imitated by an appropriate term of this sequence).*

In order to prove this it is convenient to put first

$$\mathrm{gd}(u,k,t) = \mathrm{rem}(1+kt, u)$$

and to show that gd has the same property as Gd if we allow ourselves to choose $(u,t) \in \mathbb{Z}^2$. After this we could put

$$\mathrm{Gd}(k,y) = \mathrm{gd}(t_1^{(2)}(y), k, t_2^{(2)}(y)),$$

where $t^{(2)} : \mathbb{Z}^+ \to \mathbb{Z}^2$ is the isomorphism of Lemma 3.14. Getting rid of the auxiliary parameter u in $\mathrm{Gd}(k,t)$ (in comparison with $\mathrm{gd}(u,k,t)$) causes no essential problems.

3.3.13 Discussion of the Properties of Enumerable Sets

Theorem 3.12 of §3.3.8 shows that if E is enumerable, then there exists a program "generating" E (cf. §3.3.6). Indeed, let E be the projection onto the first n coordinates of the 1–level of a primitively recursive function $f(x_1, \ldots, x_n, y)$. The program "generating" E should check one–by–one the vectors $(x_1, \ldots, x_n, y)$, say, in Kantor's order; it should compute f and output $(x_1, \ldots, x_n)$ if and only if $f(x_1, \ldots, x_n, y) = 1$. Since f is primitively recursive, the generating program will sooner or later write down each element of E, and no other element. It cannot stop forever on elements not belonging to E. However, if E were empty we could never find this out just by waiting.

The set $E \subset \mathbb{Z}^n$ is called *solvable*, if it and its complement are enumerable. Intuitively this means that that there is a program which decides for any element of $\mathbb{Z}^n$ whether it belongs to E or not. These sets can be characterized as being the level sets of general recursive (everywhere defined recursive) functions, or as the sets whose characteristic function is recursive. In order to establish these properties, the following result is used.

Proposition 3.17. *A partial function g from $\mathbb{Z}^n$ to $\mathbb{Z}^+$ is partially recursive iff its graph is enumerable.*

3.4 Diophantineness of a Set and algorithmic Undecidability

3.4.1 Algorithmic undecidability and unsolvability

Before explaining how to prove that the classes of Diophantine and of ennumerable sets are the same, we first give some interesting applications of this theorem. It is known from logic that there are sets which are ennumerable but not solvable. Combining this fact with Matiyasevich's theorem (see Theorem 3.13) and the Church thesis, we deduce that Hilbert's tenth problem (see section 3.1.2) is undecidable, see [Mat04].

First of all, every natural number is a sum of four squares (Lagrange's theorem, see Part I, section 1.2.6). The solvability of the equation $f(x_1, \ldots, x_n) = 0$ in $\mathbb{Z}^n$ is therefore equivalent to the solvability of the equation

$$f\left(1 + \sum_{i=1}^{4} y_{i1}^2, \ldots, 1 + \sum_{i=1}^{4} y_{in}^2\right) = 0$$

in $\mathbb{Z}^{4n}$. It is thus sufficient to establish the algorithmic undecidability of the class of questions whether equations have solutions in $\mathbb{Z}^n$. Let $E \subset \mathbb{Z}^+$ be ennumerable but not solvable. We represent it as the 0–level set of a polynomial $f_t = f(t; x_1, \cdots, x_n) = 0$, $f \in \mathbb{Z}[t; x_1, \cdots, x_n]$. The equation $f_{t_0} = 0$; $t_0 \in \mathbb{Z}^+$ is solvable iff $t_0 \in E$. According to a general principle (the Church thesis), intuitive computability is equivalent to partial recursivity of a function. This implies that the corresponding class of problems for the family $\{f_t\}$ is algorithmically decidable, iff the characteristic function of E is computable. However this is not the case by the choice of E: although E is ennumerable, its complement is not.

Thus the question of solvability in integers is *undecidable* even for an appropriate one parameter family of equations. The number of variables, or more generally the codimension of the projection can be reduced to 9 (Yu. I. Matiyasevich). The precise minimum is still unknown, although this is a very intriguing problem.

3.4.2 Sketch Proof of the Matiyasevich Theorem

One introduces temporarily a class of sets, intermediate between the ennumerable and Diophantine sets. In order to define this class, consider the map which takes a subset $E \subset \mathbb{Z}^n$ to a new subset $F \subset \mathbb{Z}^n$ defined by the following law:

$$(x_1, \ldots, x_n) \in F \Longleftrightarrow \forall k \in [1, x_n]$$
$$(x_1, \ldots, x_{n-1}, k) \in E.$$

We shall say in this case that F is obtained from E by use of the restricted generality quantor on the nth coordinate. The restricted generality quantor is defined analogously on any other coordinate.

Definition–Lemma. *Consider the following three classes of subsets of $\mathbb{Z}^n$ for any n:*

I. Projections of the level sets of primitively recursive functions.
II. The smallest class of sets containing the level sets of polynomials with integral coefficients, which is closed under the operations of finite direct sum, finite union, finite intersection, projection and the restricted generality quantor.
III. Projections of the level sets of polynomials with integral coefficients.

Then

a) Class I coincides with the class of ennumerable sets, and Class III with the class of Diophantine sets. The sets of the class II will be called D–sets.
b) The following inclusions hold: $I \supset II \supset III$.

The final steps in the proof of the Matiyasevich theorem consist of reductions similar to those described above. The crucial part is the proof that the class of Diophantine sets is closed under the use of the restricted generality quantor. Here one makes use of the Diophantine representations of concrete sets from §3.2, in order to check that application of Gödel's function does not damage the Diophantineness.

Note that B. Poonen studied in [Po03] Hilbert's tenth problem for large subrings of $\mathbb{Q}$ in connection with Mazur's conjecture on varieties over $\mathbb{Q}$ whose real topological closure of rational points has infinitely many components (no such varieties are known to this point). For the field of rational numbers Hilbert's tenth problem is a major open question. In trying to answer it two general methods have been used: one is to study the similar question in other global fields (such as fields of rational functions $\mathbb{F}_q(t)$ over finite fields) and try to transfer the methods to $\mathbb{Q}$; the other is to try to prove it for ever larger subrings of $\mathbb{Q}$. Relations of this problem with arithmetic and algebraic geometry were studied in [DLPvG], see also [Shl03].

4

Arithmetic of algebraic numbers

4.1 Algebraic Numbers: Their Realizations and Geometry

4.1.1 Adjoining Roots of Polynomials

The idea to extend the field of rational numbers owes a lot to various attempts to solve some concrete Diophantine equations. The use of irrational numbers which are roots of polynomials with rational coefficients often makes it possible to reduce such equations to more convenient forms. An intriguing example of this is the study of the Fermat equation (cf. [BS85], [Pos78], [Edw77], [Rib79]):

$$x^n + y^n = z^n \quad (n > 2). \tag{4.1.1}$$

The unsolvability of (4.1.1) in non–zero integers for $n > 2$ is now established in the work of Wiles [Wi] and Wiles-Taylor [Ta-Wi] on the Shimura-Taniyama-Weil conjecture and Fermat's Last Theorem, see chapter 7. Wiles used various sophisticated techniques and ideas due to himself and a number of other mathematicians (K.Ribet, G.Frey, Y.Hellegouarch, J.–M.Fontaine, B.Mazur, H.Hida, J.–P.Serre, J.Tunnell, ...). This genuinely historic event concludes a whole epoque in number theory.

Notice that before the work of Wiles, it was known from results due to Faltings (see chapter 5, §5.5) that the number of primitive solutions (i.e. such that $\mathrm{GCD}(x, y, z) = 1$) is finite for each $n > 2$. If n is an odd integer then the left hand side transforms into the following product:

$$\prod_{k=0}^{n-1}(x + \zeta^k y) = z^n, \tag{4.1.2}$$

where $\zeta = \exp(2\pi i/n)$ is a primitive n^{th} root of unity. If we suppose that the ring $R = \mathbb{Z}[\zeta]$ has unique factorization of elements, then by studying the divisibility properties of the left hand side of (4.1.1) one can prove that (4.1.1) has no solutions in integers not dividing n (this is the first case of the Fermat conjecture: $n \nmid xyz$) (Kummer). However, this unique factorization property is far from being always satisfied: J.M.Masley and H.L.Montgomery (cf. [MM76]) have found all n with this property; it turns out that there are altogether 29 such numbers, and the primes among them are $n = 3, 5, 7, 11, 13, 17, 19$. Notice that before the work of Wiles, the validity of the first case of Fermat's Last theorem has been established for infinitely many primes ([AdHB85], [Fou85], [GM]).

Let α be a complex root of an irreducible polynomial $f(x) = x^n + a_{n-1}x^{n-1} + \dots a_1x + a_0 \in \mathbb{Q}[x]$ with rational coefficients $a_i \in \mathbb{Q}$. If $k = \mathbb{Q}(\alpha)$ is the smallest field containing α then each of its elements β has the form: $\beta = r(\alpha)$, where $r(x) \in \mathbb{Q}[x]$ is a polynomial of degree $\deg r(x) < n$, and the arithmetical operations in $\mathbb{Q}(\alpha)$ are the same as those with residues mod f in the ring of polynomials $\mathbb{Q}[x]$.

In other words, there is an isomorphism between k and the quotient ring $\mathbb{Q}[x]/(f)$, and k is an n–dimensional vector space over $\mathbb{Q}$ (with basis $1, \alpha, \dots, \alpha^{n-1}$). A choice of basis gives another realization of elements of k as $n \times n$ square matrices: to an element β one attaches the matrix of the linear transformation $\varphi_\beta : x \mapsto \beta x$ (with respect to the chosen basis). For the basis $\{1, \alpha, \dots, \alpha^{n-1}\}$ the endomorphism φ_α is described by the matrix (sometimes called the *adjoint matrix*):

$$A_\alpha = \begin{pmatrix} 0 & 0 & \dots & 0 & -a_0 \\ 1 & 0 & \dots & 0 & -a_1 \\ 0 & 1 & \dots & 0 & -a_2 \\ \dots & \dots & \dots & \dots & \dots \\ 0 & 0 & \dots & 1 & -a_{n-1} \end{pmatrix},$$

and the smallest subring of the matrix algebra $M_n(\mathbb{Q})$ containing A_α can be identified with k. Each element $\beta \in k$ is a root of the characteristic polynomial of the endomorphism φ_β, and its determinant and trace are denoted $\mathrm{N}\beta$ and $\mathrm{Tr}\beta$. These are called the *norm* and the *trace* of β. The bilinear form $B : k \times k \to \mathbb{Q}$ defined by $B(u, v) = \mathrm{Tr}(uv)$ is non-degenerate. An element β is called *integral* if all of the coefficients b_i of its characteristic polynomial

$$\det(X \cdot 1_n - \varphi_\beta) = X^n + b_{n-1}X^{n-1} + \dots + b_0 \in \mathbb{Q}[X]$$

are integers. This condition is equivalent to saying that the ring $\mathbb{Z}[\beta]$ is a finitely generated Abelian group. The set of all integers of k will be denoted by $\mathcal{O} = \mathcal{O}_k$. This is a free $\mathbb{Z}$–module (a free Abelian group) with a basis $\omega_1, \cdots, \omega_n$. The determinant of the bilinear form $B(u, v)$ with respect to such a basis is called the discriminant of k, and is denoted by $D = D_k$. This is independent of the choice of basis of $\mathcal{O}_k$.

The idea of symbolically manipulating the roots of polynomials has lead to the theory of algebraic extensions of arbitrary fields, for which one may repeat the above constructions. If $k \subset K$ are two fields and the dimension $[K : k]$ is finite, then for any $\beta \in K$ one defines analogously $\mathrm{N}_{K/k}(\beta)$ and $\mathrm{Tr}_{K/k}(\beta)$. The claim that the form $B(u, v) = \mathrm{Tr}_{K/k}(uv)$ is non-degenerate is one of the definitions of a separable extension. If this is the case one can always find an element $\gamma \in K$ such that $K = k(\gamma)$ (this statement is known as the *primitive element* theorem) (cf. [La65], [Sha87]).

Adjoining the roots of all the irreducible polynomials in $k[X]$ to the ground field k leads to the construction of an *algebraic closure* $\overline{k}$ of k. This is a field, uniquely defined by k upto isomorphism, which consists of elements algebraic over k, and which is also *algebraically closed.* This means that every polynomial $f(X) \in \overline{k}[K]$ with $\deg f > 0$ has a root $\alpha \in \overline{k}$. When we write $\overline{\mathbb{Q}}$ we often mean the complex realization of this field as the set of all complex numbers $\alpha \in \mathbb{C}$ which are roots of polynomials with rational coefficients.

4.1.2 Galois Extensions and Frobenius Elements

(cf. [La65], [LN83]). In general let K/k be a finite separable extension, $k \subset K \subset \overline{k}$. Then K/k is called a Galois extension if for every embedding $\lambda : K \to \overline{k}$ over k (i.e. $\lambda(x) = x$ for $x \in k$) one has $\lambda(K) = K$. In this case the automorphisms $\lambda : K \to K$ over k form a group $G(K/k) = \mathrm{Aut}(K/k)$ of order n which is called the *Galois group.* In what follows the action of $\sigma \in G(K/k)$ on $x \in K$ will be denoted either by x^σ, or by $\sigma(x)$ so that the composition law is $(\tau\sigma)(x) = \tau(\sigma(x))$, $x^{\tau\sigma} = (x^\sigma)^\tau$ (a left action of $G(K/k)$ on K).

Theorem 4.1 (Main Theorem of Galois Theory). *There is a one–to–one correspondence between subgroups $H \subset G(K/k)$ and intermediate fields L with $k \subset L \subset K$. This correspondence is defined by the following law:*

$$H \mapsto K^H = \{x \in K \mid x^\sigma = x \text{ for all } \sigma \in H\},$$

$$L \mapsto H_L = \{\sigma \in G(K/k) \mid x^\sigma = x \text{ for all } x \in L\}.$$

The normal subgroups $H \triangleleft G(K/k)$ correspond exactly to the Galois subextensions L/k, and for such subgroups or extensions we have $G(L/k) = G(K/k)/H_L$.

Example 4.2 (Finite Fields.). Let $K = \mathbb{F}_q$ be a finite field with q elements. Then $q = p^f$ and $\mathbb{F}_q$ is a vector space of dimension f over the prime subfield $\mathbb{F}_p = \mathbb{Z}/p\mathbb{Z}$ (cf. [LN83]). For any integer $r > 0$ the algebraic closure $\overline{\mathbb{F}}_q$ contains exactly one extension of $\mathbb{F}_q$ of degree r:

$$\mathbb{F}_{q^r} = \{x \in \overline{\mathbb{F}}_q \mid x^{q^r} = x\},$$

so that $x^{q^r-1} = 1$ for all elements of the multiplicative group $\mathbb{F}_{q^r}^\times$. The extension $\mathbb{F}_{q^r}/\mathbb{F}_q$ is therefore a Galois extension, and its Galois group is cyclic of order r:

$$G(\mathbb{F}_{q^r}/\mathbb{F}_q) = \{1, \mathrm{Fr}_q, \mathrm{Fr}_q^2, \cdots, \mathrm{Fr}_q^{r-1}\},$$

where $\mathrm{Fr}_q(x) = x^q$ is the Frobenius automorphism.

Example 4.3 (Cyclotomic Fields.). Let ζ_m be a primitive root of unity of degree m. Then the field $K_m = \mathbb{Q}(\zeta_m)$ contains all roots of the polynomial $X^m - 1 = \prod_{i=0}^{m-1}(X - \zeta_m^i)$ and is therefore a Galois extension. If $\sigma \in G(K_m/\mathbb{Q})$ then the element $\sigma\zeta_m$ must also be a primitive m^{th} root of unity, so that $\sigma\zeta_m = \zeta_m^a$ for some a with $(a, m) = 1$. If ζ_m^k is another m^{th} root of unity then $\sigma(\zeta_m^k) = \zeta_m^{ak}$. Hence the correspondence $\sigma \mapsto a(\bmod\ m)$ produces a canonical map $G(K_m/\mathbb{Q}) \to (\mathbb{Z}/m\mathbb{Z})^\times$ which is in fact an isomorphism. In order to prove this it suffices to show that the cyclotomic polynomial

$$\Phi_m(X) = \prod_{\substack{i=1 \\ (i,m)=1}}^{m-1} (X - \zeta_m^i)$$

is irreducible over $\mathbb{Q}$. First we see that $X^m - 1 = \prod_{d|m} \Phi_d(X)$, and hence $\Phi_m(X) = \prod_{d|m}(X^{m/d} - 1)^{\mu(d)} \in \mathbb{Z}[X]$ (where $\mu(d)$ is the *Möbius function* of d). The irreducibility is established by reducing the polynomials modulo p: $\mathbb{Z}[X] \to \mathbb{F}_p[X]$: $f(X) \mapsto \overline{f}(X) \in \mathbb{F}_p[X]$. One applies the properties of the Frobenius endomorphism $\overline{f}(X) \to \overline{f}(X)^p = \overline{f}(X^p) \in \mathbb{F}_p[X]$ in the ring $\mathbb{F}_p[X]$. Suppose that $\Phi_m(X)$ is not irreducible and let

$$\Phi_m(X) = f_1(X) \cdot \dots \cdot f_r(X)$$

be the decomposition of Φ_m as a product of irreducible polynomials in $\mathbb{Z}[X]$. We show that for all $a \bmod m$ with $(a, m) = 1$, $f_1(\zeta_m) = 0$ implies $f_1(\zeta_m^a) = 0$. We use the existence of a prime p such that $p \equiv a \bmod m$. The polynomial $X^m - 1$ is coprime to its derivative mX^{m-1} in $\mathbb{F}_p[X]$ since $p \not| m$. Hence the polynomials $\overline{f}_1(X), \dots, \overline{f}_r(X)$ are pairwise coprime.

If $f_1(\zeta_m^a) \neq 0$ then we have $f_j(\zeta_m^a) = 0$ for some $j \neq 1$, which implies $f_j(\zeta_m^p) = 0$. Hence $f_1(X)$ has a common factor with $f_j(X^p)$. In fact since f_1 is irreducible, it must divide $f_j(X^p)$. Therefore $\overline{f}_1(X)$ divides $\overline{f}_j(X^p) = \overline{f}_j(X)^p$. This contradicts the fact that $\overline{f}_1(X)$ and $\overline{f}_j(X)$ are coprime.

Note that we do not need to assume the existence of a p such that $p \equiv a \bmod m$. We could instead consider the decomposition $a = p_1^{\alpha_1} \cdot \dots \cdot p_s^{\alpha_s}$ and study all the reductions mod p_i, $i = 1, \dots, s$ (cf. [BS85], [La65], [Chev40], [La78b], [Wash82]).

Recall that a *Dirichlet character* χ modulo m is a homomorphism $\chi : (\mathbb{Z}/m\mathbb{Z})^\times \to \mathbb{C}^\times$. These are often regarded as a functions on $\mathbb{Z}$ such that $\chi(x) = \chi(x \bmod m)$ if $(x, m) = 1$, and $\chi(x) = 0$ if $(x, m) > 1$ (see Part I, §2.2.2). According to what we have proved, there is a canonical isomorphism $G(K_m/\mathbb{Q}) \cong (\mathbb{Z}/m\mathbb{Z})^\times$. Hence a Dirichlet character defines a homomorphism $\rho_\chi : G(\overline{\mathbb{Q}}/\mathbb{Q}) \to \mathbb{C}^\times$ by means of the projection $G(\overline{\mathbb{Q}}/\mathbb{Q}) \to G(K_m/\mathbb{Q})$.

Theorem 4.4 (Theorem of Kronecker–Weber). *For any homomorphism $\rho : G(\overline{\mathbb{Q}}/\mathbb{Q}) \to \mathbb{C}^\times$ of finite order there exists a Dirichlet character χ such that $\rho = \rho_\chi$*

(cf. [Sha51], [AT51], [Chev40]) .

The theorem of Kronecker–Weber can be restated as saying that any Galois extension $K/\mathbb{Q}$ whose Galois group $G(K/\mathbb{Q})$ is commutative (i.e. any Abelian extension) is contained in a cyclotomic extension.

A remarkable fact is that the elements of the Galois group $G(K_m/\mathbb{Q})$ correspond to prime numbers (more precisely $p \bmod m$ for $p \nmid m$). The deepest results of algebraic number theory are related to generalizations of the Kronecker–Weber theorem. For example, Deligne and Serre have shown that there exists a correspondence between two–dimensional irreducible complex representations $\rho : G(\overline{\mathbb{Q}}/\mathbb{Q}) \to \mathrm{GL}_2(\mathbb{C})$ such that $\det\rho = \rho_\chi$ for an odd character χ, and *primitive cusp forms of weight one* (cf. Chap. 6, §6.4, the *theorem of Deligne–Serre*). It is conjectured that this correspondence is one–to–one, and therefore gives a two–dimensional analogue of the Kronecker–Weber theorem.

4.1.3 Tensor Products of Fields and Geometric Realizations of Algebraic Numbers

In order to obtain a convenient geometric realization of an algebraic number field k, we use the tensor product $k \otimes \mathbb{R}$. Constructions involving tensor products of fields are frequently used in algebraic number theory, and for this reason we begin with a general result on these products.

Theorem 4.5 (Theorem on Tensor Products of Fields). *Let K/k be a finite separable extension, $K = k(\gamma)$, and let L/k be another extension, and suppose that*

$$K \cong k[X]/(f_\gamma(X)), \quad f_\gamma(X) = \prod_{i=1}^{m} g_i(X)$$

is the decomposition as a product of irreducible polynomials in the ring $L[X]$. Then there is a ring isomorphism

$$K \otimes_k L \cong \prod_{i=1}^{m} L_i,$$

where $L_i \cong L[X]/(g_i(X))$ are finite extensions of L containing K under the embeddings $\lambda_i : K \hookrightarrow L_i$ defined by

$$\lambda_i(r(\gamma)) = r(X) \bmod g_i(X).$$

(cf. [CF67], [Chev40]).

The proof of this theorem is similar to that of the *Chinese remainder theorem*. The elements $r(\gamma) \otimes_k l$ with $l \in L$, $r(X) \in k[X]$ generate the whole ring $K \otimes_k L$, and the isomorphism is given by

$$r(\gamma) \otimes_k l \mapsto (lr(X) \bmod g_1(X), \cdots, lr(X) \bmod g_m(X)).$$

Corollary 4.6. *Let $\beta \in K$. If $f_\beta(X) \in k[X]$ is its characteristic polynomial in the extension K/k and $f_{\beta,i}(X) \in L[X]$ are its characteristic polynomials in the extensions L_i/L, then*

$$f_\beta(X) = \prod_{i=1}^{m} f_{\beta,i}.$$

In particular, we have

$$\mathrm{N}_{K/k}(\beta) = \prod_{i=1}^{m} \mathrm{N}_{L_i/L}(\lambda_i(\beta)), \tag{4.1.3}$$

$$\mathrm{Tr}_{K/k}(\beta) = \sum_{i=1}^{m} \mathrm{Tr}_{L_i/L}(\lambda_i(\beta)). \tag{4.1.4}$$

If we take $\overline{k}$ for L, then $m = n$ and $\lambda_1, \ldots, \lambda_n$ are all possible embeddings $\lambda_i : K \to \overline{k}$,

$$f_\beta(X) = (X - \lambda_1(\beta)) \cdot \cdots \cdot (X - \lambda_n(\beta)).$$

Hence for any $\beta \in K$ we have

$$\mathrm{N}_{K/k}(\beta) = \prod_{i=1}^{n} \lambda_i(\beta), \quad \mathrm{Tr}_{K/k}(\beta) = \sum_{i=1}^{n} \lambda_i(\beta). \tag{4.1.5}$$

By putting $L = \mathbb{R}$, $K = \mathbb{Q}(\gamma)$ and $k = \mathbb{Q}$, we obtain a geometric realization of the algebraic numbers. Let $f_\gamma(X) = (X - \gamma_1) \cdot \cdots \cdot (X - \gamma_{r_1}) \cdot (X^2 + \alpha_1 X + \beta_1) \cdot \cdots \cdot (X^2 + \alpha_{r_2} X + \beta_{r_2})$ be the decomposition of the minimal polynomial $f_\gamma(X) \in \mathbb{Q}[X]$ of γ into irreducible polynomials over $\mathbb{R}$. Then

$$K \otimes_k \mathbb{R} = \mathbb{Q}(\gamma) \otimes \mathbb{R} \cong \mathbb{R}^{r_1} \times \mathbb{C}^{r_2} \tag{4.1.6}$$

(this is an $\mathbb{R}$–algebra isomorphism), or $\mathbb{Q}(\gamma) \otimes \mathbb{R} \cong \mathbb{R}^n$ as a real vector space, so that $n = r_1 + 2r_2$. Let $\lambda_1, \cdots, \lambda_{r_1}, \cdots, \lambda_{r_1+r_2}$ be the embeddings of 4.1.2. Then the tuple

$$\lambda = (\lambda_1, \cdots, \lambda_{r_1}, \cdots, \lambda_{r_1+r_2})$$

defines an embedding of K into $\mathbb{R}^n$, and any embedding of K into $\mathbb{C}$ is one of the following

$$\lambda_1, \cdots, \lambda_{r_1}, \lambda_{r_1+1}, \overline{\lambda}_{r_1+1}, \cdots, \lambda_{r_1+r_2}, \overline{\lambda}_{r_1+r_2}.$$

A *lattice* M in a vector space $\mathbb{R}^n$ is by definition a discrete subgroup $M \subset \mathbb{R}^n$ such that the quotient group $\mathbb{R}^n/M$ is compact (in the natural topology). Every lattice is a free Abelian group generated by a basis $e_1, \ldots, e_n$ of $\mathbb{R}^n$.

If $\mathcal{O}$ is the ring of integers in K, then one verifies that its image $M = \lambda(\mathcal{O}) \subset \mathbb{R}^n$ is a lattice, and

$$D_K = (-4)^{r_2}\mathrm{vol}(\mathbb{R}^n/\lambda(\mathcal{O})), \tag{4.1.7}$$

where D_K is the *discriminant* of K, and $\mathrm{vol}(\mathbb{R}^n/\lambda(\mathcal{O}))$ is the volume of the fundamental parallelogram $\{\sum_{i=1}^n x_i e_i \mid 0 \le x_i \le 1\}$ of the lattice $\mathcal{O} = \langle e_1, \dots, e_n\rangle$ with respect to the usual Lebesgue measure on $\mathbb{R}^n$.

For example, let $K = \mathbb{Q}(\alpha)$ be a quadratic field, where $\alpha^2 = d$ for some square free integer d. Then a calculation of the characteristic polynomial of a typical element $\beta = a + b\alpha$ (where a and b are rational numbers) shows that $\mathcal{O} = \mathcal{O}_K = \mathbb{Z}[\omega]$ where

$$\omega = \frac{1+\alpha}{2} \text{ and } D_K = d \text{ for } d \equiv 1(\mathrm{mod}\ 4),$$

$$\omega = \alpha \text{ and } D_K = 4d \text{ for } d \equiv 2, 3(\mathrm{mod}\ 4).$$

If d is positive then the geometric realization of the number $\beta = a + b\alpha$ will be the point $\lambda(\beta) = (a + b\sqrt{d}, a - b\sqrt{d})$. In the case of an imaginary quadratic field ($d < 0$) the geometric realization of the number $\beta = a + b\alpha$ will be the point $(a + ib\sqrt{|d|})$ in the complex plane. Since $\mathbb{Z}[\omega] = \langle 1, \omega\rangle$ we have for positive d

$$\mathrm{vol}^2(\mathbb{R}^2/\lambda(\mathbb{Z}[\omega])) = \begin{cases} \sqrt{d} & \text{if } d \equiv 1 \bmod 4, \\ 2\sqrt{d} & \text{if } d \equiv 2, 3 \bmod 4, \end{cases}$$

and for negative d

$$\mathrm{vol}^2(\mathbb{C}/\mathbb{Z}[\omega]) = \begin{cases} \frac{\sqrt{|d|}}{2} & \text{if } |d| \equiv 3 \bmod 4, \\ \sqrt{|d|} & \text{if } |d| \equiv 1, 2 \bmod 4. \end{cases}$$

Figures 4.1 and 4.2 illustrate the lattices of integers in the quadratic fields $\mathbb{Q}(\sqrt{-1})$ and $\mathbb{Q}(\sqrt{2})$.

4.1.4 Units, the Logarithmic Map, and the Regulator

In the ring $\mathbb{Z}$ there are only two invertible elements (units): 1 and –1. The *group of units*, i.e. invertible elements of the ring of integers $\mathcal{O}_K$ of a number field K has a less trivial structure. However, this group can be completely described. One uses the notation $E_K = \mathcal{O}_K^\times$.

Some interesting arithmetical problems can be reduced to finding elements of E_K. For example, consider Pell's equation (see Part I, section 1.2.5)

$$x^2 - dy^2 = 1 \tag{4.1.8}$$

(where d is a square free positive integer).

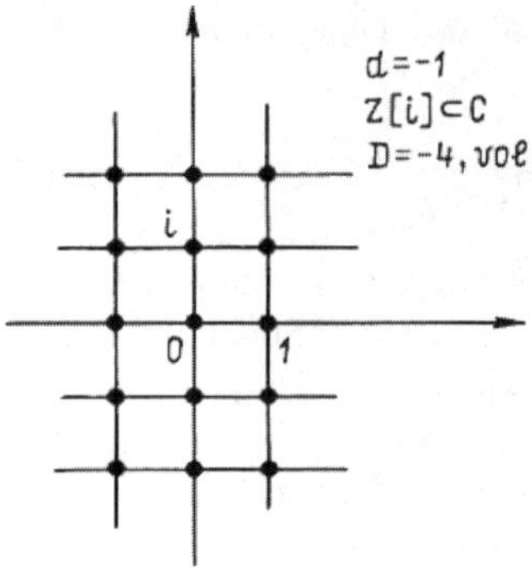

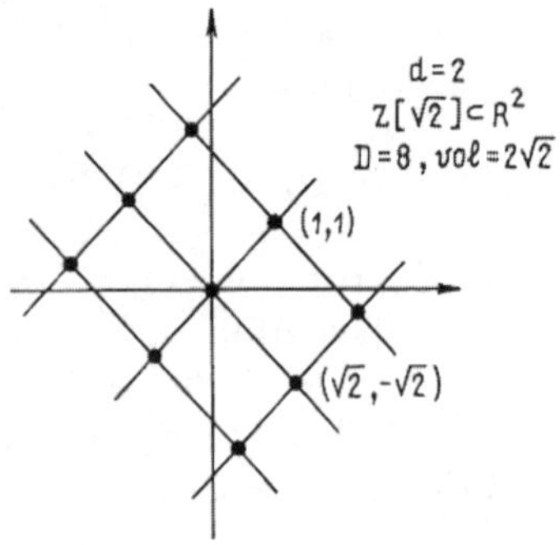

Fig. 4.1.

Fig. 4.2.

Note that if $\beta \in \mathcal{O}_K^\times$ then $\mathrm{N}\beta$ and $\mathrm{N}\beta^{-1} = \mathrm{N}(\beta^{-1})$ are rational integers, hence $\mathrm{N}\beta = \pm 1$. Conversely, any solution to (4.1.8) in integers x, y produces a unit $\beta = x + y\alpha$ in the real quadratic field $k = \mathbb{Q}(\alpha)$, $\alpha^2 = d$ since $\mathrm{N}\beta = (x + y\sqrt{d})(x - y\sqrt{d}) = x^2 - dy^2$. On the other hand for all $\beta \in \mathcal{O}_K$ with $\mathrm{N}\beta = \pm 1$ we have that $\beta \in \mathcal{O}_K^\times$. It follows from a general theorem of Dirichlet on the structure of E_K for an algebraic number field K (the Dirichlet unit theorem), that for $K = \mathbb{Q}(\sqrt{d})$ one has $E_K = \{\pm\varepsilon^n \mid n \in \mathbb{Z}\}$. Here ε is a *fundamental unit* (which can be uniquely defined by the condition that $\lambda_1(\varepsilon) = a + b\sqrt{d}$ is minimal with $\lambda_1(\varepsilon) > 1$). The set of solutions to (4.1.8) can be identified with a subgroup of E_K of the form $\{\pm\varepsilon_0^n \mid n \in \mathbb{Z}\}$, where $\varepsilon_0 = x_0 + y_0\sqrt{\alpha}$ corresponds to the minimal solution $\lambda(\varepsilon_0) = x_0 + y_0\sqrt{d} > 1$.

In order to describe the structure of E_K in the general case, one uses the embedding $\lambda : K \to K \otimes \mathbb{R} \cong \mathbb{R}^{r_1} \times \mathbb{C}^{r_2}$ and the following logarithmic map $l : (\mathbb{R}^{r_1} \times \mathbb{C}^{r_2})^\times \to \mathbb{R}^{r_1+r_2}$ where for $i \le r_1$ by definition $l_i(x) = \log|x|$, $l_i : \mathbb{R}^\times \to \mathbb{R}$, and for $i > r_1$ $l_i(x) = \log|x|^2$, $l_i : \mathbb{C}^\times \to \mathbb{R}$. Under the map $l \circ \lambda$, multiplication in K becomes addition in $\mathbb{R}^{r_1+r_2}$. If $x \in K$ then in view of (4.1.3) we know that

$$\mathrm{N}x = \lambda_1(x) \cdot \dots \cdot \lambda_{r_1}(x)\lambda_{r_1+1}(x)\overline{\lambda_{r_1+1}(x)} \cdot \dots \cdot \lambda_{r_1+r_2}(x)\overline{\lambda_{r_1+r_2}(x)}.$$

Hence

$$\sum_{i=1}^{r_1+r_2} l_i(\lambda_i(x)) = \log|\mathrm{N}x|.$$

In particular, the image $l\lambda(\mathcal{O}_K^\times)$ of $\mathcal{O}_K^\times$ lies in the hyperplane

$$V = \left\{ (x_1, \dots, x_{r_1+r_2}) \in \mathbb{R}^{r_1+r_2} \;\middle|\; \sum_{i=1}^{r_1+r_2} x_i = 0 \right\} \quad V \cong \mathbb{R}^r, \quad r = r_1 + r_2 - 1.$$

The kernel of the map $l : (K \otimes \mathbb{R})^\times \to \mathbb{R}^{r_1+r_2}$ is the following compact set

$$\{\pm 1\}^{r_1} \times S^{r_2} \subset \mathbb{R}^{r_1} \times \mathbb{C}^{r_2} \cong \mathbb{R}^n,$$

where $S = \{z \in \mathbb{C} \mid |z| = 1\}$ is the unit circle. We see that the logarithmic map provides an effective way of drawing the units: the kernel of $l\lambda : E_K \to \mathbb{R}^{r_1+r_2}$ consists of only a finite number of elements (the roots of unity in K). Dirichlet's theorem says that the image $l\lambda(E_K)$ is a complete lattice in $V \cong \mathbb{R}^r$ (where $r = r_1 + r_2 - 1$). In other words, one can find elements $\varepsilon_1, \ldots, \varepsilon_r \in E_K$ such that any unit $\varepsilon \in E_K$ can be uniquely represented in the form

$$\varepsilon = \eta\varepsilon_1^{n_1} \cdot \ldots \cdot \varepsilon_r^{n_r}$$

where $n_i \in \mathbb{Z}$ and η is a root of unity in K. In particular, $\varepsilon_1, \ldots, \varepsilon_r \in E_K$ are multiplicatively independent:

$$l\lambda(\varepsilon_1), \ldots, l\lambda(\varepsilon_r)$$

form a basis of the hyperplane V. Consider now the volume $\mathrm{vol}(V/l\lambda(E_K))$ of a fundamental parallelogram for the lattice of units (with respect to the measure on V induced by Lebesgue measure on $\mathbb{R}^n$). The number $R_K = \mathrm{vol}(V/l\lambda(E_K))/\sqrt{r+1}$ is called the *regulator* of K and is equal to the absolute value of the determinant

$$\begin{vmatrix} l_1\lambda_1(\varepsilon_1) & l_2\lambda_2(\varepsilon_1) & \cdots & l_{r_1+r_2}\lambda_{r_1+r_2}(\varepsilon_1) \\ \cdots & \cdots & \cdots & \cdots \\ l_1\lambda_1(\varepsilon_r) & l_2\lambda_2(\varepsilon_r) & \cdots & l_{r_1+r_2}\lambda_{r_1+r_2}(\varepsilon_r) \\ (r_1+r_2)^{-1} & (r_1+r_2)^{-1} & \cdots & (r_1+r_2)^{-1} \end{vmatrix}.$$

4.1.5 Lattice Points in a Convex Body

We now describe a general geometric idea, on which the proof of the Dirichlet's theorem, and some other interesting facts (such as bounds for discriminants and class numbers) is based.

Theorem 4.7 (Minkowski's Lemma on a Convex Body). *Let M be a lattice in $\mathbb{R}^n$, $\Delta = \mathrm{vol}(\mathbb{R}^n/M)$, and let $X \subset \mathbb{R}^n$ be a centrally-symmetric convex body of finite volume $v = \mathrm{vol}(X)$. If $v > 2^n\Delta$, then there exists $0 \neq \alpha \in M \cap X$.*

Proof. In order to prove the lemma, it is convenient to consider the lattice $2M \subset \mathbb{R}^n$ whose fundamental parallelotope has volume $\mathrm{vol}(\mathbb{R}^n/2M) = 2^n\Delta$. Then under the natural projection of $X \subset \mathbb{R}^n$ onto a fundamental parallelopiped $\mathbb{R}^n/2M$ there will be overlaps in the image of X, because the volume of X is bigger than the volume of a fundamental parallelepiped. Hence there exist two different points $z_1, z_2 \in X$, $z_1 \neq z_2$ such that $z_1 \cong z_2 \bmod 2M$, i.e. $(z_1 - z_2)/2 \in M$. The proof follows: the point $(z_1 - z_2)/2 \neq 0$ belongs to X in view of its convexity and central symmetry, since $(z_1 - z_2)/2 = (z_1 + (-z_2))/2$ (if $z \in X$ then $-z \in X$).

Here are some examples of convex bodies to which we can apply Minkowski's lemma. Let $x^0 = (x_1^0, \dots, x_{r_1+r_2}^0) \in K\otimes\mathbb{R}$, $|\mathrm{N}(x^0)| = \prod_{i=1}^{r_1} |x_i^0| \prod_{j=1}^{r_2} |x_{r_1+j}^0|^2 \neq 0$. Put

$$W(x^0) = \{x \in K \otimes \mathbb{R} \mid |x_i| < |x_i^0|,\ i = 1, \dots, r_1 + r_2\}.$$

For a positive integer a we put

$$U(a) = \left\{ x \in K \otimes \mathbb{R} \;\middle|\; \sum_{i=1}^{r_1} |x_i| + 2\sum_{j=1}^{r_2} |x_{r_1+j}| < a \right\}.$$

A calculation of these volumes shows that

$$\mathrm{vol}(W(x^0)) = 2^{r_1}\pi^{r_2}|\mathrm{N}(x^0)|, \quad \mathrm{vol}(U(a)) = 2^{r_1}\left(\frac{\pi}{2}\right)^{r_2}\frac{a^n}{n!}. \tag{4.1.9}$$

Applying Minkowski's lemma to the lattice $M = \lambda(\mathcal{O}_K)$ and these bodies (where $\Delta = 2^{-r_2}\sqrt{|D_K|}$ by (4.1.7)), we see that

a) for arbitrary constants $c_i > 0$ $(i = 1, \dots, r_1 + r_2)$ satisfying the condition $\displaystyle\prod_{i=1}^{r_1} c_i \prod_{j=1}^{r_2} c_{r_1+j}^2 > \left(\frac{2}{\pi}\right)^{r_2} \sqrt{|D_K|}$ there exists a non zero element $\alpha \in \mathcal{O}_K$ such that

$$|\lambda_i(\alpha)| < c_i \quad (i = 1, \dots, r_1 + r_2); \tag{4.1.10}$$

it suffices to take $x^0 \in K \otimes \mathbb{R}$ with $|x_i^0| = c_i$ $(i = 1, \dots, r_1 + r_2)$ and $\alpha \in W(x^0)$;

b) for $a \geq \left(n!\left(\frac{4}{\pi}\right)^{r_2}\sqrt{|D_K|}\right)^{1/n}$ there exists $\beta \in \mathcal{O}_K$, $\beta \neq 0$ from $U(a)$, such that

$$\sum_{i=1}^{r_1} |\lambda_i(\beta)| + 2\sum_{j=1}^{r_2} |\lambda_{r_1+j}(\beta)| < a,$$

hence in view of the inequality between the arithmetic and geometric means we have the estimate

$$|\mathrm{N}(\beta)| < \left(\frac{4}{\pi}\right)^{r_2}\frac{n!}{n^n}\sqrt{|D_K|} \quad (|\mathrm{N}(\beta)| \geq 1). \tag{4.1.11}$$

From (4.1.11) follows the estimate for the discriminant:

$$|D_K| > \left(\frac{\pi}{4}\right)^{2r_2}\frac{n^{2n}}{(n!)^2} = \left(\frac{\pi}{4}\right)^{2r_2}\frac{1}{2\pi}ne^{2n-\theta/6n} \quad (0 < \theta < 1)$$

showing that $|D_K|$ grows with n.

Some other remarkable consequences of Minkowski's lemma are:

Theorem 4.8 (Hermite's Theorem, (1863)). *There are only finitely many algebraic number fields with a given discriminant.*

Theorem 4.9 (Minkowski's Theorem, (1890)). *If $K \neq \mathbb{Q}$ then $|D_K| > 1$.*

For the proofs of these theorems cf. [Wei74a].

From the above estimate for the discriminant it follows also that for large n one has $|D_K|^{1/n} > (7.3)^{r_1/n}(5.9)^{r_2/n}$. However nowadays much stronger estimates for discriminants are known: $|D_K|^{1/n} > (188)^{r_1/n}(41)^{r_2/n}$ (for large n), cf. [Odl75] , [Kuz84]. The latter are deduced from analytic properties of the Dedekind zeta–function via explicit formulae (cf. §6.2.3 and §6.2.5).

4.1.6 Deduction of Dirichlet's Theorem From Minkowski's Lemma

Consider the hypersurface $T_c = \{x \in K \otimes \mathbb{R} \mid |\mathrm{N}x| = c\}$ for a fixed $c > 0$. Under the logarithmic map this becomes the affine hyperplane

$$V_{\log c} = \left\{ y \in \mathbb{R}^{r_1+r_2} \;\middle|\; \sum_{i=1}^{r_1+r_2} y_i = \log c \right\}.$$

The *group of units* E_K acts on T_c by multiplication with $\lambda(\varepsilon)$, $\varepsilon \in E_K$. Under the logarithmic map the action of ε becomes a translation by the vector $l\lambda(\varepsilon)$, which maps $V_{\log c}$ into itself. The number of orbits of this action on $T_c \cap \lambda(\mathcal{O}_K)$ is *finite* for any fixed c. Indeed it suffices to show that if $\mathrm{N}(\alpha) = \mathrm{N}(\beta) = c \in \mathbb{Z}$ and $\alpha \equiv \beta (\bmod c)$ in the ring $\mathcal{O}_K$ then $\alpha/\beta \in E_K$. In order to see this, notice that α divides its norm $\mathrm{N}\alpha = c$. Hence the number $\frac{\beta}{\alpha} = 1 + \frac{\beta-\alpha}{\alpha}$ belongs to $\mathcal{O}_K$. Similarly, $\frac{\alpha}{\beta} \in \mathcal{O}_K$, hence $\frac{\alpha}{\beta} \in E_K = \mathcal{O}_K^\times$.

We now use the results of §4.1.5, and choose some $c > \left(\frac{2}{\pi}\right)^{r_2} \sqrt{|D_K|}$. Then for any element $x \in T_c$ one can find an element $\alpha \in \mathcal{O}_K$ such that $\lambda(\alpha) \in W(x)$. We use this fact to show that the quotient group $V/l\lambda(E_K)$ is *compact*. It suffices to show that $V = V_0$ can be covered by translations of a bounded set by vectors $l\lambda(\varepsilon)$, $\varepsilon \in E_K$. In turn, this is implied by the analogous statement for any hyperplane parallel to V, for example of the type $V_{\log c}$ instead of V_0. For any $\alpha \in \mathcal{O}_K$, $\alpha \neq 0$, consider the set $Y_c(\alpha) \subset V_{\log c}$ consisting of all $y = l(x) \in V_{\log c}$ such that $\lambda(\alpha) \in W(x)$. Then $Y_c(\alpha)$ are all bounded, $Y_c(\alpha\varepsilon) = Y_c(\alpha) + l\lambda(\varepsilon)$ for $\varepsilon \in E_K$, and Minkowski's lemma implies that any $y \in V_{\log c}$ is contained in some $Y_c(\alpha)$. On the other hand, we know that there are only finitely many classes of $\alpha \in \mathcal{O}_K$ with $|\mathrm{N}(\alpha)| < c$ modulo the action of E_K. If $\{\alpha_i\}$ is a finite system of representatives of these classes, then the desired compact set can be defined to be the union $\cup Y_c(\alpha_i)$. This proves the *compactness* statement; *discreteness* is implied by the analogous fact for the lattice $\lambda(\mathcal{O}_K)$, and the fact that the logarithmic map restricted to any hypersurface T_c is a surjective open map onto $V_{\log c}$.

4.2 Decomposition of Prime Ideals, Dedekind Domains, and Valuations

4.2.1 Prime Ideals and the Unique Factorization Property

The original purpose of Dedekind's theory of ideals was to extend the results of Kummer on Fermat's theorem to a larger class of exponents. Let R be a commutative ring with unity. An *ideal* α of R is by definition an additive subgroup $\alpha \subset R$ such that $R\alpha \subset \alpha$. An ideal $\alpha \neq R$ is called prime iff $ab \in \alpha$ implies $a \in \alpha$ or $b \in \alpha$ (i.e. the factor ring R/α has no zero–divisors). An ideal of the type $\alpha = (a) = Ra$ for $a \in R$ is called a principal ideal. The notation $(a_i)_{i\in I}$ denotes the smallest ideal containing all $a_i \in R$, $(i \in I)$. An element $\pi \in R$ is called prime iff $\pi = ab$ implies that either a or b is invertible (i.e. a unit) in R. The reason for the lack of uniqueness of factorization into prime elements in R, is related to the fact that the ideal (π) generated by a prime element π is not always prime.

Example 4.10. Let $R = \mathbb{Z}[\sqrt{-5}]$ then there are two essentially different factorizations into prime elements:

$$21 = 3 \cdot 7 = (1 + 2\sqrt{-5}) \cdot (1 - 2\sqrt{-5}).$$

A simple check shows that none of the divisors of two different factors in this identity belong to R. However, the uniqueness of factorization can be restored if we pass from prime elements to prime ideals. Indeed, the following ideals are *prime*:

$$\mathfrak{p}_1 = (3, \sqrt{-5} - 1), \quad \mathfrak{p}_2 = (3, \sqrt{-5} - 2),$$
$$\mathfrak{p}_3 = (7, \sqrt{-5} - 3), \quad \mathfrak{p}_4 = (7, \sqrt{-5} - 4).$$

This is implied by the decompositions:

$$X^2 + 5 \equiv (X - 1)(X - 2)(\text{mod } 3), \quad X^2 + 5 \equiv (X - 3)(X - 4)(\text{mod } 7),$$

for example,

$$R/\mathfrak{p}_1 = \mathbb{Z}[X]/(3, X - 1, X^2 + 5) \cong \mathbb{F}_3[X]/(X - 1) \cong \mathbb{F}_3,$$

in view of the identity $(X - 1, X^2 + 5) = X - 1$ in $\mathbb{F}_3[X]$. Analogously one proves the decompositions

$$(3) = \mathfrak{p}_1 \cdot \mathfrak{p}_2, \quad (7) = \mathfrak{p}_3 \cdot \mathfrak{p}_4, \quad (1 + 2\sqrt{-5}) = \mathfrak{p}_1 \cdot \mathfrak{p}_3, \quad (1 - 2\sqrt{-5}) = \mathfrak{p}_2 \cdot \mathfrak{p}_4,$$

and the factorization $(21) = \mathfrak{p}_1\mathfrak{p}_2\mathfrak{p}_3\mathfrak{p}_4$ is the unique decomposition as a product of four ideals. The ideals (3), (7), $(1 + 2\sqrt{-5})$, $(1 - 2\sqrt{-5})$ are not prime.

A *Dedekind domain* is by definition a commutative associative ring with identity, in which the factorization of non-zero ideals into prime ideals is unique. This is equivalent to R being a *Noetherian* (every ideal being finitely generated), *integrally closed* (containing every element of its field of fractions which is integral over R) ring, all of whose non–zero prime ideals are maximal (i.e. $R/\mathfrak{p}$ is a field).

One can prove that the ring $\mathbb{Z}[\sqrt{-5}]$ of our example is a Dedekind domain. From the given characterization it follows that for a given number field K, $[K : \mathbb{Q}] < \infty$, the ring of integers $\mathcal{O}_K$ is a Dedekind domain. It also follows that no proper subring of $\mathcal{O}_K$ with the same field of fractions can be a Dedekind ring, since it cannot be integrally closed. For example, the ring $\mathbb{Z}[\sqrt{5}]$ is not a Dedekind ring: the ideal $(1 - \sqrt{5})$ cannot be decomposed into a product of prime ideals. However the bigger ring $\mathbb{Z}[\frac{1-\sqrt{5}}{2}] = \mathcal{O}_K$, $K = \mathbb{Q}(\sqrt{5})$ is a Dedekind ring. Thus one can build a good divisibility theory in this class of rings by replacing elements α by the corresponding ideals and using prime ideals rather than prime elements. However, the class of Dedekind rings is quite narrow, and a good divisibility theory can be built in a much larger class of rings. For example, in the polynomial ring $k[x_1, x_2, \ldots, x_n]$ over a field k one has unique factorization of elements, and the prime elements here are the irreducible polynomials. On the other hand, the existence and uniqueness of factorization of ideals into prime ideals does not hold in this ring. For instance, the ideal $(x^2, y) \subset k[x, y]$ does not have such a decomposition. This last example explains particularly Kronecker's mistrust of the prime ideals of Dedekind. Kronecker himself began developing a different theory of divisibility, based on valuations. This is described below (§4.2.5 and §4.3). The history of the controversy between Kronecker and Dedekind is nicely presented by H.Weyl (cf. [Wey40]).

Fractional ideals. Let $\mathcal{O}_K$ be the ring of all integers in a number field K, $[K : \mathbb{Q}] < \infty$. A *fractional ideal* is by definition a non-zero $\mathcal{O}_K$–submodule $\mathfrak{a} \subset K$ such that $\alpha\mathfrak{a} \subset \mathcal{O}_K$ for some $\alpha \in K^\times$. The properties of Dedekind domains imply that together with a fractional ideal $\mathfrak{a}$, the $\mathcal{O}_K$–submodule $\mathfrak{a}^{-1} = \{x \in K \mid x\mathfrak{a} \subset \mathcal{O}_K\}$ will also be a fractional ideal. If $\mathfrak{a}$ and $\mathfrak{b}$ are fractional ideals, then $\mathfrak{a}\mathfrak{b}$ is also a fractional ideal. Thus the fractional ideals form a multiplicative group I_K whose identity element is $\mathcal{O}_K$. Since $\mathcal{O}_K$ is a Dedekind domain, it follows that I_K is a free Abelian group in which the prime ideals $\mathfrak{p} \subset \mathcal{O}_K$ form a basis: every $\mathfrak{a} \in I_K$ can be uniquely written in the form:

$$\mathfrak{a} = \mathfrak{p}_1^{n_1} \cdot \cdots \cdot \mathfrak{p}_k^{n_k} \quad (n_i \in \mathbb{Z}).$$

The *norm* $\mathrm{N}\mathfrak{a}$ of an integral ideal $\mathfrak{a} \subset \mathcal{O}_K$ is defined to be the number of elements of the corresponding factor ring: $\mathrm{N}\mathfrak{a} = \mathrm{Card}(\mathcal{O}_K/\mathfrak{a})$, and the norm of an arbitrary fractional ideal $\mathfrak{a} \in I_K$ is defined by multiplicativity. If $\mathfrak{a} = (\alpha)$ is a principal ideal, then $\mathrm{N}((\alpha)) = |\mathrm{N}\alpha| = |\mathrm{N}_{K/\mathbb{Q}}\alpha|$: multiplication by α defines an endomorphism of the lattice $\mathcal{O}_K$, and one easily verifies that the absolute value of its determinant coincides with the index of its image: $(\mathcal{O}_K : (\alpha)) = \mathrm{N}((\alpha))$.

4.2.2 Finiteness of the Class Number

To each element $\alpha \in K^\times$ one can associate $(\alpha) \in I_K$, so that we have a homomorphism $K^\times \stackrel{\mathrm{div}}{\to} I_K$. The image of this homomorphism is called the *group of principal ideals*, and is denoted by P_K. The quotient group $Cl_K = I_K/P_K$ is called the *ideal class group*. The following result is another corollary of Minkowski's lemma.

Theorem 4.11. *The group Cl_K is finite.*

The order $|Cl_K| = h_K$ is called the *class number* of K.

In order to prove the theorem we note that each ideal class can be represented by an integral ideal (replacing if necessary $\mathfrak{a}$ by $M\mathfrak{a}$ with an appropriate integer M, and so getting rid of denominators). According to Minkowski's lemma (see §4.1.5) there exists a non–zero element $\alpha \in \mathfrak{a}$ such that $|\mathrm{N}\alpha| < \left(\frac{2}{\pi}\right)^{r_2} \sqrt{|D_K|}\mathrm{N}\mathfrak{a}$. We have $\alpha\mathcal{O}_K \subset \mathfrak{a}$ because $\mathfrak{a}$ is an ideal, i.e. $\mathcal{O}_K \subset \alpha^{-1}\mathfrak{a}$. We see now that the index $(\alpha^{-1}\mathfrak{a} : \mathcal{O}_K) = (\mathcal{O}_K : \alpha\mathfrak{a}^{-1})$ is bounded by the constant $\left(\frac{2}{\pi}\right)^{r_2} \sqrt{|D_K|}$, because

$$(\mathcal{O}_K : \alpha\mathfrak{a}^{-1}) = |\mathrm{N}(\alpha)|\mathrm{N}\mathfrak{a}^{-1} < \left(\frac{2}{\pi}\right)^{r_2} \sqrt{|D_K|}.$$

If $\mathfrak{a}'$ is an arbitrary fractional ideal containing $\mathcal{O}_K$ and $(\mathfrak{a}' : \mathcal{O}_K) = r$ then $r^{-1}\mathcal{O}_K \supset \mathfrak{a}' \supset \mathcal{O}_K$. But it is obvious that the number of intermediate ideals $\mathfrak{a}'$ between $r^{-1}\mathcal{O}_K$ and $\mathcal{O}_K$ is finite. The theorem follows, in view of the fact that r can take only a finite number of values.

As we shall see below, this theorem and Dirichlet's unit theorem not only have similar proofs, but can be incorporated as parts of a more general result on the structure of the idele class group (cf. [Chev40], [Wei74a]).

The class number plays an exceptionally important role in number theory. For example the statement $h_K = 1$ is equivalent to saying that $\mathcal{O}_K$ is a unique factorization domain. Another example is that the theorem of Kummer from §4.1.1 on the first case of Fermat's Last Theorem can be extended to all prime exponents n with the property that h_K is not divisible by n, where $K = \mathbb{Q}(\exp(2\pi i/n))$ is the corresponding cyclotomic field.

There have been a number of experimental and empirical observations of class groups of number fields made over the years. H. Cohen and H. W. Lenstra, Jr. in [CoLe83] introduced a heuristic principle that succeeded in predicting the statistical distribution of ideal class groups of imaginary quadratic number fields and totally real abelian number fields. Many numerically verified observations are a precise consequence of the Cohen-Lenstra conjecture, cf. e.g. [Lee02], where a relation with Leopoldt's Spiegelungssatz (cf. [Leo58]) is discussed.

4.2.3 Decomposition of Prime Ideals in Extensions

If K is a number field with ring of integers $\mathcal{O}_K$, and p is a prime number, then $(p) = p\mathcal{O}_K$ can be decomposed into a product of prime ideals of $\mathcal{O}_K$:

$$(p) = \mathfrak{p}_1^{k_1}\mathfrak{p}_2^{k_2} \cdot \dots \cdot \mathfrak{p}_s^{k_s}. \tag{4.2.1}$$

The form of the decomposition (4.2.1) for primes p is one of the most important characteristics of K; if say $K/\mathbb{Q}$ is a Galois extensions, then K is uniquely determined by the set of primes p satisfying $(p) = \mathfrak{p}_1\mathfrak{p}_2 \cdot \dots \cdot \mathfrak{p}_n$, where $n = [K : \mathbb{Q}]$ (the product of n distinct primes). If this is the case for p, we say that p *splits completely* in K. For a general number field it is difficult to determine the precise form of the decomposition (4.2.1) for all p. This problem is related to the deepest questions of algebraic number theory ("non–commutative class field theory", see §6.4). However, for Abelian extensions K, i.e. Galois extensions $K/\mathbb{Q}$ whose Galois group $G(K/\mathbb{Q})$ is commutative, this decomposition is known. We shall give the precise form of the decomposition for quadratic fields $K = \mathbb{Q}(\sqrt{d})$ and cyclotomic fields $K = \mathbb{Q}(\sqrt[m]{1})$. This is done by a general method, applicable to any extension $R \subset S$ of commutative rings, where it is supposed that S is a finitely generated R–module. In this case each element $\alpha \in S$ is a root of a normalized (monic) polynomial $f(X) \in R[X]$. For example, one could take $f(X) = X^n + a_{n-1}X^{n-1} + \dots + a_0$, $a_i \in R$ (the characteristic polynomial). Let $\mathfrak{p}$ be a maximal ideal in R. Denote by $\overline{\alpha}$ the image of α in the quotient ring $S/\mathfrak{p}S$.

Theorem 4.12 (Theorem on the Decomposition of a Maximal Ideal). *Suppose that for an element $\alpha \in S$ one has $S/\mathfrak{p}S = (R/\mathfrak{p})[\overline{\alpha}]$ and $n = \deg f_\alpha(X) = \dim_{R/\mathfrak{p}} S/\mathfrak{p}S$. Choose normalized polynomials $g_1(X), \dots, g_r(X) \in R[X]$ such that*

$$f_\alpha(X) \equiv g_1(X)^{e_1} \cdot \dots \cdot g_r(X)^{e_r} (\mathrm{mod}\ \mathfrak{p}R[X]) \tag{4.2.2}$$

where $g_i(X) (\mathrm{mod}\ \mathfrak{p}R[X])$ are distinct and irreducible in $(R/\mathfrak{p})[X]$. Then the ideals $\mathfrak{P}_i = (\mathfrak{p}, g_i(\alpha))$ are maximal, and the following decomposition holds:

$$\mathfrak{p}S = \mathfrak{P}_1^{e_1} \cdot \dots \cdot \mathfrak{P}_r^{e_r}. \tag{4.2.3}$$

The maximality of $\mathfrak{P}_i$ follows from the isomorphism:

$$S/\mathfrak{P}_i \cong R[X]/(g_i(X), \mathfrak{p}) \cong (R/\mathfrak{p})[X]/(g_i(X))$$

and from the irreducibility of $g_i(X) (\mathrm{mod}\ \mathfrak{p}R[X])$; the decomposition (4.2.3) is deduced from an analogue of the theorem on tensor products of fields, see 4.1.2 (or from the Chinese Remainder theorem):

$$S/\mathfrak{p}S \cong S \otimes_R (R/\mathfrak{p}) \cong \prod_{i=1}^{r} (R/\mathfrak{p})[X]/(g_i(X)^{e_i}).$$

Example 4.13. a) Quadratic Fields (see [BS85], Chapter 2). For a quadratic extension $K = Q(\sqrt{d})$ ($d \in \mathbb{Z}$ being square free), $\mathcal{O} = \mathcal{O}_K = \mathbb{Z}[\omega]$ where

$$\omega = \frac{1+\sqrt{d}}{2}, f_\omega(X) = X^2 - X + (d-1)/4 \text{ and } D_K = d \text{ for } d \equiv 1 (\text{mod } 4),$$

$$\omega = \sqrt{d} \text{ and } D_K = 4d \text{ for } d \equiv 2, 3 (\text{mod } 4).$$

The result on the decomposition of primes can be conveniently stated in terms of the *quadratic character* χ_K of K. By definition χ_K is the unique primitive Dirichlet character of order 2 modulo $|D_K|$ such that $\chi(-1) = \text{sgn } D_K$. It can be written explicitly as follows

$$\chi_K(x) = \begin{cases} \left(\frac{x}{|d|}\right), & \text{if } d \equiv 1 (\text{mod } 4) \\ (-1)^{(x-1)/2} \left(\frac{x}{|d|}\right), & \text{if } d \equiv 3 (\text{mod } 4) \\ (-1)^{(x^2-1)/8+(x-1)(d'-1)/4} \left(\frac{x}{|d'|}\right), & \text{if } d = 2d', d' \equiv 1 (\text{mod } 2). \end{cases}$$

Then p decomposes in $\mathcal{O}_K$ as follows:

$$p\mathcal{O}_K = \begin{cases} \mathfrak{p}\mathfrak{p}', \ \mathfrak{p} \neq \mathfrak{p}', \text{ and } \mathrm{N}\mathfrak{p} = \mathrm{N}\mathfrak{p}' = p & \text{for } \chi_K(p) = 1, \\ \mathfrak{p}, \ \mathrm{N}\mathfrak{p} = p^2 \text{ (i.e. } p \text{ remains prime)} & \text{for } \chi_K(p) = -1, \\ \mathfrak{p}^2, \ \mathrm{N}\mathfrak{p} = p & \text{for } \chi_K(p) = 0. \end{cases}$$

In order to prove these decompositions one applies the above theorem with $R = \mathbb{Z}$, $S = \mathcal{O}_K$, $\alpha = \omega$, using the decomposition of the corresponding quadratic polynomial $f_\omega(X) \bmod p$, which either has two distinct roots over $\mathbb{F}_p$, or is irreducible, or has a double root over $\mathbb{F}_p$ in the cases when $\chi_K(p) = 1$, $\chi_K(p) = -1$ or $\chi_K(p) = 0$ respectively. This result can be elegantly rewritten as an identity for the Euler factors of the Dedekind zeta–function (cf. §6.2.3 below):

$$\prod_{\mathfrak{p}|(p)} (1 - \mathrm{N}\mathfrak{p}^{-s}) = (1 - p^{-s})(1 - \chi_K(p)p^{-s}) \quad (s \in \mathbb{C}). \tag{4.2.4}$$

Example 4.14. b) Cyclotomic fields. $K = K_m = \mathbb{Q}(\zeta_m)$. We use the fact $\mathcal{O}_K = \mathbb{Z}[\zeta_m]$. Consider the extension $\mathbb{Z}[\zeta_m] \supset \mathbb{Z}$, and take for $f_\alpha(X)$ the cyclotomic polynomial $\Phi_m(X)$ (see 4.1.2). The proof that $\mathcal{O}_K$ coincides with $\mathbb{Z}[\zeta_m]$ is rather fine but elementary; it is based on a calculation of the discriminant of $R = \mathbb{Z}[\zeta_m]$ which turns out to be equal to

$$(-1)^{\varphi(m)/2} m^{\varphi(m)} / \left(\prod_{p|m} p^{\varphi(m)/(p-1)} \right),$$

see [BS85].

4.2.4 Decomposition of primes in cyslotomic fields

Theorem 4.15. *a) Let $p \nmid m$, then*

$$pR = \mathfrak{p}_1 \cdot \dots \cdot \mathfrak{p}_r, \quad \mathrm{N}\mathfrak{p}_i = p^f,$$

where $\mathfrak{p}_i \subset R$ are distinct prime ideals, and the number f is equal to the order of $p \bmod m$ in $(\mathbb{Z}/m\mathbb{Z})^\times$, $f \cdot r = \varphi(m)$.

b) If $m = p_1^{\alpha_1} \cdot \dots \cdot p_s^{\alpha_s}$ then

$$p_i R = (\mathfrak{p}'_1 \cdot \dots \cdot \mathfrak{p}'_{r'})^{\varphi(p_i^{\alpha_i})}, \quad \mathrm{N}\mathfrak{p}'_i = p^{f'},$$

where $\varphi(p_i^{\alpha_i}) = p_i^{\alpha_i - 1}(p_i - 1)$, f' is equal to the order of the element $p_i \bmod mp_i^{-\alpha_i}$ in $\mathbb{Z}/(mp_i^{-\alpha_i})\mathbb{Z}$, and $f' \cdot r' = \varphi(mp_i^{-\alpha_i})$.

Proof. Note first that for prime ideals $\mathfrak{p} \subset \mathcal{O}_K$ the number $f = \log_p \mathrm{N}\mathfrak{p}$ coincides with the degree of the corresponding extension of residue fields: $f = [(\mathcal{O}_K/\mathfrak{p}) : \mathbb{F}_p]$, and thus f is the order of the Frobenius automorphism $x \mapsto x^p$, generating the cyclic Galois group $G((\mathcal{O}_K/\mathfrak{p})/\mathbb{F}_p)$. Applying the theorem on the decomposition of maximal ideals we see that it suffices to find the form of the decomposition of the cyclotomic polynomial $\Phi_m(X) \bmod p$ in $\mathbb{F}_p[X]$ into irreducible polynomials.

It follows also that the form of the decomposition depends in this case only on $p \bmod m$. In particular, p splits completely in $K \iff p \equiv 1 \bmod m$. A useful observation is that the decomposition of (p) in $\mathcal{O}_{K_m}$ is fully determined by the action of the Frobenius endomorphism Fr_p on the finite ring $\mathcal{O}_K/(p)$, so that in the case $p \nmid m$ this endomorphism may be regarded as the element of the Galois group $G(K_m/\mathbb{Q})$:

$$(\mathrm{Fr}_p : \zeta_m \mapsto \zeta_m^p) \iff p \bmod m \in (\mathbb{Z}/m\mathbb{Z})^\times = G(K_m/\mathbb{Q}).$$

It is useful for further applications to reformulate theorem 4.15 using the Dirichlet characters $\chi : (\mathbb{Z}/m\mathbb{Z})^\times \to \mathbb{C}^\times$. The *conductor* of χ is by definition the least positive integer $m(\chi)$ such that χ can be defined modulo $m(\chi)$, i.e. to which χ factors through the natural projection

$$(\mathbb{Z}/m\mathbb{Z})^\times \xrightarrow{\mathrm{pr}} (\mathbb{Z}/m(\chi)\mathbb{Z})^\times \xrightarrow{\chi_0} \mathbb{C}^\times.$$

The corresponding character $\chi_0 \bmod m(\chi)$ is called the primitive Dirichlet character associated with χ. Theorem 4.15 is equivalent to the following identity (cf. §6.2.3 below):

$$\prod_{\mathfrak{p}|(p)} (1 - \mathrm{N}\mathfrak{p}^{-s}) = \prod_{\chi \bmod m} (1 - \chi_0(p)p^{-s}) \quad (s \in \mathbb{C}). \tag{4.2.5}$$

Indeed, the theorem implies that the left hand side has the form $(1 - p^{-fs})^r$ for $p \not| m$, and $(1 - p^{-f's})^{r'}$ for $p|m$, where f is equal to the order of $p \bmod m$

in $(\mathbb{Z}/m\mathbb{Z})^\times$, $f \cdot r = \varphi(m)$, f' is equal to the order of the element p_i mod m' (with $m' = mp_i^{-\alpha_i}$) in $\mathbb{Z}/m'\mathbb{Z}$, and $f' \cdot r' = \varphi(m')$. It remains to verify the equation

$$(1 - T^f)^r = \prod_{\chi \bmod m} (1 - \chi(p)T). \tag{4.2.6}$$

Let μ_f be the group of roots of unity of degree f, then $1 - T^f = \prod_{\mu_f}(1 - \zeta T)$. Equations (4.2.5) and (4.2.6) follow from the fact that for any $\zeta \in \mu_f$ there are exactly r characters $\chi(\text{mod } m)$ such that $\chi(p) = \zeta$ (cf. [BS85], [La70], [Se70]).

4.2.5 Prime Ideals, Valuations and Absolute Values

An alternative approach to the theory of divisibility has arisen from the notion of the π-order $\text{ord}_\pi a$ of an element $a \neq 0$, $a \in R$ for a prime element π of a unique factorization domain R: here $\text{ord}_\pi a$ is defined to be the largest exponent of π dividing a in R, so that there is a decomposition: $a = \varepsilon\pi_1^{k_1} \cdot \dots \cdot \pi_r^{k_r}$, in which $k_i = \text{ord}_{\pi_i} a$, $\varepsilon \in R^\times$ is a unit.

The function ord_π can be uniquely extended to the *field of fractions* K of R as a homomorphism $\text{ord}_\pi : K^\times \to \mathbb{Z}$ with the following properties:

1) $\forall a, b \in K^\times$ $\text{ord}_\pi(ab) = \text{ord}_\pi a + \text{ord}_\pi b$,
2) $\forall a, b \in K^\times$ $\text{ord}_\pi(a + b) \geq \min(\text{ord}_\pi a, \text{ord}_\pi b)$,
3) a divides b in $R \iff \forall \pi$ $\text{ord}_\pi a \leq \text{ord}_\pi b$,
4) $\pi R = \{a \in R \mid \text{ord}_\pi a > 0\}$ is a prime ideal of R,
5) $R = \{x \in K^\times \mid \forall \pi \;\; \text{ord}_\pi x \geq 0\} \cup \{0\}$.

Generalizing, for an arbitrary field K the notion of a valuation v is introduced as a function $v : K^\times \to \mathbb{Z}$ satisfying the conditions

1) $\forall a, b \in K^\times$ $v(ab) = v(a) + v(b)$,
2) $\forall a, b \in K^\times$ $v(a + b) \geq \min(v(a), v(b))$.

More often one uses instead of v a multiplicative absolute value: for a fixed ρ, $0 < \rho < 1$ put $|x|_{\rho,v} = \rho^{v(x)}$, $|0|_{\rho,v} = 0$.

Definition 4.16. *An absolute value* $|\cdot|$ *of a field* K *is a real–valued function* $x \mapsto |x|$ *with non–negative values, such that*

1) $\forall a, b \in K^\times$ $|a \cdot b| = |a| \cdot |b|$,
2) $\forall a, b \in K^\times$ $|a + b| \leq |a| + |b|$,
3) $|x| = 0 \iff x = 0$.

An absolute value is called non–Archimedean iff instead of 2) the following stronger inequality is satisfied

2′) $\forall a, b \in K^\times \quad |a+b| \le \max(|a|, |b|)$.

Thus the function $|\cdot|_{\rho,v}$ is a non–Archimedean absolute value. An absolute value of the type $|\cdot|_{\rho,v}$ is called a discrete absolute value. An example of such an absolute value is given by the p–adic absolute value $|a/b|_p = p^{\mathrm{ord}_p b - \mathrm{ord}_p a}$ $(a, b \in \mathbb{Z})$ of the field $\mathbb{Q}$. The usual absolute value $|x|$ of $x \in \mathbb{Q} \subset \mathbb{R}$ is an Archimedean absolute value of $\mathbb{Q}$.

If $|\cdot|$ is a non-Archimedean absolute value of K, then the subset $\mathcal{O} = \{x \in K \mid |x| \le 1\}$ is a ring with a unique maximal ideal $\mathfrak{p} = \{x \in K \mid |x| < 1\}$. Such rings are called valuation rings. For the discrete absolute value $|\cdot| = |\cdot|_{\rho,v}$ corresponding to a valuation v, the notation $R_{(v)} = \mathcal{O}$, $\mathfrak{p}_{(v)} = \mathfrak{p}$ is used, and $\mathfrak{p}_{(v)}$ is a principal ideal generated by any $\pi \in K$ such that $v(\pi) = 1$.

Now one can define a *divisibility theory* on an *integral domain* R with field of fractions K with the help of a family of valuations $\Sigma = \{v\}$ such that the following properties are satisfied:

1) a divides b in $R \iff \forall v \in \Sigma,\ v(a) \le v(b)$;
2) for all $a \in K^\times$ one has $v(a) = 0$ for all but a finite number of $v \in \Sigma$;
3) the set $R_{(v)} = \{x \in K \mid v(x) \ge 0\} \cup \{0\}$ uniquely determines v;
4) $R = \cap_{v \in \Sigma} R_{(v)}$.

If such a family Σ exists then the group of divisors $\mathcal{D} = \mathcal{D}_\Sigma$ is defined to be the free Abelian group with basis Σ. Its elements are written additively as finite formal sums $\sum_i k_i v_i$ or multiplicatively $\prod_i \mathfrak{p}_{v_i}^{k_i}$, where only finitely many of the k_i are non zero. The following homomorphism is defined

$$\mathrm{div} : K^\times \to \mathcal{D}, \quad \mathrm{div}(x) = \prod_{v \in \Sigma} \pi_v^{v(x)}.$$

This homomorphism is called a *divisor map* on R.

The class of *rings with a divisibility theory* is larger than the class of Dedekind rings, and it admits a purely algebraic characterization as the class of Krull rings. Notice that in order to construct valuations, not all of the prime ideals of the ring are used. If we try to define for a prime ideal $\mathfrak{p} \subset R$ a valuation v by putting for $a \in R$, $v(a) = \min\{n \ge 0 \mid a \in \mathfrak{p}^n\}$, then we succeed only when the localization $R_\mathfrak{p}$ of R with respect to $\mathfrak{p}$ is a Noetherian, integrally closed ring with a unique maximal ideal, where

$$R_\mathfrak{p} = \{x = a/b \mid a, b \in R, b \notin \mathfrak{p}\}.$$

The idea of using valuations rather than prime ideals, which arose from the study of algebraic numbers, has turned out to be very fruitful in algebraic geometry. In turn, developments in algebraic geometry have lead to a number of inventions in number theory (cf. Chapters 5 and 6).

To conclude this section we remark that all absolute values of $\mathbb{Q}$ either have the form $|x|^\alpha$ $(0 < \alpha < 1,\ |x|$ being the usual absolute value of $x \in \mathbb{Q} \subset \mathbb{R})$, or have the form $|x|_p^\alpha$ $(\alpha > 0$, where $|x|_p$ is the p–adic absolute value of $x \in \mathbb{Q})$. This result is due to Ostrowski, cf. [BS85], [Chev40].

4.3 Local and Global Methods

4.3.1 p–adic Numbers

The idea of extending the field $\mathbb{Q}$ appears in algebraic number theory in various different guises. For example, the embedding $\mathbb{Q} \subset \mathbb{R}$ often gives useful necessary conditions for the existence of solutions to Diophantine equations over $\mathbb{Q}$ or $\mathbb{Z}$. The important feature of $\mathbb{R}$ is its completeness: every Cauchy sequence $\{\alpha_n\}_{n=1}^{\infty}$ in $\mathbb{R}$ has a limit α (a sequence is called Cauchy if for any $\varepsilon > 0$ we have $|\alpha_n - \alpha_m| < \varepsilon$ whenever n and m are greater than some large $N = N(\varepsilon)$). Also, every element of $\mathbb{R}$ is the limit of some Cauchy sequence $\{\alpha_n\}_{n=1}^{\infty}$ with $\alpha_n \in \mathbb{Q}$.

An analogous construction exists using the p–adic absolute value $|\cdot|_p$ of $\mathbb{Q}$ (see §2):

$$|\cdot|_p : \mathbb{Q} \to \mathbb{R}_{\geq 0} = \{x \in \mathbb{R} \,|x \ \geq 0\}$$
$$|a/b|_p = p^{\mathrm{ord}_p b - \mathrm{ord}_p a}, \quad |0|_p = 0,$$

where $\mathrm{ord}_p a$ is the highest power of p dividing the integer a. This general construction of "adjoining the limits of Cauchy sequences" to a field k with an absolute value $|\cdot|$ leads to a completion of k. This completion, often denoted $\hat{k}$, is complete, and contains k as a dense subfield with respect to the extended absolute value $|\cdot|$, [BS85], [Kob80].

As was noted at the end of §2, all absolute values of $\mathbb{Q}$ are equivalent either to the usual Archimedean absolute value, or to the p–adic absolute value. Thus any completion of $\mathbb{Q}$ is either $\mathbb{R}$, or $\mathbb{Q}_p$, the field of p-adic numbers, i.e. the completion of the field of rational numbers $\mathbb{Q}$ with respect to the p-adic absolute value. Using the embeddings $\mathbb{Q} \hookrightarrow \mathbb{R}$ and $\mathbb{Q} \hookrightarrow \mathbb{Q}_p$ (for all primes p) many arithmetical problems can be simplified. An important example is given by the following *Minkowski–Hasse theorem* [BS85], [Cas78], [Chev40]: the equation

$$Q(x_1, x_2, \ldots, x_n) = 0, \tag{4.3.1}$$

given by a quadratic form $Q(x_1, x_2, \ldots, x_n) = \sum_{i,j} a_{ij} x_i x_j$, $a_{ij} \in \mathbb{Q}$ has a non–trivial solution in rational numbers, iff it is non–trivially solvable over $\mathbb{R}$ and over all $\mathbb{Q}_p$. There are very effective tools for finding solutions in $\mathbb{Q}_p$. These tools are somewhat analogous to those for $\mathbb{R}$ such as the "Newton - Raphson algorithm", which in the p–adic case becomes *Hensel's lemma.*

The simplest way to define the p–adic numbers is to consider expressions of the type

$$\alpha = a_m p^m + a_{m+1} p^{m+1} + \ldots, \tag{4.3.2}$$

where $a_i \in \{0, 1, \ldots . p-1\}$ are digits to the base p, and $m \in \mathbb{Z}$. It is convenient to write down α as a sequence of digits, infinite to the left:

$$\alpha = \begin{cases} \cdots a_{m+1}a_m \overbrace{000\ldots0}^{m-1 \text{ zeros}}{}_{(p)}, & \text{if } m \geq 0, \\ \cdots a_1 a_0.a_{-1} \cdots a_{m(p)}, & \text{if } m < 0. \end{cases}$$

These expressions form a field, in which algebraic operations are executed in the same way as for natural numbers $n = a_0 + a_1p + \ldots a_rp^r$, written as sequences of digits to the base p. Consequently, this field contains all the natural numbers and hence all rational numbers. For example,

$$-1 = \frac{p-1}{1-p} = (p-1) + (p-1)p + (p-1)p^2 + \cdots = \cdots (p-1)(p-1)_{(p)};$$

$$\frac{-a_0}{p-1} = a_0 + a_0p + a_0p^2 + \cdots = \cdots a_0a_0a_{0\,(p)}.$$

For $n \in \mathbb{N}$ the expression for $-n = n \cdot (-1)$ of type (4.3.2) is obtained if we multiply the above expressions for n and for -1. Generally, for $\alpha \in \mathbb{Q}$ write $\alpha = c - \frac{a}{b}$, where $a, c \in \mathbb{Z}$, $b \in N$, $0 \leq a < b$, i.e. a/b is a proper fraction. Then by an elementary theorem of Euler, $p^{\varphi(b)} - 1 = bu$, $u \in \mathbb{N}$. Hence

$$-\frac{a}{b} = \frac{au}{p^{\varphi(b)} - 1},$$

and $au < bu = p^r - 1$, $r = \varphi(b)$. Now let au be written to the base p as $a_{r-1} \cdots a_{0(p)}$, then the expression of type (4.3.2) for α is obtained as the sum of the expression for $c \in \mathbb{N}$ and

$$-\frac{a}{b} = \cdots a_0 \overbrace{a_{r-1} \cdots a_0}^{r \text{ digits}} \overbrace{a_{r-1} \cdots a_0}^{r \text{ digits}}{}_{(p)}.$$

For example, if $p = 5$,

$$\frac{9}{7} = 2 - \frac{5}{7} = 2 + \frac{5 \cdot 2232}{1 - 5^6} \quad c = 2\ a = 5,\ b = 7,$$

so that

$$2232 = 32412_{(5)} = 3 \cdot 5^4 + 2 \cdot 5^3 + 4 \cdot 5^2 + 1 \cdot 5 + 2,$$

thus

$$\frac{9}{7} = \cdots \overbrace{324120}\overbrace{324120}324122_{(5)}.$$

It is easy to verify that the completion of $\mathbb{Q}$ with respect to the p–adic metric $|\cdot|_p$ can be identified with the described field of p–adic expansions (4.3.2), where $|\alpha|_p = p^m$ for α as in (3.2) with $a_m \neq 0$ (see Koblitz N. (1980)).

It is curious to compare the expansions (4.3.2) infinite to the left with the ordinary expansions of real numbers $\alpha \in \mathbb{R}$, infinite to the right:

$$\alpha = a_m a_{m-1} \cdots a_0.a_{-1} \cdots = a_m 10^m + a_{m-1}10^{m-1} + \cdots a_0 + a_{-1}10^{-1} + \cdots,$$

where $a_i \in \{0, 1, \cdots, 9\}$ are digits, $a_m \neq 0$. These expansions to any natural base lead to the same field $\mathbb{R}$. Also, a given α can possess various expressions of this type, e.g. $2.000\cdots = 1.999\cdots$. However, in the p–adic case the expressions (4.3.2) are uniquely determined by α. This fact provides additional comfort when calculating with p–adic numbers.

The field $\mathbb{Q}_p$ is a *complete metric space* with the topology generated by the "open discs":

$$U_a(r) = \{x \mid |x - a| < r\} \quad (x,\ a \in \mathbb{Q}_p,\ r > 0)$$

(or "closed discs" $D_a(r) = \{x \mid |x - a| \leq r\}$). From the topological point of view, the sets $U_a(r)$ and $D_a(r)$ are both open and closed in $\mathbb{Q}_p$.

An important topological property of $\mathbb{Q}_p$ is its *local compactness*: all discs of finite radius are compact. The easiest way to show this is to consider any sequence $\{\alpha_n\}_{n=1}^{\infty}$ of elements $\alpha_n \in D_a(r)$ and to construct a limit point. Such a point may be found step–by–step using the *p–adic digits* (4.3.2). One knows that the number of digits "after the point" is bounded on any finite disc. In particular, the disc

$$\mathbb{Z}_p = D_0(1) = \{x \mid |x|_p \leq 1\} = \left\{x = a_0 + a_1 p + a_2 p^2 + \cdots\right\}$$

is a compact topological ring, whose elements are called p–adic integers. $\mathbb{Z}_p$ is the closure of $\mathbb{Z}$ in $\mathbb{Q}_p$. The ring $\mathbb{Z}_p$ is local, i.e. it has only one maximal ideal $p\mathbb{Z}_p = U_0(1)$ with residue field $\mathbb{Z}_p/p\mathbb{Z}_p = \mathbb{F}_p$. The set of invertible elements (units) of $\mathbb{Z}_p$ is

$$\mathbb{Z}_p^\times = \mathbb{Z}_p \backslash p\mathbb{Z}_p = \{x \mid |x|_p = 1\} = \left\{x = a_0 + a_1 p + a_2 p^2 + \cdots \mid a_0 \neq 0\right\}.$$

For each $x \in \mathbb{Z}_p$ its Teichmüller representative

$$\omega(x) = \lim_{n \to \infty} x^{p^n}$$

is defined. This limit always exists and satisfies the relations: $\omega(x)^p = \omega(x)$, $\omega(x) \equiv x \bmod p$. For example, if $p = 5$, we have

$$\begin{aligned}
&\omega(1) = 1;\\
&\omega(2) = 2 + 1\cdot 5 + 2\cdot 5^2 + 1\cdot 5^3 + 3\cdot 5^4 \cdots;\\
&\omega(3) = 3 + 3\cdot 5 + 2\cdot 5^2 + 3\cdot 5^3 + 1\cdot 5^4 + \cdots;\\
&\omega(4) = 4 + 4\cdot 5 + 4\cdot 5^2 + 4\cdot 5^3 + 4\cdot 5^4 + \cdots = -1;\\
&\omega(5) = 0.
\end{aligned}$$

The ring $\mathbb{Z}_p$ can also be described as the projective limit

$$\varprojlim_n \mathbb{Z}/p^n\mathbb{Z}$$

of rings $A_n = \mathbb{Z}/p^n\mathbb{Z}$ with respect to the homomorphisms $\varphi_n : A_n \to A_{n-1}$ of reduction modulo p^{n-1}. The sequence

$$\cdots \xrightarrow{\varphi_{n+1}} A_n \xrightarrow{\varphi_n} A_{n-1} \xrightarrow{\varphi_{n-1}} \cdots \xrightarrow{\varphi_3} A_2 \xrightarrow{\varphi_2} A_1 \tag{4.3.3}$$

forms a *projective system* indexed by positive integers $n \geq 1$. The projective limit of the system is defined as a ring

$$\varprojlim_n A_n$$

with the following universal property: there are uniquely defined projections

$$\pi_n : \varprojlim_n A_n \to A_n$$

such that for an arbitrary ring B and a system of homomorphisms $\psi_n : B \to A_n$ compatible with each other under the condition: $\psi_{n-1} = \varphi_n \circ \psi_n$ for $n \geq 2$, there exists a unique homomorphism $\psi : B \to A$ such that $\psi_n = \pi_n \psi$ (cf. [Kob80], [Se70]). Note that the uniqueness of A is implied from its existence by abstract nonsense. Hence for the ring $\mathbb{Z}_p$ it suffices to define the projections $\pi_n : \mathbb{Z}_p \to \mathbb{Z}/p^n\mathbb{Z}$, and to check the universal property using digits as in (4.3.2).

Analogously,

$$\mathbb{Z}_p^\times = \varprojlim_n (\mathbb{Z}/p^n\mathbb{Z})^\times,$$

and one can describe the structure of the multiplicative group $\mathbb{Q}_p^\times$.

Put $\nu = 1$ for $p > 2$ and $\nu = 3$ for $p = 2$, and define

$$U = U_p = \{x \in \mathbb{Z}_p | x \equiv 1 \bmod p^\nu\}.$$

Then there is an isomorphism $U \xrightarrow{\sim} \mathbb{Z}_p$ from the multiplicative group U_p to the additive group $\mathbb{Z}_p$, which is given by combining the natural homomorphism

$$U \xrightarrow{\sim} \varprojlim_n U/U^{p^n}$$

with the special isomorphisms

$$\alpha_{p^n} : U/U^{p^n} \xrightarrow{\sim} \mathbb{Z}/p^n\mathbb{Z},$$

given by

$$\alpha_{p^n}((1+p^\nu)^a) = a \bmod p^n \quad (a \in \mathbb{Z}). \tag{4.3.4}$$

One easily verifies that (4.3.4) is well defined and gives the desired isomorphism. Therefore, the group U is a topological cyclic group, and $1+p^\nu$ can be taken as its generator. Another proof of this fact is obtained using the power series

$$\log(1+x) = \sum_{n=1}^{\infty} (-1)^{n+1} \frac{x^n}{n}.$$

which defines an isomorphism from U onto $p\mathbb{Z}_p$

One has the following decompositions

$$\mathbb{Q}_p^\times = p^\mathbb{Z} \times \mathbb{Z}_p^\times, \quad \mathbb{Z}_p^\times \cong (\mathbb{Z}/p^\nu\mathbb{Z})^\times \times U. \tag{4.3.5}$$

4.3.2 Applications of p–adic Numbers to Solving Congruences

The first appearances of p–adic numbers, in papers by Hensel, were related to the problem of finding solutions to congruences modulo p^n. An application of this method by his student H.Hasse to the theory of quadratic forms has lead to an elegant reformulation of this theory, without the use of considerations over the residue rings $\mathbb{Z}/p^n\mathbb{Z}$. These considerations are tiring because of the zero–divisors in $\mathbb{Z}/p^n\mathbb{Z}$. From the above presentation of $\mathbb{Z}_p$ as the projective limit

$$\varprojlim_n \mathbb{Z}/p^n\mathbb{Z}$$

it follows that for $f(x_1, \dots, x_n) \in \mathbb{Z}_p[x_1, \dots, x_n]$, the congruences

$$f(x_1, \dots, x_n) \equiv 0(\text{mod } p^n)$$

are solvable for all $n \geq 1$ iff the equation

$$f(x_1, \dots, x_n) = 0$$

is solvable in p–adic integers. Solutions in $\mathbb{Z}_p$ can be obtained using the following p–adic version of the *"Newton - Raphson algorithm"*.

Theorem 4.17 (Hensel's Lemma). *Let $f(x) \in \mathbb{Z}_p[x]$ be a polynomial in one variable x, $f'(x) \in \mathbb{Z}_p[x]$ its formal derivative, and suppose that for some $\alpha_0 \in \mathbb{Z}_p$ the initial condition*

$$|f(\alpha_0)/f'(\alpha_0)^2|_p < 1 \tag{4.3.6}$$

is satisfied.

Then there exists a unique $\alpha \in \mathbb{Z}_p$ such that

$$f(\alpha) = 0, \quad |\alpha - \alpha_0| < 1.$$

We prove this by induction using the sequence of "successive approximations":

$$\alpha_n = \alpha_{n-1} - \frac{f(\alpha_{n-1})}{f'(\alpha_{n-1})}.$$

Taking into account the formal Taylor expansion of $f(x)$ at $x = \alpha_{n-1}$ one shows that this sequence is Cauchy, and its limit α has all the desired properties (cf. [CF67], [BS85], [Se70]).

For example, if $f(x) = x^{p-1} - 1$, then any $\alpha_0 \in \{1, 2, \dots, p-1\}$ satisfies the condition $|f(\alpha_0)|_p < 1$ At the same time $f'(\alpha_0) = (p-1)\alpha_0^{p-2} \not\equiv 0 \bmod p$, hence the initial condition (3.6) is satisfied. The root α coincides then with the uniquely defined Teichmüller representative of α_0: $\alpha = \omega(\alpha_0)$.

The method described is applicable to polynomials in many variables, although for more than one variable the p-adic solution is not unique (cf. [BS85], [Kob80], [Se70]).

Another interesting application of Hensel's Lemma is related to describing the squares of the field $\mathbb{Q}_p$: for an arbitrary

$$\alpha = p^m \cdot v \in \mathbb{Q}_p \quad (m \in \mathbb{Z},\ v \in \mathbb{Z}_p^\times),$$

the property that α is a square is equivalent to saying that

a) for $p > 2$, $m \in 2\mathbb{Z}$, and $\overline{v} = v \bmod p \in (\mathbb{Z}/p\mathbb{Z})^{\times 2}$ (i.e. $\left(\frac{\overline{v}}{p}\right) = 1$, where $\left(\frac{\overline{v}}{p}\right)$ is the Legendre symbol (see §1.1.5));
b) for $p = 2$, $m \in 2\mathbb{Z}$ and $v \equiv 1 \bmod 8$.

The solvability of $x^2 = \alpha$ in $\mathbb{Q}_p$ under conditions a) and b) is implied by Hensel's Lemma, and the necessity of these conditions is deduced more trivially from considerations modulo p and modulo 8.

As a corollary we give the following description of the quotient group $\mathbb{Q}_p^\times/\mathbb{Q}_p^{\times 2}$

a) for $p > 2$ it is isomorphic to $\mathbb{Z}/2\mathbb{Z} \times \mathbb{Z}/2\mathbb{Z}$ with the system of coset representatives $\{1, p, v, pv\}$, $\left(\frac{\overline{v}}{p}\right) = -1$;
b) for $p = 2$ it is isomorphic to $\mathbb{Z}/2\mathbb{Z} \times \mathbb{Z}/2\mathbb{Z} \times \mathbb{Z}/2\mathbb{Z}$ with the system of coset representatives $\{\pm 1, \pm 5, \pm 2, \pm 10\}$.

4.3.3 The Hilbert Symbol

In this subsection we allow $p = \infty$, in which case we write $\mathbb{Q}_\infty$ for the field of real numbers $\mathbb{R}$. The *Hilbert symbol* (or *norm residue symbol*)

$$(a, b) = \left(\frac{a, b}{p}\right) = \left(\frac{a, b}{p}\right) = (a, b)_p$$

is defined for $a, b \in \mathbb{Q}_p^\times$ by

$$(a, b) = \begin{cases} 1, & \text{if the form } ax^2 + by^2 - z^2 \text{ has a non–trivial zero in } \mathbb{Q}_p; \\ -1, & \text{otherwise.} \end{cases}$$

It is clear that (a, b) depends only on a and b modulo squares. There is a asymmetric form of the definition, namely $(a, b) = 1$ iff

$$a = z^2 - by^2 \text{ for some } y,\ z \in \mathbb{Q}_p. \tag{4.3.7}$$

Indeed, from (4.3.7) it follows that $(1, y, z)$ is a non–trivial zero of the quadratic form $ax^2 + by^2 - z^2$. Conversely, if (x_0, y_0, z_0) is a non–trivial zero, then one can obtain all other zeros using a geometric trick in which one draws secants from the point (x_0, y_0, z_0) in all directions given by vectors with coordinates in $\mathbb{Q}_p$ (see §1.2.3). Using this method we may reduce to the case $x_0 \neq 0$. Then $(y_0/x_0, z_0/x_0)$ satisfies (4.3.7).

Local properties of the Hilbert symbol:

a) $(a, b) = (b, a);$ (4.3.8)

b) $(a_1 a_2, b) = (a_1, b)(a_2, b), \quad (a, b_1 b_2) = (a, b_1)(a, b_2);$ (4.3.9)

c) if $(a, b) = 1$ for all b, then $a \in \mathbb{Q}_p^{\times 2}$; (4.3.10)

d) $(a, -a) = 1$ for all a; (4.3.11)

e) if $p \neq 2, \infty$ and $|a|_p = |b|_p = 1$, then $(a, b) = 1$. (4.3.12)

In particular for a fixed b, the a for which $(a, b) = 1$ form a multiplicative group. Equation (4.3.7) expresses the fact that a is a norm from the quadratic extension $\mathbb{Q}_p(\sqrt{b})/\mathbb{Q}_p$ (cf. [BS85], [Cas78], [Chev40], [Se70]).

A calculation of the Hilbert symbol makes it possible to solve completely the "global" question on the existence of non–trivial rational zeros of quadratic forms (in view of the *Minkowski–Hasse theorem*). If, say

$$Q(x, y, z) = ax^2 + by^2 + cz^2 \quad (a, b, c \in \mathbb{Q}, \ c \neq 0), \tag{4.3.13}$$

then (4.3.13) has a non–trivial zero over $\mathbb{Q}$ iff $(-a/c, -b/c)_p = 1$ for all p including $p = \infty$. This criterion is very effective because for almost all p we have $|a|_p = |b|_p = 1$, whence $(a, b)_p = 1$ for $p \neq 2, \infty$ in view of (3.8e). We give a table of the values of $(a, b)_p$:

Table 4.1. The Hilbert symbol for $p > 2$. Here v denotes an element $v \in \mathbb{Z}$ such that $\left(\frac{v}{p}\right) = -1$, and $\varepsilon = 1$ iff $-1 \in \mathbb{Q}_p^{\times 2}$ (i.e. iff $p \equiv 1 \bmod 4$). Otherwise $\varepsilon = -1$

b \ a	1	v	p	pv
1	$+1$	$+1$	$+1$	$+1$
v	$+1$	$+1$	-1	-1
p	$+1$	-1	ε	$-\varepsilon$
pv	$+1$	-1	$-\varepsilon$	ε

A global property of the Hilbert symbol (the product formula). Let $a, b \in \mathbb{Q}^\times$. Then $(a, b)_p = 1$ for almost all p and

$$\prod_{p \text{ including } \infty} (a, b)_p = 1. \tag{4.3.14}$$

Formula (4.3.14) is equivalent to the *quadratic reciprocity law* (see part I, §1.1.5). Indeed, by (4.3.12) one has $|a|_p = |b|_p = 1$ for all but a finite number of p, hence $(a,b)_p = 1$ for $p \neq 2, \infty$ in view of (4.3.12). Denote the left hand side of (4.3.14) by $f(a,b)$, then by (4.3.9) one has

$$f(a_1 a_2, b) = f(a_1, b) f(a_2, b),$$
$$f(a, b_1 b_2) = f(a, b_1) f(a, b_2),$$

and one verifies that $f(a,b) = 1$ when a and b run through the set of generators of the group $\mathbb{Q}^\times$: -1, 2, $-q$ an odd prime.

Table 4.2. The Hilbert symbol for $p = 2$.

b \ a	1	5	−1	−5	2	10	−2	−10
1	+1	+1	+1	+1	+1	+1	+1	+1
5	+1	+1	+1	+1	−1	−1	−1	−1
−1	+1	+1	−1	−1	+1	+1	−1	−1
−5	+1	+1	−1	−1	−1	−1	+1	+1
2	+1	−1	+1	−1	+1	−1	+1	−1
10	+1	−1	+1	−1	−1	+1	−1	+1
−2	+1	−1	−1	+1	+1	−1	−1	+1
−10	+1	−1	−1	+1	−1	+1	+1	−1

In what follows we shall need an analogous product formula for the normalized absolute values $|\cdot|_p$.

The product formula for absolute values. Let $a \in \mathbb{Q}^\times$. Then $|a|_p = 1$ for all but a finite number of p, and

$$\prod_{p \text{ including } \infty} |a|_p = 1. \tag{4.3.15}$$

Indeed, if $a \in \mathbb{Q}^\times$, then

$$a = \pm \prod_{p \neq \infty} p^{v_p(a)},$$

where $v_p(a) \in \mathbb{Z}$ and $v_p(a)$ for all but a finite number of p. The product formula now follows from the identities:

$$|a|_p = p^{-v_p(a)} \text{ (for } p \neq \infty),$$

$$|a|_\infty = \prod_{p \neq \infty} p^{v_p(a)}.$$

In §4.3.6 we discuss the global properties of absolute values in more detail.

4.3.4 Algebraic Extensions of $\mathbb{Q}_p$, and the Tate Field

If K is a finite algebraic extension of $\mathbb{Q}_p$ then K is generated over $\mathbb{Q}_p$ by some primitive element $\alpha \in K$. The element α is a root of an irreducible polynomial of degree $d = [K : \mathbb{Q}_p]$,

$$f(x) = x^d + a_{d-1}x^{d-1} + ... + a_0 \in \mathbb{Q}_p[x].$$

The absolute value $|\cdot|_p$ has a unique extension to K defined by

$$|\beta|_p = (|\mathrm{N}_{K/\mathbb{Q}_p}(\beta)|_p)^{1/d}, \tag{4.3.16}$$

where $\mathrm{N}_{K/\mathbb{Q}_p}(\beta) \in \mathbb{Q}_p$ is the algebraic norm of the element $\beta \in K$. Formula (4.3.16) defines a unique extension of $|\cdot|_p$ to the algebraic closure $\overline{\mathbb{Q}}_p$ of $\mathbb{Q}_p$. The uniqueness of this extension can easily be deduced from the local compactness of K as a finite–dimensional $\mathbb{Q}_p$–vector space: all of its norms over $\mathbb{Q}_p$ are equivalent (the same thing happens for $\mathbb{R}^n$). It then follows from the multiplicativity of absolute values that any two must coincide.

The function ord_p can then also be extended to $\overline{\mathbb{Q}}_p$ by setting $\mathrm{ord}_p\alpha = \log_p|\alpha|_p$. Formula (4.3.16) implies that $\mathrm{ord}_pK^\times$ is an additive subgroup of $\frac{1}{d}\mathbb{Z}$. Hence $\mathrm{ord}_pK^\times = \frac{1}{e}\mathbb{Z}$ for some positive integer e dividing d. We shall call e the *ramification index* of the extension $K/\mathbb{Q}_p$.

Put

$$\mathcal{O}_K = \{x \in K |\, |x|_p \le 1\}\ ,\ \mathfrak{p}_K = \{x \in K |\, |x|_p < 1\}. \tag{4.3.17}$$

Then $\mathfrak{p}_K$ is the maximal ideal in $\mathcal{O}_K$ and the residue field $\mathcal{O}_K/\mathfrak{p}_K$ is a finite extension of degree f of $\mathbb{F}_p$. One has the relation $d = e \cdot f$, in which f is called *the inertial degree* of the extension. For each $x \in \mathcal{O}_K$ its *Teichmüller representative* is defined by

$$\omega(x) = \lim_{n\to\infty} x^{p^{fn}},\quad \omega(x) \equiv x \pmod{\mathfrak{p}_K}, \tag{4.3.18}$$

and satisfies the equation

$$\omega(x)^{p^f} = \omega(x).$$

The map ω provides a homomorphism from the group of invertible elements

$$\mathcal{O}_K^\times = \mathcal{O}_K \setminus \mathfrak{p}_K = \{x \in K |\, |x|_p = 1\}$$

of $\mathcal{O}_K$ onto the group of roots of unity of degree $p^f - 1$ in K, denoted by μ_{p^f-1}. One also has an isomorphism

$$(\mathcal{O}_K/\mathfrak{p}_K)^\times \xrightarrow{\sim} \mu_{p^f-1} \subset \mathcal{O}_K^\times. \tag{4.3.19}$$

The structure of the multiplicative group $K^\times$ can be described analogously to (4.3.5): if $[K : \mathbb{Q}_p] = d$, then

$$K^\times = \pi^{\mathbb{Z}} \times \mathcal{O}_K^\times, \quad \mathcal{O}_K^\times \cong (\mathcal{O}_K/\mathfrak{p}_K)^\times \times U_K, \tag{4.3.20}$$

where π is a generator of the principal ideal $\mathfrak{p}_K = \pi\mathcal{O}_K$ (i.e. any element $\pi \in K^\times$ with $\mathrm{ord}_p\pi = 1/e$),

$$U_K = \{x \in \mathcal{O}_K^\times | \, |x-1|_p < 1\} = D_1(1^-; K).$$

The structure of the group U_K is then described as a direct product of d copies of the additive group $\mathbb{Z}_p$ and a finite group consisting of all p–power roots of unity contained in K.

Example 4.18. If $e = 1$ then the extension K is called *unramified.* In this case $f = d$ and the Teichmüller representatives generate K over $\mathbb{Q}_p$. Therefore

$$K = \mathbb{Q}_p(1^{1/N}), \quad N = p^d - 1.$$

On the other hand, if $e = d$ then the extension K is called *totally ramified.* For example, if ζ is a primitive root of unity of degree p^n, then $\mathbb{Q}_p(\zeta)$ is totally ramified of degree $d = p^n - p^{n-1}$, and we have that

$$\mathrm{ord}_p(\zeta - 1) = \frac{1}{p^n - p^{n-1}}. \tag{4.3.21}$$

The Tate Field. For purposes of analysis it is convenient to embed $\mathbb{Q}_p$ into a bigger field, which is complete both in the topological and in the algebraic sense. This field is constructed as the completion $\mathbb{C}_p = \widehat{\overline{\mathbb{Q}}}_p$ of an algebraic closure $\overline{\mathbb{Q}}_p$ of $\mathbb{Q}_p$ with respect to the unique absolute value satisfying the condition $|p|_p = \frac{1}{p}$. The proof that $\mathbb{C}_p$ is algebraically closed is not difficult.

We shall use the notation

$$\mathcal{O}_p = \{x \in \mathbb{C}_p | \, |x|_p \le 1\}, \quad \mathfrak{p} = \{x \in \mathbb{C}_p | \, |x|_p < 1\}.$$

Note that the $\mathcal{O}_p$ and $\mathfrak{p}$ are no longer compact, so the field $\mathbb{C}_p$ is not locally compact. We also have that $\mathcal{O}_p/\mathfrak{p} = \overline{\mathbb{F}}_p$ is an algebraic closure of $\mathbb{F}_p$.

4.3.5 Normalized Absolute Values

If F is a locally compact field, then its topology can by given by an absolute value. This fact is deduced from the existence of a *Haar measure* μ on a locally compact group G, i.e. a measure invariant under group shifts $x \mapsto gx$ $(x, g \in G)$:

$$\int_G f(x)\, d\mu(x) = \int_G f(x)\, d\mu(gx) \equiv \int_G f(g^{-1}x)\, d\mu(x)$$

for all integrable functions $f : G \to \mathbb{R}$. This measure is defined uniquely up to a multiplicative constant. However, we do not need a general construction of $d\mu$ (cf. [Wei40]), and we point out only some concrete examples.

If $G = \mathbb{R}$ (the additive group) then $d\mu(x) = dx$ (Lebesgue measure), and $d(x+a) = dx$, $a \in \mathbb{R}$. If $G = \mathbb{R}^\times$ (the multiplicative group), then $d\mu = \frac{dx}{x}$.

If $G = \mathbb{C}$, $z = x + iy \in \mathbb{C}$, then $d\mu = dx\,dy$.

If $K/\mathbb{Q}_p$ is an extension of degree d, and $q = p^f$ is the number of elements of the residue field $\mathcal{O}_K/\mathfrak{p}_K$, then the measure $d\mu$ on the additive group K is uniquely determined by the number $\int_{\mathcal{O}_K} d\mu = \mu(\mathcal{O}_K) = c > 0$; one has $\mu(a + \mathfrak{p}_K) = cq^{-1}$, because the measures of all of the sets $a + \mathfrak{p}_K$ are equal and $\mathcal{O}_K = \bigcup_{a \bmod \mathfrak{p}_K} (a + \mathfrak{p}_K)$. More generally, for all $n \in \mathbb{Z}$ and $a \in K$ one has

$$\mu(a + \mathfrak{p}_K^n) = cq^{-n}. \tag{4.3.22}$$

Any measure $d\mu$ on the additive group of a locally compact field F defines an absolute value $\|\cdot\| : F \to \mathbb{R}_{\geq 0}$: for $a \in F^\times$ the number $\|a\|$ is defined as the multiple, by which the two Haar measures $d\mu(x)$ and $d\mu(ax)$ on F differ:

$$\mu(aU) = \|a\|\mu(U), \tag{4.3.23}$$

where U is an open subset of positive measure, $\mu(U) = \int_U d\mu(x)$. The multiplicativity property

$$\|\alpha\beta\| = \|\alpha\| \cdot \|\beta\| \quad (\alpha, \beta \in F^\times) \tag{4.3.24}$$

follows immediately from definition (4.3.23). If the topology of F is *non–discrete*, i.e. not all subsets are open, then one verifies, that discs of finite radius $\mathcal{D}_a(r) = \{x \in F \mid \|x - a\| \leq r\}$ are compact, and the function $\|\cdot\|$ is continuous. Hence this function is bounded on such discs. In particular,

$$\|1 + \alpha\| \leq C \text{ for } \|\alpha\| \leq 1 \tag{4.3.25}$$

for a positive constant $C \geq 1$. From (4.3.25) it follows that

$$\forall \alpha, \beta \in F \quad \|\alpha + \beta\| \leq C\max(\|\alpha\|, \|\beta\|) \tag{4.3.26}$$

which is weaker than that in the definition of an absolute value from §4.2. These functions are called generalized absolute values. If for example $F = \mathbb{C}$, and $U = \{z = x+iy \in \mathbb{C} \mid |z| = 1\}$, then $\mu(wU) = |w|^2\mu(U)$, where $|w|^2 = w\overline{w}$, and (4.3.26) is satisfied with $C = 4$. However, if for all $n \in \mathbb{N}$ one has $\|n\| \leq 1$, then $C = 1$, so that $\|\cdot\|$ is a *non–Archimedean absolute value.*

In particular, for an extension $K/\mathbb{Q}_p$ with $[K : \mathbb{Q}_p] = d$ put

$$U = \mathcal{O}_K, \quad \alpha = \pi^m v \quad (m \in \mathbb{Z}, v \in \mathcal{O}_K^\times),$$

where π is a uniformizing element, $\mathfrak{p}_K = (\pi)$. We have $\|\alpha\| = q^{-m} = p^{-fm}$. Since $p = \pi^e u$ for some $u \in \mathcal{O}_K^\times$, we obtain

$$\|p\| = \mu(p\mathcal{O}_K)/\mu(\mathcal{O}_K) = |\mathcal{O}_K/p\mathcal{O}_K|^{-1} = p^{-d}.$$

This proves the formula $d = e \cdot f$.

4.3.6 Places of Number Fields and the Product Formula

We shall call two (generalized) absolute values $\|\cdot\|_1$ and $\|\cdot\|_2$ of a field F *equivalent* if $\|x\|_1 = \|x\|_2^c$ for all $x \in F$ and for a constant $c > 0$. A class of equivalent absolute values is called a *place* of F, and it will be denoted by v. The symbol F_v denotes the corresponding completion (with respect to one of the equivalent absolute values in v).

The *theorem of Ostrowski* (see §4.2) says that every place of $\mathbb{Q}$ is either $v = p$ (p a prime), or $v = \infty$. If the place v is non–Archimedean, then we let the same symbol v denote the valuation of F normalized by the condition $v(F^\times) = \mathbb{Z}$.

We list places of finite extensions F of $\mathbb{Q}$. To do this we construct all possible extensions to F of absolute values on $\mathbb{Q}$, since the restriction to $\mathbb{Q}$ of any absolute value on F is an absolute value of $\mathbb{Q}$. More generally, let F/k be a finite separable extension of k with an absolute value $|\cdot|_v$ (for example, $k = \mathbb{Q}$ and $v = p$ or $v = \infty$); $f(x) \in k[x]$ the irreducible polynomial of degree $n = [F : k]$ of a primitive element α for F over k, and let

$$f(x) = \prod_{j=1}^{m} g_j(x) \quad (g_j(x) \in L[x]) \tag{4.3.27}$$

be the decomposition of $f(x)$ into polynomials irreducible over L, where $L = k_v$ is the completion of k with respect to v.

In view of the theorem on tensor products of fields (see §4.1.3), there is a ring isomorphism

$$F \otimes_k L \cong \prod_{j=1}^{m} L_j, \tag{4.3.28}$$

where $L_j \cong L[x]/(g_j(x))$ is the finite extension of L containing F via $\lambda_j : F \hookrightarrow F \otimes_k L \to L_j$.

In §4.3.4 we saw that there exists a unique absolute value on L_j extending $|\cdot|_v$ from $L = k_v$, where it is canonically defined as on the completion. Let us denote this extended absolute value on L_j by the same symbol $|\cdot|_v$ and define an absolute value $|\cdot|_{v,j}$ on F using the embedding λ_j by putting

$$|\beta|_{v,j} = |\lambda_j(\beta)|_v. \tag{4.3.29}$$

It is not difficult to verify that all the $|\cdot|_{v,j}$ are different, and that they are the only extensions of $|\cdot|_v$ from k to F, such that (4.3.28) becomes an isomorphism of topological rings. Thus there are no more than $n = [F : k]$ extensions of an absolute value $|\cdot|_v$ of k to F. These extensions are described explicitly by (4.3.29), assuming one knows the decomposition (4.3.27). Formula (4.3.16) shows that

$$|\lambda_j(\beta)|_v = \sqrt[n_j]{|\mathrm{N}_{L_j/L}(\lambda_j(\beta))|_v},$$

where $n_j = [L_j : L] = \deg g_j(x)$ is the local degree.

To obtain the normalized absolute value $\|\cdot\|_{v,j}$ we put for $\beta \in F^\times$

$$\|\beta\|_{v,j} = |\mathrm{N}_{L_j/L}(\lambda_j(\beta))|_v.$$

Then for all $\beta \in F^\times$ one has:

$$\prod_{j=1}^{m} \|\beta\|_{v,j} = |\mathrm{N}_{F/k}(\beta)|_v. \tag{4.3.30}$$

This follows from $\mathrm{N}_{F/k}(\beta) = \prod_{j=1}^{m} \mathrm{N}_{L_j/L}(\lambda_j(\beta))$ in view of §4.1.3. *Product Formula for Normalized Absolute Values.* Let $k/\mathbb{Q}$ be a finite extension, $\alpha \in k^\times$, and let $|\cdot|_v$ run through the normalized absolute values of k. Then $|\alpha|_v = 1$ for all but a finite number of v, and the following product formula holds

$$\prod_{v} |\alpha|_v = 1. \tag{4.3.31}$$

This is easily deduced from formula (4.3.30), in which we put $k/\mathbb{Q}$ instead of F/k and notice that $\mathrm{N}_{k/\mathbb{Q}}(\alpha) \in \mathbb{Q}^\times$. It then suffices to apply the already proven product formula for $\mathbb{Q}$, see (4.3.15).

Global Fields. We use the term "global field" to refer to either a finite extension of $\mathbb{Q}$ (an algebraic number fields) or a finite, separable extension of $\mathbb{F}_q(t)$, where $\mathbb{F}_q$ is the field with q elements and t is a (transcendental) variable (a function field with positive characteristic) [AW45], [AT51] , [Wei74a], [CF67], [BoCa79].

In every global field there is a product formula and a similar classification of the normalized absolute values. Many problems concerning integers have natural analogies in function fields. These analogies can sometimes be more successfully treated using methods of algebraic geometry, and they provide a rich source of intuition for the number field case (see §4.5, §5.2, §6.5, and Introductory survey to Part III).

4.3.7 Adeles and Ideles

The Ring of Adeles.

In arithmetical questions the ring $\mathbb{Z}$ is often considered as a lattice in $\mathbb{R}$, i.e. a discrete subgroup of the additive group of the locally compact field $\mathbb{R}$ with compact quotient group $\mathbb{R}/\mathbb{Z}$, the quotient being isomorphic to a circle. It turns out, that for a global field k one can canonically construct the "smallest" locally compact ring $\mathbb{A}_k$, containing k as a lattice. This means that k is a discrete subring in $\mathbb{A}_k$ with compact additive quotient group $\mathbb{A}_k/k$. The ring $\mathbb{A}_k$, which is called the *ring of adeles* is constructed using all the embeddings $k \hookrightarrow k_v$, where v runs through the set $\Sigma = \Sigma_k$ of all places of k. One defines

$\mathbb{A}_k$ to be the subring of the product $\prod_{v\in\Sigma} k_v$ consisting of all infinite vectors $\alpha = (\alpha_v)_{v\in\Sigma}$, $\alpha_v \in k_v$ such that $\alpha_v \in \mathcal{O}_v$ for all but a finite number of v. In view of §4.3.6 the number of Archimedean places does not exceed $n = [k : \mathbb{Q}]$. Hence all but a finite number of places are non–Archimedean, and the compact subring $\mathcal{O}_v \subset k_v$ is defined to be the valuation ring of v):

$$\mathbb{A}_k = \tag{4.3.32}$$
$$\left\{ \alpha = (\alpha_v) \in \prod_{v\in\Sigma} k_v \;\middle|\; \alpha_v \in \mathcal{O}_v \text{ for all but a finite number of } v \right\}.$$

One gives $\mathbb{A}_k$ the topology generated by the open subsets of the type

$$W_S = \prod_{v\in S} W_v \times \prod_{v\notin S} \mathcal{O}_v, \tag{4.3.33}$$

where S runs through all finite subsets $S \subset \Sigma$, and W_v are open subsets in k_v. The set W_S is compact (has compact closure) if all the W_v are bounded. Hence $\mathbb{A}_k$ is a locally compact topological ring in which k is embedded diagonally

$$k \ni \alpha \mapsto (\cdots, \alpha, \alpha, \cdots)_{v\in\Sigma} \in \mathbb{A}_k \subset \prod_{v\in\Sigma} k_v$$

(note that in view of §4.3.6 $|\alpha|_v = 1$ for all but a finite number of $v \in \Sigma$). It is interesting to note that the product $\prod_{v\in\Sigma} k_v$ is too big to be locally compact: by definition of the product topology, the projection of any open subset $U \subset \prod_{v\in\Sigma} k_v$ onto k_v coincides with k_v for almost all v, thus U would never be compact having non–compact image under a continuous map (projection). The above construction of $\mathbb{A}_k$ is called the *restricted topological product* of the topological spaces k_v with respect to the compact subspaces $\mathcal{O}_v$ defined for all but a finite number of indices v. The convergence of a sequence $\{\alpha\}_{n=1}^{\infty}$, $\alpha_n = (\alpha_{v,n})_v \in \mathbb{A}_k$ to $\beta = (\beta_v) \in \mathbb{A}_k$ means that for any $\varepsilon > 0$ and any finite set $S \subset \Sigma$ there exist $N \in \mathbb{N}$ such that

$$1)\ \forall n > N\ \forall v \notin S\ \alpha_{n,v} - \beta_v \in \mathcal{O}_v,$$
$$2)\ \forall n > N\ \forall v \in S\ |\alpha_{n,v} - \beta_v|_v < \varepsilon.$$

Every principal adele α, i.e.

$$\alpha = (\cdots, \alpha, \alpha, \cdots)_v \in k \subset \mathbb{A}_k \tag{4.3.34}$$

can be separated from the rest of k by a neighborhood of type (4.3.33) with $S = \{v \in \Sigma \mid \alpha \notin \mathcal{O}_v\}$. Hence k is discrete in $\mathbb{A}_k$. The compactness of the quotient group $\mathbb{A}_k/k$ has an explanation via the Pontryagin duality theory of locally compact commutative topological groups: $\mathbb{A}_k/k$ is isomorphic to the group $\hat{k}$ of all characters of k. Recall that for a locally compact group G its group of continuous characters

$$\hat{G} = \mathrm{Hom}_{\mathrm{contin}}(G, S^1) \tag{4.3.35}$$

(where $S^1 = \{z \in \mathbb{C}^\times \mid |z| = 1\}$) is again a locally compact group in the natural topology of the character group; one always has $G^{\wedge\wedge} = G$, and for any exact sequence

$$1 \to G_1 \to G \to G_2 \to 1$$

with continuous homomorphisms, the dual sequence for characters is exact:

$$1 \to \hat{G}_2 \to \hat{G} \to \hat{G}_1 \to 1.$$

By the association $G \mapsto \hat{G}$, finite groups remain finite; discrete groups become compact groups (and conversely), and for a connected group G its dual $\hat{G}$ is torsion free. If $H \subset G$ is a closed subgroup, then its annihilator

$$H^\perp = \left\{\chi \in \hat{G} \mid \chi(H) = 1\right\} \tag{4.3.36}$$

is isomorphic to $(G/H)\hat{}$.

In the simplest example $\mathbb{Z} \subset \mathbb{R}$ one has $\mathbb{Z}^\wedge \cong S^1$, $S^{1\wedge} \cong \mathbb{Z}$, and the group $\mathbb{R}$ is self-dual: $\mathbb{R}^\wedge \cong \mathbb{R}$ (the number $t \in \mathbb{R}$ corresponds to the character $(x \mapsto e^{2\pi ixt})$, so that $\mathbb{Z}^\perp \cong \mathbb{Z}$.

One can verify that the additive group $\mathbb{A}_k$ is self dual, and $\alpha \in \mathbb{A}_k$ corresponds to the character $(\beta \mapsto \chi(\alpha\beta)) \in \hat{\mathbb{A}}_k$, where χ is a non–trivial additive character of $\mathbb{A}_k$ satisfying $\chi(k) = 1$, so that $k \cong k^\perp = (\mathbb{A}_k/k)^\wedge$.

Consider in detail the case $k = \mathbb{Q}$, and the ring $\mathbb{A} = \mathbb{A}_\mathbb{Q}$. For $\alpha = (\alpha_v)_v \in \mathbb{A}$ the fractional parts $\{\alpha_v\}$ are defined (for $v = p$ one uses the p-adic expansion (4.3.2) to define $\{\alpha_v\} = a_{-1}p^{-1} + \cdots + a_m p^m$ for $m < 0$). Then for all but a finite number of v we have that $\{\alpha_v\} = 0$, and $\{\alpha\} = \sum_{v \neq \infty}\{\alpha_v\}$ is a rational number. The character χ can be defined by the formula

$$\beta \mapsto \exp(-2\pi i\{\beta_\infty\}) \cdot \prod_{v \neq \infty} \exp(2\pi i\{\beta_v\}), \tag{4.3.37}$$

and for each $\beta \in \mathbb{Q}^\times$ one has $\chi(\beta) = 1$.

For each component v the character $\chi_v : \mathbb{Q}_v^\times \to S^1$ is defined by

$$\chi_v(\beta) = \exp(2\pi i\{\beta\})$$

($\beta \in \mathbb{Q}_v$), which provides the self-duality of the locally compact field $\mathbb{Q}_v$ ($v = p, \infty$) in a similar way: an element $t \in \mathbb{Q}_v$ corresponds to the character $x \mapsto \chi_v(tx)$. This also gives us a description of the quotient group

$$\mathbb{A}/\mathbb{Q} \cong \mathbb{R}/\mathbb{Z} \times \prod_p \mathbb{Z}_p, \tag{4.3.38}$$

which is easily seen by subtracting from an adele α its fractional part

$$\{\alpha\} = \sum_{v \neq \infty} \{\alpha_v\} \in \mathbb{Q}. \tag{4.3.39}$$

The quotient group $\mathbb{A}/\mathbb{Q}$ is compact by the *theorem of A.N. Tychonov* on products of compact spaces. For a number field k it is useful to consider the isomorphism of topological rings

$$\mathbb{A}_k \cong k \otimes \mathbb{A}_{\mathbb{Q}}, \tag{4.3.40}$$

which implies an isomorphism of additive groups $\mathbb{A}_k^{(+)} \cong (\mathbb{A}_{\mathbb{Q}}^{(+)})^n$, where $n = [k : \mathbb{Q}]$, and also statements on the discreteness of k in $\mathbb{A}_k$ and on the compactness of the quotient group $\mathbb{A}_k/k$. One verifies easily that an analogous isomorphism takes place for an arbitrary extension of global fields F/k:

$$\mathbb{A}_F \cong F \otimes_k \mathbb{A}_k. \tag{4.3.41}$$

The Idele Group

(cf. [Chev40], [Wei74a]). The set of all invertible elements of a ring R forms a multiplicative group $R^\times$. If R is topological, the topology on $R^\times$ is defined by means of the embedding $x \mapsto (x, x^{-1})$ $(R \to R \times R)$ so that the inversion map $x \mapsto x^{-1}$ is continuous. The *idele group* J_k of a global field k is the topological group $\mathbb{A}_k^\times$ of invertible elements of the ring $\mathbb{A}_k$. The group J_k coincides with the *restricted topological product* of the locally compact groups $k_v^\times$ with respect to the compact subgroups $\mathcal{O}_v^\times$ defined for non–Archimedean places $v \in \Sigma$.

4.3.8 The Geometry of Adeles and Ideles

The embedding of k into its ring of adeles $\mathbb{A}_k$ is reminiscent of the geometric interpretation of the ring of integers $\mathcal{O} = \mathcal{O}_k$ of k as a lattice in the $\mathbb{R}$–algebra

$$k_\infty = k \otimes \mathbb{R} \cong \prod_{v|\infty} k_v \cong \mathbb{R}^{r_1} \times \mathbb{C}^{r_2}, \quad \lambda : \mathcal{O} \hookrightarrow k_\infty. \tag{4.3.42}$$

This analogy goes much further. Consider a Haar measure μ on the locally compact additive group $\mathbb{A}_k$; this measure can be defined on the open subsets W_S of type (4.3.33) by

$$\mu(W_S) = \prod_{v \in S} \mu_v(W_v), \tag{4.3.43}$$

where $\mu_v(\mathcal{O}_v) = 1$ for $v \not| \infty$ (i.e. for non–Archimedean v); for Archimedean places one normalizes the measure as follows:

$$\mu_v = \begin{cases} dx \quad \text{(Lebesgue measure)} & \text{if } k_v \cong \mathbb{R}, \\ 2dx\, dy = |dz \wedge d\overline{z}| & \text{if } z = x + iy \in k_v \cong \mathbb{C}. \end{cases}$$

If $\beta = (\beta_v) \in J_k$ is an idele, then its *module* is defined to be the multiplicative constant $|\beta|$, by which the Haar measures $\mu(x)$ and $\mu(\beta x)$ on $\mathbb{A}_k$ differ:

$$\mu(\beta x) = |\beta| \cdot \mu(x). \tag{4.3.44}$$

It follows from the description (4.3.43) of μ that $|\beta| = \prod_v |\beta_v|_v$, where $|\cdot|_v$ is the normalized absolute value from the class of a place $v \in \Sigma$, which for Archimedean places is given by the following:

$$|x|_v = \begin{cases} |x| \quad \text{(the usual absolute value)} & \text{if } k_v \cong \mathbb{R}, \\ |z|^2 = z\overline{z} & \text{if } z = x + iy \in k_v \cong \mathbb{C}. \end{cases}$$

On the compact quotient group $\mathbb{A}_k/k$ we define a measure μ by means of a general notion of fundamental domain: if Γ is a discrete subgroup of a locally compact group G, then a fundamental domain X for G modulo Γ is a complete set of coset representatives for (left) cosets G/Γ, which has some additional measurability properties. By restricting the Haar measure α of G onto the subset X, one obtains a uniquely defined measure on G/Γ, which is denoted by the same letter, and $\alpha(G/\Gamma) = \alpha(X)$.

In order to construct a fundamental domain X for $\mathbb{A}_k/k$ we choose a $\mathbb{Z}$–basis $\omega_1, \cdots, \omega_n$ of the free Abelian group $\mathcal{O} \subset k$ of algebraic integers in k. This is also a basis of the vector space $k_\infty = k \otimes \mathbb{R}$ over $\mathbb{R}$, and it defines an isomorphism $\theta : \mathbb{R}^n \xrightarrow{\sim} k_\infty$ by the formula

$$\theta((u_1, \ldots, u_n)) = \sum_{i=1}^{n} u_i \omega_i.$$

Denote by I the interval $0 \le t < 1$ in $\mathbb{R}$. Then $\theta(I^n)$ is a fundamental parallelogram for the lattice $\mathcal{O}$ in k_∞ (see 1.3). Now take X to be the set

$$X = \theta(I^n) \times \prod_{v \nmid \infty} \mathcal{O}_v \tag{4.3.45}$$

(a *fundamental domain for k in $\mathbb{A}_k$*). To prove that X is a fundamental domain, we note that $k_\infty + k$ is dense in $\mathbb{A}_k$. This statement is known as the *approximation theorem* and it is a version of the *Chinese remainder theorem* (cf. §1.1.5). Moreover, $k_\infty \times \prod_v \mathcal{O}_v$ is an open subgroup in $\mathbb{A}_k$, hence for any $x \in \mathbb{A}_k$ there exists $\eta \in k$ such that

$$x - \eta \in k_\infty \times \prod_v \mathcal{O}_v.$$

The condition that another element $\eta' \in k$ has the same property is equivalent to saying that $\eta - \eta' \in \mathcal{O}_v$ for all non–Archimedean places v, i.e. that $\eta - \eta' \in \mathcal{O}_k$. Thus by an appropriate choice of η we may assume that the y_∞–coordinate of $y = x - \eta$ belongs to $\theta(I^n)$; therefore $y_\infty = \theta(u)$, $u \in I^n$, where u is uniquely determined. This establishes the statement.

The first application of the measure constructed on $\mathbb{A}_k/k$ is a simple proof of the product formula (4.3.31): if $\beta \in k^\times \subset J_k$ is a principal idele, then $\beta k^\times = k^\times$ in J_k, and multiplication by β defines a homeomorphism of $\mathbb{A}_k/k$ with itself, hence the Haar measures $\mu(x)$ and $\mu(\beta x)$ on $\mathbb{A}_k/k$ must coincide, i.e. by (4.3.44) we see that

$$|\beta| = \prod_v |\beta_v|_v = \frac{\mu(\beta\mathbb{A}_k/k)}{\mu(\mathbb{A}_k/k)} = 1.$$

Let us calculate the measure $\mu(\mathbb{A}_k/k)$. The form of the fundamental domain X constructed reduces this calculation to the problem of determining the volume of the fundamental parallelogram $\theta(I^n)$ in k_∞. This volume was already found in §4.1.3, (4.1.7). We obtain

$$\mu(\mathbb{A}_k/k) = |D_k|^{1/2}, \tag{4.3.46}$$

where $D_k = \det(\mathrm{Tr}(\omega_i\omega_j))$ is the discriminant of k. Here we have taken into account that the measure $\prod_{v|\infty} \mu_v$ on k_∞ differs by a multiple of 2 from the Lebesgue measure on those components v such that $k_v \cong \mathbb{C}$, when

$$d\,\mu_v(z) = 2dx\,dy = |dz \wedge d\bar{z}| \text{ for } z = x + iy \in k_v \cong \mathbb{C}.$$

Consider the constant

$$C = \left(\frac{2}{\pi}\right)^{r_2} \sqrt{|D_k|}. \tag{4.3.47}$$

This number is important for finding non–zero points β in the lattice $k \subset \mathbb{A}_k$ belonging to a *parallelotope*, i.e. to a set of the form

$$V(c) = \{x = (x_v)_v \in \mathbb{A}_k \mid \forall v \in \Sigma_k \;\; |x_v|_v \le c_v\}, \tag{4.3.48}$$

such that $c = (c_v)$ is an infinite tuple of positive constants defined for the places v of k, all but a finite number of which are 1.

Lemma 4.19 (Blichfeldt). *Assume that for the numbers c_v we have that*

$$\prod_v c_v > C = \left(\frac{2}{\pi}\right)^{r_2} \sqrt{|D_k|}.$$

Then there exist $\beta \in k^\times \cap V(c) \subset \mathbb{A}_k$.

Proof. Consider the auxiliary parallelotope $V(c')$, $c' = (c'_v)_v$, $v \in \Sigma$, where

$$\begin{cases} c'_v = c_v, & \text{if } v \text{is non-Archimedean} \\ c'_v = c_v/2, & \text{if } k_v \cong \mathbb{R} \\ c'_v = c_v/4, & \text{if } z = x + iy \in k_v \cong \mathbb{C}. \end{cases}$$

Then one can calculate the measure of $V(c')$:

$$\mu(V(c')) = \left(\frac{\pi}{2}\right)^{r_2} \prod_v c_v > \sqrt{|D_k|}.$$

In other words, the measure of $V(c')$ is bigger than that of the fundamental domain for $\mathbb{A}_k/k$, hence there exist two distinct points y and $y' \in V(c')$ whose images modulo k coincide, i.e. $y-y' \in k^\times$. We obtain for the number $\beta = y-y'$ the following estimates:

$$|\beta|_v \leq \begin{cases} \max(|y_v|_v, |y'_v|_v) \leq c_v & \text{if } v \text{ is non-Archimedean,} \\ 2\max(|y_v|_v, |y'_v|_v) \leq c_v & \text{if } k_v \cong \mathbb{R}, \\ 4\max(|y_v|_v, |y'_v|_v) & \text{if } z = x+iy \in k_v \cong \mathbb{C}, \end{cases}$$

proving the lemma.

We now turn our attention to the structure of the idele group. Consider the homomorphism $|\cdot|_v : J_k \to \mathbb{R}_+^\times$, which takes $y = (y_v)_v \in J_k$ to $|y| = \prod_v |y_v|_v$. Denote by J_k^1 its kernel, then J_k^1 is a closed subgroup, and in view of the product formula (3.27) we have that $k^\times \subset J_k^1$. The following theorem is one of the most important facts in algebraic number theory.

Theorem 4.20. *The quotient group $J_k^1/k^\times$ is compact.*

The proof relies on Blichfeldt's lemma, and is very similar to the proof of Dirichlet's unit theorem, and the deduction of the latter from Minkowski's lemma. One can show that this theorem is equivalent to the conjunction of Dirichlet's unit theorem and the finiteness of the ideal class group (see §4.1.6 and §4.2.2). These two statements can be easily deduced from the above theorem as follows:

The Divisor Map. Let I_k be the group of fractional ideals (divisors), i.e. the free Abelian group generated by the set of non–Archimedean places of k. Define

$$\operatorname{div} : J_k \longrightarrow I_k, \quad \operatorname{div}((x_v)) = \sum_{v \nmid \infty} v(x_v) \cdot v, \tag{4.3.49}$$

where v denotes as agreed above the valuation of k normalized by the condition $v(k^\times) = \mathbb{Z}$. Note that $\operatorname{div}(J_k) = I_k$, and that changing only the Archimedean component $x_\infty = (x_v)_{v|\infty}$ of an idele x does not change $\operatorname{div}(x)$. Note also that $\operatorname{div}(k^\times) = P_k$ is the subgroup of principal ideals in the discrete group I_k. Hence we have a continuous epimorphism $\operatorname{div} : J_k^1/k^\times \to I_k/P_k = Cl_k$ of a compact group onto a discrete group. The image is both compact and discrete, and is therefore finite.

The Logarithmic Map and S–Units. Let $S \subset \Sigma$ be a finite set of places containing the set Σ_∞ of all Archimedean places. The set of elements $\eta \in k^\times$ satisfying $|\eta|_v = 1$ for all $v \notin S$ forms a multiplicative group, which is denoted by E_S and is called the group of S–units.

Theorem 4.21 (Theorem on S–Units). *The group E_S is the direct sum of a finite cyclic group and a free Abelian group of rank $s-1$, where $s = \mathrm{Card}\, S$ is the number of places in S.*

(cf. [La70]). The proof of this theorem is similar to that of Dirichlet's unit theorem (see §4.1.4). One considers the logarithmic map

$$l : J_k \longrightarrow \underbrace{\mathbb{R} \oplus \cdots \oplus \mathbb{R}}_{s \text{ times}} \tag{4.3.50}$$

(where $\mathbb{R}$ is the additive group of real numbers), defined by

$$l((x_v)_v) = (\cdots, \log |x_v|_v, \cdots)_{v \in S}.$$

This map is continuous, and its image contains a basis of the vector space $\mathbb{R}^s$ (if $S = \Sigma_\infty$ then l is an epimorphism).

With the help of (4.3.49) and (4.3.50) it is not difficult to describe fundamental domains for $k^\times$ in J_k and J_k^1 (cf. [Wei74a], pp.137–139]). One can calculate the volume $\tilde{\gamma}(J_k^1/k^\times)$ with respect to the Haar measure $\tilde{\gamma}$ on $J_k^1/k^\times$. We normalize the measure $\tilde{\gamma}$ by using the decomposition:

$$J_k/k^\times \cong J_k^1/k^\times \times \mathbb{R}_+^\times, \quad \gamma = \left(\tilde{\gamma} \times \frac{dx}{|x|}\right), \tag{4.3.51}$$

in which

$$\gamma = \prod_v \gamma_v$$

is the Haar measure on J_k, normalized as follows:

$$\begin{cases} \gamma_v(\mathcal{O}_v^\times) = 1 & \text{if } v \text{ is non-Archimedean,} \\ d\gamma_v(x) = |x|^{-1} dx & \text{if } k_v \cong \mathbb{R}, \\ d\gamma_v(z) = |z\overline{z}|^{-1} |dz \wedge d\overline{z}| = 2 dx\, dy & \text{if } z = x + iy \in k_v \cong \mathbb{C}. \end{cases}$$

Then the following formula holds:

$$\tilde{\gamma}(J_k^1/k^\times) = \frac{2^{r_1}}{(2\pi)^{r_2}} hR_k, w, = \kappa_k, \tag{4.3.52}$$

where $h = |Cl_k|$ is the class number of k; R_k is the regulator, and $w = w_k$ is the number of roots of unity in k, see 4.1.3. This formula means that for any positive number $m > 1$ in $\mathbb{R}$ the subset $C(m)$ of $J_k/k^\times$ defined by $C(m) = \{x \in J_k/k^\times \mid 1 \le |x| \le m\}$ has measure

$$\gamma(C(m)) = \kappa_k \log m. \tag{4.3.53}$$

The quantities $R = R_k$, $h = h_k$ and $D = D_k$ turn out to be the most important constants characterizing a number field k. These quantities occur together in formulae (4.3.52) and (4.3.53) for the volumes of fundamental domains, and are not independent. According to a deep result of Brauer and Siegel (cf. [La70], [La83]), one knows that for a sequence of number fields k_m of degrees $n_m = [k_m : \mathbb{Q}]$ satisfying the condition $n_m/\log|D_{k_m}| \to 0$ as $m \to \infty$, the following asymptotic relation holds

$$\log(h_{k_m} \cdot R_{k_m}) \sim \log(|D_{k_m}|)^{1/2}. \tag{4.3.54}$$

The idele class group

$$C_k = J_k/k^\times$$

plays a key role in classifying the Abelian extensions of k (class field theory), cf. §4.4.

If $k = \mathbb{Q}$ then there are isomorphisms

$$J_\mathbb{Q}/\mathbb{Q}^\times \cong \mathbb{R}_+^\times \times \prod_p \mathbb{Z}_p^\times, \tag{4.3.55}$$

$$J_\mathbb{Q}^1/\mathbb{Q}^\times \cong \prod_p \mathbb{Z}_p^\times, \tag{4.3.56}$$

which are easily established by dividing an idele $\alpha \in J_\mathbb{Q}$ by its (multiplicative) divisor $\mathrm{div}(\alpha) = \prod_p p^{v_p(\alpha_p)}$, which in this situation turns out to be a positive rational number. As a result one obtains the element $\alpha \cdot \mathrm{sign}(\alpha_\infty) \cdot \mathrm{div}(\alpha)^{-1}$, which belongs to the right hand side of (4.3.55).

4.4 Class Field Theory

4.4.1 Abelian Extensions of the Field of Rational Numbers

(cf. [AT51], [Chev40], [Wei74a]). One of the central objects of algebraic number theory is the full Galois group $G = G(\overline{\mathbb{Q}}/\mathbb{Q})$ of $\overline{\mathbb{Q}}$ over $\mathbb{Q}$, together with its subgroups $H \subset G$ of finite index, which correspond to finite extensions k of $\mathbb{Q}$:

$$H = G_k = G(\overline{\mathbb{Q}}/k) \subset G.$$

From the topological point of view G is a compact, totally disconnected group, with the topology of a profinite group (the projective limit of its finite quotient groups):

$$G \cong \varprojlim_k G/G_k = \varprojlim_k G(k/\mathbb{Q}),$$

where G_k are normal subgroups which are both closed and open, as they correspond to finite Galois extensions $k/\mathbb{Q}$.

Class field theory provides a purely arithmetical description of the maximal Abelian (Hausdorff) quotient group $G_k^{\mathrm{ab}} = G_k/G_k^c$, where G_k^c is the closure of the commutator subgroup of G_k. Moreover, one has this description both for algebraic number fields and for function fields (global fields of positive characteristic). One form of this description of G_k^{ab} is given by a calculation of all characters (one–dimensional complex representations) of the full Galois group G_k.

The topological structure of infinite Galois groups is similar to that of locally compact analytic Lie groups over p–adic fields such as $\mathrm{SL}_n(\mathbb{Q}_p)$, $\mathrm{Sp}_n(\mathbb{Q}_p)$ etc. The use of analytic methods such as the representation theory of Lie groups and Lie algebras, has developed drastically in recent decades. These techniques are related to non–commutative generalizations of class field theory (see §6.5). We first describe the group $G_{\mathbb{Q}}^{\mathrm{ab}}$ starting from the Kronecker–Weber theorem, which says that every Abelian extension k of $\mathbb{Q}$ (i.e. an extension whose Galois group $G(k/\mathbb{Q})$ is Abelian) is contained in a cyclotomic field $K_m = \mathbb{Q}(\zeta_m)$, where ζ_m is a primitive root of unity of degree m (see §4.1.2). There is an isomorphism

$$\psi_m : (\mathbb{Z}/m\mathbb{Z})^\times \longrightarrow G_m = G(K_m/\mathbb{Q}), \tag{4.4.1}$$

which associates to a residue class $a \pmod{m} \in (\mathbb{Z}/m\mathbb{Z})^\times$, $(a,m) = 1$ an automorphism $\sigma = \sigma_a = \psi_m(a) \in G_m$ given by the condition $\zeta_m^\sigma = \zeta_m^a$. The arithmetical isomorphism (4.4.1) makes it possible to regard Dirichlet characters $\chi : (\mathbb{Z}/m\mathbb{Z})^\times \to \mathbb{C}^\times$ as one–dimensional representations

$$\rho_\chi : G \xrightarrow{\pi_m} G_m \xrightarrow{\psi_m} (\mathbb{Z}/m\mathbb{Z})^\times \xrightarrow{\chi} \mathbb{C}^\times, \tag{4.4.2}$$

where $G \xrightarrow{\pi_m} G_m$ is the natural homomorphism restricting the action of the Galois automorphisms to the subfield K_m; $\rho_\chi = \chi \circ \psi_m \circ \pi_m$. Hence each character $\rho : G \to \mathbb{C}^\times$ has the form $\rho = \rho_\chi$ for some χ. For example a quadratic extension $k = \mathbb{Q}(\sqrt{d})$ is contained in the cyclotomic extension $\mathbb{Q}(\zeta_{|D|})$, where D is the discriminant of k. This is easily shown using Gauss sums: for the quadratic character $\chi = \chi_k$ of k we have $G(\chi)^2 = D$. Hence $G(\chi) = \pm\sqrt{D} \in K_{|D|}$. Recall that χ is a primitive quadratic character modulo $|D|$ which is uniquely determined by the condition $\chi(-1) = \text{sign } D$. The field k corresponds by Galois theory to the subgroup $\text{Ker}\rho \subset G(K_{|D|}/\mathbb{Q})$ of index 2, $\rho = \rho_{\chi_k}$.

By the Kronecker–Weber theorem, the maximal Abelian extension $\mathbb{Q}^{\text{ab}}$ can be described as the union of all K_m, and its Galois group coincides with the projective limit of the groups $G_m \cong (\mathbb{Z}/m\mathbb{Z})^\times$, that is

$$G^{\text{ab}} \cong \varprojlim_m (\mathbb{Z}/m\mathbb{Z})^\times,$$

where the limit is taken over the system of natural projection homomorphisms

$$(\mathbb{Z}/m_1\mathbb{Z})^\times \to (\mathbb{Z}/m_2\mathbb{Z})^\times$$

for m_2 dividing m_1. Hence the group G^{ab} coincides with the group $\prod_p \mathbb{Z}_p^\times$ of invertible elements of the ring $\hat{\mathbb{Z}} = \prod_p \mathbb{Z}_p$ (the profinite completion of the ring of integers).

A more invariant formulation of this isomorphism is based on the introduction of the ring of adeles $\mathbb{A}$ and its multiplicative group $J = \mathbb{A}^\times$, the ideles of $\mathbb{Q}$ (see §4.3.7 and §4.3.8). The group J consists of all infinite vectors

$$\alpha = (\alpha_\infty; \alpha_2, \alpha_3, \ldots, \alpha_p, \ldots) \in \mathbb{R}^\times \times \prod_p \mathbb{Q}_p^\times,$$

such that $\alpha_p \in \mathbb{Z}_p^\times$ for all but a finite number of p. The quotient $\mathbb{A}^\times/U_1$ is discrete. According to (4.3.55) we have

$$J/\mathbb{Q}^\times \cong \mathbb{R}_+^\times \times \prod_p \mathbb{Z}_p^\times,$$

where $\mathbb{R}_+^\times$ is the multiplicative group of all positive real numbers.

The group G^{ab} is therefore isomorphic to the quotient of $J/\mathbb{Q}^\times$ by the connected component of 1:

$$G^{\text{ab}} \cong \prod_p \mathbb{Z}_p^\times \cong J/\mathbb{R}_+^\times\mathbb{Q}^\times. \tag{4.4.3}$$

The important feature of this isomorphism is that the elements of G_m, and hence of G^{ab}, have an arithmetical nature; they correspond to prime numbers. Namely, a prime p not dividing m corresponds to its Frobenius element $\sigma = \sigma_p : \zeta_m \mapsto \zeta_m^p$. The set of all primes corresponding to a fixed element $\sigma \in G_m$ is infinite by *Dirichlet's theorem on primes in arithmetical progressions.* This set coincides with the set of primes of type $p = a + km$ $(k \in \mathbb{Z})$, where $\sigma = \psi_m(a)$.

The automorphism σ is called the Frobenius automorphism (and denoted Fr_p or Frob_p) for the following reason: if we consider the ring $\mathcal{O}_m = \mathbb{Z}[\zeta_m]$ of all integers in K_m then in the reduction $\mathcal{O}_m/p\mathcal{O}_m$ we have $\mathrm{Fr}_p(x) = x^p$, i.e. Fr_p acts as the Frobenius automorphism. The way that p splits into prime ideals in $\mathcal{O}_m$ depends only on the image of p in the Galois group $G_m \cong (\mathbb{Z}/m\mathbb{Z})^\times$ (see §4.1.2). The idea of associating a Galois automorphism to a prime number (or prime ideal) leads to the isomorphism (4.4.3), in which to Fr_p one associates the class of the idele

$$\pi_p = (1; 1, \cdots, 1, p, 1, \cdots) \text{ in } J/(\mathbb{R}_+ \times \mathbb{Q}^\times).$$

The field K_m corresponds to the open subgroup

$$U_m = \mathbb{R}_+^\times \times \prod_{p|m} (1 + m\mathbb{Z}_p)^\times \times \prod_{p \not| m} \mathbb{Z}_p^\times \subset J,$$

so that $G_m \cong J/U_m\mathbb{Q}^\times$, [La73/87]. This formulation of the result is very easy to extend to the general case of the group G_k^{ab} for arbitrary global fields k. Note that the set of all primitive Dirichlet characters can be identified with the discrete group of all characters of finite order of the idele class group C_k ($k = \mathbb{Q}$) using the projection

$$J/\mathbb{Q}^\times \cong \mathbb{R}_+ \times \prod_p \mathbb{Z}_p^\times \to G_m.$$

Such characters are all trivial on the connected component of the identity. Abelian extensions of $\mathbb{Q}$ correspond bijectively to open subgroups of $J/\mathbb{R}_+^\times\mathbb{Q}^\times$, and any such group is the intersection of the kernels of a finite number of Dirichlet characters.

4.4.2 Frobenius Automorphisms of Number Fields and Artin's Reciprocity Map

Let K be an algebraic number field, $[K : \mathbb{Q}] = n$, Σ_K^0 the set of all finite places of K (normalized discrete valuations which correspond to prime ideals $\mathfrak{p}_v \neq 0$ in the ring of integers $\mathcal{O}_K$ of K);

$$\mathfrak{p}_v = \{x \in \mathcal{O}_K \mid |x|_v < 1\}.$$

The residue field $k(v) = \mathcal{O}_K/\mathfrak{p}_v$ is finite, having $\mathrm{N}v = p_v^{\deg v}$ elements, where $p_v = \mathrm{Char}\ k(v)$ is the characteristic and $\deg v = f_v$ is the degree of the extension (or *inertial degree*) of $k(v)$ over $\mathbb{F}_{p_v}$. The absolute value is normalized by the condition

$$v(x) = -\log_{\mathrm{N}v} |x|_v \quad (|x|_v = \mathrm{N}v^{-v(x)}). \tag{4.4.4}$$

The *ramification index* e_v of v is the number $v(p_v)$. With this notation one has the following decomposition $p\mathcal{O}_K = \prod_{v, v(p)>0} \mathfrak{p}_v^{e_v}$.

Let L/K be a finite Galois extension with Galois group $G(L/K)$, and let w be a place of L, which extends a fixed place v of K. Define the action of the group $G(L/K)$ on the places $w \in \Sigma_L$ by $w \mapsto \sigma w$,

$$|x|_{\sigma w} = |x^{\sigma^{-1}}|_w.$$

If v and w are non–Archimedean, $\mathfrak{p}_v$ and $\mathfrak{P}_w$ being the corresponding prime ideals, then σw corresponds to the ideal $\mathfrak{P}_{\sigma w} = \mathfrak{P}_w^\sigma$. A Galois automorphism $\sigma \in G(L/K)$ induces an isomorphism of the completions $\sigma : L_w \xrightarrow{\sim} L_{\sigma w}$ as normed vector spaces over K_v.

The *decomposition group* G_w is introduced as the subgroup

$$G_w = \{\sigma \in G(L/K) \mid \sigma w = w\} \subset G(L/K). \tag{4.4.5}$$

By definition we have that

$$G_{\tau w} = \{\sigma \in G(L/K) \mid \sigma \tau w = \tau w\} = \tau G_w \tau^{-1}.$$

On the other hand, it is immediate from the explicit construction of the extensions of places, that $G(L/K)$ acts transitively on the set of places of L lying over a fixed place v of K. Hence all the corresponding subgroups G_w are conjugate [Wei74a].

The *inertia group* $I_w \subset G_w$ is by definition the kernel of the natural homomorphism $G_w = G(L_w/K_v) \to G(l(w)/k(v))$ where $l(w)$ denotes the residue field of the place w. The quotient group $G_w/I_w \cong G(l(w)/k(v))$ is generated by the Frobenius automorphism: $G(l(w)/k(v)) = \langle \mathrm{Fr}_w \rangle$, $\mathrm{Fr}_w(x) = x^{\mathrm{N}v}$. The place w is called unramified iff $I_w = \{1\}$; in this case one has $G_w = \langle \mathrm{Fr}_w \rangle$. It follows from the definitions that $\mathrm{Fr}_{\tau w} = \tau^{-1}\mathrm{Fr}_w \tau$, so that the conjugacy class of Fr_w in $G(L/K)$, if defined, can depend only on v. It turns out that all but a finite number of places are unramified; for such places we put

$$F_{L/K}(v) = \text{ (the conjugacy class of } \mathrm{Fr}_w \text{ for } w|v). \tag{4.4.6}$$

If $G(L/K)$ is commutative, then the right hand side of (4.4.6) consists of one element.

The *Artin reciprocity law* tells us where the Frobenius elements $F_{L/K}$ are situated in a *commutative* Galois group $G(L/K)$. Let S be a finite set of places of K, including all Archimedean places and those places ramified in the extension L/K. Denote by I^S the free Abelian (multiplicative) group generated by the elements $\mathfrak{p}_v$ for $v \notin S$. Then the association $v \mapsto F_{L/K}(v) \in G(L/K)$ extends to a homomorphism

$$F_{L/K} : I^S \longrightarrow G(L/K), \tag{4.4.7}$$

which is called *Artin's reciprocity map*,

$$F_{L/K}\left(\prod_{v\notin S}\mathfrak{p}_v^{n_v}\right)=\prod_{v\notin S}F_{L/K}(v)^{n_v}. \tag{4.4.8}$$

Class field theory gives an explicit description of the kernel of the Artin reciprocity map (4.4.7) (see section §4.4.5 below). The statement that (4.4.7) is surjective was established first, and it could be deduced from the general *Chebotarev density theorem*, which is a far–reaching generalization of Dirichlet's theorem on primes in arithmetical progressions (cf. [Chebo25], [Se70], [Se68a], [Chev51]).

Let P be a subset of the set Σ_K^0 of all non–Archimedean places of K. For any integer $x \geq 1$ denote by $a_x(P)$ the number of places $v \in P$ such that $\mathrm{N}v \leq x$. We say that P has density $a \geq 0$ if the limit exists

$$\lim_{x\to\infty}\frac{a_x(P)}{a_x(\Sigma_K^0)}=a. \tag{4.4.9}$$

Not every set of places has a density. For example, if $K=\mathbb{Q}$ and P is the set of primes whose first digit is equal to 1, then P does not have a density.

By the prime number theorem one has $a_x(\Sigma_K^0)\sim x/\log x$, hence the condition (4.4.9) is equivalent to the following asymptotic expression

$$a_x(P)=a\frac{x}{\log x}+o\left(\frac{x}{\log x}\right). \tag{4.4.10}$$

4.4.3 The Chebotarev Density Theorem

Theorem 4.22. *Let L/K be a finite extension of a number field K, and X a subset of $G(L/K)$, invariant under conjugation. Denote by P_X the set of places $v \in \Sigma_K^0$ unramified in L such that the classes of Frobenius elements of these places belong to X: $F_{L/K}(P_X)\subset X$. Then the set P_X has a density, which is equal to* $\mathrm{Card}\,X/\mathrm{Card}\,G(L/K)$.

The proof is based on analytic methods; the notion of the analytic density of P is introduced as the limit

$$\lim_{s\to 1+}\frac{\sum_{v\in P}\mathrm{N}v^{-s}}{\log\left(\frac{1}{s-1}\right)}. \tag{4.4.11}$$

Proving the existence of and calculating this limit for $P=P_X$ can be done with the help of the *Artin L–functions* (see §6.2.2); the density statement in the above sense (4.4.9) can then be deduced (cf. [Chev40], [La70]).

4.4.4 The Decomposition Law and the Artin Reciprocity Map

If L/K is an Abelian extension, then the decomposition of $\mathfrak{p}_v$ in $\mathcal{O}_L$ is completely determined by the order f of the element $F_{L/K}(v)\in G(L/K)$: in this case $\mathfrak{p}_v = \mathfrak{P}_{w_1} \cdot \dots \cdot \mathfrak{P}_{w_s}$, where $s = (G(L/K) : \langle\sigma\rangle)$ and $f = f(w_i/v) = \deg w_i/\deg v = [l(w_i) : k(v)]$ is the *relative residue field degree*. This fact is deduced from the transitivity of the action of the Galois group $G(L/K)$ on the set of places w dividing v. In particular, the place v *splits completely* (i.e. $f = 1$ and v is unramified) iff $F_{L/K}(v) = 1 \in G(L/K)$.

Theorem 4.22 shows that a finite Galois extension $G(L/K)$ is uniquely determined (in a fixed algebraic closure $\overline{K}$) by the set $\mathrm{Spl}_{L/K}$ of places which split completely in L/K. The *Artin reciprocity law* gives us amongst other things a description of this set when L/K is Abelian. For non–Abelian extensions there are only some special cases when $\mathrm{Spl}_{L/K}$ is known. However these examples provide a basis for quite general conjectures (the *Langlands program*, see §6.5. These conjectures determine nowadays one of the main directions in modern algebraic number theory.

4.4.5 The Kernel of the Reciprocity Map

In order to formulate the main result on the kernel of the reciprocity map (4.4.7) we recall that the relative norm $\mathrm{N}_{L/K}(w)$ of a non–Archimedean place w is defined as $\mathfrak{p}_v^{f(w/v)}$ (or, in additive terms, as $f(w/v)\cdot v$), where

$$f(w/v) = \deg w/\deg v = [l(w) : k(v)] = \log_{\mathrm{N}v} \mathrm{N}w$$

is the *relative degree of residue fields.* Also, consider the *divisor map* (see (4.3.49))

$$\mathrm{div}_S : K^\times \to I^S, \quad \mathrm{div}_S(a) = \prod_{v\notin S} \mathfrak{p}_v^{v(a)} \in I^S,$$

where S is the set of all Archimedean places and places ramified in L/K.

Let L/K be an Abelian extension of K, $\mathfrak{f} = \prod_v \mathfrak{p}_v^{r(v)}$ an ideal in $\mathcal{O}_K$, divisible by sufficiently high powers $r(v)$ of the prime ideals ramified in L. For each Archimedean place $v \in \Sigma_K^\infty$ we fix an embedding

$$a \mapsto a^{(v)} \in \mathbb{C}, \quad K \hookrightarrow K_v \subset \mathbb{C},$$

which induces v, and let

$$\Sigma_{L/K}^\infty = \{v \in \Sigma_K^\infty \mid K_c \cong \mathbb{R}, L_w \cong \mathbb{C} \text{ for } w|v\}.$$

Define the subgroups $P_{L/K}(\mathfrak{f}), \mathfrak{N}_{L/K}(\mathfrak{f}) \subset I^S$ by

$$P_{L/K}(\mathfrak{f}) = \left\{\mathrm{div}_S(a) \mid a \in K^\times,\ a \equiv 1 \bmod \mathfrak{f},\ \forall v \in \Sigma_{L/K}^\infty\ a^{(v)} > 0\right\}, \tag{4.4.12}$$

$$\mathfrak{N}_{L/K}(\mathfrak{f}) = \langle \mathrm{N}_{L/K}(w)\rangle_{w\notin S}, \tag{4.4.13}$$

the latter being the subgroup generated by the relative norms of prime divisors of those places v (or ideals $\mathfrak{p}_v$) which are unramified in L/K.

Theorem 4.23 (The Artin Reciprocity Law). *Let L/K be an Abelian extension. Then*

$$\mathrm{Ker}F_{L/K} = P_{L/K}(\mathfrak{f}) \cdot \mathfrak{N}_{L/K}(\mathfrak{f}). \tag{4.4.14}$$

Corollary 4.24 (Description of the Galois group). *For an Abelian extension L/K the reciprocity map (4.4.7) induces an isomorphism*

$$G(L/K) \cong I^S/(P_{L/K}(\mathfrak{f})\mathfrak{N}_{L/K}(\mathfrak{f})).$$

4.4.6 The Artin Symbol

Consider the group J_K of ideles, and define a surjective homomorphism

$$(\cdot, L/K) : J_K \to G(L/K), \quad s \mapsto (s, L/K) \tag{4.4.15}$$

with the help of the reciprocity map (4.4.7). For an arbitrary $s \in J_K$ let us choose a principal idele $\alpha \in K^\times$ such that $|\alpha s_v - 1|_v < \varepsilon$ for $v \in S$ and sufficiently small $\varepsilon > 0$. Define the S–divisor (cf. (4.3.49)) by

$$\mathrm{div}(\alpha s) = \prod_v \mathfrak{p}_v^{v(\alpha s_v)} \in I^S.$$

Then the *Artin symbol* $(s, L/K) = \psi_{L/K}(s)$ is defined by the formula

$$(s, L/K) = \psi_{L/K}(s) \overset{\mathrm{def}}{=} F_{L/K}(\mathrm{div}(\alpha s)). \tag{4.4.16}$$

We stress that (4.4.16) is defined in terms of ideles, and in order to show that (4.4.16) is well defined it is essential that the reciprocity law in terms of ideals (4.4.14) is satisfied. Indeed, the condition on α in (4.4.16) is satisfied if $\mathrm{div}(\alpha) \in P_{L/K}(\mathfrak{f})$ with an appropriate choice of $\mathfrak{f}$. Now the reciprocity law transforms into the statement that $\mathrm{Ker}\psi_{K/L}$ coincides with $K^\times \mathrm{N}_{L/K} J_L$, where $\mathrm{N}_{L/K} J_L$ is the subgroup of relative norms of ideles from J_L:

$$\mathrm{N}_{L/K}((\beta_w)_w) = \left(\prod_{w|v} \mathrm{N}_{L_w/K_v}(\beta_w) \right)_v . \tag{4.4.17}$$

Hence the Artin symbol $\psi_{L/K}$ in (4.4.16) is defined for *idele classes* $s \in C_K = J_K/K^\times$. Furthermore $F_{L/K}(v) = \psi(s(v))$, where $s(v)$ is the idele class of $(\cdots, 1, \pi_v, 1, \cdots)$ for a local uniformizer $\pi_v \in K_v^\times$, i.e. an element with the condition $v(\pi_v) = 1$. The homomorphism $\psi_{L/K} : C_K \to G(L/K)$ is continuous and its kernel is both open and closed, again in view of (4.4.14).

4.4.7 Global Properties of the Artin Symbol

Let H be a subgroup of a finite group G. Then the *transfer homomorphism* (or *Verlagerung*)

$$\mathrm{Ver} : G/[G,G] \longrightarrow H/[H,H], \tag{4.4.18}$$

is defined by $\mathrm{Ver}(g[G,G]) = \prod_{r\in R} h(g,r)$ where r runs through a system of representatives R of left cosets G/H and $h(g,r) \in H$ is defined by the condition $gr = \widetilde{gr}h(g,r)$ ($\widetilde{gr} \in R$ being the representative of gr in R).

1) There is a one–to–one correspondence between open subgroups $U \subset C_K$ and finite Abelian extensions L/K, such that the symbol (4.4.16) induces an isomorphism
$$C_K/U \stackrel{\sim}{\to} G(L/K),$$
and U coincides with the norm subgroup $U = \mathrm{N}_{L/K}(C_L)$ (see (4.4.17)).
2) Let K'/K be an arbitrary finite extension. Then for $\alpha \in C_{K'}$ the following equation holds
$$(\mathrm{N}_{K'/K}(\alpha), L/K) = \overline{(\alpha, LK'/K')}. \tag{4.4.19}$$
3) Let L'/K be a finite Galois extension, L/K the maximal Abelian subextension of L'/K, and K' a subextension of L'/K over which L' is Abelian. Then
$$(\alpha, L'/K') = \mathrm{Ver}(\alpha, L/K), \tag{4.4.20}$$
where Ver is the transfer (4.4.18).
4) Let L'/K be a finite Galois subextension of L/K. Then for all $\alpha \in C_K$ the following equation holds
$$(\alpha, L'/K) = \overline{(\alpha, L/K)}. \tag{4.4.21}$$
5) Let σ be an isomorphism of K onto σK, $\sigma \in \mathrm{Aut}\,\overline{K}$. Then for all $\alpha \in C_K$ the equation holds
$$(\sigma\alpha, \sigma L/\sigma K) = \sigma(\alpha, L/K)\sigma^{-1}. \tag{4.4.22}$$

The bar in the above formulae denotes the restriction to a subfield (cf. [Chev40], [Koch70], [AT51], [Wei74a]).

These properties make it possible to extend the definition of the Artin symbol to infinite Abelian extensions L/K. Consider the correspondence

$$s \mapsto (s, L/K) = \varprojlim_{\nu}(s, L_\nu/K), \tag{4.4.23}$$

where L_ν/K runs through all finite subextensions of L/K. It follows from 4) that this is well defined and one has a map from C_K to $G(L/K)$ with dense image. Taking into account the one–to–one correspondence between subgroups of finite index of C_K and $G(K^{\rm ab}/K)$, we see that $G(K^{\rm ab}/K)$ is isomorphic to the profinite completion of C_K, which is in fact with the quotient of C_K by its connected component. The thus constructed reciprocity homomorphism satisfies the properties 2), 4) and 5).

Note that a new approach to the class field theory was developped by J.Neukirch in Chapters 4–6 of [Neuk99].

From a profinite group G endowed with a surjective continuous homomorphism $d : G \to \widehat{\mathbb{Z}}$ and a G-module A endowed with a "Henselian valuation with respect to d", one constructs in an elementary way the *reciprocity homomorphisms*; if A satisfies the so-called *class field axiom* (a statement involving zeroth and (-1)st cohomology of A), then the reciprocity homomorphisms are isomorphisms. From this abstract class field theory, both local and global class field theory are deduced, and the classical formulation of global class field theory, using ray class groups instead of the idèle class group, is presented as well. Neukirch's approach minimizes the cohomological tools needed to construct class field theory.

4.4.8 A Link Between the Artin Symbol and Local Symbols

Suppose that we already know the existence of the Artin symbol on ideles (4.4.16). For a finite Abelian extension L/K, a non–Archimedean place v of K and an extension w of v to L, consider the completions K_v and L_w, and the decomposition group

$$G_v \subset G(L/K), \quad G_v \cong G(L_w/K_v),$$

which in the Abelian case does not depend on the choice of w. Consider the embedding $i_v : K_v^\times \hookrightarrow J_K$, and the projection onto the v–component $J_v : J_K \to K_v^\times$, where i_v maps $x \in K_v^\times$ onto the element of J_K, whose v–component is equal to x, and whose other components are all 1. Put

$$\psi_v = \psi_{L/K} \circ i_v = (\cdot, L_w/K_v)_v. \tag{4.4.24}$$

Then one verifies that the image of ψ_v belongs to the decomposition group G_v. The homomorphism $\psi_v : K_v^\times \to G_v$ is called the *local Artin homomorphism* (or the *norm residue homomorphism*). If $x = (x_v) \in J_K$, then the following decomposition holds

$$\psi_{L/K}(x) = \prod_v \psi_v(x_v), \tag{4.4.25}$$

where

$$x = \lim_S \left(\prod_{v \in S} i_v(x_v) \right)$$

(the limit is taken over an increasing family of places of K). The product (4.4.25) is actually finite: if a component x_v is a v–unit and v is unramified, then x_v is a norm in the extension L_w/K_v: for some $y_w \in L_w$ one has $x_v = \mathrm{N}_{L_w/K_v} y_w$. The existence of y_w is established by Hensel's lemma (see 4.3.2).

Thus the knowledge of all *local Artin maps* ψ_v is equivalent to the knowledge of the *global Artin map* $\psi_{L/K}$. In classical work on class field theory the local reciprocity maps were studied using the global theory; in particular, it was shown that these local maps depend only on the local extensions L_w/K_v, and are independent on a global extension L/K from which they are obtained. In this sense, modern expositions of class field theory (for example, in [Chev40] , [Wei74b]) differ from classical texts: first one gives a purely local and independent construction of maps

$$\theta_v : K^\times \to G_v = G(L^v/K_v), \tag{4.4.26}$$

where L^v is a finite extension of K_v. Then one proves that the product $\prod_v \theta_v$ has the properties which uniquely characterize the homomorphism $\psi_{L/K}$. The most important part of the proof consists of verifying the product formula

$$\theta_v(a) = 1 \text{ for all } a \in K^\times. \tag{4.4.27}$$

In the case of a quadratic extension $L = K(\sqrt{b})$ the image $\theta_v(a)$ belongs to $\{\pm 1\} = G(L/K)$, and coincides with the Hilbert symbol, defined in §4.3.3. The product formula is equivalent to the quadratic reciprocity law of Gauss, which thus becomes a special case of the general reciprocity law (4.4.14).

The construction of the map (4.4.26) for an arbitrary Abelian extension L^v/K_v is usually carried out using methods of *Galois cohomology theory* (see §4.5, and [Se63], [Se64], [Chev40], [Koch70] [Koch97]). A more direct construction of θ_v was suggested by [Haz78] , [Iw86], based on an explicit analysis of cohomological constructions in low dimensions, compare with the approach of J.Neukirch cf. [Neuk99]).

4.4.9 Properties of the Local Symbol

The properties of the local symbol

$$\theta_v = \psi_v = (\cdot, L_w/K_v) : K_v^\times \to G_v$$

are completely analogous to the corresponding properties 1) to 5) from §4.4.7, replacing $C_K = J_K/K^\times$ by $K_v^\times$, $G(L/K)$ by G_v and $\psi_{L/K}$ by θ_v. Also, the homomorphism θ_v maps the group of units $U_v = \mathcal{O}_v^\times$ of K_v onto the inertia group $I^w \subset G_v$. If L_w/K_v is unramified, then for all $\alpha \in K_v^\times$ one has

$$\theta_v(\alpha) = \mathrm{Fr}_v^{v(\alpha)}$$

where $\mathrm{Fr}_v \in G_v$ is the Frobenius element of the extension, and the valuation v of K_v is normalized by the condition $v(K_v^\times) = \mathbb{Z}$. In the same way as for

$\psi_{L/K}$ the local symbol can be generalized for infinite Abelian extensions, and thus one obtains a reciprocity map $\theta_v : K_v^\times \to G(K_v^{\rm ab}/K_v)$, where $K_v^{\rm ab}$ is the maximal Abelian extension of K_v. The Galois group can then be described as follows:

$$G(K_v^{\rm ab}/K_v) = (K_v^\times)^\wedge \cong \hat{\mathbb{Z}} \times \mathcal{O}_v^\times \tag{4.4.28}$$

where $^\wedge$ denotes the profinite completion. Under the isomorphism (4.4.28) the Galois group $G(K_v^{\rm nr}/K_v) = G(\overline{\mathbb{F}}_q/\mathbb{F}_q)$ of the maximal unramified extension $K_v^{\rm nr}$ of K_v becomes $\hat{\mathbb{Z}}$, and the inertia group $I_w = I^v$ maps isomorphically onto the whole group of units $\mathcal{O}_v^\times$:

$$\theta_v : \mathcal{O}_v^\times \xrightarrow{\sim} I^v \tag{4.4.29}$$

(the field $K_v^{\rm nr}$ can be defined as the maximal extension of K, for which the extension of the valuation v satisfies the property $v(K_v^\times) = \mathbb{Z}$).

Below we give a remarkable explicit construction of the maximal Abelian extension $K_v^{\rm ab}$ of a local non–Archimedean field K_v, generalizing the contruction of $\mathbb{Q}_p^{\rm ab}$ by adjoining roots of unity to $\mathbb{Q}_p$.

4.4.10 An Explicit Construction of Abelian Extensions of a Local Field, and a Calculation of the Local Symbol

(cf. [LT65], [Se63], [Chev40], [Ha50], [Sha50], [CW77], [Koly79]). Consider first the field $\mathbb{Q}_p$ as a model example. Any Abelian extension of this field is contained in a cyclotomic extension, i.e. $\mathbb{Q}_p^{\rm ab} = \mathbb{Q}_p(W_\infty)$, where $W_\infty = \cup_{n>1} W_n$, $W_n = \{\zeta \in \overline{\mathbb{Q}}_p \mid \zeta^n = 1\}$, W_∞ is the set of all roots of unity from $\overline{\mathbb{Q}}_p$. Let $W_{p^\infty} = \cup_{m\geq 0} W_{p^m}$ be the subset of all roots of unity of p–power order, and $V_\infty = \cup_{p\nmid n} W_n$ the subset of roots of unity of order not divisible by p. Then

$$W_\infty = V_\infty \times W_{p^m}, \quad \mathbb{Q}_p(W_\infty) = \mathbb{Q}_p(V_\infty)\cdot \mathbb{Q}_p(W_{p^\infty}),$$

and the following decomposition takes place:

$$G(\mathbb{Q}_p^{\rm ab}/\mathbb{Q}_p) \cong G(\mathbb{Q}_p(V_\infty)/\mathbb{Q}_p) \times G(\mathbb{Q}_p(W_{p^\infty})/\mathbb{Q}_p). \tag{4.4.30}$$

Here $\mathbb{Q}_p(V_\infty)$ is the maximal unramified extension (see the example from 4.3.4), for which $v_p(\mathbb{Q}_p(V_\infty)^\times) = \mathbb{Z}$ and

$$G(\mathbb{Q}_p(V_\infty)/\mathbb{Q}_p) \cong G(\overline{\mathbb{F}}_p/\mathbb{F}_p) \cong \hat{\mathbb{Z}} = \langle {\rm Fr}_p\rangle^\wedge. \tag{4.4.31}$$

The field generated by $W_{p^\infty} = \cup_{m\geq 0} W_{p^m}$ is the union of all the totally ramified extensions of $\mathbb{Q}_p$. The *Galois group* $G(\mathbb{Q}_p(W_{p^\infty})/\mathbb{Q}_p)$ can be described by means of its action on the set W_{p^∞} of all roots of unity of p–power order. In order to do this we note the isomorphisms

$$\operatorname{End} W_{p^\infty} \cong \mathbb{Z}_p, \quad \operatorname{Aut} W_{p^\infty} \cong \mathbb{Z}_p^\times,$$

in which a p–adic number $\alpha = a_0 + a_1 p + a_2 p^2 + \cdots \in \mathbb{Z}_p$ in its digital form (4.3.2) corresponds to the endomorphism $[\alpha] : \zeta \mapsto \zeta^\alpha$ for $\zeta \in W_{p^\infty}$:

$$\zeta^\alpha \stackrel{\text{def}}{=} \zeta^{a_0 + a_1 p + a_2 p^2 + \cdots + a_{m-1} p^{m-1}} \text{ if } \zeta \in W_{p^m} \subset W_{p^\infty}.$$

From the action of the Galois group on W_{p^∞} one obtains a homomorphism

$$\delta_p : G(\mathbb{Q}_p(W_{p^\infty})/\mathbb{Q}_p) \xrightarrow{\sim} \operatorname{Aut} W_{p^\infty} \cong \mathbb{Z}_p^\times, \tag{4.4.32}$$

which is a one–dimensional p–adic Galois representation (the cyclotomic representation), and (4.4.32) is an isomorphism.

It turns out that the local symbol $\theta_p(\alpha) = (\alpha, \mathbb{Q}_p^{\mathrm{ab}}/\mathbb{Q}_p) \in G(\mathbb{Q}_p^{\mathrm{ab}}/\mathbb{Q}_p)$ for an element $\alpha = p^m u$ ($m \in \mathbb{Z}$, $u \in \mathbb{Z}_p^\times$) can be described using isomorphisms (4.4.31), (4.4.32):

$$\theta_p(\alpha) = \begin{cases} \mathrm{Fr}_p^m & \text{on the subfield } \mathbb{Q}_p(V_\infty), \\ [u^{-1}] & \text{on the subfield } \mathbb{Q}_p(W_{p^\infty}). \end{cases}$$

We now reformulate this in a manner more suitable for generalization. Consider the sets $E_{p^\infty} = \cup_{m \geq 0} E_{p^m}$, where

$$E_{p^m} = \{w = \zeta - 1 \mid \zeta \in W_{p^\infty}\}.$$

These sets are groups with respect to the group law

$$w_1 \circ w_2 := w_1 + w_2 + w_1 w_2 \quad (w_1, w_2 \in E_{p^\infty}),$$

and for all $w \in E_{p^\infty}$ one has $|w|_p < 1$. The set E_p consists of all roots of the polynomial

$$f_p(X) = (X+1)^p - 1 = pX + \binom{p}{2} X^2 + \cdots + X^p,$$

which becomes irreducible after division by X according to Eisenstein's irreducibility criterion. Its roots therefore generate a field $\mathbb{Q}_p(E_p)$ of degree $p-1$ over $\mathbb{Q}_p$.

Now consider the iterations of the polynomial $f_p(X)$:

$$f_{p^2}(X) = f_p(f_p(X)) = ((X+1)^p - 1)^p - 1,$$
$$\cdots \quad \cdots \quad \cdots$$
$$f_{p^m}(X) = f_{p^{m-1}}(f_p(X)).$$

The group E_{p^m} coincides with the set of all roots of the polynomial $f_{p^m}(X)$, and this is isomorphic to $p^{-m}\mathbb{Z}_p/\mathbb{Z}_p$. Under this isomorphism the obvious inclusions $E_{p^m} \subset E_{p^{m+1}}$ become the natural embeddings $p^{-m}\mathbb{Z}_p/\mathbb{Z}_p \subset p^{-m-1}\mathbb{Z}_p/\mathbb{Z}_p$, and we see that $E_{p^\infty} \cong \mathbb{Q}_p/\mathbb{Z}_p$. From this it follows that

End $E_\infty \cong \mathbb{Z}_p$. We have $\mathbb{Q}_p(E_\infty) = \mathbb{Q}_p(W_{p^\infty})$, and the isomorphism (4.4.32) takes the form

$$\delta_p : G(\mathbb{Q}_p(E_{p^\infty})/\mathbb{Q}_p) \xrightarrow{\sim} \text{Aut } E_{p^\infty} \cong \mathbb{Z}_p^\times. \tag{4.4.33}$$

Now let K_v be an arbitrary finite expension of $\mathbb{Q}_p$ with valuation ring $\mathcal{O}_v$, maximal ideal $\mathfrak{p}_v = (\pi)$ and $q = |\mathcal{O}_v/\mathfrak{p}_v|$. Here π is a uniformizing element, i.e. $v(\pi) = 1$. There is an analogous construction of the maximal Abelian extension of K_v. Consider the polynomial

$$f_\pi(X) = \pi X + X^q. \tag{4.4.34}$$

It follows as before by Eisenstein's criterium that $f_\pi(X)/X$ is irreducible. Define recursively the iterations

$$f_{\pi^m}(X) = f_{\pi^{m-1}}(f_\pi(X)), \quad m \geq 1.$$

Then the sets of roots

$$W_{f,m} = \{x \in \overline{K}_v \mid f_{\pi^m}(x) = 0\} \tag{4.4.35}$$

of the polynomials $f_{\pi^m}(X)$ form an increasing sequence:

$$W_{f,m} \subset W_{f,m+1},$$

and there is a natural group structure on (4.4.35) such that $W_{f,m}$ is isomorphic to $\mathfrak{p}_v^{-m}/\mathcal{O}_v$ ($\cong \mathcal{O}_v/\mathfrak{p}_v^m$). The inclusions $W_{f,m} \subset W_{f,m+1}$ become the natural embeddings $\mathfrak{p}_v^{-m}/\mathcal{O}_v \subset \mathfrak{p}_v^{-m-1}/\mathcal{O}_v$. Thus we obtain a group, which is analogous to the group of all roots of unity of p–power order:

$$W_{f,\infty} = \bigcup_{m\geq 1} W_{f,m} \text{ is isomorphic to } K_v/\mathcal{O}_v. \tag{4.4.36}$$

There is a natural action of elements $a \in \mathcal{O}_v = \text{End}(K_v/\mathcal{O}_v)$ on $W_{f,m}$ for which the equation $[\pi]_f(x) = f_\pi(x)$ holds. This action will be denoted by $[a]_f : x \mapsto [a]_f x$. The action of the Galois group on the roots of the polynomials $f_{\pi^m}(X)$ provides us with a representation analogous to (4.4.32):

$$G(\overline{K}_v/K_v) \to \text{Aut } W_{f,\infty} \cong \mathcal{O}_v^\times. \tag{4.4.37}$$

Denote by K_π the field which corresponds to the kernel of the homomorphism (4.4.37). Then K_π is an Abelian extension of K_v in view of the isomorphism

$$\delta_v : G(K_\pi/K_v) \xrightarrow{\sim} \mathcal{O}_v^\times, \tag{4.4.38}$$

and we obtain the following explicit description of the Abelian extensions of K_v:

$$K_v^{\text{ab}} = K_v^{\text{nr}} \cdot K_\pi,$$

where $K_v^{\mathrm{nr}} = K_v(V_\infty)$ is the maximal unramified extension of K_v (V_∞ is the group of all roots of unity of degree not divisible by p),

$$G(K_v^{\mathrm{nr}}/K_v) \cong G(\overline{\mathbb{F}}_q/\mathbb{F}_q) \cong \hat{\mathbb{Z}} = \langle \mathrm{Fr}_q \rangle^\wedge. \qquad (4.4.39)$$

The field $K_\pi = \bigcup_{m \geq 1} K_v(W_{f,m})$ is the union of all totally ramified Abelian extensions of K_v.

The norm residue symbols can then be described as follows:

1) for $u \in \mathcal{O}_v^\times$ the element $\theta_v(u) = (u, K_\pi/K_v)_v$ acts on $W_{f,\infty}$ via $[u^{-1}]_f$;
2) the norm residue symbol of $(\pi, K_\pi/K_v)_v$ is equal to 1.
3) the symbol $\theta_v(\alpha)$ for $\alpha = \pi^m$ ($m \in \mathbb{Z}$, $u \in \mathcal{O}_v^\times$) acts on K_v^{nr} as $\mathrm{Fr}_q^m \in G(K_v^{\mathrm{nr}}/K_v)$.

A remarkable feature of the construction of the group law on the set $W_{f,\infty}$ is that the field K_π is independent of the choice of uniformizer π and of the polynomial $f(X) \in \mathcal{O}_v[X]$, which need only satisfy the following requirements:

$$f(X) \equiv \pi X \text{ (modulo degree 2 polynomials)}, \qquad (4.4.40)$$

$$f(X) \equiv X^q (\mathrm{mod}\ \pi). \qquad (4.4.41)$$

Moreover, instead of a polynomial $f(X)$ one may use any element of the set $\mathcal{F}_\pi$ of power series $f(X) \in \mathcal{O}_v[[X]]$ satisfying the above conditions (4.4.40), (4.4.41).

The above group law is constructed in the theory of *Lubin–Tate formal groups.*

4.4.11 Abelian Extensions of Number Fields

For the field of rational numbers $\mathbb{Q}$ the theorem of Kronecker–Weber (see §4.1.2) gives an explicit description of all Abelian extensions with the help of the action of the Galois group on roots of unity, which may be regarded as certain special values of the exponential function: $\zeta_m = \exp(2\pi i/m)$. An analogous theory exists also over an imaginary quadratic field $K = \mathbb{Q}(\sqrt{d})$, whose Abelian extensions are constructed with the help of the action of the Galois group $G(\overline{K}/K)$ on the points of finite order of an elliptic curve with complex multiplication (more precisely, on the coordinates of these points, see §5.4 of Chapter 5). This description is essentially the content of the *theory of complex multiplication.* In more classical terms, Abelian extensions of an imaginary quadratic field are described by means of the special values of elliptic functions and the j–invariants corresponding to lattices with complex multiplication. The Galois action on these values is explicitly described in terms of the arithmetic of the imaginary quadratic ground field (this was Kronecker's "Jugendtraum" ("dream of youth"), cf. a nice book by S.Vladut,

[Vla91]) The content of Hilbert's famous twelfth problem is to give an explicit description of all Abelian extensions of an arbitrary number field K, $[K:\mathbb{Q}] < \infty$ using special values of certain special functions (such as the exponential function or elliptic functions), and by means of the Galois actions on these values.

Some progress has been made in solving this problem for the so–called CM–fields K. These are totally imaginary quadratic extensions $K = F(\sqrt{-a})$ of totally real fields F: F is a number field generated by a root of a polyomial which splits as a product of linear factors over $\mathbb{R}$, and $a \in F$ is totally positive (positive in each real embedding of F). This multi–dimensional complex multiplication theory is based on the study of Abelian varieties with complex multiplication by elements of K. For a real quadratic field K a description of certain Abelian extensions of K is given by Shimura's theory of "real multiplication". However, in these cases the situation is less satisfactory than for $\mathbb{Q}$ or for an imaginary quadratic field K, since these constructions do not give all Abelian extensions of the ground field K. A completely different situation takes place in the function field case, when K is a finite, separable extension of $\mathbb{F}_q(T)$. Here there is a complete description of all Abelian extensions of K in terms of the *elliptic modules* of V.G.Drinfel'd (and in terms of elliptic functions in positive characteristic attached to these modules, [Dr]). This result gives an illustrative example of analogy between numbers and functions.

The idea of describing extensions of K via the action the Galois group $G(\overline{K}/K)$ on certain groups and other algebraic objects has turned out to be very fruitful. Many examples of constructions of Abelian and non–Abelian extensions of a ground field K are based on this idea. A complete classification of all these extensions in terms of Galois representations and in terms of certain objects of analysis and algebraic geometry (automorphic forms and motives) is an important aim in Langlands far-reaching program, see §6.5.

In a new book [Yos03] the main object are special values given by the exponential of the derivative at $s = 0$ of the partial zeta function of a certain ideal-class c attached to a number field F. Such a special value is an important invariant which conjecturally gives a unit of an abelian extension of a number field and should give an answer to *Hilbert's 12-th problem.*

Let F be a totally real field, i.e. $F \otimes \mathbb{R} \cong \mathbb{R}^n$ as an $\mathbb{R}$-algebra. According to Shintani, the special values at non–positive integers of the partial zeta function

$$\zeta_F(\mathfrak{a}, \mathfrak{f}, s) = \sum_{\substack{I \in \mathfrak{a} \\ I \subset \mathcal{O}_F}} N(I)^{-s}$$

are rational numbers which can be expressed in terms of certain generating functions which generalize the generating function of Bernoulli numbers, see [Shin76].

These generating functions are associated to "cone decompositions" in $F \otimes \mathbb{R}$ which are defined non–canonicaly (see also [Hi93], Chapter I).

Conjectural absolute periods are described in [Yos03] in terms of a function constructed geometrically from cone decomposions, whose principial term is given by periodes of abelian varieties of CM-type for an arbitrary CM-field, explicifying a conjecture of Colmez, [Colm93].

Our knowledge of the nature of CM-periods is quite limited, the only essential fact in the case $F = \mathbb{Q}$ is the classical *Chowla – Selberg formula,*

$$\pi p_K(id, id)^2 \sim \prod_{a=1}^{d-1} \Gamma\left(\frac{a}{d}\right)^{w\chi(a)/2h} = d \cdot \exp\left(\frac{L'(0,\chi)}{L(0,\chi)}\right),$$

where K is an imaginary quadratic field of discriminant $-d$, w is the number of roots of unity in K, h the class number of K, and χ the Dirichlet character corresponding to K [ChSe68]. Also, using the jacobian of the Fermat curve, G.W.Anderson found that the CM-periods in the case of a cyclotomic field are linked with the logarithmic derivative of Dirichlet's L-functions at $s = 0$:

$$p_K(id, \sigma) \sim \pi^{-\mu(\sigma)/2} \prod_{\eta \in \hat{G}_-} \exp\left(\frac{\eta(\sigma)}{[K:\mathbb{Q}]} \frac{L'(0,\eta)}{L(0,\eta)}\right).$$

where η is a Dirichlet character, $\mu(c) = \pm 1$ or 0 (see [Ande82]).

The conjectures of Harold Stark were made in the 1970's and 80's (see in [St71-80]) concerning the values at $s = 1$ and $s = 0$ of complex Artin L series attached to Galois extensions of number fields K/F. A systematic approach to the Stark conjectures was presented in the book by Tate, cf. [Ta84]. In the most general terms these conjectures concern the special values of Artin L-functions of number fields and their analogies, relating them to certain "regulators of S-units" and analogous objects.

More precisely the Conjecture S (on units) discussed in Chapter II of [Yos03], is formulated in terms of the *partial zeta function*

$$\zeta_F(s, \sigma) = \sum_{\mathfrak{A} \subset \mathcal{O}_F, \left(\frac{K/F}{\mathfrak{A}}\right) = \sigma} N(\mathfrak{A})^{-s}$$

attached to an element σ of the Galois group $\mathrm{Gal}(K/F)$ of an abelian extension K of a totally real ground field F ramified over only one infinite place of F (here $\left(\frac{K/F}{\mathfrak{A}}\right)$ denotes the Artin symbol). Conjecture S says that there exists a unit ϵ in K such that

$$\zeta_F'(0, \sigma) = \epsilon^\sigma$$

for every $\sigma \in \mathrm{Gal}(K/F)$. A method is given, which derives Conjecture S on units from the following Conjecture on Galois action:

$$\left(\frac{c(\chi)}{R(\chi)}\right)^\sigma = \frac{c(\chi^\sigma)}{R(\chi^\sigma)}.$$

Here $\chi \in \hat{G}$ denotes a (non-Abelian) irreducible character of a Galois extension K of F with the finite Galois group $G = \mathrm{Gal}(K/F)$, $c(\chi)$ the leading term of the Taylor expansion of its Artin L-function at $s = 0$:

$$L(s, \chi, K/F) = c(\chi)s^{r(\chi)} + \mathcal{O}(s^{r(\chi)+1}), \quad 0 \leq r(\chi) \in \mathbb{Z}$$

and $R(\chi)$ denotes a generalized regulator which is the determinant of a matrix of size $r(\chi)$ whose entries are linear combinations of the absolute values of units of K with algebraic coefficients.

In recent years there was an important developpement in the study of "class invariants" of ray classes in a totally real field F, and in their arithmetical interpretations (which include generalizations of Stickelbergers's theorem, generalization of Dedekind sums as certain cocycles etc.) A recent interpretation of *Stark's conjectures* using notions of Noncommutative geometry was given in [Man02], [Man02a].

4.5 Galois Group in Arithetical Problems

4.5.1 Dividing a circle into n equal parts

The problem of dividing a circle into n equal parts (cf. [Gau], [Gin85]) has a geometric form. However its solution, given by Gauss, was based essentially on arithmetical and algebraic considerations. The construction of the regular 17–gon was the first mathematical invention of Gauss, written in his diary on March 30th 1796, one month before his 19th birthday. Previously one could only construct triangles, squares, pentagons, 15–gons, and all those n–gons which are obtained from these by doubling the number of sides. From the algebraic point of view, the construction of a regular n–gon is equivalent to constructing the roots of unity of degree n on the complex plane, i.e. the solutions to the equation

$$X^n - 1 = 0, \tag{4.5.1}$$

which have the form

$$\varepsilon_k = \cos\frac{2\pi k}{n} + i\sin\frac{2\pi k}{n} = \exp\left(\frac{2\pi i k}{n}\right), \quad k = 0, 1, \dots, n-1. \tag{4.5.2}$$

Assuming that the segment of length one is given, we can construct using ruler-and-compass methods all new segments whose length is obtained from the lengths of given segments using the operations of addition, subtraction, multiplication, division and extraction of the square root. Through a sequence of these operations one may construct any number belonging to any field L, which is a union of a tower of quadratic extensions

$$L = L_m \supset L_{m-1} \supset \cdots L_1 \supset L_0 = \mathbb{Q}, \tag{4.5.3}$$

where $L_{i+1} = L_i(\sqrt{d_i})$, $d_i \in L_i$. It is not difficult to prove that no other points of the complex plane can be constructed starting from the point $z = 1$ and using only ruler-and-compass methods. In order to construct $z = \alpha$ (if this is possible) one constructs the corresponding tower of type (4.5.3) for the field L, generated by all the roots of the minimal polynomial $f(X) \in \mathbb{Q}[X]$ of α (the decomposition field of $f(X)$). By Galois theory, to a quadratic extension $L_1/\mathbb{Q}$ corresponds a subgroup $G_1 = G(L/L_1)$ of index two in the Galois group $G_0 = G(L/\mathbb{Q})$ (the group of Galois symmetries of the polynomial $f(X)$). The action of the subgroup G_1 partitions the set of all roots of $f(X)$ into two parts, such that the sum of all elements of each part belongs to L_1 and generates this field, being invariant under under automorphisms in G_1. In the next step each of these two parts is divided into two further parts using the action on the roots by elements of $G_2 = G(L/L_2)$, which is of index 2 in G_1 etc.. This process continues until we obtain the subset of roots consisting of only one element $z = \alpha$.

For example, for the root of unity $\alpha = \varepsilon_1$ from (4.5.2) the corresponding irreducible polynomial $f(X)$ is the cyclotomic polynomial $\Phi_n(X)$, whose roots ε_k $((k,n) = 1)$ are primitive roots of unity; the Galois symmetries have the form

$$\sigma_a : \varepsilon_k \mapsto \varepsilon_k^a = \varepsilon_{ka \bmod n} \quad (a \in (\mathbb{Z}/n\mathbb{Z})^\times).$$

For $n = 5$ one has $G_0 = \{\sigma_1, \sigma_2, \sigma_3, \sigma_4\}$ and the subgroup $G_1 = \{\sigma_1, \sigma_4\}$ partitions the set of primitive roots into the parts $\{\varepsilon_1, \varepsilon_4\}$ and $\{\varepsilon_2, \varepsilon_3\}$. One has $\Phi_5(x) = x^4 + x^3 + x^2 + x + 1$. Hence

$$\varepsilon_1^2 + \varepsilon_1 + 1 + \varepsilon_1^{-1} + \varepsilon_1^{-2} = 0.$$

By putting $u = \varepsilon_1 + \varepsilon_1^{-1} = \varepsilon_1 + \varepsilon_4$ we obtain the equation

$$u^2 + u - 1 = 0, \quad \varepsilon_1 + \varepsilon_4 = \frac{-1 + \sqrt{5}}{2}, \quad \varepsilon_2 + \varepsilon_3 = \frac{-1 - \sqrt{5}}{2},$$

which gives the desired construction of the regular pentagon.

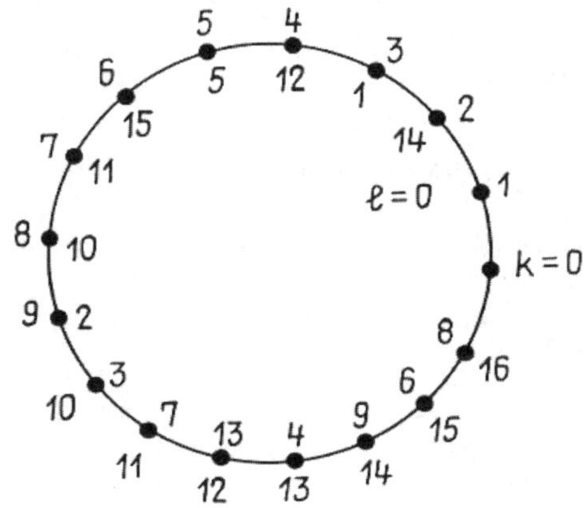

Fig. 4.3.

In the case $n = 17$ Gauss' intuition led him to the correct partition of the roots of $\Phi_{17}(x) = x^{16} + x^{15} + \cdots + x + 1$ given by Galois symmetries (Galois theory had not yet been discovered!). The group of symmetries $G_0 \cong (\mathbb{Z}/17\mathbb{Z})^\times$ is a cyclic group of order 16 with a generator $3 \bmod 17$ (a primitive root), and Gauss's idea was to use a more convenient indexing system for the roots (see Fig. 13). Let us assign to the root ε_k the new number l (the notation $\varepsilon_{[l]}$) defined by the condition $k \equiv 3^l \bmod 17, l = 0, 1, \ldots, 15$, and let T_l denote the automorphism σ_k. Then

$$T_l\varepsilon_{[m]} = \varepsilon_{[m+l]} \quad (m, l \bmod 16). \tag{4.5.4}$$

The corresponding subgroups have the form

$$G_0 = \{T_0, T_1, \cdots, T_{15}\}, \quad G_1 = \{T_0, T_2, \cdots, T_{14}\},$$

$$G_2 = \{T_0, T_4, \cdots, T_{12}\}, \quad G_3 = \{T_0, T_8\}.$$

We now show how the idea described above works in this case. First of all note that

$$\varepsilon_1 + \varepsilon_2 + \cdots + \varepsilon_{16} = \varepsilon_{[0]} + \varepsilon_{[1]} + \cdots + \varepsilon_{[15]} = -1 \tag{4.5.5}$$

(the sum of a geometric progression). Denote by $\sigma_{m,r}$ the sum of $\varepsilon_{[l]}$ with l, congruent to r modulo m. We thus obtain

$$\sigma_{2,0} = \varepsilon_{[0]} + \varepsilon_{[2]} + \cdots + \varepsilon_{[14]} = \sum_{T_l \in G_1} T_l\varepsilon_{[0]},$$

$$\sigma_{2,1} = \varepsilon_{[1]} + \varepsilon_{[3]} + \cdots + \varepsilon_{[15]} = \sum_{T_l \in G_1} T_l\varepsilon_{[1]}.$$

Identity (4.5.5) implies

$$\sigma_{2,0} + \sigma_{2,1} = -1,$$

and by termwise multiplication we find that

$$\sigma_{2,0} \cdot \sigma_{2,1} = 4(\varepsilon_{[0]} + \varepsilon_{[1]} + \ldots \varepsilon_{[15]}) = -4.$$

Now using Viète's formulae, we may express $\sigma_{2,0}$ and $\sigma_{2,1}$ as the roots of the quadratic equation $x^2 + x - 4 = 0$:

$$\sigma_{2,0} = \frac{\sqrt{17}-1}{2}, \quad \sigma_{2,1} = \frac{-\sqrt{17}-1}{2},$$

which generate the field $L_1 = \mathbb{Q}(\sqrt{17})$. We distinguish the two roots by the condition that $\sigma_{2,0} > \sigma_{2,1}$; in each of these fields the roots arise together with their conjugates. In the first case we have to add and to multiply the real parts of the numbers ε_1, ε_2, ε_4, ε_8 and in the second case we do the same for ε_3, ε_5, ε_6, ε_7. In a similar way we have that $\sigma_{4,0} + \sigma_{4,2} = \sigma_{2,0}$, $\sigma_{4,1} + \sigma_{4,3} = \sigma_{2,1}$, and the multiplication using (4.5.4) shows that $\sigma_{4,0} \cdot \sigma_{4,2} = \sigma_{2,0} + \sigma_{2,1} = -1$. Hence $\sigma_{4,0}$ and $\sigma_{4,2}$ are roots of the equation $x^2 + \sigma_{2,0} + 1 = 0$ which generates the field L_2:

$$\sigma_{4,0} = \frac{1}{4}\left(\sqrt{17} - 1 + \sqrt{34 - 2\sqrt{17}}\right),$$

$$\sigma_{4,2} = \frac{1}{4}\left(\sqrt{17} - 1 - \sqrt{34 - 2\sqrt{17}}\right).$$

In the same way we see that

$$\sigma_{4,1} = \frac{1}{4}\left(-\sqrt{17} - 1 + \sqrt{34 + 2\sqrt{17}}\right),$$

$$\sigma_{4,2} = \frac{1}{4}\left(-\sqrt{17} - 1 - \sqrt{34 + 2\sqrt{17}}\right).$$

An analogous argument shows that

$$\begin{aligned}
\sigma_{8,0} &= \varepsilon_{[0]} + \varepsilon_{[8]} \\
&= 2\cos\frac{2\pi}{17} \\
&= (1/2)\cdot\sqrt{(\sigma_{4,0})^2 - 4\sigma_{4,1})} \\
&= (1/8)\cdot(\sqrt{17} - 1 + \sqrt{34 - 2\sqrt{17}}) \\
&\quad + (1/4)\cdot\sqrt{17 + 3\sqrt{17} - \sqrt{170 + 38\sqrt{17}}},
\end{aligned}$$

which completes the construction.

In the general case of an n–gon with $n = 2^r p_1^{r_1}\cdot\ldots\cdot p_s^{r_s}$, where p_i are odd primes, we have that

$$G(\mathbb{Q}(\zeta_n)/\mathbb{Q}) \cong (\mathbb{Z}/n\mathbb{Z})^\times.$$

By considering the tower (4.5.3) of quadratic extensions, we see that the possibility of constructing the regular n–gon is equivalent to the condition that the number

$$|(\mathbb{Z}/n\mathbb{Z})^\times| = \varphi(n) = 2^{r-1}(p_1 - 1)p_1^{r_1-1}\cdot\ldots\cdot(p_s - 1)p_1^{r_s-1}$$

is a power of 2. This holds precisely when $n = 2^r p_1 \cdot \ldots \cdot p_s$, where the p_i are primes such that $p_i = 2^{m_i} + 1$. It follows from Lagrange's theorem applied to the cyclic group $(\mathbb{Z}/p_i\mathbb{Z})^\times$, that m_i divides $p_i - 1$. Hence m_i is also a power of 2. The construction is therefore only possible for $n = 2^r p_1 \cdot \ldots \cdot p_s$ where p_i are *Fermat primes* $p_i = 2^{2^{t_i}} + 1$, which were discussed in Part I, §1.1.2. The proof of the latter statement was not published by Gauss: "Although the framework of our treatise does not allow us to proceed with this proof, we think that it is necessary to point out this fact, in order to prevent somebody else from wasting his time, by attempting to find some other cases, which are not given by our theory."

4.5.2 Kummer Extensions and the Power Residue Symbol

(see [CF67], [Chev40], [Koch70]). Let K be a field containing a primitive root of unity ζ of degree m, where m is a fixed positive integer not divisible by the characteristic of K. One may show that cyclic extensions L/K of degree dividing m coincide with the so–called Kummer extensions of type $K(\sqrt[m]{a})/K$ $(a \in K)$. In applications K will be either a number field or a completion of one. Any extension L/K containing a root α of $X^m = a$, also contains all the other roots $\zeta\alpha, \ldots, \zeta^{m-1}\alpha$ of this polynomial. Let σ be an element of the Galois group $G(K(\sqrt[m]{a})/K)$. If we fix a root α then the automorphism σ is completely determined by the image of α under the action of σ: $\alpha^\sigma = \zeta^b\alpha$. In particular, if α is an element of order m in the multiplicative group $K^\times/K^{\times m}$ then $X^m - a$ is irreducible and a^r is an mth power iff $m|r$. In this case the assignement $\sigma \mapsto b \bmod m$ provides an isomorphism of the Galois group $G(K(\sqrt[m]{a})/K)$ with the cyclic group $\mathbb{Z}/m\mathbb{Z}$.

Now let L be an arbitrary cyclic extension of degree m of K. We shall construct explicitly an element $a \in K$ such that $L = K(\sqrt[m]{a})/K$. Let σ be a generator of the cyclic group $G(L/K)$ and let $L = K(\gamma)$ for some primitive element $\gamma \in L$. Then the elements $\gamma, \gamma^\sigma, \ldots, \gamma^{\sigma^{m-1}}$ form a basis of L over K. Consider the sum

$$\beta = \sum_{s=0}^{m-1} \zeta^s \gamma^{\sigma^s}. \tag{4.5.6}$$

Then $\beta^\sigma = \zeta^{-1}\beta$, and $\beta \neq 0$ since the elements $\gamma, \gamma^\sigma, \gamma^{\sigma^{m-1}}$ are linearly independent over K. Thus $\beta^m \in K$ and $\beta^r \notin K$ for $0 < r < m$, i.e. $a = \beta^m$ is an element of order m in the quotient group $K^\times/K^{\times m}$ and the above argument shows that the field $K(\beta)$ is a cyclic extension of degree m contained in L and is therefore equal to $L = K(\sqrt[m]{a})$. In a similar way we can check that two extensions $K(\sqrt[m]{a})/K$ and $K(\sqrt[m]{b})/K$ coincide iff $a = b^r c^m$ for some $c \in K$ and $r \in \mathbb{Z}$ such that $(r, m) = 1$. These statements can be unified into one statement by saying that for a given field $K \supset \mu_m$ and its Galois group $G_K = G(\overline{K}/K)$ there is the isomorphism

$$K^\times/K^{\times m} \cong \mathrm{Hom}(G_K, \mu_m), \tag{4.5.7}$$

where $\mu_m = \{\zeta \in \overline{K} \mid \zeta^m = 1\}$ and Char $K \not| m$. In order to construct (4.5.7) for a given $a \in K^\times$ choose $\gamma \in \overline{K}^\times$ with the condition $\gamma^m = a$, and for $\sigma \in G_K$ the formula $\varphi_a(\sigma) = \gamma^\sigma/\gamma$ defines then a homomorphism $\varphi_a : G_K \to \mu_m$. The fact that this map defines a homomorphism (4.5.7) is deduced from *Hilbert's Theorem 90* on the cohomology of the multiplicative group: $H^1(G_K, K^\times) = \{1\}$ (see §4.5.3).

Now let K be a number field, $\mu_m \subset K$, $\mathfrak{p} = \mathfrak{p}_v$ a prime divisor attached to a non–Archimedean place v of K. The decomposition of $\mathfrak{p}$ in the extension $K(\sqrt[m]{a})/K$ is reduced to study of the extension $K_v(\sqrt[m]{a})/K_v$ of the local field K_v (by the construction of extensions of absolute values, see §4.3.6). One can assume that a belongs to the ring $\mathcal{O}_K$ of integers of K, and that $\mathfrak{p} \not| ma$. Then the decomposition of the *maximal ideal* $\mathfrak{p} \subset \mathcal{O}_K$ is determined by the decomposition of the polynomial $X^m - a(\bmod \mathfrak{p})$ over the field $\mathcal{O}_K/\mathfrak{p}$ (by the lemma in §4.2.3). This decomposition is a product of pairwise coprime irreducible factors of degree f, where f is the degree of the residue field extension: the least positive integer f such that the congruence $a^f \equiv x^m (\bmod \mathfrak{p})$ is solvable in $\mathcal{O}_K/\mathfrak{p}$. Under our assumptions the ideal $\mathfrak{p}$ is *unramified* in $L = K(\sqrt[m]{a})$

and $\mathfrak{p} = \mathfrak{P}_{w_1} \cdot \ldots \cdot \mathfrak{P}_{w_r}$ $(f \cdot r = m)$. In particular, $\mathfrak{p}$ splits completely iff $f = 1$, i.e. iff the congruence $x^m \equiv a \bmod \mathfrak{p}$ is solvable.

We now define the *power residue symbol.* In order to do this denote by S the set of places of K which either divide m or are Archimedean. For elements $a_1, \ldots, a_l \in K^\times$ denote by $S(a_1, \ldots, a_l)$ the union of S and the set of places v for which $|a_i|_v \neq 1$ for some i. For $a \in K^\times$ and a place $v \notin S(a)$ define the power residue symbol $\left(\frac{a}{v}\right) \in \mu_m$ by

$$\sqrt[m]{a}^{F_{L/K}(v)} = \left(\frac{a}{v}\right) \sqrt[m]{a}, \tag{4.5.8}$$

where $L = K(\sqrt[m]{a})$ and $F_{L/K}(v) \in G(L/K)$ is the global *Artin symbol,* cf. §4.4.6. The number $\left(\frac{a}{v}\right) \in \mu_m$ does not depend on a choice of $\sqrt[m]{a}$, and one verifies that

$$\left(\frac{aa'}{v}\right) = \left(\frac{a}{v}\right)\left(\frac{a'}{v}\right) \quad (v \notin S(a, a')). \tag{4.5.9}$$

According to the definition of $F_{L/K}(v)$ as a Frobenius element, the identity (4.5.8) is equivalent to the congruence $\sqrt[m]{a}^{\mathrm{N}v-1} \equiv \left(\frac{a}{v}\right) (\bmod\ \mathfrak{p}_v)$, which implies that $m|(\mathrm{N}v - 1)$ and

$$a^{(\mathrm{N}v-1)/m} \equiv \left(\frac{a}{v}\right) (\bmod\ \mathfrak{p}_v), \tag{4.5.10}$$

(the *generalized Euler criterium*) since the group $(\mathcal{O}_v/\mathfrak{p}_v)^\times$ is cyclic of order $\mathrm{N}v - 1$. For an arbitrary divisor $\beta = \prod_{v \notin S(a)} \mathfrak{p}_v^{n(v)} \in I^{S(a)}$ put

$$\left(\frac{a}{\beta}\right) = \prod_{v \notin S(a)} \left(\frac{a}{v}\right)^{n(v)}.$$

Then we have that

$$\sqrt[m]{a}^{F_{L/K}(\beta)} = \left(\frac{a}{\beta}\right) \sqrt[m]{a}, \tag{4.5.11}$$

where $L = K(\sqrt[m]{a})$, $F_{L/K}(\beta) \in G(L/K)$ is the global *Artin symbol,* and the following equation holds:

$$\left(\frac{a}{\beta\beta'}\right) = \left(\frac{a}{\beta}\right)\left(\frac{a}{\beta'}\right) \quad (\beta, \beta' \in I^{S(a)}). \tag{4.5.12}$$

For any prime divisor $v \notin S(a)$ the following statements are equivalent:

1) $\left(\frac{a}{v}\right) = 1$;
2) the congruence $x^m \equiv a (\bmod\ \mathfrak{p}_v)$ is solvable for some $x \in \mathcal{O}_v$;
3) the equation $x^m = a$ is solvable for some $x \in K_v$

A solution in 2) can be lifted to a solution in 3) by Hensel's lemma, see §4.3.2. For an integral ideal $\beta \subset \mathcal{O}_K$ the value of $\left(\frac{a}{\beta}\right)$ depends only on a mod β as long as $a \in \mathcal{O}_K$. Thus the following character of order m is defined

$$\chi_\beta : (\mathcal{O}_K/\beta)^\times \to \mu_m, \quad \chi_\beta(a) = \left(\frac{a}{\beta}\right). \tag{4.5.13}$$

The cubic reciprocity law. Let $K = \mathbb{Q}(\zeta_3) = \mathbb{Q}(\sqrt{-3})$, $m = 3$. Then $\mathcal{O}_K = \mathbb{Z}[\zeta_3]$ is a principal ideal domain and if $\mathfrak{p} = \mathfrak{p}_v = (\pi)$ for a prime element π, then we shall use the notation $\left(\frac{a}{\pi}\right)$ instead of $\left(\frac{a}{\mathfrak{p}}\right)$. Call a prime element *primary* if $\pi \equiv 2 \bmod 3$, i.e. either $\pi = q$ is a rational prime number, $q \equiv 2 (\mathrm{mod}\ 3)$, or $\mathrm{N}\pi = p \equiv 1 (\mathrm{mod}\ 3)$, $\pi \equiv 2 \bmod 3$. One easily verifies that among the generators of an ideal $\mathfrak{p}$, $\mathfrak{p} \not| 3$ there is exactly one primary element. Let $\mathfrak{p}_1 = (\pi_1)$ and $\mathfrak{p}_2 = (\pi_2)$, where π_1 and π_2 are coprime primary elements such that $\mathrm{N}\mathfrak{p}_1 \neq \mathrm{N}\mathfrak{p}_2 \neq 3$. Then the following "reciprocity law" holds:

$$\left(\frac{\pi_1}{\pi_2}\right) = \left(\frac{\pi_2}{\pi_1}\right). \tag{4.5.14}$$

The biquadratic reciprocity law. Let $m = 4$, $K = \mathbb{Q}(i)$ and $\mathcal{O}_K = \mathbb{Z}[i]$, the ring of *Gaussian integers.* We shall call $\alpha \in \mathcal{O}_K$ *primary* if $\alpha \equiv 1 (\mathrm{mod}\ (1+i)^3)$. Then one verifies that in any prime ideal $\mathfrak{p}$, $\mathfrak{p} \not| 2$ one can choose a unique primary generator. If $\mathfrak{p}_1 = (\pi_1)$ and $\mathfrak{p}_2 = (\pi_2)$, where π_1 and π_2 are coprime primary elements, then the following reciprocity law holds:

$$\left(\frac{\pi_1}{\pi_2}\right) = \left(\frac{\pi_2}{\pi_1}\right)(-1)^{((\mathrm{N}\pi_1 - 1)/4)\cdot((\mathrm{N}\pi_2 - 1)/4)}. \tag{4.5.15}$$

4.5.3 Galois Cohomology

The group cohomology theory provides a standard method of obtaining arithmetical information from Galois groups, acting on various objects: algebraic numbers, idele classes, points of algebraic varieties and algebraic groups etc. (cf. [Se58], [Se63], [Se64], [Chev40], [Ire82], [Koch70], [Koly88] [Wei74a]). Let G be a finite (or profinite) group acting on a G–module A (endowed with the discrete topology). The cohomology groups of G with coefficients in A are defined with the help of the complex of cochains. Consider the following Abelian groups:

$$C^0(G, A) = A,$$

and for $n \geq 1$

$$C^n(G, A) = \{f : G \times \cdots \times G \to A \mid f \text{ is continuous}\}$$

(the addition of functions is pointwise and the continuity of $f \in C^n(G, A)$ means that the function $f(g_1, \dots, g_n)$ depends only on a coset of g_i modulo some open subgroup of G).

The formula

$$\begin{aligned}(d_n f)(g_1, \dots, g_{n+1}) =& g_1 f(g_2, \dots, g_{n+1}) \\ &+ \sum_{i=1}^{n} (-1)^i f(g_1, \dots, g_i g_{i+1}, \dots, g_{n+1}) \\ &+ (-1)^{n+1} f(g_1, \dots, g_n), \end{aligned} \tag{4.5.16}$$

defines a homomorphism $d_n : C^n(G, A) \to C^{n+1}(G, A)$, such that $d_n \circ d_{n+1} = 0$.

The group $Z^n(G, A) = \mathrm{Ker} d_n$ is called the group of n*–cocycles*, and the group $B^n(G, A) = \mathrm{Im} d_{n-1}$ is called the group of n*–coboundaries*. The property $d_n \circ d_{n+1} = 0$ implies that $B^n(G, A) \subset Z^n(G, A)$. The cohomology groups are then defined by

$$H^n(G, A) = B^n(G, A)/Z^n(G, A) = \begin{cases} \mathrm{Ker} d_n / \mathrm{Im} d_{n-1} & \text{for } n \geq 1; \\ \mathrm{Ker} d_0 & \text{for } n = 0. \end{cases} \tag{4.5.17}$$

If $n = 0$ then

$$H^0(G, A) = A^G = \{a \in A \mid ga = a \text{ for all } g \in G\}. \tag{4.5.18}$$

For $n = 1$ we call a continuous map $f : G \to A$ a scew–homomorphism iff for all $g_1,\ g_2 \in G$ one has

$$f(g_1 g_2) = f(g_1) + g_1 f(g_2). \tag{4.5.19}$$

One says that a scew–homomorphism *splits*, iff for a fixed $a \in A$ it can be written in the form $f(g) = a - ga$. The group $H^1(G, A)$ can be identified with the quotient group of the group of all scew–homomorphisms modulo the subgroup formed by all split scew–homomorphisms. If the action of G on A is trivial then $H^1(G, A)$ coincides with the group of all (continuous) homomorphisms from G into A.

For $n = 2$ the elements of $H^2(G, A)$ correspond bijectively to equivalence classes of extensions of G by A. Consider an extension

$$0 \to A \to \tilde{G} \to G \to 1. \tag{4.5.20}$$

For all $g \in G$ choose a lift $\tilde{g}$ in $\tilde{G}$ (i.e. choose a section $g \mapsto \tilde{g}$ of the projection $\tilde{G} \to G$). Define $f : G \times G \to A,\ f(g_1, g_2) \in A$ by

$$\tilde{g}_1 \cdot \tilde{g}_2 = f(g_1, g_2) \widetilde{g_1 g_2}.$$

Then the function f is a 2–cocycle of G with values in A. If we change our choice of representatives $\tilde{g}$ (i.e. the choice of section $G \to \tilde{G}$), then f is altered by a coboundary. Hence the class of f depends only on the extension (4.5.20). The group $H^2(G, \mathbb{C}^\times)$ is called also the *Schur multiplier* of G. Let L/K be a Galois extension with Galois group $G = G(L/K)$. Then $L^\times$ is a G–module and $H^2(G, L^\times)$ can be interpreted as the *Brauer group*, see §4.5.5.

For the action of the Galois group $G = G(L/K)$ on $L^\times$ one has the following fundamental theorem.

Theorem 4.25 (Hilbert's Theorem 90).

$$H^1(G(L/K), L^\times) = \{1\}.$$

The idea of the proof of this theorem is the same as in the description of all cyclic extensions of K in §4.5.2. Let $f : G \to L^\times$ be an arbitrary scew–homomorphism, $f \in Z^1(G(L/K), L^\times)$. In multiplicative notation this means that for all $g,\ h \in G$ we have $f(h)^g = f(gh)/f(g) \in L^\times$. We shall find an element $b \in L^\times$ such that for all $g \in G$ one has $f(g) = b/b^g$. In order to do this choose a primitive element γ in the extension L/K, so that the elements γ^g $(g \in G)$ form a normal basis of L over K. Then the element

$$b = \sum_{h \in G} f(h)\gamma^h \in L \tag{4.5.21}$$

is not equal to zero. We apply to both sides of (4.5.21) an element $g \in G$. Then

$$\begin{aligned} b^g &= \sum_{h \in G} f(h)^g (\gamma^h)^g \\ &= \sum_{h \in G} f(h)^g \gamma^{gh} \\ &= f(g)^{-1} \sum_{h \in G} f(gh)\gamma^{gh} \\ &= f(g)^{-1} b \end{aligned}$$

(by the formula of the (left) action of G on $L^\times$: $(\gamma^h)^g = \gamma^{gh}$ for $g, h \in G$). This method of taking the average is also known as the construction of the *Lagrange resolution* in the theory of solvable extensions of fields.

Properties of cohomology groups.

1) For an arbitrary exact sequence of G–modules

$$0 \longrightarrow A \longrightarrow B \longrightarrow C \longrightarrow 0$$

the following long exact sequence of cohomology groups is defined:

$$\begin{aligned} &0 \longrightarrow H^0(G, A) \longrightarrow H^0(G, B) \longrightarrow H^0(G, C) \xrightarrow{\Delta_0} H^1(G, A) \\ &\longrightarrow H^1(G, B) \longrightarrow H^1(G, C) \xrightarrow{\Delta_1} H^2(G, A) \longrightarrow \cdots H^n(G, A) \longrightarrow \\ &\longrightarrow H^n(G, B) \longrightarrow H^n(G, C) \xrightarrow{\Delta_n} H^{n+1}(G, A) \longrightarrow \cdots \end{aligned} \tag{4.5.22}$$

Example 4.26. Kummer theory. Let K be a field containing the group μ_m of all roots of unity of degree m in $\overline{K}$. Assume further that Char K does not divide m. For an arbitrary Galois extension L/K with Galois group $G = G(L/K)$ the map $x \mapsto x^m$ defines a homomorphism of G–modules: $\nu : L^\times \longrightarrow L^\times$. For $L = \overline{K}$, and $G = G_K$, one has the following exact sequence

$$1 \longrightarrow \mu_m \longrightarrow \overline{K}^\times \stackrel{\nu}{\longrightarrow} \overline{K}^\times \longrightarrow 1.$$

Passing to cohomology groups (4.5.22) we obtain the following long exact sequence

$$\begin{aligned} &H^0(G_K, \mu_m) \longrightarrow H^0(G_K, \overline{K}^\times) \stackrel{\nu}{\longrightarrow} H^0(G_K, \overline{K}^\times) \longrightarrow \\ &H^1(G_K, \mu_m) \longrightarrow H^1(G_K, \overline{K}^\times) \stackrel{\nu}{\longrightarrow} H^1(G_K, \overline{K}^\times) \longrightarrow \cdots . \end{aligned} \tag{4.5.23}$$

Since the group G_K acts trivially on μ_m, it follows that $H^1(G_K, \mu_m)$ coincides with the group $\mathrm{Hom}(G_K, \mu_m)$. The group $H^0(G_K, \overline{K}^\times)$ is the subgroup of all fixed points of the Galois action, i.e. $H^0(G_K, \overline{K}^\times) = \overline{K}^{\times G_K(\overline{K}/K)} = K^\times$. Also, $H^0(G_K, \mu_m) = \mu_m$, and $H^1(G_K, \overline{K}^\times) = \{1\}$ by Hilbert's theorem 90. We thus have the following exact sequence

$$1 \longrightarrow \mu_m \longrightarrow K^\times \stackrel{\nu}{\longrightarrow} K^\times \longrightarrow \mathrm{Hom}(G_K, \mu_m) \longrightarrow 1,$$

which is equivalent to the isomorphism of Kummer:

$$K^\times / K^{\times m} \cong \mathrm{Hom}(G_K, \mu_m).$$

2) Let H be an open normal subgroup in G and A a G–module. Then one has the following "inflation - restriction" exact sequence:

$$0 \longrightarrow H^1(G/H, A^H) \stackrel{\mathrm{Inf}}{\longrightarrow} H^1(G, A) \stackrel{\mathrm{Res}}{\longrightarrow} H^1(H, A), \tag{4.5.24}$$

in which Inf denotes the inflation homomorphism, which is defined by "inflating" a cocycle f on G/H with values in $A^H \subset A$ to a cocycle $\overline{f}$ on G; and Res is the restriction homomorphism given by restricting cocycles on G to the subgroup H.

3) $\cup$–*products.* Let A, B, C be three G–modules, for which some G–invariant pairing $\circ : A \times B \to C$ is given (i.e. for all $g \in G$, $a \in A$, $b \in B$ we have that $g(a \circ b) = ga \circ gb$). For example, if $A = B = C$ is a ring on which the group G acts trivially, then the multiplication in A is such a pairing. Any pairing $A \times B \to C$ induces for every $n \geq 0$ and $m \geq 0$ a bilinear map

$$H^n(G, A) \times H^m(G, B) \to H^{n+m}(G, C), \tag{4.5.25}$$

which is called $\cup$–product. This is defined on cocycles by the following rule. If $f \in C^n(G, A)$, $f' \in C^m(G, B)$ then the cochain

$$(f \circ f')(g_1, \cdots, g_{n+m}) = f(g_1, \cdots, g_n) \circ (g_1 \dots g_n) f'(g_{n+1}, \cdots, g_{n+m})) \tag{4.5.26}$$

turns out to be a cocycle, as can be seen from the following equation:

$$d_{n+m}(f \circ f') = d_n f \circ f' + (-1)^n f \circ d_m f'.$$

The $\cup$–product (4.5.25) is well defined by the formula

$$\overline{f} \cup \overline{f'} = \overline{f \circ f'} \in H^{n+m}(G, C).$$

One has the equation

$$\alpha \cup \Delta_m \beta = (-1)^n \Delta_{n+m}(\alpha \cup \beta), \tag{4.5.27}$$

where Δ_m is the "connecting homomorphism" of the long exact sequence (4.5.22). If $A = B = C$ is a commutative ring on which G acts trivially, then for all $\alpha \in H^n(G, A)$, $\beta \in H^m(G, A)$ one has

$$\alpha \cup \beta = (-1)^{nm} \beta \cup \alpha. \tag{4.5.28}$$

4.5.4 A Cohomological Definition of the Local Symbol

Let K be a finite extension of the field $\mathbb{Q}_p$ of p–adic numbers. The local Artin symbol is a homomorphism

$$\theta : K^\times \to G_K^{\mathrm{ab}} = \varprojlim_L G(L/K) \tag{4.5.29}$$

from the multiplicative group of K to the Galois group of the maximal Abelian extension (the union of all finite Abelian extensions L/K) of K. This homomorphism was described in §4.4 using powerful global methods – the *Artin reciprocity law*. However, the local symbol can be defined purely locally. With this approach the global reciprocity law can then be deduced from the properties of the local symbols by proving the product formula (4.3.31).

We shall define for a given $\alpha \in K^\times$ the image $\theta(\alpha) = \theta_{L/K}(\alpha) \in G(L/K)$ (in a finite extension L/K) using the characters $\chi \in \mathrm{Hom}(G(L/K), \mathbb{Q}/\mathbb{Z})$. Note that the element $\theta(\alpha)$ of the finite Abelian group $G(L/K)$ is completely determined by the values $\chi(\theta(\alpha))$ for all characters χ of $G(L/K)$. For the trivial $G(L/K)$-module $\mathbb{Q}/\mathbb{Z}$ we have:

$$\mathrm{Hom}(G(L/K), \mathbb{Q}/\mathbb{Z}) = H^1(G(L/K), \mathbb{Q}/\mathbb{Z}),$$

and there is an exact sequence

$$0 \to \mathbb{Z} \to \mathbb{Q} \to \mathbb{Q}/\mathbb{Z} \to 0,$$

which gives rise to the isomorphism

$$\Delta_1 : H^1(G(L/K), \mathbb{Q}/\mathbb{Z}) \xrightarrow{\sim} H^2(G(L/K), \mathbb{Z}). \qquad (4.5.30)$$

The latter is found by considering the long exact sequence (4.5.22), and using the fact that all the higher cohomology groups of the divisible group $\mathbb{Q}$ are trivial: $H^i(G(L/K), \mathbb{Q}) = \{0\}$ for $i \geq 1$.

As we have seen in §4.5.3, $H^1(G(L/K), L^\times) = \{1\}$. Moreover, the following fundamental facts on the cohomology groups of the multiplicative group are known:

a) $H^3(G(L/K), L^\times) = \{1\}$ (L is a local field)
b) There exists an embedding

$$\mathrm{inv}_K : H^2(G(L/K), L^\times) \to \mathbb{Q}/\mathbb{Z}. \qquad (4.5.31)$$

The image of an element $\beta \in H^2(G(L/K), L^\times)$ under this embedding is called the *invariant* of β. For a finite extension L/K the group $H^2(G(L/K), L^\times)$ is cyclic of order $[L : K]$.

Now consider the pairing

$$L^\times \times \mathbb{Z} \to L^\times \quad ((x, m) \mapsto x^m).$$

This induces a $\cup$–product in the cohomology groups

$$H^0(G(L/K), L^\times) \times H^2(G(L/K), \mathbb{Z}) \to H^2(G(L/K), L^\times).$$

Recall that $H^0(G(L/K), L^\times) = K^\times$. For $\Delta_1\chi \in H^2(G(L/K), \mathbb{Z})$, we have

$$\alpha \cup \Delta_1\chi \in H^2(G(L/K), L^\times)$$

for $\alpha \in K^\times$. Define for each character χ,

$$\chi(\theta_{L/K}(\alpha)) = \mathrm{inv}_K(\alpha \cup \Delta_1\chi). \qquad (4.5.32)$$

This determines $\theta_{L/K}(\alpha)$ as a well defined element of $G(L/K)$. Passing to the projective limit in (4.5.29), we obtain an element

$$\theta(\alpha) = \lim_{L} \theta_{L/K}(\alpha) \in G_K^{\mathrm{ab}}.$$

To do this we need the following compatibility property. Consider a tower of (Abelian) Galois extensions $K \subset L' \subset L$ and let $G = G(L/K)$, $H = G(L/L')$. Let χ' be a character of G. Then if $\alpha \in K^\times$ induces an element $s_\alpha = \theta_{L/K}(\alpha) \in G$ and the element $s'_\alpha \in G/H$ under the projection $G \to G/H$, then we have that $\chi(s_\alpha) = \chi'(s'_\alpha)$. This follows from the definition $\chi(s_\alpha) = \mathrm{inv}_K(\alpha \cup \Delta_1\chi)$ together with the fact that the inflation map takes χ' (respectively, $\Delta_1\chi'$) to the character χ (respectively, to $\Delta_1\chi$), using the commutative diagram

$$\begin{array}{ccc} H^2(G/H, L'^{\times}) & \xrightarrow{\mathrm{Inf}} & H^2(G, L^{\times}) \\ & {\scriptstyle \mathrm{inv}_K} \searrow \quad \swarrow {\scriptstyle \mathrm{inv}_K} & \\ & \mathbb{Q}/\mathbb{Z} & \end{array} \tag{4.5.33}$$

The map inv_K will be defined in the next subsection via the *Brauer group*. The above compatibility property will also be discussed there. This compatibility property is very important, since it makes it possible to define the symbol (4.5.29).

If the field K contains a primitive root of unity ζ_m of degree m, then the *power residue symbol* (α, β) of degree m can be defined for α, $\beta \in K^{\times}$ by the condition

$$\theta_{L/K}(\beta) \cdot \sqrt[m]{\alpha} = (\alpha, \beta) \cdot \sqrt[m]{\alpha}, \tag{4.5.34}$$

where $L = K(\sqrt[m]{\alpha})$ is a cyclic extension and $\theta_{L/K}(\beta)$ is the local symbol (4.5.32). The values of (α, β) are roots of unity of degree m, and they satisfy the following conditions:

1) $(\alpha\alpha', \beta) = (\alpha, \beta)(\alpha', \beta)$;
2) $(\alpha, \beta\beta') = (\alpha, \beta)(\alpha, \beta')$;
3) $(\alpha, \beta)(\beta, \alpha) = 1$;
4) if $(\alpha, \beta) = 1$ for all $\beta \in K^{\times}$ then $\alpha \in K^{\times m}$;
5) $(\alpha, \beta) = 1$ iff β is a norm of an element in the extension $K(\sqrt[m]{\alpha})/K$.

The power residue symbol symbol can be interpreted as a $\cup$–product in certain one–dimensional cohomology groups, cf. [Koch70]. An explicit calculation of this symbol is given in [Koly79], [Sha50].

4.5.5 The Brauer Group, the Reciprocity Law and the Minkowski–Hasse Principle

Recall first some basic facts about the Brauer group of an arbitrary field K (see [Man70b], [Man72b], [Se63], [Se86] [Chebo49]).

A finite dimensional algebra A over K is called a *simple central algebra* over K, if there exist $n \geq 1$ such that $A \otimes \overline{K} \cong \mathrm{M}_n(\overline{K})$, where M_n denotes the $n \times n$–matrix algebra and $\overline{K}$ is an algebraic closure of K. The tensor product induces a commutative semigroup structure on the set of simple central K–algebras (modulo isomorphism). The following equivalence relation turns this set into a group: we say that an algebra A is equivalent to an algebra B, if there exist $m, n \geq 1$ such that $A \otimes \mathrm{M}_m(K)$ is isomorphic to $B \otimes \mathrm{M}_n(K)$. All matrix algebras are equivalent to each other, and they form the identity class of algebras. The class of the algebra A°, inverse to A (i.e. consisting of the same elements and having the same addition but the opposite order of multiplication), is the inverse of A in the group structure induced by the tensor product. To see this, consider the canonical map $A \otimes A^\circ \to \mathrm{End}_K(A)$ (endomorphisms of the linear space A), which assignes to an element $x \otimes y \in A \otimes A^\circ$ the multiplication by x on the left, followed by the multiplication by y on the right. The kernel of this map is trivial, since $A \otimes A^\circ$ is simple, and the dimension of $A \otimes A^\circ$ coincides with the dimension of $\mathrm{End}_K(A)$, i.e. with $(\dim A)^2$. Hence the map is an isomorphism, so $A \otimes A^\circ$ is isomorphic to $\mathrm{End}_K(A) \cong \mathrm{M}_{\dim A}(K)$.

The group of classes of central simple algebras over K is called the *Brauer group* of K and is denoted by Br K. We shall now describe the Brauer group in cohomological terms.

Let L/K be an extension of K. It is called a *splitting field* of a K-algebra A iff $A \otimes_K L \cong \mathrm{M}_n(L)$. Equivalent algebras have the same splitting fields. Let $\mathrm{Br}(K, L)$ be the subset of the Brauer group, consisting of those classes of K-algebras which split over L. This is a subgroup. Now asume that L/K is a Galois extension with Galois group $G = G(L/K)$. One has the following fundamental isomorphism:

$$\mathrm{Br}(K, L) \cong H^2(G, L^\times). \tag{4.5.35}$$

This isomorphism can be constructed in various ways; we point out one of these, the so–called construction of "scew–products". This method consists of explicitly constructing a central simple algebra over K from a given "factor set", i.e. from a cocycle $\{a_{g,h}\} \in Z^2(G, L^\times)$. The algebra is construced as follows:

$$A = \bigoplus_{g \in G} L e_g,$$

with multiplication given by

$$e_g e_h = a_{g,h} e_{gh} \text{ for all } g, h \in G,$$

$$e_g a = g(a) e_g \text{ for all } g \in G.$$

Its dimension over K is obviously equal to $[L : K]^2$. We omit to verify the various necessary properties of the construction; note only that the associativity of A is equivalent to the fact that the cochain of *structural constants* is actually a cocycle.

The condition that A splits over L has important arithmetical implications. Put $N = n^2$ and choose a basis $\{a_1, \ldots, a_N\}$ of A over K. If we use the isomorphism

$$F : A \otimes_K \overline{K} \xrightarrow{\sim} \mathrm{M}_n(\overline{K}), \tag{4.5.36}$$

then all of the elements $a = \sum_{i=1}^N x_i a_i \in A$ $(x_i \in K)$ become matrices $F(a) \in \mathrm{M}_n(\overline{K})$. Then it is not difficult to check that the maps

$$\tau(a) = \mathrm{Tr}(F(a)), \quad \nu(a) = \det(F(a))$$

are polynomial functions of $x_1, \ldots, x_N$ with coefficients in the ground field K. These maps are called respectively the *reduced trace* and the *reduced norm* of the element $a \in A$, cf. [Wei74a]:

$\tau(a) = l_A(x_1, x_2, \ldots, x_N)$ a linear form,
$\nu(a) = \Phi_A(x_1, x_2, \ldots, x_N)$ a homogeneous polynomial of degree n.

Since $F(ab) = F(a)F(b)$ by the isomorphism (4.5.36), $\nu(ab) = \nu(a)\nu(b)$. However, in case the algebra A is a division algebra, notice that each non–zero element of A is invertible. Thus the form Φ_A has no non–trivial zero over K. On the other hand if $A \otimes_K L \cong \mathrm{M}_n(L)$, then Φ_A does have a non–trivial zero over L; under this isomorphism the solutions to the equation

$$\Phi_A(x_1, \ldots, x_N) = 0 \quad (x_i \in L) \tag{4.5.37}$$

correspond exactly to degenerate matrices.

We now describe the local invariant (cf. [Chev40], [Se63])

$$\mathrm{inv}_K : \mathrm{Br}\ K \longrightarrow \mathbb{Q}/\mathbb{Z} \tag{4.5.38}$$

in the case when K is a finite extension of $\mathbb{Q}_p$. Let A be a central division algebra (a scew–field) over the field K, $[A : K] = n^2$. The valuation $v = v_K$ of K has a unique extension to a valuation v_A of A, coinciding with v_K on the center of A. For example, one can first extend v over local fields $K(\alpha)$ for $\alpha \in A$ and then use the compatibility of these continued absolute values (in view of the uniqueness property of continuations of absolute values to finite extensions of a local field). Considering the reduction of the algebra A modulo the valuation v_A one checks that A contains a maximal commutative subfield L unramified over the center K, and an element $\delta \in \mathrm{Br}\ K$ corresponding to A splits over L, i.e. $\delta \in H^2(G(L/K), L^\times)$. A maximal unramified extension L may not be unique in A, but all these extensions are conjugate in view of the *theorem of Skolem–Noether.* This theorem states that each automorphism of L in A over K is induced by an inner automorphism of A. Consequently, there exists an element $\gamma \in A$ such that $\gamma L \gamma^{-1} = L$ and the inner automorphism $x \mapsto \gamma x \gamma^{-1}$, restricted to the subfield L, coincides with the Frobenius automorphism $\mathrm{Fr}_{L/K}$. Moreover, the element γ is uniquely defined upto a factor from $L^\times$. Let $v_A : A^\times \to \frac{1}{n}\mathbb{Z}$ be an extension of v_K onto A. Then one can define $\mathrm{inv}_K \delta$ as the image of $v_A(\gamma)$ in the group $(\frac{1}{n}\mathbb{Z})/\mathbb{Z} \subset \mathbb{Q}/\mathbb{Z}$. This definition may be restated, taking into account the fact that the map $x \mapsto \gamma^n x \gamma^{-n}$ is equal to $\mathrm{Fr}^n_{L/K}$ and is thus the identity (since $n = [L : K]$). It therefore follows that the element γ^n commutes with all elements of L and $\gamma^n = c \in L^\times$. This gives us

$$v_A(\gamma) = \frac{1}{n} v_A(\gamma^n) = \frac{1}{n} v_A(c) = \frac{1}{n} v_L(c). \tag{4.5.39}$$

Thus we have that

$$\mathrm{inv}_K \delta = i/n \quad (c = \pi_L^i u),$$

where $u \in \mathcal{O}_L^\times$, π_L is a uniformizing element in L, i.e. $v_L(\pi_L) = 1$, $v_L(u) = 0$.

Passing to the global case, we consider a Galois extension of number fields L/K with Galois group $G = G(L/K)$. Let $G^v \subset G$ denote the decomposition group of an extension w of a place v to L. If the extension L/K is Abelian then we know that the group G^v is uniquely determined by v (cf. §4.4). The inclusion $L \to L_w$ induces a homomorphism

$$\varphi_v : H^2(G, L^\times) \longrightarrow H^2(G^v, L_w^\times). \tag{4.5.40}$$

One verifies that for an element $\alpha \in H^2(G, L)$ the images $\varphi_v \alpha$ vanish for almost all v (all but a finite number): if a cocycle $\{a_{g,h}\} \in Z^2(G, L^\times)$ representing α satisfies the condition $a_{g,h} \in \mathcal{O}_w^\times$ and the extension L_w/K_v is unramified, then $H^i(G^v, \mathcal{O}_v^\times) = 0$ for $i \geq 1$. This fact is deduced from the exact sequence of cohomology groups obtained from the short exact sequence

$$1 \longrightarrow \mathcal{O}_w^\times \longrightarrow L_w^\times \longrightarrow \mathbb{Z} \longrightarrow 0.$$

This is actually a version of Hensel's lemma , cf. §4.3.2.

Thus there exists a well defined map

$$H^2(G, L^\times) \longrightarrow \bigoplus_v H^2(G^v, L_w^\times) \tag{4.5.41}$$

where w is a fixed continuation of a place v and the summaton runs through all places v of K. In this situation the local invariants

$$\mathrm{inv}_{K_v} : H^2(G^v, L_w^\times) \longrightarrow \mathbb{Q}/\mathbb{Z}$$

induce a map

$$\bigoplus_v H^2(G^v, L_w^\times) \longrightarrow \mathbb{Q}/\mathbb{Z}, \tag{4.5.42}$$

which is defined to be the sum of all the local invariants.

The *Minkowski–Hasse Local–Global Principle* states that that the sequence

$$0 \longrightarrow H^2(G, L^\times) \longrightarrow \bigoplus_v H^2(G^v, L_w^\times) \longrightarrow \mathbb{Q}/\mathbb{Z}, \tag{4.5.43}$$

obtained from (4.5.41) and (4.5.42), is exact.

This exact sequence (4.5.43) plays a key role in many arithmetical questions. For example, the statement that (4.5.41) is an embedding is equivalent to saying that for the reduced norm

$$\nu(a) = \Phi_A(x_1, x_2, \ldots, x_N) \tag{4.5.44}$$

the Minkowski–Hasse principle holds, i.e. the form $\nu(a) = \Phi_A(x_1, x_2, \ldots, x_{n^2})$ has a non–trivial zero over L iff it has a non–trivial zero over each completion of L.

The exactness in the middle term $\oplus_v H^2(G^v, L_w^\times)$ describes completely the classes of cenral simple algebras A which split over L. They correspond bijectively to tuples of numbers $i(v)$, $0 \le i(v) < n$, the sum of which is divisible by n; for some algebra A with the class $\delta \in H^2(G, L^\times)$ one has $\mathrm{inv}_{K_v}\varphi_v(\delta) = i(v)/n \in \frac{1}{n}\mathbb{Z}/\mathbb{Z}$.

Finally, the statement that for $\delta \in H^2(G, L^\times)$ one always has

$$\sum_v \mathrm{inv}_{K_v}(\varphi_v(\delta)) = 0 \in \mathbb{Q}/\mathbb{Z},$$

is essentially equivalent to the product formula for local symbols (4.5.38), and to the *global reciprocity law.*

Indeed, if $\alpha = (\alpha_v)_v \in J_K$ is an idele, then the *global Artin symbol* $\theta(\alpha) \in G_K^{\mathrm{ab}}$ is defined as the limit $\theta(\alpha) = \lim_S \prod_{v \in S} \theta_v(\alpha_v)$ where the product is finite, and the local symbols are defined by the condition

$$\chi(\theta_v(\alpha_v)) = \mathrm{inv}_{K_v}(\alpha \cup \Delta_1 \chi) \tag{4.5.45}$$

(see (4.5.32)) for all characters $\chi \in H^1(G_{K_v}^{\mathrm{ab}}, \mathbb{Q}/\mathbb{Z})$.

If $\alpha \in K^\times$, i.e. $\alpha_v = \alpha \in K^\times$ for all v, then for all characters $\chi \in H^1(G_K^{\mathrm{ab}}, \mathbb{Q}/\mathbb{Z})$ one has

$$\chi\left(\prod_v \theta_v(\alpha_v)\right) = \sum_v \mathrm{inv}_{K_v}(\alpha \cup \Delta_1 \chi) = 0,$$

since the element

$$\alpha \cup \Delta_1 \chi \in H^2(G_K^{\mathrm{ab}}, \mathbb{Q}/\mathbb{Z})$$

belongs to the global Brauer group.

In the case when the extension L/K is cyclic, one can construct using purely cohomological methods a canonical isomorphism

$$H^2(G(L/K), L^\times) \cong K^\times / \mathrm{N}_{L/K} L^\times \tag{4.5.46}$$

and the exact sequence (4.5.43) implies the following:

Theorem 4.27 (Hasse's Theorem on Norms). *If $a \in K^\times$ and L/K a cyclic extension, then $a \in \mathrm{N}_{L/K} L$ if and only if $a \in \mathrm{N}_{L_w/K_v} L_w$ for all places v of K.*

In particular, let G be the group of order 2, so that $L = K(\sqrt{b})$. Then $\mathrm{N}_{L/K}(x + y\sqrt{b}) = x^2 - by^2$. Hence a can be represented by the form $x^2 - by^2$ over K iff it can be represented by it everywhere locally, i.e. over every completion of K. This implies that a quadratic form $Q(x, y, z)$ in three variables over K

has a non–trivial zero over K iff it has a non–trivial zero over every completion of K. Passing to arbitrary n we obtain the Minkowski–Hasse theorem, which states that a quadratic form has a non–trivial zero over K iff it has a non–trivial zero everywhere locally, cf. [Chev40], [Cas78].

It was pointed out to us by B.Moroz (MPIM-Bonn), that Hasse's Theorem on Norms may hold for some non-cyclic extensions, providing interesting examples of the validity of the Minkowski–Hasse principle, and Theorem 6.11 of [PlRa83] at p.309 gives an interesting cohomological condition for Hasse's Theorem on Norms to hold.

5

Arithmetic of algebraic varieties

5.1 Arithmetic Varieties and Basic Notions of Algebraic Geometry

5.1.1 Equations and Rings

(cf. [Sha88], [Sha87], [Bou62]). The machinery of algebraic geometry uses commutative rings instead of equations. Replacing a system of equations by a ring is similar to replacing an algebraic number given as a root of a polynomial by the corresponding field (or ring) extension. Consider a system of equations

$$X : F_i(T_j) = 0 \quad (i \in I, j \in J).$$

Here I and J are index sets; the T_j are independent variables; F_i are polynomials from the ring $K[T_j]$ and K is a commutative ring. We shall say that X is defined over K. Now the question arises, which objects should be called solutions of the system X? There is an obvious definition: it is a family $(t_j), j \in J$, of elements of K such that $F_i(t_j) = 0$ for all $i \in I$. However, this definition is too restrictive. We could also be interested in solutions not belonging to K, for example the complex roots of a polynomial with rational coefficients. In general, consider a K–algebra L.

5.1.2 The set of solutions of a system

Definition 5.1. *An L–valued solution of X is a family $(t_j), j \in J$ of elements of L such that $F_i(t_j) = 0$ for all $i \in I$. The set of all such solutions is denoted $X(L)$.*

Since every ring is a $\mathbb{Z}$–algebra, if X is defined over $\mathbb{Z}$ then we can consider its solutions with values in any ring. Let $f : L_1 \to L_2$ be a K–algebra homomorphism, ie. a homomorphism of rings and of K–modules. Then for any L_1–valued solution (t_j) of X, $(f(t_j))$ is an L_2–valued solution. Hence f induces a map $X(L_1) \to X(L_2)$.

5.1.3 Example: The Language of Congruences

Let n be an integer of the form $4m+3$. Here is the classical proof that n is not a sum of two integral squares: if it were then there would be a solution to the congruence $T_1^2+T_2^2 \equiv 3 \bmod 4$, whereas a short case–by–case check shows that this is unsolvable. From our new viewpoint this argument can be rephrased as follows. Let X denote the equation $T_1^2+T_2^2-n=0$ $(K=\mathbb{Z})$. We want to prove that $X(\mathbb{Z})=\emptyset$. Consider $\mathbb{Z}/4\mathbb{Z}$ as $\mathbb{Z}$–algebra via the reduction homomorphism $\mathbb{Z}\to\mathbb{Z}/4\mathbb{Z}$. There is then an induced map $X(\mathbb{Z})\to X(\mathbb{Z}/4\mathbb{Z})$. If $X(\mathbb{Z})$ were non–empty, $X(\mathbb{Z}/4\mathbb{Z})$ would also be non–empty, which is false. In general, for any system X over $\mathbb{Z}$, if $X(L)$ is empty for some algebra L, then $X(\mathbb{Z})$ is empty. In practice one usually tests for solutions in the finite rings $\mathbb{Z}/m\mathbb{Z}$ and the real numbers $\mathbb{R}$. A more satisfactory theoretical formulation uses p–adic fields and the ring of adèles (see Chapter 4, §4.3).

5.1.4 Equivalence of Systems of Equations

Definition 5.2. *Two systems of equations X and Y with one and the same family of indeterminates over a ring K are called equivalent if $X(L)=Y(L)$ for each K–algebra L. Among all systems equivalent to a given one X, there is a largest one. Its left hand sides form the ideal P generated in $K[T_j]$ by the $F_i(T_j)$. In order to see that this is equivalent to X, it suffices to take $L=K[T_j]/P$.*

5.1.5 Solutions as K-algebra Homomorphisms

We summarize the results of our discussion. Starting with the system X as above, we construct the algebra $A=K[T_j]/P$. Then for any K–algebra L we have a natural identification

$$X(L)=\mathrm{Hom}_K(A,L).$$

The system X is called solvable, if $X(L)$ is non–empty for some non–trivial (that is, with $0\neq 1$) K–algebra L. One sees that X is solvable iff 1 is not contained in P.

We have established the equivalence of two languages: systems of equations up to equivalence and algebras with a marked family of generators. Forgetting about the generators, we identify further those systems of equations that are related by invertible changes of variables. Each element of A can play the role of an indeterminate in a suitable system. The value taken by this indeterminate at a given solution is equal to its image with respect to the homomorphism $A\to L$ corresponding to this solution.

In classical algebraic geometry, an (affine) algebraic variety over an algebraically closed field $K=\overline{K}$ is defined to be the set $Z\subset K^n$ of common zeroes of a system of polynomials

$$F_i(T_1, \ldots, T_n) \in K[T_1, \ldots, T_n].$$

The ring of regular algebraic functions on Z is by definition,

$$A = K[Z] = K[T_1, \ldots, T_n]/P_Z,$$

where P_Z is the ideal consisting of all polynomials vanishing on Z. Obviously, A is a finitely generated K–algebra without nilpotents. Conversely any such algebra is of the type $K[Z]$.

The abstract notion of a scheme allows us to consider an arbitrary commutative ring A as a set of functions on a space $\mathrm{Spec}(A)$.

5.1.6 The Spectrum of A Ring

Definition 5.3. *The set of all prime ideals of a (commutative) ring A (distinct from A) is called the spectrum of A and is denoted* $\mathrm{Spec}(A)$. *An element $x \in \mathrm{Spec}(A)$ is called a point of the spectrum; the corresponding ideal is denoted $p_x \subset A$.*

Recall that an ideal $p \subset A$ is prime iff the quotient ring A/p has no zero divisors. We shall denote the field of fractions of A/p_x by $R(x)$.

5.1.7 Regular Functions

Each element f of A defines a function on $\mathrm{Spec}(A)$ whose value at a point x is the residue class $f(x) = f \bmod p_x$ considered as an element of $R(x)$. Two distinct elements of A may take the same values at all points of the spectrum. This happens iff their difference belongs to the intersection of all prime ideals of A, i.e. to the ideal of all nilpotent elements of A (cf. [Bou62], [SZ75]). For this reason, the rings of functions of classical algebraic geometry usually contained no nilpotents. However, this restriction is unnatural even in many classical situations, since nilpotents arise geometrically when an algebraic variety depending on a parameter degenerates in a certain way (e.g. a polynomial acquires multiple roots). For this reason nilpotents are allowed in modern algebraic geometry, and all elements of A are thought of as pairwise distinct regular functions on the spectrum.

We now define a canonical topology on $\mathrm{Spec}(A)$. A minimal consistency requirement of this topology with a given set of functions is that the vanishing sets of all functions are closed.

5.1.8 A Topology on Spec(A)

For any subset $E \subset A$, denote by $V(E) \subset \mathrm{Spec}(A)$ the set of all points $x \in \mathrm{Spec}(A)$ for which $f(x) = 0$ for all $f \in E$. The family $\{V(E)\}$ consists of all closed sets of a topology on $\mathrm{Spec}(A)$ called the Zariski, or spectral topology.

Each ring homomorphism $\varphi : A \to B$ induces a continuous map

$$^a\varphi : \mathrm{Spec}(B) \to \mathrm{Spec}(A).$$

By definition for $y \in \mathrm{Spec}(B)$, we have

$$p_{^a\varphi(y)} = \varphi^{-1}(p_y).$$

Each set $V(E)$ is itself a prime spectrum: $V(E)$ can be identified with $\mathrm{Spec}(A/P_E)$ where P_E is the ideal generated by E. This identification is induced by the canonical homomorphism

$$A \to A/P_E.$$

There is also an important basis of open subsets of $\mathrm{Spec}(A)$ consisting of the sets $D(f) = \mathrm{Spec}(A[1/f])$ for $f \in A$. In fact for each $E \subset A$ we have $\mathrm{Spec}(A)\backslash V(E) = \cup_{f\in E} D(f)$.

The spectra $\mathrm{Spec}(A)$ have very non–classical topologies. As a rule, these spaces are not separable. The closure of any point $x \in \mathrm{Spec}(A)$, can be described as follows:

$$\overline{\{x\}} = \bigcup_{E\subset p_x} V(E) = V\Big(\bigcup_{E\subset p_x} E\Big) = V(p_x) = \{y \in \mathrm{Spec}(A)\ ,\ p_y \supset p_x\}.$$

In particular this space is isomorphic to $\mathrm{Spec}(A/p_x)$, so only the points corresponding to the maximal ideals are closed. If $y \in \overline{\{x\}}$, one sometimes says that y is a specialization of x; this is equivalent to $p_x \subset p_y$. If A has no zero divisors then the ideal $(0) \in \mathrm{Spec}(A)$ corresponds to the generic point of $\mathrm{Spec}(A)$, whose closure coincides with the whole spectrum. One can imagine that the points of $\mathrm{Spec}(A)$ have different depths which can be, loosely speaking, measured by the number of specializations of the generic point necessary to reach a given point. This idea leads to one of the definitions of dimension in algebraic geometry. A sequence $x_0, x_1, \dots, x_n$ of points of a topological space X is called a chain of length n beginning at x_0 and ending at x_n if $x_i \neq x_{i+1}$ and x_{i+1} is a specialization of x_i for all i. The dimension $\dim(X)$ is defined to be the maximal length of such chains.

For example in $X = \mathrm{Spec}\ K[T_1, \dots, T_n]$ (where K is a field) there is a chain $(0) \subset (T_1) \subset \dots \subset (T_1, \dots, T_n)$, so $\dim(X) \geq n$. Similarly, $\dim\ \mathrm{Spec}\ \mathbb{Z}[T_1, \dots, T_n] \geq n + 1$ because there is a chain

$$(0) \subset (p) \subset (p, T_1) \subset (p, T_1, T_2) \subset \dots \subset (p, T_1, T_2, \dots, T_n).$$

Actually, in both cases the strict equality holds.

Passing to the closures instead of the points themselves, one can say that this is a variant of the old "definition" of dimension due to Euclid: points are boundaries of curves, curves are boundaries of surfaces, surfaces are boundaries of solids.

Arithmetical intuition is greatly enhanced when one considers *rings of arithmetical type* (that is, quotient rings $\mathbb{Z}[T_1, \dots, T_n]/P$) and their spectra as analogues of algebraic varieties over fields.

This is in the spirit of the general analogy between numbers and functions. For example, integral extensions of rings correspond to coverings of complex varieties, in particular *Riemann surfaces.* More precisely, let $\varphi : R \subset S$ be an integral extension, so that S is a finitely generated R–module. Then the corresponding contravariant map ${}^a\varphi : \mathrm{Spec}(S) \to \mathrm{Spec}(R)$ is surjective, and its restriction to the subset $\mathrm{Spm}(S)$ of maximal ideals (closed points) is also surjective (cf. [Sha88]).

For $x \in \mathrm{Spec}(R)$, the fiber $({}^a\varphi)^{-1}(x)$ can be described as $\mathrm{Spec}(S/\varphi(p_x)S)$. The structure of the fibers over closed points is described by a decomposition theorem. In particular ${}^a\varphi$ is called unramified at $x \in \mathrm{Spm}(R)$ if $S/\varphi(p_x)S$ has no nilpotents, and is therefore a direct sum of fields.

Example 5.4. Figure 5.1 depicts $\mathrm{Spec}(\mathbb{Z}[i])$ as a covering of $\mathrm{Spec}(\mathbb{Z})$ (cf. [Sha88]). The generic point ω' of $\mathrm{Spec}(\mathbb{Z}[i])$ projects onto the generic point ω of $\mathrm{Spec}(\mathbb{Z})$. The other points are closed. A closed point of $\mathrm{Spec}(\mathbb{Z})$ is essentially a prime p. The fiber $({}^a\varphi)^{-1}((p))$ consists of the prime ideals of $\mathbb{Z}[i]$ dividing p. They are *principal.* There are two of them if $p \equiv 1 (\mathrm{mod}\ 4)$; otherwise there is one. Only 2 is ramified (of multiplicity two).

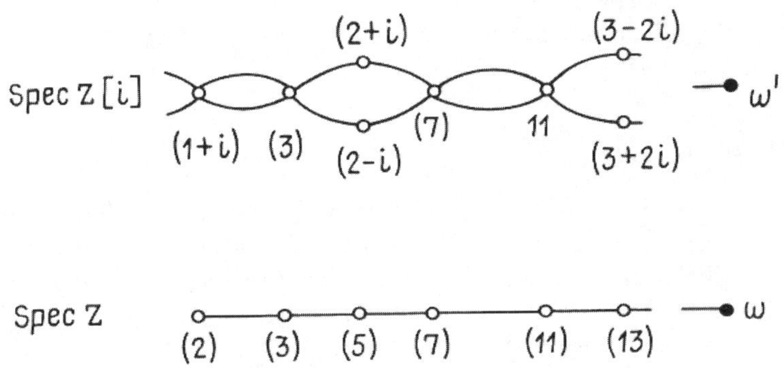

Fig. 5.1.

Notice that $\mathrm{Spec}(\mathbb{Z})$ and $\mathrm{Spec}(\mathbb{Z}[i])$ are one–dimensional (as are algebraic curves). More precisely, $\mathrm{Spec}(\mathbb{Z})$ should be thought of as being an analogue of the affine line, that is, the projective line minus one point. (We shall later explain how one "compactifies" $\mathrm{Spec}(\mathbb{Z})$ by adding the arithmetical infinity). This analogy can be illustrated by two deep theorems of algebraic number theory. The first is *Minkowski's theorem* that $\mathbb{Q}$ has no proper unramified extensions. The second theorem is *Hermite's theorem* that $\mathbb{Q}$ (or any finite extension of $\mathbb{Q}$) has only a finite number of extensions with given ramification points and bounded degree.

These arithmetical facts have their geometric counterparts in the theory of Riemann surfaces: the Riemannian sphere has no non–trivial unramified coverings, and the number of coverings (up to isomorphism) of a given compact Riemann surface X, which are unramified outside of a given finite set of points and have a fixed degree, is finite. To prove these statements, one can use the following formula due to Hurwitz. Let $f : Y \to X$ be a covering of Riemann surfaces; g_X, g_Y their genera and e_P the ramification index of f at a point $P \in Y$. Then

$$2g_Y - 2 = \deg(f)(2g_X - 2) + \sum_{P \in Y} (e_P - 1). \tag{5.1.1}$$

Alongside this one uses an explicit description of the fundamental group $\pi_1(X \backslash S)$ of a Riemann surface with a finite set of points S removed. This group has only finitely many subgroups of a given index.

A more sophisticated version of this analogy (dealing with algebraic curves over number fields instead of finite extensions) was developed by I. R. Shafarevich in his Stockholm ICM talk (cf. [Sha62]). The *finiteness conjectures* stated in this talk prompted a wealth of research which eventually lead to the proof of all these conjectures as well as the Mordell conjecture on the finiteness of the number of rational points on any curve of genus $g > 1$ over a number field ([Fal83], see also §5.5).

5.1.9 Schemes

The notion of a *scheme* is basic to algebraic geometry. An affine scheme is essentially a pair $(\mathrm{Spec}(A), A)$, where A is a commutative ring. More precisely, it is a topological space $\mathrm{Spec}(A) = X$, endowed with a sheaf of local rings $\mathcal{O}_X$ whose ring of sections over an open set $D(f)$ is $A[f^{-1}]$. A *general scheme* X is a topological space X with a structure sheaf $\mathcal{O}_X$ such that $(X, \mathcal{O}_X)$ is locally (in a neighbourhood of each point) isomorphic to an affine scheme (see [Ha77], [Sha88]).

Schemes form a category. Morphisms of affine schemes are defined to correspond bijectively to the homomorphisms of the commutative rings. Morphisms of schemes are given by such homomorphisms locally.

For a commutative ring K, one can define a K–scheme as a morphism $X \to \mathrm{Spec}(K)$. In the category of K–schemes, morphisms should be compatible with the structural morphisms to $\mathrm{Spec}(K)$. Every affine scheme defining X has locally a canonical structure as the spectrum of a K–algebra.

A scheme is called *irreducible* if its topological space is irreducible, i.e. if it is not a non–trivial union of two closed subspaces.

We shall say that X is a scheme of *geometric type* if it can be covered by a finite number of spectra of rings of finite type over a field K. Similarly we say that X is a scheme of *arithmetic type* if it can be covered by a finite number of spectra of rings of finite type over $\mathbb{Z}$.

These two classes have a non–empty intersection consisting of geometric schemes over finite fields $\mathbb{F}_q$. They were and are a standard testing ground for various conjectures in which geometric and arithmetical intuitions are combined. We shall repeatedly turn to this class of schemes. In particular, if $X \to \mathrm{Spec}(\mathcal{O}_K)$ is a scheme of arithmetic type over the ring of integers $\mathcal{O}_K$ of a number field K, we can define for every prime ideal $\mathfrak{p} \subset \mathcal{O}_K$ the reduction $X \bmod \mathfrak{p}$. This is a scheme over the finite field $\mathcal{O}_K/\mathfrak{p}$. Considerable arithmetical information concerning X is encoded in the set all reductions $X \bmod \mathfrak{p}$.

For a scheme X of one of these types, dim X is defined to be the maximal length of a chain

$$Z_0 \subset Z_1 \subset \ldots \subset Z_n, \quad Z_i \neq Z_{i+1},$$

consisting of irreducible subspaces of X. If X is itself irreducible, with generic point x having residue field $R(x)$, then dim X coincides with the so-called Kronecker dimension of $R(x)$, that is, the transcendence degree of $R(x)$ over its prime subfield, enlarged by one if Char $R(x) = 0$. In particular,

$$\dim\ \mathbb{A}^n_{\mathbb{Z}} = \dim\ \mathrm{Spec}\ \mathbb{Z}[x_1, \ldots, x_n] = n + 1.$$

Example 5.5. The projective space $\mathbb{P}^n_K$ over a ring K Consider the polynomial ring $S = K[T_0, \ldots, T_n]$ graded by total degree $S = \oplus_{d \geq 0} S_d$. Put $S_+ = \oplus_{d>0} S_d$. This is a graded ideal. Define $\mathrm{Proj}(S)$ to be the set of all homogeneous prime ideals of S which do not contain S_+. It is a topological space, whose closed subspaces are the sets

$$V(\mathfrak{a}) = \{\mathfrak{p} \in \mathrm{Proj}(S) \mid \mathfrak{p} \supset \mathfrak{a}\}$$

where $\mathfrak{a}$ is a homogeneous ideal of S. In order to turn $\mathrm{Proj}(S)$ into a scheme, put

$$A_i = K[T_0/T_i, \ldots, T_n/T_i], \quad A_{ij} = A_i[T_i/T_j].$$

We can identify $\mathrm{Spec}(A_i)$ with an open subset of $\mathrm{Proj}(S)$ in such a way that

$$\mathrm{Spec}(A_i) \cap \mathrm{Spec}(A_j) = \mathrm{Spec}(A_{ij}).$$

The structure sheaves can also be glued together in a coherent way. As a result, $\mathrm{Proj}(A)$ becomes a K–scheme $\mathbb{P}^n_K$ which is called the projective space over K.

5.1.10 Ring-Valued Points of Schemes

Let $X \to \mathrm{Spec}(K)$ be a K–scheme and L a K–algebra. We define an L–point of X (over K) to be a morphism $\mathrm{Spec}(L) \to X$ over K. Denote the set $\mathrm{Mor}_K(\mathrm{Spec}(L), X)$ of L–points by $X(L)$. If L is a field, we call these points geometric.

Example 5.6. a) Let $X = \mathbb{A}^n_{\mathbb{Z}}$, $K = \mathbb{Z}$, $L = \mathbb{F}_q$. Then an L–point

$$\mathbb{Z}[T_1, \ldots, T_n] \to \mathbb{F}_q$$

is an n–tuple

$$(t_1, \ldots, t_n) \in \mathbb{F}_q^n.$$

Hence Card $X(\mathbb{F}_q) = q^n$.

b) Let $X = \mathbb{P}^n_{\mathbb{Z}}$, $K = \mathbb{Z}$, $L = \mathbb{Z}/N\mathbb{Z}$. An element of $X(\mathbb{Z}/N\mathbb{Z})$ is a class of $(n+1)$–tuples $(t_0 : \ldots : t_n) \in (\mathbb{Z}/N\mathbb{Z})^{n+1}$ such that at least one of the coordinates is invertible. Two tuples are equivalent iff their coordinates differ by a common factor in $(\mathbb{Z}/N\mathbb{Z})^\times$. The ith coordinate is invertible precisely when the point lies in Spec(A_i) (cf. §5.1.9). It is not difficult to count the total number of $\mathbb{Z}/N\mathbb{Z}$–points:

$$\text{Card } \mathbb{P}^n_{\mathbb{Z}}(\mathbb{Z}/N\mathbb{Z}) = N^n \prod_{p|N} \frac{p^{n+1}-1}{(p-1)p^n} = \frac{N^{n+1}}{\varphi(N)} \prod_{p|N} (1 - p^{-(n+1)}). \quad (5.1.2)$$

5.1.11 Solutions to Equations and Points of Schemes

Solving a Diophantine equation or a system of equations is the same as finding a point in a scheme of arithmetic type. In fact, a family of polynomials over a ring K,

$$F_i(T_1, \ldots, T_n) \in K[T_1, \ldots, T_n] \quad (i \in I)$$

generates an ideal $\mathfrak{a} \subset K[T_1, \ldots, T_n]$ and for any K–algebra L, the L–points of the affine scheme Spec$(K[T_1, \ldots, T_n]/\mathfrak{a})$ correspond bijectively to the solutions of $F_i = 0$ in L^n.

If the F_i are homogeneous then we may consider the corresponding projective scheme

$$\text{Proj}(K[T_1, \ldots, T_n]/\mathfrak{a})$$

and its points. For a general algebra L, the relation between L–points and solutions here is somewhat complicated. For example if L is the ring of integers in an algebraic number field, then the set of L–points of $\mathbb{P}^n_{\mathbb{Z}}$ is related to the ideal class group of L. However when L is a field, the L–points correspond to non–zero L–valued solutions upto a homogeneity factor.

Projective space over a field can be obtained from the affine space by adding the hyperplane at infinity. Intuitively the transition to projective schemes is a kind of compactification. For this reason, projective schemes and varieties possess many nice geometric properties which play an important role in arithmetical investigations.

5.1.12 Chevalley's Theorem

Theorem 5.7. *Let X be a subscheme of $\mathbb{P}^n_K$ over a finite field $K = \mathbb{F}_q$ defined by an equation $F(T_0, \ldots, T_n) = 0$, where F is a form of degree d and $n > d$. Then the set $X(\mathbb{F}_q)$ (of projective solutions) is non–empty.*

cf. ([BS85], [War36]).

Denote by N_F the number of solutions of $F = 0$ in $\mathbb{F}_q^{n+1}$, i.e. the number of $\mathbb{F}_q$–points of the corresponding *affine* scheme. We shall prove that $p|N_F$ where p is the prime dividing q. Since $F(0, \ldots, 0) = 0$, this shows that there must also be a non–zero solution.

Obviously $1 - F(T)^{q-1}$ is equal to $1 \in \mathbb{F}_q$ at the points of (the cone over) X and 0 elsewhere. Therefore,

$$\bar{N}_F = N_F \bmod p = \sum_{t \in \mathbb{A}^{n+1}(\mathbb{F}_q)} (1 - F(t)^{q-1}). \tag{5.1.3}$$

We now expand the right hand side of (5.1.3) into a sum of monomials. Most of these will add up to zero. More precisely,

$$\sum_{t \in \mathbb{F}_q^{n+1}} t_0^{i_0} t_1^{i_1} \ldots t_n^{i_n} = 0 \tag{5.1.4}$$

unless all the i_k are non-zero and are divisible by $q - 1$. This can be checked for $n = 0$ directly, and then for general n by expanding the sum in (5.1.4) into the product of $n + 1$ factors.

If a monomial $T_0^{i_0} \ldots T_n^{i_n}$ appears in the expansion of $1 - F(T)^{q-1}$, then necessarily $i_j < q - 1$ for at least one j; otherwise we would have $(q - 1) \deg F(T) \geq n(q - 1)$ contradicting the assumption that $d < n$. Hence finally $\bar{N}_F = 0$, so $p|N_F$.

5.1.13 Some Geometric Notions

In this subsection, we shall briefly review some notions of algebraic geometry over fields, which will be used later. For a detailed treatment we refer the reader to the volumes of this series devoted to algebraic geometry.

i) *Irreducible components.* Every K–variety (a geometric scheme over a field K without nilpotent elements in its structure sheaf) is a finite union of its irreducible components. After a finite algebraic extension of the base field K, an irreducible variety may become reducible (its components may form an orbit with respect to the action of an appropriate Galois group). A variety which remains irreducible after any algebraic extension of the ground field, is called absolutely (or geometrically) irreducible. Each irreducible variety has a well defined dimension.

ii) *Singular points.* A point $x \in V$ can be singular or non–singular (regular). Amongst the many equivalent definitions of regularity, the following is probably the shortest: x is regular iff the completion of the local ring $\mathcal{O}_x$ (with respect to the $\mathfrak{m}_x$*–adic topology* where $\mathfrak{m}_x$ is the maximal ideal) is isomorphic to a ring of formal power series over $k(x) = \mathcal{O}_x/\mathfrak{m}_x$. The regular points form a Zariski open subset of V. If V is given by a homogeneous equation $F(x_1, \ldots, x_n) = 0$ in a projective space, one can obtain additional equations for the subvariety of singular points by putting $\partial F(x)/\partial x_i = 0$.

Intersection points. A point of intersection of two irreducible components is always singular.

Genus and singular points. The existence of singular points can drastically change both the geometry and the arithmetical properties of a variety. For example, a non–singular cubic curve in a projective plane has genus one; its set of rational points over, say, $\mathbb{Q}$ is quite small (cf. §5.3 below). When such a curve acquires a double point, the genus of its non–singular model becomes zero, and its set of rational points becomes much larger.

iii) *Embeddins and heights.* A variety V given abstractly by an affine atlas and gluing rules may or may not be embeddable in a projective space. A variety which is given as a subvariety of a projective space admits in general many more inequivalent embeddings. A choice of such an embedding (if it exists at all) is an extremely important additional structure. In geometry, it allows one to use various induction techniques (fibration by hyperplane sections etc.). In algebra, it governs most of the sheaf cohomology calculations via various finiteness and vanishing results. In arithmetic the choice of an embedding leads to the notion of the height of a rational point, which is used in most of the quantitative problems of the Diophantine geometry.

Divisors and Invertible sheafs. We therefore say a few words about divisors and invertible sheaves, the universally used geometric notions which generalize the ideas of a hyperplane sections and a projective embeddings.

Cartier divisors. Let V be a variety. A (Cartier) divisor on V is given in an affine atlas $V = \cup U_i$ by a family of elements $\{f_i\}$, where f_i is a rational function on U_i. On the intersection $U_i \cap U_j$, we require that $f_i = u_{ij} f_j$ for some regular, regularly invertible function u_{ij}. Two families $\{f_i\}$, $\{g_i\}$ determine the same divisor if $f_i = u_i g_i$ for all i, where u_i is a regular and regularly invertible function on U_i. The divisors form a group $\mathrm{Div}(V)$ under the natural composition: $\{f_i\}\{g_i\} = \{f_i g_i\}$. Every hyperplane section is a divisor. If all the $\{f_i\}$ are regular, the divisor is said to be effective.

Picard group. An invertible sheaf on V is a locally free, one dimensional $\mathcal{O}_V$–module $\mathcal{L}$. The set of all such sheaves upto isomorphism forms a group $\mathrm{Pic}(V)$ with respect to the tensor product. Every divisor D defines an invertible sheaf $\mathcal{O}(D)$: its sections over U_i can be identified with elements of $f_i \mathcal{O}_{U_i}$. Vice versa, a meromorphic section of $\mathcal{L}$ defines a divisor D and an identification $\mathcal{L} \cong \mathcal{O}(D)$. In this way, we have a surjective homomorphism $\mathrm{Div}(V) \to \mathrm{Pic}(V)$.

Ample sheaves. A projective space has a canonical invertible sheaf $\mathcal{O}(1)$. Each morphism $\varphi : V \to \mathbb{P}^n$ determines the invertible sheaf $\mathcal{L} = \varphi^*(\mathcal{O}(1))$. The sheaves $\mathcal{L}$ obtained from the closed embeddings φ are called *very ample.* $\mathcal{L}$ is called *ample* if some positive power of it is very ample.

iv) *Canonical sheaf.* If V is non–singular, one can define the locally free $\mathcal{O}_X$–module of 1–forms Ω^1_X whose rank is $d = \dim(V)$. Its dth exterior power ω_V is called the *canonical sheaf* of V. Its numerical properties have a very strong influence on the arithmetical properties of V (cf. the next section). For $V = \mathbb{P}^n$ we have $\omega_V = \mathcal{O}(-n-1)$, so ω_V^{-1} is ample. Simultaneously the set of rational points could not be larger. When ω_V becomes ample, one conjectures that most rational points are concentrated on a proper Zariski closed subvariety.

5.2 Geometric Notions in the Study of Diophantine equations

5.2.1 Basic Questions

Consider a finite system of polynomial equations over $\mathbb{Z}$. As was explained in §5.1, such a system defines an arithmetic scheme X, its set of integral points $X(\mathbb{Z})$ and sets $X(L)$ for more general rings L, for example, rings of integers $\mathcal{O}$ of algebraic number fields.

Let X be a smooth projective algebraic variety over a number field K with the maximal order $\mathcal{O} = \mathcal{O}_K$. In this case the K-points of X coincide with its $\mathcal{O}$-points, so we shall speak about $X(K)$ rather than $X(\mathcal{O})$.

In number theory, one is interested in properties of K-rational points $X(K)$ on X. In algebraic geometery one studies the properties of $X(\mathbb{C})$ considered as a topological space, analytic manifold, or algebraic variety (or, more generally, one studies $X(L)$ for various algebraically closed fields L). Geometric methods in the theory of Diophantine equations are used in order to relate the geometry of $X(\mathbb{C})$ to the arithmetical properties of $X(K)$.

The relevance of such methods is most evident for congruences, or, more generally, varieties over finite fields. A.Weil in his famous note (cf. [Wei49]) formulated several conjectures concerning the numbers of points of such schemes and suggested that there should exist a cohomology theory in finite characteristic such that a Lefschetz type theorem in this theory would imply (a part of) these conjectures. A.Grothendieck and his collaborators developed such a cohomology theory, and P.Deligne accomplished the realization of Weil's programme by proving the Weil–Riemann conjecture in full generality. In Chapter 6, §6.1 we briefly describe these results.

In this section we survey some known connections between geometry and arithmetic over number fields.

A) Is $X(K)$ non-empty?
B) Is $X(K)$ finite or infinite? Is it dense in X?
C) If $X(K)$ is infinite, what is the order of growth of

$$N(H;B) := \text{Card}\ \{x \in X(K) | H(x) \leq B\}?$$

Here H is a (exponential) certain "height" function, e.g. in fixed coordinates, for $X \subset \mathbb{P}^n$,

$$H(x_0, \ldots, x_n) = \prod_{v \in \text{Val}(K)} \max_i(|x_i|_v).$$

D) Can one, at least in some sense, describe the set $X(K)$ as a finitely generated structure?

For any of these questions one may also be interested in algorithmic solutions. Matiyasevich's theorem is, however, a strong indication that one will not be able to answer these questions for *all* varieties. Instead, one could try to prove conditional statements of the type "if $X(\mathbb{C})$ has such-and-such geometric properties (is a one-dimensional irreducible non-singular variety, projective algebraic group, flag space ...), then $X(K)$ has such-and-such arithmetical properties (is finite, finitely generated; $N(H;B)$ grows as a power of $B\ldots$)". One expects that in the *stable* range, allowing for a finite extension of K and restricting to a Zariski open subset U of X, there is a relation between the set of rational points on U and geometric invariants of X.

Below we shall briefly discuss some results of the latter type, grouping them around questions A) – D).

5.2.2 Geometric classification

One of the main geometric invariants of a smooth projective variety X is its canonical class K_X (see §5.1). Algebraic varieties can be classified, very roughly, according to the ampleness of the anticanonical class $-K_X$, resp. K_X. Varieties with ample $-K_X$ are called *Fano*, with ample K_X - varieties of *general type* and *intermediate type* varieties, otherwise. In finer classification theories, and in many arithmetic applications, one has to allow "mild" singularities and to introduce further invariants such as Kodaira dimension, cones of *effective* and *ample* divisors etc.

In dimension one the above classification coincides with the topological classification of Riemann surfaces: genus 0, ≥ 2, resp. 1. Fano varieties in dimension two are called *Del Pezzo* surfaces. Over an algebraically closed field, these are:

$$\mathbb{P}^2, \mathbb{P}^1 \times \mathbb{P}^1, S_d$$

where S_d is the blowup of $\mathbb{P}^2$ at $9-d$ points, and the degree $d = 1, \ldots, 8$. Surfaces of intermediate type include: abelian surfaces and their quotients, K3 surfaces and Enriques surfaces. The classification of Fano varieties in dimension three was a major achievement by Iskovskikh and Mori–Mukai, cf. [Isk77], [Isk78], [MoMu84], [MoMu03], completing the work of the Italian school, notably G. Fano. Examples are cubics, quartics or double covers of $\mathbb{P}^3$ ramified in a surface of degree 6. Interesting three-dimensional varieties of intermediate type are Calabi-Yau threefolds. One knows that in every dimension, the number of families of Fano varieties is finite.

Fano varieties are, in some sense, similar to projective spaces. As we have seen, Del Pezzo surfaces over $\mathbb{C}$ are birational to $\mathbb{P}^2$. Generally, Fano varieties have the following properties, quite important for arithmetic applications: through every point in X there is a rational curve of anticanonical degree $\leq \dim X + 1$, defined over $\mathbb{C}$, and any two points can be connected by a chain of rational curves. However, it is unknown whether or not all Fano threefolds are dominated by a projective space.

5.2.3 Existence of Rational Points and Obstructions to the Hasse Principle

Let X be an algebraic variety over a number field K. An obvious necessary condition for $X(K)$ to be non-empty is that $X(K_v) \neq \emptyset$, for every completion K_v of K. If this condition is also sufficient we say that X satisfies the *Hasse (or Minkowski–Hasse) principle.*

Using the circle method one can prove the Hasse principle for complete intersections in projective spaces whose dimension is sufficiently large with respect to the degree. B.J. Birch (1962) has proved the following general result.

Let $X \subset \mathbb{P}^{n-1}$ be given by r equations. Assume that the dimension of the subvariety of singular points of X is less than

$$n - 1 - r(r+1)(d-1)2^{d-1}.$$

Then X satisfies the Hasse principle. In particular, it holds for

a) quadrics of dimension ≥ 3 (with number of variables $n \geq 5$);
b) intersections of two quadrics of dimension ≥ 10 ($n \geq 13$);
c) cubic hypersurfaces of dimension ≥ 15 ($n \geq 17$).

One conjectures that this is true for $n \geq 9$ in case b) and $n \geq 10$ in case c); this last conjecture was proved, over $\mathbb{Q}$, by Ch. Hooley (cf. [H88]).

The best known results for the case b) are due to J.-L. Colliot-Thélène, J.-J. Sansuc, and P. Swinnerton-Dyer.

Of course, the case of quadrics is classical. For cubic forms in 3 and 4 variables the Hasse principle may fail. A conceptual approach to *higher* obstructions to the existence of rational points was proposed in [Man70a], [Man72b]. It is based on the Hasse–Minkowski principle for the Brauer group over a number field (see in §4.5 of the previous Chapter, the exact sequence (4.5.43)), and Grothendieck's generalization of the Brauer group for schemes. One has the following diagram

$$\begin{array}{ccccccccc}
 & & \mathrm{Br}(X) & \longrightarrow & \oplus_v \mathrm{Br}(X_v) & & & & \\
 & & \downarrow{\scriptstyle x} & & \downarrow{\scriptstyle (x_v)_v} & & & & \\
0 & \longrightarrow & \mathrm{Br}(K) & \longrightarrow & \oplus_v \mathrm{Br}(K_v) & \xrightarrow{\sum_v \mathrm{inv}_v} & \mathbb{Q}/\mathbb{Z} & \longrightarrow & 0
\end{array}$$

In detail, if X is a scheme over a field K, an element $a \in \mathrm{Br}(X)$ is represented by a family of semi-simple algebras parametrized by X. In particular, for any extension field $L \supset K$ and an L-point $x \in X(L)$, one has a natural specialization $a(x) \in \mathrm{Br}(L)$, with obvious functorial properties. Assume that $X(K_v) \neq \emptyset$ for all v and that for every $(x_v)_v \in X(\mathbb{A})$, where $\mathbb{A}$ is the adèle ring of K, there exists an $a \in \mathrm{Br}(X)$ such that

$$\sum_v \mathrm{inv}_v(a(x_v)) \neq 0.$$

Then $(x_v)_v$ cannot belong to $X(K)$ and one says that X has a non-trivial Brauer–Manin obstruction to the Hasse principle.

One of the simplest examples in which the Brauer–Manin obstruction is non-trivial, is furnished by the projective cubic surface X over $\mathbb{Q}$:

$$z(x+z)(x+2z) = \prod_{i=1}^{3}(x + \theta^{(i)}y + \theta^{(i)2}z),$$

where $\theta^{(i)}$ are the three roots of

$$\theta^3 + 7(\theta+1)^2 = 0$$

(this example is due to Swinnerton-Dyer). Its set of adèlic points is non-empty.

A local analysis shows that one can construct two elements a_1, a_2 of the Brauer group of this surface with the following properties:

i) if $v \neq 7$, the local invariants of $a_i(x_v)$ vanish for every $x_v \in X(\mathbb{Q}_v)$;
ii) for every $x_7 \in X(\mathbb{Q}_7)$, either $\mathrm{inv}_7(a_1(x_7)) \neq 0$, or $\mathrm{inv}_7(a_2(x_7)) \neq 0$.

Hence the Hasse principle fails for this surface.

J.-L. Colliot-Thélène, J.-J. Sansuc and D. Kanevsky have compiled a table of diagonal cubic surfaces $ax^3 + by^3 + cz^3 + du^3 = 0$ with integral coefficients in the range $[-500, 500]$ having rational points everywhere locally, for which the Brauer–Manin obstruction vanishes. A computer search has shown that all these surfaces have rational points. One might therefore conjecture that the vanishing of this obstruction implies the existence of a rational point for all diagonal cubic surfaces, or perhaps all non-singular cubic surfaces, or even all non-singular rational surfaces (i.e. those admitting a birational parametrization by two independent parameters over $\mathbb{C}$). This conjecture has been proved for the so called generalized Chatelet surfaces given by an equation of the form $y^2 - az^2 = P(x)$, where a is not a square and P is a polynomial of degree three or four.

The Brauer–Manin obstruction has been thoroughly investigated for three classes of varieties:

i) rational surfaces;
ii) principal homogeneous spaces of linear algebraic groups, especially algebraic tori;
iii) principal homogeneous spaces of elliptic curves and more generally Abelian varieties.

Historically, iii) was the first example. However, it appeared in a different form in the theory of the Shafarevich–Tate group, whose classical definition will be given in the next section. The connection with the Brauer–Manin obstruction is explained in [Man70a].

J.-L. Colliot-Thélène and J.-J. Sansuc have developed a geometric version of this obstruction, which is called the descent obstruction.

Assume that for a variety X over K we have somehow constructed a family of dominating morphisms $f_i : Y_i \to X$ such that $X(K) = \bigcup f_i(Y_i(K))$. Then one can establish that $X(K)$ is empty by showing that for each Y_i there exists a completion $K_{v(i)}$ such that $Y_i(K_{v(i)}) = \emptyset$. On the other hand, if $X(K)$ is non-empty, and the Y_i are in some sense simpler than X, e.g. rational, one obtains an explicit description of the set $X(K)$.

Colliot-Thélène and Sansuc have developed a systematic way of constructing such families, based on the notion of a *torsor*. They have shown that for non-singular rational varieties these descent families have the following properties:

a) The descent obstruction vanishes iff the Brauer–Manin obstruction vanishes.
b) The Brauer–Manin obstructions for the descent varieties Y_i vanish.

Using the machinery of torsors, Skorobogatov cf. [Sko99] constructed an example of a surface with trivial Brauer–Manin obstruction and not satisfying the Hasse principle. The surface in question has a nontrivial fundamental group and the obstruction may be interpreted via non-abelian descent (see also [Sko01] and [HaSk02]).

5.2.4 Finite and Infinite Sets of Solutions

Once it is established that the set of rational points $X(K)$ on an algebraic variety X over a number field K is not empty one could ask whether this set is finite or, for example, dense in X. First of all, let us describe the results in the case of smooth projective curves:

i) Let X be a curve of genus zero. The X satisfies the Hasse principle. More precisely, X can be given by a homogeneous quadratic equation in $\mathbb{P}^2_K$:

$$aX^2 + bY^2 + cZ^2 = 0$$

and the local conditions can be checked algorithmically. If $X(K) \neq \emptyset$ then X is isomorphic to $\mathbb{P}^1_K$, so that $X(K) = K \cup \{\infty\}$ is Zariski dense in X.
ii) If X is of genus 1 then $X(K)$ can be empty, finite or infinite. Even over $\mathbb{Q}$, one does not know a provably correct algorithm allowing to distinguish between these cases. However, there are algorithms that work in practice. In [Man71] an algorithm was suggested to answer the finite/infinite question when it is known that $X(K)$ is non-empty. If one assumes certain general conjectures on elliptic curves (the Birch–Swinnerton-Dyer conjecture and the Shimura–Taniyama–Weil conjecture, cf. the next section, and [Man71]) then one can deduce the correctness of this algorithm. Moreover, $X(L)$ always becomes infinite over an appropriate finite extension of K.

iii) If X is of genus > 1 then $X(K)$ is always finite. This is the famous *Mordell conjecture*, proved by G. Faltings. For more details see the following three sections.

Note that this rather distinct arithmetic behaviour is well aligned with the classification of one-dimensional algebraic varieties recalled in Section 5.2.2. One hopes that this property persists in higher dimensions as well. Bombieri, for surfaces, and Lang–Vojta, in general, conjectured that rational points on varieties of general type are always contained in proper subvarieties, i.e., they are never Zariski dense. If true, this conjecture would have remarkable consequences: non-uniqueness of the Brauer–Manin obstruction for general hypersurfaces [SarWa], *uniform* (in terms of K) upper bounds for the number of rational points on curves of genus $g \geq 2$ etc., cf. [CHM97].

One may ask for a converse to this conjecture. Note that some care is necessary. First of all, one has to allow finite extensions of the ground field (already conics may have no rational points at all). Thus we ask for *potential density*, i.e., Zariski density after a finite extension of the ground field. Secondly, X may not be of general type while admitting an étale cover which dominates a variety of general type. In this case, rational points on X cannot be Zariski dense. In dimension two, rational points are potentially dense on

i) Del Pezzo surfaces;
ii) abelian, Enriques and K3 surfaces with an elliptic fibration or an infinite automorphism group [BoTs99].

All Fano threefolds, except double covers of $\mathbb{P}^3$ ramified in a surface of degree 6, are known to satisfy potential density [HaTsch], [BoTs2000].

To get an idea how these results are proved, assume that there is a non-trivial rational map $f : C \to X$, where C is a curve of genus 0 or 1 with $C(K)$ infinite. Then, of course, $X(K)$ is also infinite. Families of such embedded curves can often be constructed by geometric methods.

Examples.

a) Every $a \in K^*$ can be represented in an infinite number of ways as a sum of three cubes in K. In fact, one representation is given by the identity

$$a = \left(\frac{1}{3^2a^2 + 3^4a + 3^6}\right)^3 ((a^3 - 3^6)^3 + (-a^3 + 3^5a + 3^6)^3 + (3^3a^2 + 3^5a)^3).$$

The geometric picture is as follows: For any non-singular cubic surface X and any point $x \in X(K)$, denote by C_x the intersection of X with the tangent plane to X at x. If x does not belong to a line in X then C_x is a plane cubic curve with a double point at x. Hence it has genus zero and a rational point (cf. Part I, §1.3). (This argument must be modified in certain degenerate cases).

b) Euler conjectured in 1769 that the equation

$$x^4 + y^4 + z^4 = u^4 \tag{5.2.1}$$

has no non-trivial integral solutions. This conjecture was disproved by N.D. Elkies (1988). He found a solution

$$2682440^4 + 15365639^4 + 18796760^4 = 20615673^4$$

and proved that there are in fact infinitely many solutions by constructing an elliptic curve lying on (5.2.1) with infinitely many points. Potential density of rational points on this quartic, and in fact on any smooth quartic surface containing a line, has been proved in [HaTsch], [BoTs99]. Let us sketch the geometry behind this result: let X be a quartic surface with a line ℓ and consider the one-parameter family of hyperplanes $\mathbb{P}^2 \subset \mathbb{P}^3$ containing ℓ. For each $\mathbb{P}^2$ in this family, its intersection with X is a curve of degree 4 containing ℓ. The residual curve to ℓ is a plane curve of degree 3 intersecting ℓ in three points. Thus X admits a fibration over $\mathbb{P}^1$ with generic fiber a curve of genus 1 and a *rational* trisection ℓ. Generically, this implies that rational points on X are Zariski dense already over the ground field. In some degenerate situtions one has to pass to a finite extension to insure Zariski density of rational points.

c) Of course, we can find even more points on X if we manage to construct maps $\mathbb{P}^n \to X$ or $A \to X$, where A is an Abelian variety with large $A(K)$ etc. Many geometric methods for such constructions are known. For example, the diagonal quartic threefold

$$x^4 + y^4 + z^4 + t^4 + u^4 = 0$$

is *geometrically* unirational, i.e., dominated by $\mathbb{P}^3$. It is unknown whether or not every smooth quartic threefold is unirational.

d) Here is another general method of proving that $X(K)$ is infinite: if X has an infinite automorphism group G, an orbit Gx of a point x can be infinite. Examples of K3 surfaces with infinite automorphism groups are hypersurfaces of degree $(2,2,2)$ in $\mathbb{P}^1 \times \mathbb{P}^1 \times \mathbb{P}^1$.

5.2.5 Number of points of bounded height

Let us start with a heuristic argument. Consider a system of equations

$$F_i(x_0, \ldots, x_n) = 0, \quad i = 1, \ldots, r, \tag{5.2.2}$$

where F_i is a form of degree d_i with integral coefficients. Put

$$N(B) = \text{Card}\,\{(x_0, \ldots, x_n) \in \mathbb{Z}^{n+1} \mid H(x) := \max(|x_i|) \leq B\}.$$

To guess the order of growth of $N(B)$, we may argue as follows. First note that there are about B^{n+1} points in $\mathbb{Z}^{n+1}$ whose heights are $\leq B$. Secondly

F_i takes roughly B^{d_i} values at these points. Assuming that the probability of taking the zero value is about B^{-d_i}, and that these events are independent for different i's, we get

$$N(B) \cong B^{n+1-\sum d_i}. \tag{5.2.3}$$

The power on the right hand side of (5.2.3) has a nice geometric interpretation: if the projective variety X defined by (5.2.2) is a non-singular complete intersection, then its anticanonical sheaf $-K_X$ is given by the following formula:

$$-K_X \cong \mathcal{O}(n+1-\sum d_i),$$

where $\mathcal{O}(1)$ is induced on X by $\mathcal{O}_{\mathbb{P}^n}(1)$. Hence we can reformulate (5.2.3) in a more general and a more cautious way, taking into account various counterexamples to the over-optimistic formulation (5.2.1): we expect the order of growth of $N(B)$ to be B^a, $a > 0$ when $-K_X$ is ample and $O(B^\epsilon)$ for any $\epsilon > 0$ when K_X is ample, if one deletes from X some "point-accumulating" subvarieties, and if one passes to a sufficiently large ground field.

These conjectures were stated in a precise form by V.V. Batyrev and Yu.I. Manin. We shall add some comments without going into much detail.

a) To obtain a stable picture, allowing to involve geometric notions and constructions, we must pass to finite field extensions.
b) We should consider counting problems with respect to arbitrary invertible sheaves, not only $-K_X$. The latter could fail to be ample, for example, it could be zero.

The first step in this program is the theory of heights, going back to an old construction of A. Weil.

Let X be a projective algebraic variety over a number field K and $\mathcal{L}$ an invertible sheaf on X. Consider all completions K_v of K. Denote by $|\cdot|_v : K \to \mathbb{R}$ the local norm which is the scaling factor of an additive Haar measure with respect to multiplication by elements of K_v. We have the classical product formula $\prod_v |x|_v = 1$ for all $x \in K^\times$. If Λ is a one-dimensional vector space over K, $\| \cdot \|_v$: $\Lambda \to \mathbb{R}$ denotes a norm such that $\| a\lambda \|_v = |a|_v \| \lambda \|_v$ for all $a \in K$ and $\lambda \in \Lambda$. The invertible sheaf $\mathcal{L}$ can be considered as a family of one-dimensional spaces parametrized by X, and one can define an admissible metrization as a family of metrics $\| \cdot \|_v$ for all v, on each fiber of $\mathcal{L}$, with natural continuity properties (cf. Lang S. (1983)). Given such a metrized sheaf $\mathbf{L} = (\mathcal{L}, \| \cdot \|_v)$, the height with respect to it is a function $H_{\mathbf{L}} : X(K) \to \mathbb{R}$ defined by the following formula:

$$H_{\mathbf{L}}(x) := \prod_v \| \mathsf{s}(x) \|_v^{-1}, \tag{5.2.4}$$

where s is a local section of $\mathcal{L}$ not vanishing at x. (Its choice is irrelevant due to the product formula).

For a list of properties of heights, we refer the reader to Lang S. (1983). We mention only the following ones:

i) Up to a function of the type $\exp(O(1))$ $H_{\mathbf{L}}$ does not depend on the choice of metrization and is multiplicative in $\mathcal{L}$. We shall therefore write $H_{\mathcal{L}}$ instead, if we are interested only in questions invariant with respect to such choice.
ii) If $\mathcal{L}$ is ample and $U \subset X$ is a Zariski open subset then the number

$$N_U(\mathcal{L}; B) := \text{Card } \{x \in U(K) \mid H_{\mathcal{L}}(x) \leq B\}$$

is finite for every B.
iii) We have

$$N_{\mathbb{P}^n}(-K_X; B) = cB(1 + o(1)) \tag{5.2.5}$$

for all $n > 0$ and number fields K (this is Schanuel's theorem, Schanuel S. (1979)).

A natural generalization of (5.2.5) is the following

Conjecture 5.8 (Linear Growth – first version). Let X be smooth, with ample $-K_X$, and let r denote the rank of the Picard group of X. Then there exists a sufficiently small Zariski open subset such that for all sufficiently large ground fields one has

$$N_U(-K_X; B) = cB\log(B)^{r-1}(1 + o(1)). \tag{5.2.6}$$

Clearly, the conjecture cannot be true for varieties without rational points or for cubic surfaces containing rational lines, since each such line would already contribute about B^2 rational points to the asymptotic. These are the obvious necessary conditions.

Thus, if X is a cubic surface such that all 27 lines on X are defined over the ground field and $U \subset X$ is the complement to these lines then

$$N_U(-K_X; B) = cB\log(B)^6(1 + o(1)).$$

Lower bounds of this shape have been proved over $\mathbb{Q}$ in [SSw-D]. Non-trivial upper bounds are unknown.

Now consider the variety $X \subset \mathbb{P}^3 \times \mathbb{P}^3$ given by the $(1, 3)$-form

$$\sum_{j=0}^{3} x_j y_j^3 = 0$$

over a field K containing $\sqrt[3]{1}$. The projection to the x-coordinates exhibits X as a fibration over $\mathbb{P}^3$ with generic fiber a cubic surface. A Zariski dense set of fibers corresponds to cubic surfaces with all 27 lines defined over K, each contributing $B\log(B)^6$ to $N(B)$. However, the rank of the Picard group of X

is 2 and Conjecture 5.8 predicts $B \log(B)$ points of $-K_X$-height bounded by B, leading to a contradiction [BaTsch98a]. A refined approach to the Linear Growth conjecture, taking into account such fibration structures, is explained in [BaTsch98b].

On the other hand, varieties closely related to linear algebraic groups do satisfy Conjecture 5.8, see its refinement by Peyre [Pey95] and its generalization to arbitrary ample line bundles in [BatMan] and [BaTsch98b]. Precise asymptotics, compatible with the above conjectures are known for:

- smooth complete intersections of small degree (for example, [Bir61]);
- split smooth Del Pezzo surface of degree 5 over $\mathbb{Q}$ [dlBre02];
- generalized flag varieties [FMTsch]
- toric varieties [BaTsch98a];
- smooth equivariant compactifications of G/U - (horospherical varieties), where G is a semi-simple group and $U \subset G$ a maximal unipotent subgroup [StTsch];
- smooth equivariant compactifications of $\mathbb{G}_a^n$ [Ch-LT02];
- smooth bi-equivariant compactifications of unipotent groups [ShT04];
- wonderful compactification of some semi-simple algebraic group of adjoint type [ShT-BT04a], [ShT-BT04b].

Very little is known for general higher-dimensional varieties. Geometric arguments, based on Mori's theorem that every point on a Fano variety X lies on a rational curve of degree at most $\dim X + 1$, imply that

$$N_U(\mathcal{L}; B) > cB^{\beta(U,\mathcal{L})}$$

for any dense Zariski open subset $U \subset X$, sufficiently large K, and some positive constants $c > 0$, $\beta(U, L) > 0$. Batyrev and Manin state conjectures about the best possible values of $\beta(U, L)$ and relate them to Mori's theory.

Further developments are reflected in the book [PeyTsch01].

5.2.6 Height and Arakelov Geometry

S.Yu.Arakelov (cf. [Ara74a] and §III.2) had the brilliant idea of considering *Hermitian metrizations* of various linear objects related to algebraic varieties such as invertible sheaves, tangent bundles etc., in order to compactify arithmetic schemes over number fields at the arithmetical infinity. In particular, each curve has a well defined minimal model over $\mathcal{O}$ which is called an arithmetical surface (since we added an arithmetical dimension to the geometric one). Adding metrics at infinity to this, Arakelov developed the intersection theory of arithmetical divisors. Heights in this picture become the (exponentiated) intersection index, see [Ara74b], [La88].

This theory was vastly generalized by H.Gillet and C.Soulé [GS91], [GS92], [SABK94], following some suggestions in [Man84].

Figure 5.2 is a visualization of a minimal arithmetical surface (this notion was defined and studied by I.R.Shafarevich (cf. [Sha65], [Sha66]). Its fibers over the closed points of Spec($\mathcal{O}$) can be non-singular ("non-degenerate", or with "good reduction") or singular (having "bad reduction"). Rational points of the generic fiber correspond to the horizontal arithmetical divisors; there are also vertical divisors (components of closed fibers) and "vertical divisors at infinity" added formally, together with an *ad hoc* definition of their intersection indices with other divisors defined via Green's functions, (see §III.2).

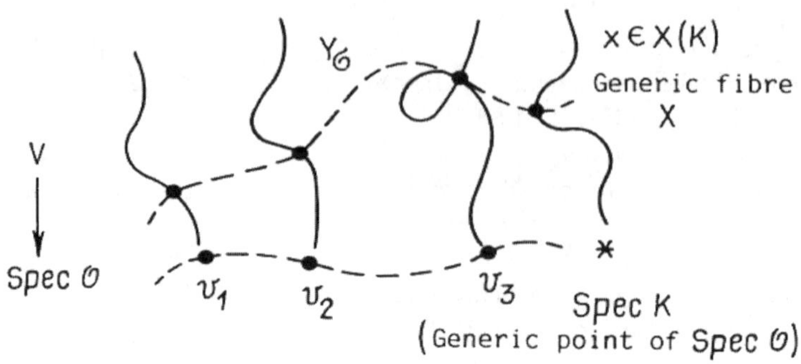

Fig. 5.2.

Arakelov's picture and the theory of heights played a prominent role in Faltings' proof and the subsequent development of his work. We postpone a more detail discussion of Arakelov geometry and its relation with Non-Commutative geometry to Chapter 8.

5.3 Elliptic curves, Abelian Varieties, and Linear Groups

5.3.1 Algebraic Curves and Riemann Surfaces

An algebraic curve is a one-dimensional algebraic variety over a field K. Usually we shall tacitly assume it to be irreducible. Every algebraic curve can be obtained by deleting a finite number of points from a projective curve. For every projective curve C, there exists a non-singular projective curve C' and a morphism $C' \to C$ which is an isomorphism outside of singular points of C. The curve C' is called a (complete) non-singular model of its function field. It is uniquely defined (upto isomorphism) by this function field.

The *genus* g of a projective non-singular curve C (and its function field) can be defined (or calculated) in many ways. Here are some of them:

i) It is the dimension of the space $\Gamma(\omega)$ of regular differential 1-forms on C (the differentials of the first kind).
ii) If $K = \mathbb{C}$ then g is the topological genus (the number of handles) of the Riemann surface $C(\mathbb{C})$ of complex points of C.
iii) Consider a projective embedding $C \subset \mathbb{P}^n_K$. In general one can take $n \geq 3$ but not $n = 2$: our curve may have no non-singular plane model. However, C always has a plane projection with only simple double points. Let d be its degree and ν the (geometric) number of double points. Then

$$g = \frac{(d-1)(d-2)}{2} - \nu.$$

The basic theorem on algebraic curves is the Riemann-Roch theorem. To state this theorem we require some definitions.

Let D be a divisor on C. It has a *degree* $\deg(D)$: a Cartier divisor on a non-singular curve can be identified with a formal linear combination of (geometric) points, and the degree is the sum of the coefficients of this linear combination.

Recall that each invertible sheaf $\mathcal{L}$ is isomorphic to a sheaf of the type $\mathcal{O}(D)$ (cf. §5.1.13). Although D is not uniquely defined by $\mathcal{L}$, its degree is. We may therefore define $\deg(\mathcal{L}) = \deg(D)$. In particular $\deg(\omega_C) = 2g - 2$, where g is the genus of C. A divisor $K = K_C$ such that $\omega \cong \mathcal{O}(K)$ is called a *canonical divisor.* A sheaf $\mathcal{L}$ is ample iff its degree is positive.

For a divisor D, put $l(D) = \dim\ \Gamma(\mathcal{O}(D))$. The Riemann – Roch theorem for curves can be stated as follows:

$$l(D) - l(K - D) = \deg(D) - g + 1. \tag{5.3.1}$$

5.3.2 Elliptic Curves

We shall call a non-singular projective curve X of genus one with a non-empty set $X(K)$ of K-points an *elliptic curve.* An elliptic curve has exactly one (upto constant factor) differential of the first kind. The divisor of this is zero. In other words $\omega_X \cong \mathcal{O}_X$. From the Riemann-Roch theorem (5.3.1) it follows that $l(D) = \deg(D)$ for $\deg(D) > 1$. We can use this to show that i) X is an algebraic group; ii) X is isomorphic to a plane cubic curve. To prove i) choose a point $o \in X(K)$. For any two points $x, y \in X(K)$ let $D = x + y - o$. Since $\deg(D) = 1$ we have $l(D) = 1$. It follows that there exists a unique (upto constant factor) function f whose divisor is $x + y - o - z$. Define $x * y := z$. One can check directly that $*$ is a commutative group law on $X(K)$ (with identity o). Actually, one can ameliorate this construction in order to define the algebraic addition law which is a morphism $* : X \times X \to X$, verifying the standard axioms.

To prove ii) choose a non-constant section $f \in \mathcal{L}(2o)$. Then f has a pole of order precisely two, since sections of $\mathcal{L}(o)$ are constants. Furthermore $l(3o) = 3$, so there is a section $h \in \mathcal{L}(3o)$ with a pole of order three at o. From f and h we can construct seven sections of $\mathcal{L}(6o)$: 1, f, f^2, f^3, h, fh, fh^2, whereas $l(6o) = 6$. Hence these seven sections are connected by a linear relation

$$a_0 + a_1 f + a_2 f^2 + a_3 f^3 + b_0 h + b_1 fh + b_2 fh^2 = 0. \qquad (5.3.2)$$

Equation (5.3.2) defines a smooth affine cubic curve. Its projective completion is a non-singular projective plane model Y of X. The identity point $o \in X(K)$ corresponds to the infinite point $(0:1:0)$ of Y, and the group law $*$ becomes the law described in Part I, §1.3.2 in terms of secants and tangents.

Making additional linear changes of variables we may reduce (5.3.2) to the following (Weierstrass) normal forms over a field K:

$$y^2 + a_1 xy + a_3 y = x^3 + a_2 x^2 + a_4 x + a_6,$$

where $a_1, a_2, a_3, a_4, a_6 \in K$ and

$$\Delta = -b_2^2 b_8 - 8b_4^3 - 27b_6^2 + 9b_2 b_4 b_6 \neq 0,$$

where

$$b_2 = a_1^2 + 4a_2, \quad b_4 = 2a_4 + a_1 a_3, \quad b_6 = a_3^2 + 4a_6.$$

The notation $j = \dfrac{c_4^3}{\Delta}$ is used, where

$$c_4 = b_2^2 - 24b_4, c_6 = -b_2^3 + 36b_2 b_4 - 216b_6.$$

Then this equation can be further simplified using the transformation $x \mapsto u^2 x' + r$, $y \mapsto u^3 y' + su^2 x' r + t$ in order to obtain the following (cf. [Ta73], [Kob87] :

1) For $p \neq 2, 3$:

$$y^2 = x^3 + a_4 x + a_6 \text{ with } \Delta = -16(4a_4^3 + 27a_6^2) \neq 0. \qquad (5.3.3)$$

2) For $p = 2$ we have the condition $j = 0$ is equivalent to $a_1 = 0$, and the equation transforms as follows: if $a_1 \neq 0$ (i.e. $j \neq 0$), then choosing suitably r, s, t we can achieve $a_1 = 1$, $a_3 = 0$, $a_4 = 0$, and the equation takes the form

$$y^2 + xy = x^3 + a_2x^2 + a_6, \tag{5.3.4}$$

with the condition of smoothness given by $\Delta \neq 0$. Suppose next that $a_1 = 0$ (i.e. $j = 0$), then the equation transforms to

$$y^2 + a_3y = x^3 + a_4x + a_6, \tag{5.3.5}$$

and the condition of smoothness in this case is $a_3 \neq 0$.

3) For $p = 3$:

$$y^2 = x^3 + a_2x^2 + a_4x + a_6, \tag{5.3.6}$$

(here multiple roots are again disallowed).

The proper Weierstrass form (in the case (5.3.3)) is

$$y^2 = 4x^3 - g_2x - g_3. \tag{5.3.7}$$

The *discriminant*

$$\Delta = g_2^3 - 27g_3^2 \tag{5.3.8}$$

does not vanish. The coefficients g_2 and g_3 are defined upto the substitution $g_2 \mapsto u^4g_2$, $g_3 \mapsto u^6g_3$ with $u \in K$. The modular, or absolute invariant j of our elliptic curve is defined to be

$$j = 2^6 3^3 \frac{g_2^3}{g_2^3 - 27g_3^2} = 1728\frac{g_2^3}{\Delta}. \tag{5.3.9}$$

Two elliptic curves have the same absolute invariant iff they become isomorphic over an algebraic closure of the ground field K. The classical Weierstrass form (5.3.7) emerged in the theory of complex parametrizations of the complex elliptic curves.

The Riemann surface $E(\mathbb{C})$ of an elliptic curve E defined over $\mathbb{C}$, is a complex torus, that is, a quotient $\mathbb{C}/\Lambda$ where Λ is a lattice

$$\Lambda = \{z = n_1 + n_2\tau | n_1, n_2 \in \mathbb{Z}, \mathrm{Im}(\tau) > 0\}. \tag{5.3.10}$$

The connection between this analytic description of E and an algebraic one is based on the identification of rational functions on E with Λ–periodic meromorphic functions on $\mathbb{C}$, i.e. elliptic functions.

Weierstrass considered the following basic functions:

$$\wp(z) = \wp(z, \Lambda) = \frac{1}{z^2} + \sum_{\omega \in \Lambda}{}' \left(\frac{1}{(z-\omega)^2} - \frac{1}{\omega^2} \right) \tag{5.3.11}$$

(prime denoting $\omega \neq 0$);

$$\wp'(z) = \wp'(z, \Lambda) = -2 \sum_{\omega \in \Lambda} \frac{1}{(z-\omega)^2}. \tag{5.3.12}$$

These series converge absolutely outside Λ and define elliptic functions. The set of all elliptic functions with periods Λ forms a field which is generated over $\mathbb{C}$ by $\wp(z)$ and $\wp'(z)$. These two functions are related by the equation

$$\wp'(z)^2 = 4\wp(z)^3 - g_2\wp(z) - g_3, \tag{5.3.13}$$

where

$$g_2 = 60 \sum_{\omega \in \Lambda}{}' \frac{1}{\omega^4}, \quad g_3 = 140 \sum_{\omega \in \Lambda}{}' \frac{1}{\omega^6}. \tag{5.3.14}$$

Now if an elliptic curve $E_\tau \subset \mathbb{P}^2_{\mathbb{C}}$ is defined by the equation (5.3.7) with g_2 and g_3 from (5.3.14), we can define a map

$$\mathbb{C}/\Lambda \xrightarrow{\sim} E_\tau(\mathbb{C}) \tag{5.3.15}$$

for which $z \mapsto (\wp(z) : \wp'(z) : 1)$ when z is not in Λ. The point 0 is mapped to the infinite point (0:1:0).

The map (5.3.15) is a complex analytic isomorphism. In order to define its inverse, consider the differential of the first kind

$$dx/y = dx/\sqrt{4x^3 - g_2x - g_3} \tag{5.3.16}$$

on the Riemann surface $E_\tau(\mathbb{C})$. We integrate this form over a path joining a fixed initial point (say, o) with a variable point.

The integral depends on the choice of path, but its image in $\mathbb{C}/\Lambda$ is determined only by the endpoints.

According to a classical theorem due to Jacobi, the discriminant $\Delta = \Delta(\tau)$ of E_τ can be expressed via $\Lambda = \Lambda_\tau$ as

$$\Delta = (2\pi)^{12} q \prod_{m=1}^{\infty} (1 - q^m)^{24} = (2\pi)^{12} \sum_{n=0}^{\infty} \tau(n) q^n \tag{5.3.17}$$

for all $\tau \in \mathbb{C}$ with $\mathrm{Im}(\tau) > 0$, $q = \exp(2\pi i \tau)$. The function $\tau(n)$ is called *Ramanujan's function.* Its first few values are

$$\tau(1) = 1, \quad \tau(2) = -24, \quad \tau(3) = 252, \quad \tau(4) = -1472.$$

The absolute invariant of E_τ is by definition

$$j(\tau) = \frac{1728 g_3(\tau)^3}{\Delta(\tau)} = q^{-1} + 744 + \sum_{n=1}^{\infty} c(n) q^n, \tag{5.3.18}$$

where $c(1) = 196884$, $c(2) = 21493760, \dots$. One can prove that $j(\tau)$ takes all complex values, which shows that every elliptic curve over $\mathbb{C}$ is isomorphic to E_τ for an appropriate τ.

Two curves $E_\tau, E_{\tau'}$ are isomorphic iff $\tau' = \frac{a\tau+b}{c\tau+d}$ for some matrix $\begin{pmatrix} a & b \\ c & d \end{pmatrix} \in \mathrm{SL}(2, \mathbb{Z})$. In fact, a complex analytic isomorphism $\mathbb{C}/\Lambda \xrightarrow{\sim} \mathbb{C}/\Lambda'$ is necessarily induced by multiplication by some $u \in \mathbb{C}^\times$. Therefore, $\Lambda_\tau = u\Lambda_{\tau'}$, so that $(u, u\tau')$ is a basis of Λ_τ and $u = c\tau+d$, $u\tau' = a\tau+b$. The linear transformation is unimodular because $(1, \tau')$, $(u, u\tau')$ and $(1, \tau)$ all define the same orientation of $\mathbb{C}$. We therefore have

$$g_2(\tau') = u^4 g_2(\tau), \quad g_3(\tau') = u^6 g_3(\tau). \tag{5.3.19}$$

To sum up, isomorphism classes of elliptic curves over $\mathbb{C}$ correspond bijectively to points of the quotient space $\mathbb{H}/\Gamma$, where $\mathbb{H}$ is the upper half plane

$$\mathbb{H} = \{\tau \in \mathbb{C} | \mathrm{Im}(\tau) > 0\}, \tag{5.3.20}$$

and the modular group

$$\Gamma = \mathrm{SL}(2, \mathbb{Z}) \tag{5.3.21}$$

acts on $\mathbb{H}$ by fractional linear transformations. The isomorphism $\mathbb{C}/\Lambda_\tau \xrightarrow{\sim}$

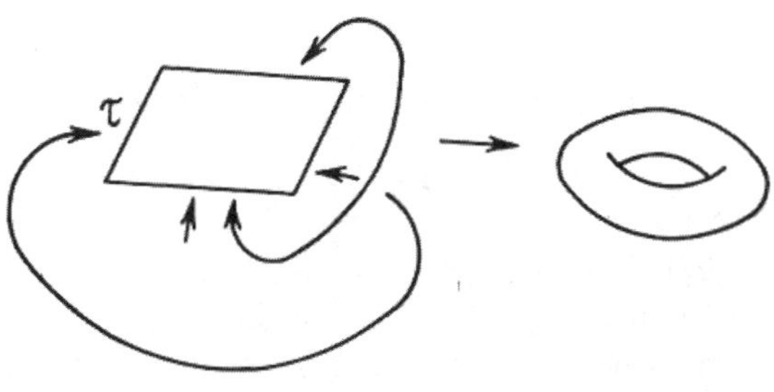

Fig. 5.3.

$E_\tau(\mathbb{C})$ is also compatible with the natural group structures. In terms of elliptic functions, this is reflected in the *addition theorem for elliptic functions*:

$$\wp(z_1 + z_2) = -\wp(z_1) - \wp(z_2) + \frac{1}{4}\left(\frac{\wp'(z_1) - \wp'(z_2)}{\wp(z_1) - \wp(z_2)}\right)^2. \tag{5.3.22}$$

In terms of the coordinates (x, y) satisfying (5.3.7), we have

$$x_3 = -x_1 - x_2 + \frac{1}{4}\left(\frac{y_1 - y_2}{x_1 - x_2}\right)^2,$$

where

$$P_1 = (x_1, w_1), \quad P_2 = (x_2, w_2), \quad P_3 = P_1 * P_2 = (x_3, w_3).$$

Topologically, $\mathbb{C}/\Lambda$ is a surface of genus one. It can be obtained from the parallelogram $\{u_1 + u_2\tau \mid 0 \le u_1, u_2 \le 1\}$ by identifying the opposite sides (cf. Fig. 5.3).

Points of finite order. Let E be an elliptic curve defined over a field K. For an integer N denote by E_N the kernel of the map which multiplies each point by N:

$$N_E : E(\overline{K}) \to E(\overline{K}), \quad N_E(t) = Nt. \tag{5.3.23}$$

If E is defined over $\mathbb{C}$ then the isomorphism $\mathbb{C}/\Lambda \cong E(\mathbb{C})$ shows that

$$E_N \cong \mathbb{Z}/N\mathbb{Z} \times \mathbb{Z}/N\mathbb{Z}.$$

In fact E_N corresponds to the subgroup $\frac{1}{N}\Lambda/\Lambda \subset \mathbb{C}/\Lambda$. For example 2-torsion points are represented by 0, $1/2$, $\tau/2$, $(1+\tau)/2$. It follows that (5.3.15) maps $1/2$, $\tau/2$, $(1+\tau)/2$ onto $(x_i, 0)$ for $i = 1, 2, 3$, where x_i are the roots of the polynomial $4x^3 - g_2x - g_3$. In other words,

$$\wp'(1/2) = \wp'(\tau/2) = \wp'((1+\tau)/2) = 0,$$

and

$$4x^3 - g_2x - g_3 = 4(x - e_1)(x - e_2)(x - e_3),$$

where

$$e_1 = \wp'(1/2), \quad e_2 = \wp'(\tau/2), \quad e_3 = \wp'((1+\tau)/2).$$

The 3-torsion points have a nice geometric interpretation: they are the *points of inflection* of the projective Weierstrass model.

For any ground field K, the morphism N_E has degree N^2, and if $(\mathrm{char}\, K, N) = 1$, we still have

$$E(\overline{K})_N \cong \mathbb{Z}/N\mathbb{Z} \times \mathbb{Z}/N\mathbb{Z}. \tag{5.3.24}$$

However for $\mathrm{char}(K) = p$ and $N = p^m$ we have

$$E(\overline{K})_N \cong (\mathbb{Z}/p^m\mathbb{Z})^{\gamma_E}, \tag{5.3.25}$$

where $\gamma_E = 0$ or 1, (cf. [La73/87], [La88]).

Assume that $(\mathrm{char}(K), N) = 1$. The field $K(E_N)$, generated by the coordinates of all points of E_N, is a Galois extension of K and the action of $\mathrm{Gal}(\overline{K}/K)$ on $E(\overline{K})_N \cong \mathbb{Z}/N\mathbb{Z} \times \mathbb{Z}/N\mathbb{Z}$ determines a representation

$$\rho_N : \mathrm{Gal}(\overline{K}/K) \to \mathrm{GL}_2(\mathbb{Z}/N\mathbb{Z}) \tag{5.3.26}$$

whose image is isomorphic to $\mathrm{Gal}(K(E_N)/K)$. The field $K(E_N)$ can be regarded as an analog of the cyclotomic field $K(\zeta_N)$. However it is not in general Abelian; only for the so called *complex multiplication curves* is the analogy really far-reaching. This is a basic example of the kind of construction made in Abelian and non–Abelian class field theory.

It is known that the representation $\det(\rho_N)$ is the cyclotomic character of $G_K = \mathrm{Gal}(\overline{K}/K)$; that is, it corresponds to the action of G_K on the group μ_N of Nth roots of unity. Actually these roots are contained in $K(E_N)$, and E_N is endowed with a canonical non–degenerate alternating *Weil pairing*

$$e_N : E(\overline{K})_N \times E(\overline{K})_N \to \mu_N. \tag{5.3.27}$$

compatible with the action of G_K. This is defined purely algebraically, with the help of the functions f_P, $P \in E(K)$ such that $\mathrm{div}(f) = NP - No$. Calculating the pairing for an elliptic curve E over $\mathbb{C}$, given by a period lattice Λ, we obtain

$$e_N((a+b\tau)/N, (c+d\tau)/N) = \exp(2\pi i(ad-bc)/N). \tag{5.3.28}$$

5.3.3 Tate Curve and Its Points of Finite Order

(see [Ta74], [He97], p.343). Let us write again the Weierstrass equation for

$$\mathbb{C}/(2\pi i)\Lambda \xrightarrow{\sim} \mathbb{C}^\times/\langle q^{\mathbb{Z}}\rangle \quad (u \mapsto \exp(u))$$

in the following form

$$Y^2 = 4X^3 - \frac{E_4}{12}X + \frac{E_6}{216} \quad (X = \wp(2\pi i u, (2\pi i)\Lambda),\ X = \wp'(2\pi i u, (2\pi i)\Lambda)), \omega = 2\pi i\, du,$$

using the Eisenstein series (see also in §6.3.2, (6.3.4)):

$$\begin{aligned} G_k(\tau) &= \sum_{m_1,m_2\in\mathbb{Z}}{}' (m_1+m_2\tau)^{-k} \qquad\qquad (5.3.29)\\ &= \frac{2(2\pi i)^k}{(k-1)!}\left[-\frac{B_k}{2k} + \sum_{n=1}^{\infty}\sigma_{k-1}(n)\exp(2\pi i n\tau)\right] = \\ &= \frac{2(2\pi i)^k}{(k-1)!}\cdot\left(-\frac{B_k}{2k}\right)\left[1 - \frac{2k}{B_k}\sum_{n=1}^{\infty}\sigma_{k-1}(n)\exp(2\pi i n\tau)\right] = -\frac{(2\pi i)^k B_k}{k!}E_k, \end{aligned}$$

where the prime denoting $(m_1, m_2) \neq (0,0)$,

$$E_k(\tau) = 1 - \frac{2k}{B_k}\sum_{n=1}^{\infty}\sigma_{k-1}(n)\exp(2\pi i n\tau),$$

$\sigma_{k-1}(n) = \sum_{d|n} d^{k-1}$ and B_k is the kth *Bernoulli number*. In particular we have that

$$12(2\pi i)^4 g_2(\tau) = E_4 = 1 + 240\sum_{n=1}^{\infty} \sigma_3(n)q^n \quad (q = \exp(2\pi i\tau),$$

$$-216(2\pi i)^6 g_3(\tau) = E_6 = 1 - 504\sum_{n=1}^{\infty} \sigma_5(n)q^n.$$

Let us pass to the variables x, y via the substitution

$$X = x + \frac{1}{12}, Y = x + 2y,$$

in order to obtain a new equation of this curve (with coefficients in $\mathbb{Z}[[q]]$):

$$Tate(q) : y^2 + xy = x^3 + B(q)x + C(q),$$

where

$$B(q) = -5\left(\frac{E_4 - 1}{2}40\right) = -5\sum_{n=1}^{\infty} \sigma_3(n)q^n = -5\sum_{n=1}^{\infty} \frac{n^3q^n}{1-q^n}, \qquad (5.3.30)$$

$$C(q) = \frac{-5\left(\frac{E_4-1}{240}\right) - 7\left(\frac{E_6-1}{-504}\right)}{12} = -\frac{1}{12}\sum_{n=1}^{\infty} \frac{(7n^5+5n^3)q^n}{1-q^n}.$$

This equation defines an elliptic curve over the ring $\mathbb{Z}((q))$ with the canonical differential ω_{can} given by

$$\frac{dx}{2y+x} = \frac{dX}{Y}.$$

$$\begin{cases} g_2 = 60G_4 = (2\pi i)^4 \dfrac{E_4}{12} \\ g_3 = 140G_6 = -(2\pi i)^6 \dfrac{E_6}{216}. \end{cases}$$

Let $N \geq 1$ be a natural number. Let us define

$$Tate(q^N) : y^2 + xy = x^3 + B(q^N)x + C(q^N).$$

Next we put $t = \exp(2\pi iu)$, then the points of order N on $Tate(q^N)$ correspond to $t = \zeta_N^i q^j$, $(0 \leq i, j \leq N-1)$, $\zeta_N = \exp(2\pi i/N)$, and their coordonates are given by

$$\begin{cases} x(t) = \displaystyle\sum_{n\in\mathbb{Z}} \frac{q^{Nn}t}{(1-q^{Nn}t)^2} - 2\sum_{n=1}^{\infty} \frac{nq^{Nn}}{1-q^{Nn}t} \\ y(t) = \displaystyle\sum_{n\in\mathbb{Z}} \frac{q^{2Nn}t^2}{(1-q^{Nn}t)^3} + \sum_{n=1}^{\infty} \frac{nq^{Nn}}{1-q^{Nn}t}. \end{cases}$$

It is important for arithmetical applications for the Tate curve that these coordonates belong to the ring $\mathbb{Z}[\zeta_N, N^{-1}][[q]]$.

Proof uses the identity

$$\sum_{n\in\mathbb{Z}}(u+n)^{-k} = \frac{(2\pi i)^k}{(k-1)!}\sum_{n=1}^{\infty} n^{k-1}e^{2\pi inu} \quad (k\geq 2, \quad u\in\mathbb{Z}).$$

We have for the lattice $\Lambda = 2\pi i(\mathbb{Z}+\tau\mathbb{Z})$ the following equalities

$$X = \wp(2\pi iu) =$$

$$(2\pi i)^{-2}\left(u^{-2} + \sum_{m,n\in\mathbb{Z}}{}' \left((u+m\tau+n)^{-2} - (m\tau+n)^{-2}\right)\right) = \tag{5.3.31}$$

$$(2\pi i)^{-2}\left(\sum_{m\in\mathbb{Z}}\sum_{n\in\mathbb{Z}}(u+m\tau+n)^{-2} - 2\sum_{m=1}^{\infty}\sum_{n\in\mathbb{Z}}(m\tau+n)^{-2} - 2\zeta(2)\right) =$$

$$\sum_{m\in\mathbb{Z}}\sum_{n=1}^{\infty} ne^{2\pi i(u+m\tau)n} - 2\sum_{m=1}^{\infty}\sum_{n=1}^{\infty} ne^{2\pi imn\tau} + \frac{1}{12},$$

implying the above identities.

5.3.4 The Mordell – Weil Theorem and Galois Cohomology

The fundamental arithmetical property of elliptic curves defined over an algebraic number field K is the following result.

Theorem 5.9 (The Mordell–Weil Theorem). *The Abelian group $E(K)$ is finitely generated, that is*

$$E(K) \cong E(K)_{\text{tors}} \oplus \mathbb{Z}^{r_E}, \tag{5.3.32}$$

where $E(K)_{\text{tors}}$ is a finite (torsion) group, and r_E is an integer ≥ 0, called the rank of E over K.

(cf. [Wei79], [La83], [Se97] and Appendix by Yu.Manin to [Mum74]).

This theorem is proved in two steps. One first shows that $E(K)/nE(K)$ is finite (for some, or every $n\geq 2$). Then one uses a descent argument based on the following property of logarithmic heights $h(P)$:

$$h(P) \leq \text{const} + n^{-2}h(nP+P_0)$$

for a fixed P_0, variable P and a constant independent of P.

The weak finiteness theorem for $E(K)/nE(K)$ can be established by a kind of *Kummer theory* for $K(E_n)$.

Consider the extension $K(\frac{1}{n}E(K))$ of $K(E_n)$. One proves that this is a finite Abelian extension whose order divides n. This can be deduced from Hermite's theorem on the finiteness of the number of extensions of a fixed number field having prescribed degree and ramification points. In order to apply Hermite's result one must check that every ramified prime either divides n or is a point of bad reduction of E.

Now consider the exact sequence

$$0 \to E_n \longrightarrow E(\overline{K}) \xrightarrow{n} E(\overline{K}) \to 0. \qquad (5.3.33)$$

This gives rise to an exact sequence of Galois cohomology groups

$$E(K) \xrightarrow{n} E(K) \to H^1(G_K, E_n) \to H^1(G_K, E(\overline{K})) \xrightarrow{n} H^1(G_K, E(\overline{K}))$$

which can be rewritten as

$$0 \to E(K)/nE(K) \to H^1(G_K, E_n) \to H^1(G_K, E(\overline{K}))_n \to 0. \qquad (5.3.34)$$

Although the group $H^1(G_K, E_n)$ is infinite, the image of $E(K)/nE(K)$ is contained in a finite subgroup, which we shall describe in geometric terms.

An element $\alpha \in H^1(G_K, E_n)$ corresponds to an *n-fold covering* of E over K, that is to a map $a : C \to E$ of algebraic curves, which becomes isomorphic to $n : E \otimes \overline{K} \to E \otimes \overline{K}$ when the ground field K is extended to $\overline{K}$. Given such a covering, one constructs a 1–cocycle by choosing a point $P \in E(K)$, an inverse image $Q = a^{-1}P$, and a point $Q_1 \in E(\overline{K})$, which corresponds to Q under a structure isomorphism $C(\overline{K}) \cong E(\overline{K})$. Then one defines α as the class of the cocycle:

$$\sigma \mapsto \alpha_\sigma = Q_1 - Q_1^\sigma \in E_n \quad (\sigma \in G_K) \qquad (5.3.35)$$

(subtraction refers to the group law on E; we shall later on denote it by $+$ instead of $*$). Elements $\beta \in H^1(G_K, E(\overline{K}))$ are interpreted as isomorphism classes of the principal homogeneous spaces X of E over K, that is, curves X given together with group actions $E \times X \to X$ which become isomorphic to the addition morphism of E when the ground field is extended to $\overline{K}$. Given such an X, choose a point $P \in E(K)$, a point $P_1 \in X(\overline{K})$ corresponding to P under a structure isomorphism, and define a cocycle

$$\sigma \mapsto \beta_\sigma = P_1 - P_1^\sigma \in E(\overline{K}) \quad (\sigma \in G_K). \qquad (5.3.36)$$

A different choice of P leads to a cohomological cocycle. The cohomology class is trivial iff X has a rational K–point. This establishes a direct connection between Galois cohomology and Diophantine geometry.

The exact sequence (5.3.34) can be conveniently described in this setting. A point $P \in E(K)$ determines an n–covering

$$t_P n_E : E \to E, \qquad (5.3.37)$$

where t_P is the translation by P. Now choose a point $Q \in E(\overline{K})$ such that $nQ = P$. Then $t_P n = nt_Q$, so that the translation $t_Q : E \otimes \overline{K} \to E \otimes \overline{K}$ is a $\overline{K}$–isomorphism of algebraic curves, turning (5.3.37) into multiplication by n. Therefore, our n–covering becomes trivial over $K' = K(\frac{1}{n}E(K))$. Hence its class belongs to the finite subgroup

$$M_n = \mathrm{Infl}(H^1(G(K'/K), E_n)) \subset H^1(G_K, E_n),$$

whose order can be bounded in terms of the degree and ramification of K'. This finishes the proof of the weak Mordell–Weil theorem.

The descent argument proceeds as follows: choose a finite number of representatives

$$P_1, \ldots, P_s$$

of $E(K)/nE(K)$. There is a constant C such that if $h(P) > C$, then

$$h\left(\frac{1}{n}(P - P_i)\right) < h(P),$$

where P_i is congruent to P modulo n. Hence P can be represented as a linear combination of

$$P_1, \ldots, P_s$$

and points of height $<C$ whose number is finite.

The exact sequence (5.3.34) can be used to obtain upper bounds for the *rank* r_E. In fact, if $n = p$ is a prime and M is a finite subgroup of $H^1(G_K, E_p)$ containing the image of $E(K)/pE(K)$ then (5.3.34) shows that

$$r_E \leq \mathrm{rk}_{\mathbb{Z}/p\mathbb{Z}}(M) - \mathrm{rk}_{\mathbb{Z}/p\mathbb{Z}}(E(K)_p). \tag{5.3.38}$$

Any improvement on this bound would require an understanding of the cokernel of the map

$$E(K)/pE(K) \to M.$$

In order to choose a small, well–defined M, it is convenient to apply the usual local–to–global constructions. For each place v of K, choose an extension w of v to $\overline{K}$ and denote by $G_v \subset G_K$ the corresponding decomposition subgroup $G_v \cong G(\overline{K}_w/K_v)$. Then for an arbitrary G_K–module A we have restriction homomorphisms $H^i(G_K, A) \to H^i(G_v, A)$. In our setting, these fit into the commutative diagram

$$\begin{array}{ccccccccc}
0 \to & E(K)/nE(K) & \to & H^1(G_K, E_n) & \xrightarrow{\alpha} & H^1(G_K, E(\overline{K}))_n & \to 0 \\
& \downarrow & & \downarrow & & \beta = \prod_v \beta_v \downarrow & \\
0 \to & \prod_v E(K_v)/nE(K) & \to & \prod_v H^1(G_v, E_n) & \to & \prod_v H^1(G_K, E(\overline{K}_w))_n & \to 0
\end{array}$$

in which β_v denotes the composition of the restriction morphism and the morphism induced by the inclusion $E(\overline{K}) \to E(\overline{K}_w)$.

Let us consider the group

$$Ш(E,K) = \bigcup_{n\in\mathbb{N}} Ш(E,K)_n, \quad Ш(E,K)_n = \mathrm{Ker}(\beta). \tag{5.3.39}$$

This group is called *the Shafarevich–Tate group* of E over K; its interpretation in terms of the Brauer group and connection with the Brauer–Manin obstruction is explained in [Man70b]. In our setting, an element of $Ш(E,K)$ corresponds to a principal homogeneous space of E over K (up to isomorphism) which has a K_v–point in every completion of K.

The group

$$S(E,K)_n = \alpha^{-1}(Ш(E,K)_n) \tag{5.3.40}$$

(and the inductive limit of these groups over all n) is called the *Selmer group* of E. An element of $S(E,K)_n$ can be interpreted as (the class of) an n–covering $C \to E$ such that C has a K_v–point in each completion K_v of K. By definition we have an exact sequence

$$0 \to E(K)/nE(K) \to S(E,K)_n \to Ш(E,K)_n \to 0. \tag{5.3.41}$$

One can say that $Ш(E,K)$ is a cohomological obstruction to a calculation of $E(K)$. There is a conjecture that $Ш(E,K)$ is *finite.* This was proved by K.Rubin in [Rub77] for certain curves with complex multiplication, and by V.Kolyvagin (cf. [Koly88]) for a class of curves uniformized by modular curves. More recently, these results were extended to a classe of curves without complex multiplication by K.Kato, cf. [Scho98].

We shall return to this question in Chapter 6 in connection with *zeta–functions* and *modular functions.*

We now consider in more detail the properties of the height function $h_D : E(\overline{K}) \to \mathbb{R}$ corresponding to a divisor D, or, equivalently, to the invertible sheaf $\mathcal{O}(D)$ of degree d on an elliptic curve E. Since the degree of the map n_E is n^2, one can check that

$$h_D \circ n_E \sim n^2 h_D. \tag{5.3.42}$$

More precisely the following limit exists:

$$\hat{h}_D(x) = \lim_{N\to\infty} h_D(2^N x)/2^{2N}. \tag{5.3.43}$$

This limit $\hat{h}_D$ is called the *Néron-Tate height.*

If the divisor D is ample (see 5.1.13) then $\hat{h}_D$ is a quadratic form on $E(K)$, which is positive definite modulo torsion. Moreover its natural extension

$$\hat{h}_D : E(\overline{K}) \otimes_{\mathbb{Z}} \mathbb{R} \to \mathbb{R}$$

is of the form db_0, where $d = \deg(D)$ and b_0 is a positive definite quadratic form independent of D. The kernel of the natural map $E(K) \to E(K) \otimes_{\mathbb{Z}} \mathbb{R}$ is the finite torsion subgroup $E(K)_{\text{tors}}$; its image is a lattice in the r_E–dimensional Euclidean space with the scalar product

$$\langle P, Q \rangle = \frac{1}{2} \left[b_0(P+Q) - b_0(P) - b_0(Q) \right[.$$

Therefore, the region $\hat{h}_D \leq \log(B)$ in this space is a ball of radius

$$(d^{-1} \log(B))^{1/2}.$$

The number of points in this ball is asymptotically proportional to its volume, that is const $\cdot\, (\log(B))^{r/2}$. The constant in this expression depends on the volume of a fundamental domain for the lattice $E(K)$ mod torsion, that is, on the *regulator* of E over K:

$$H = H(E, K) = \det(\langle P_i, P_j \rangle)^{1/2}. \tag{5.3.44}$$

B.Mazur has proved that $\mathrm{Card}(E(\mathbb{Q})_{\text{tors}})$ is universally bounded (cf. [Maz77]). This result was extended to all number fields by L. Merel, cf. [Mer96]. Actually Mazur showed that $E(\mathbb{Q})_{\text{tors}}$ is always isomorphic to one of the following fifteen groups:

$$\mathbb{Z}/m\mathbb{Z} \;\; (m \leq 10, m = 12), \quad \mathbb{Z}/2\mathbb{Z} \times \mathbb{Z}/2\nu\mathbb{Z} \;\; (\nu \leq 4).$$

All these groups arise in this way.

It is conjectured that there are elliptic curves of arbitrarily large rank over $\mathbb{Q}$. J.–L. Mestre constructed curves of rank $r_E \geq 14$ ([Me82]), by choosing equations in such a way that their reductions modulo many primes p have as many points modulo p as possible. A concrete example of a calculation of the group $E(\mathbb{Q})$ is given in [Maz86]. Consider the curve

$$E : -206y^2 = x^3 - x^2 + 1/4$$

and three points on it

Point	x	y	Néron – Tate height
P_1	$-15/8$	$7/32$	1.52009244
P_1	$-55/8$	$43/32$	2.05430703
P_3	$-55/98$	$47/1372$	2.42706090

A descent argument shows that $r_E \leq 3$, and a height computation allows one to conclude that P_1, P_2, P_3 are linearly independent generators of $E(\mathbb{Q}) \cong \mathbb{Z}^3$; The absence of torsion can be checked by p–adic calculations.

For a given elliptic curve, the numbers $|E(\mathbb{Q})_{\rm tors}|$, r_E, $H(E,K)$, $|\text{Ш}(E,K)|$ (conjecturally finite), and the conductor (a product of primes of bad reduction) are the most important arithmetical invariants of E. Later we shall see that all these invariants are combined (partly conjecturally) in the properties of its *zeta-function* (see §6.4.4).

5.3.5 Abelian Varieties and Jacobians

(cf. [Mum74], [La58], [Wei48]). Abelian varieties are multi-dimensional generalizations of elliptic curves. By definition, an Abelian variety A over a field K is a non–singular projective variety, together with a group structure given by morphisms over K:

$$A\times A\to A\ \ ((x,y)\mapsto x+y),\qquad A\to A\ \ (x\mapsto -x).$$

One can prove that any such structure is commutative, which justifies the additive notation.

A homomorphism of Abelian varieties is a morphism $\lambda : A \to B$ of algebraic varieties which is a group homomorphism. If $\dim(A) = \dim(B)$, the surjectivity of λ is equivalent to the condition that the kernel of λ is finite. If these conditions are satisfied then λ is called an *isogeny*, and A and B are said to be isogenous.

In particular, multiplication by an integer $m_A : A \to A$, $m_A(x) = mx$, is an isogeny of degree m^{2g}, $g = \dim(A)$. If the characteristic of the ground field does not divide m, then

$$A_m = A(\overline{K})_m = \mathrm{Ker}(m_A) \cong (\mathbb{Z}/m\mathbb{Z})^{2g}.$$

In particular the action of the Galois group on A_m defines a Galois representation

$$\rho_m : G_K \to \mathrm{Aut}(A_m) \subset \mathrm{GL}_{2g}(\mathbb{Z}/m\mathbb{Z}). \tag{5.3.45}$$

These representations are the best studied examples of the general Galois actions on Grothendieck's *étale cohomology groups.* As in the case of elliptic curves, there is a non–degenerate alternating *Weil pairing*

$$e_m : A(\overline{K})_m \times A(\overline{K})_m \to \mu_m. \tag{5.3.46}$$

This is compatible with the action of the Galois group, so that

$$\mathrm{Im}(\rho_m) \subset \mathrm{GSp}_g(\mathbb{Z}/m\mathbb{Z}) \subset \mathrm{GL}_{2g}(\mathbb{Z}/m\mathbb{Z}),$$

where GSp_g is the group of symplectic matrices: for an arbitrary ring R,

$$\mathrm{GSp}_g(R) = \{M \in GL_{2g}(R) | M^t J_g M = \mu(M) J_g,\ \mu(M) \in R^\times\}, \tag{5.3.47}$$

where J_g is a standard symplectic matrix. Actually, the construction of e_m depends on the choice of a *polarization* on A (cf. below).

If A is an Abelian variety defined over $\mathbb{C}$, the complex variety $A(\mathbb{C})$ is isomorphic to a complex torus $\mathbb{C}^g/\Lambda$, where Λ is a lattice in $\mathbb{C}^g$. Not every complex torus, however, can be obtained in this way. A necessary and sufficient condition for this is the existence of an $\mathbb{R}$–valued, $\mathbb{R}$–bilinear form $E(z,w)$ with the following properties:

$$E(z,w) = -E(w,z). \tag{5.3.48}$$

$$E(z,w) \in \mathbb{Z} \text{ for all } z,w \in \Lambda. \tag{5.3.49}$$

$$E(z,iw) \text{ is an } \mathbb{R} - \text{bilinear, symmetric, positive definite form.} \tag{5.3.50}$$

Such a form E is called a *Riemannian form* on the complex torus $\mathbb{C}^g/\Lambda$. It also defines a Hermitean Riemannian form on $\mathbb{C}^g$:

$$H(z,w) = E(iz,w) + iE(z,w). \tag{5.3.51}$$

If such a form E exists at all, it is not unique. We shall say that a choice of E defines a polarization of A.

An Abelian variety together with a polarization is called *a polarized Abelian variety.*

We recall the following classification theorem for non–degenerate alternating integral forms on a lattice $\Lambda \cong \mathbb{Z}^{2g}$: for each form, there exists a basis $\{\lambda_1,\dots,\lambda_{2g}\}$ of Λ such that

$$E(\lambda_i,\lambda_j) = E(\lambda_{g+i},\lambda_{g+j}) = 0 \quad \text{for} \quad 1 \le i,j \le g,$$
$$E(\lambda_i,\lambda_{g+j}) = e_i\delta_{ij} \quad \text{for} \quad 1 \le i,j \le g,$$

where $e_1,\dots,e_g$ are natural numbers,

$$e_1|e_2,\dots,e_{g-1}|e_g.$$

Clearly,

$$\det_\Lambda(E) = (e_1e_2\dots e_g)^2.$$

A polarization with determinant 1 is called a principal polarization.

There is a totally different definition of polarization, which is purely algebraic and is valid over any ground field. Namely, consider an arbitrary projective embedding $A \hookrightarrow \mathbb{P}^N$. Call two embeddings equivalent if one can be obtained from the other by a projective transformation composed with a translation by a point of A. An equivalence class of projective embeddings defines a linear system of hyperplane sections D of A. Over the complex ground field, this gives rise to an integral 2–cohomology class of $A(\mathbb{C})$, which in turn defines a Riemannian form E, in view of the known structure of the cohomology ring of a torus. Elaborating this correspondence, one obtains the following

Definition 5.10. *An (algebraic) polarization of an Abelian variety A is a class of ample divisors $\{D\}$ up to algebraic equivalence.*

5.3.6 The Jacobian of an Algebraic Curve

([La58], [Wei48]). Let X be a non–singular projective curve over a field K. One defines in an invariant way an Abelian variety $J = J_X$, which parametrizes the invertible sheaves (or divisor classes) of degree zero on X. This Abelian variety is called *the Jacobian* of X. For $K = \mathbb{C}$, its structure is essentially described by *Abel's theorem.* Consider a divisor

$$\mathfrak{a} = \sum n_i P_i, \quad \sum n_i = 0.$$

We have $\mathfrak{a} = \partial C$ where C is a *1-chain.* Choose a basis of the differentials of the first kind

$$\{\omega_1, \ldots, \omega_g\}$$

on X, where g is the genus of X. Consider the point

$$\left(\int_C \omega_1, \ldots, \int_C \omega_g\right) \in \mathbb{C}^g.$$

Since one can replace C by a homologous 1–chain, this point is only well defined modulo the period lattice $H_1(X, \mathbb{Z})$ of our basis. Abel's theorem asserts that the map sending α to the class of this point in the torus $\mathbb{C}^g/H_1(X, \mathbb{Z})$, identifies this torus with the group $J_X(\mathbb{C})$ of all classes of divisors of degree zero.

The classical *Riemann periodicity relations* imply that the lattice $H_1(X, \mathbb{Z})$ is self-dual with respect to a canonical Hermitean metric. Hence

$$\widehat{H_1(X)} \cong \mathbb{C}^g/H_1(X, \mathbb{Z})$$

where $\widehat{H_1(X, \mathbb{Z})}$ denotes the Pontryagin character group of $H_1(X, \mathbb{Z})$. This shows that J_X can be considered as an algebraic avatar of the 1–cohomology of X.

Properties of Jacobians.

1) $\dim(J_X) = g$ (the *genus* of X).
2) J_X is an Abelian variety, and for every extension field L of K, the group $J_X(L)$ is canonically isomorphic to the group of divisor classes of degree zero on X with ground field extended to L.
3) Every morphism of curves of finite degree $f : X \to Y$ determines a functorial homomorphism $f^* : J_Y \to J_X$, corresponding to the inverse image map on divisor classes.

4) J_X has a canonical principal polarization. This has an algebraic description as the class of the *Poincaré divisor* θ. The Poincaré divisor can be defined as follows. Start with Abel's map

$$\varphi : X \to J_X, \tag{5.3.52}$$

which sends a point $x \in X(\overline{K})$ to the divisor class $cl(x - P)$, where $P \in X(\overline{K})$ is some fixed point. Consider the map

$$\psi : X^g \xrightarrow{\varphi^g} J_X^g \xrightarrow{\mu} J_X,$$

where μ is the addition map. From the Riemann–Roch theorem it follows that ψ is surjective. Put $\theta = \psi(X^{g-1})$.

Many geometric and arithmetical properties of a curve X can be read off from the properties of its Jacobian. In particular, the classical *theorem of Torelli* (cf. [Wei57]) states that X can be uniquely reconstructed from J_X together with its canonical principal polarization. Essentially this theorem was used in Faltings' theory and in earlier constructions due to A.N.Parshin and Yu.I.Zarkhin (cf. [Zar74], [Zar85], [Par71], [Par73], [PZ88]).

If X is defined over K, the Jacobian and its principal polarization are both defined over K. If X has a K–point P, the map (5.3.52) is also defined over K.

One can also prove that if X is an algebraic curve over an algebraic number field K having good reduction modulo a prime $\mathfrak{p} \subset \mathcal{O}_K$, then J_X with its canonical projective embedding (given by the divisor θ) also has good reduction.

Every Abelian variety A over a number field (or absolutely finitely generated field) K satisfies the *Mordell – Weil theorem*: $A(K)$ is a finitely generated commutative group, that is

$$A(K) \cong A(K)_{\text{tors}} \oplus \mathbb{Z}^{r_A},$$

where $A(K)_{\text{tors}}$ is finite and r_A is the rank of A over K (cf. [La83], [Se97] and Appendix by Yu.Manin to [Mum74]).

As with elliptic curves, one can define the *Selmer groups* $S(A, K)_m$ and the Shafarevich – Tate groups $Ш(A, K)$. A standard conjecture is that the latter are all finite.

Every divisor D on A determines a Néron – Tate height

$$\hat{h}_D : A(\overline{K}) \otimes \mathbb{R} \to \mathbb{R},$$

and if D (that is, $\mathcal{O}(D)$) is ample, then $\hat{h}_D$ induces a Euclidean metric on the r_A–dimensional vector space $A(K) \otimes \mathbb{R}$.

A very important role in the theory of Abelian varieties is played by the endomorphism ring $\mathrm{End}(A)$ of A (over $\overline{K}$) together with the $\mathbb{Q}$–algebra $\mathrm{End}(A) \otimes \mathbb{Q}$. It is known that this algebra is semi-simple.

The Abelian variety A is called simple if $\mathrm{End}(A)\otimes\mathbb{Q}$ is simple. A decomposition of $\mathrm{End}(A)\otimes\mathbb{Q}$ as a sum of simple algebras $R_1\oplus\cdots\oplus R_s$ corresponds to a decomposition of A into a product of simple Abelian varieties *up to isogeny*: there exists an Abelian variety

$$B = B_1 \times \cdots \times B_s$$

isogenous to A such that $\mathrm{End}(B_i)\otimes\mathbb{Q} \cong R_i$ ([La58]).

Let E be a Riemannian form corresponding to a polarization of an Abelian variety A over $\mathbb{C}$. Such a form determines a *Rosatti involution* ρ on $\mathrm{End}(A)\otimes\mathbb{Q}$ (that is, an anti-isomorphism of order 1 or 2) which verifies the relation $E(\lambda x, y) = E(x, \lambda^\rho y)$ for every $\lambda \in \mathrm{End}(A)\otimes\mathbb{Q}$. Involutions of this kind can also be defined over a ground field of finite characteristic.

Semi-simple algebras with involutions have been classified, cf. [Mum74], [Shi71].

If K is a number field, $g = 1$, then either $\mathrm{End}(A)\otimes\mathbb{Q} = \mathbb{Q}$ or $\mathrm{End}(A)\otimes\mathbb{Q}$ is an imaginary quadratic field k. In the latter case A is called an *elliptic curve with complex multiplication.* It can be represented as a complex torus $\mathbb{C}/\Lambda_\tau$ (see (5.3.15)) with $\tau \in k$, $\mathrm{Im}(\tau) > 0$.

We now sketch an analytic construction of the space $\mathcal{A}_g$ of isomorphism classes of Abelian varieties over $\mathbb{C}$ with principal polarizations. The crucial observation is that each such variety can be represented as a complex torus $\mathbb{C}^g/\Lambda_\tau$, where

$$\Lambda = \Lambda_\tau = \{n_1 + n_2\tau \mid n_1, n_2 \in \mathbb{Z}^g,\ \tau \in \mathbb{H}_g\} \tag{5.3.53}$$

and $\mathbb{H}_g$ is the *Siegel upper half space*

$$\mathbb{H}_g = \{\tau \in \mathrm{GL}_g(\mathbb{C}) \mid \mathrm{Im}(\tau) \text{ is positive definite}\}.$$

In fact, let A be an Abelian variety with a principal polarization, given as a torus $\mathbb{C}^g/\Lambda$, and a Riemannian form E on Λ with determinant 1. Choose a symplectic basis $\{\omega_1, \omega_2, \cdots, \omega_{2g}\}$ of Λ. Representing ω_i by its column of coordinates, we can construct a $(g \times 2g)$-matrix

$$\Omega = (\omega_1, \omega_2, \cdots, \omega_{2g})$$

which is called a *period matrix* of A. Put $\Omega = (\Omega_1, \Omega_2)$ where $\Omega_i \in M_g(\mathbb{C})$. From (5.3.48) and (5.3.50) it follows that

$$\Omega_2\Omega_1^t - \Omega_1\Omega_2^t = 0, \tag{5.3.54}$$

$$2i(\Omega_2\overline{\Omega}_1^t - \Omega_1\overline{\Omega}_2^t) > 0 \text{ is positive definite.}$$

Thus $\Omega_1, \Omega_2 \in \mathrm{GL}_g(\mathbb{C})$ and $\tau = \Omega_2^{-1}\Omega_1 \in \mathbb{H}_g$. From this one deduces that the complex variety $A(\mathbb{C})$ is isomorphic to the torus $\mathbb{C}^g/\Lambda_\tau$, and the initial polarization corresponds to one given by the form

$$E(x_1 + \tau y_1, x_2 + \tau y_2) = x_1^t y_2 - x_2^t y_1,$$

where $x_i, y_i \in \mathbb{R}^g$.

The varieties $\mathbb{C}^g/\Lambda_\tau$ and $\mathbb{C}^g/\Lambda_{\tau'}$ are isomorphic iff

$$\tau' = (A\tau + B)(C\tau + D)^{-1}$$

for a certain matrix $M = \begin{pmatrix} A & B \\ C & D \end{pmatrix}$ from the group

$$\mathrm{Sp}_g(\mathbb{Z}) = \left\{ M = \begin{pmatrix} A & B \\ C & D \end{pmatrix} \in \mathrm{SL}_{2g}(\mathbb{Z}) \mid M^t J_g M = M_g \right\}. \tag{5.3.55}$$

This group is called the *Siegel modular group of genus g*.

Summing up, we see that $\mathcal{A}_g$ can be described as the quotient space $\mathbb{H}_g/\mathrm{Sp}_g(\mathbb{Z})$ where M acts on $\mathbb{H}_g$ by matrix fractional linear transformations.

One can show that $\mathcal{A}_g$ is a complex analytic space of dimension $g(g+1)/2$ with a natural structure as a normal quasi–projective variety defined over $\mathbb{Q}$. A generic Abelian variety over $\mathbb{C}$ is simple, and its endomorphism ring is $\mathbb{Z}$.

There are important variations of this construction. One can consider families of pairs A, E in which $\mathrm{End}(A)$ and E verify some additional constraints, and one can supply such pairs with so called *level structures*, for example a choice of symplectic basis for the subgroup A_m of points of order m. In many situations there exist universal PEL–families (Polarization, Endomorphisms, Level), whose bases are very important algebraic varieties (*Shimura Varieties*) defined over number fields. The action of the Galois group on the Algebraic points of these varieties can be described in considerable detail.

5.3.7 Siegel's Formula and Tamagawa Measure

Algebraic groups comprise not only of Abelian varieties but also of linear groups. The latter are affine varieties, whereas Abelian varieties are projective. The arithmetic of linear groups is a well–developed chapter of algebraic geometry. For an extensive report on its qualitative aspects we refer the reader to the papers of [Pl82], and [PlRa83]. We shall describe here only classical results due to C.-L. Siegel, which were generalized and reinterpreted by Weil. These results give a quantitative form to the Minkowski-Hasse principle for quadratic forms, and lead to certain precise formulae of the kind furnished by the circle method for the principal terms of some arithmetical functions.

Siegel's formulae concern the equations

$$S[X] = T \quad (S[X] = X^t S X) \tag{5.3.56}$$

where $S \in \mathrm{M}_m(\mathbb{Q})$ and $T \in M_n(\mathbb{Q})$ are the symmetric matrix forms of $\mathbb{Q}$–rational quadratic forms q_S and q_T, the solutions X being in $\mathrm{M}_{m,n}(\mathbb{Q})$.

Let us consider in more detail the case when S and T are the matrices of integral positive definite quadratic forms corresponding to the lattices $\Lambda_S \subset$

$\mathbb{R}^m$, $\Lambda_T \subset \mathbb{R}^n$ (in the sense that q_S and q_T express the lengths of elements of Λ_S, resp. Λ_T). Then an integral solution X to (5.3.56) determines an isometric embedding $\Lambda_S \to \Lambda_T$. Denote by $\mathrm{N}(S,T)$ the total number of such maps, which is also called the number of integral representations of q_T by q_S. The *genus* of q_S is by definition the set of quadratic forms, rationally equivalent to q_S. The genus consists of a finite number of classes with respect to integral equivalence. Let I be the set of these classes. One of Siegel's formulae gives the value of a certain weighted average of the numbers $\mathrm{N}(S_x,T)$ over a set of representatives S_x for classes $x \in I$ of forms of a given genus. To be more precise, denote by $w(x)$ the order of the group of orthogonal transformations of the lattice Λ_{S_x} and define the *mass* of S by the formula

$$\mathrm{Mass}(S) = \sum_{x \in I} \frac{1}{w(x)}. \tag{5.3.57}$$

Assume that $\mathrm{N}(S_x,T) \neq 0$ for at least one x (or, equivalently, that there is an isometric embedding $\Lambda_S \otimes \mathbb{Q} \to \Lambda_T \otimes \mathbb{Q}$) and put

$$\tilde{\mathrm{N}}(S,T) = \frac{1}{\mathrm{Mass}(S)} \sum_{x \in I} \frac{\mathrm{N}(S_x,T)}{w(x)}.$$

Siegel's formula expresses this average as a product of local factors

$$\tilde{\mathrm{N}}(S,T) = c_{m-n} c_m^{-1} \alpha_\infty(S,T) \prod_p \alpha_p(S,T), \tag{5.3.58}$$

where $c_1 = 1/2$, $c_a = 1$ for $a > 1$, and the proper local factors are defined as follows. For a prime p denote by $\mathrm{N}(S,T;p^r)$ the number of solutions of the congruence

$$S[X] \equiv T \pmod{p^r} \quad (X \in \mathrm{M}_{m,n}(\mathbb{Z}/p^r\mathbb{Z})) \tag{5.3.59}$$

and introduce the *local density*

$$\alpha_p(S,T) = \lim_{r\to\infty} c_{m-n+1} \mathrm{N}(S,T;p^r)\mathfrak{p}^{rd}, \quad d = mn - n(n+1)/2 \tag{5.3.60}$$

(the expression inside the limit actually stabilizes when r is sufficiently large). One can define $\alpha_\infty(S,T)$ similarly, replacing the p-adic measure by an Archimedean measure. Consider a neighbourhood V of a matrix S in the space of symmetric matrices $\{T = (t_{i,j}) \in M_m(\mathbb{R}) \mid T^t = T\}$ with the measure given by the volume form $\alpha_T = \wedge_{i \le j} dt_{i,j}$. Put

$$U = \{X = (x_{i,j}) \in \mathrm{M}_{m,n}(\mathbb{R}) \mid X^t S X \in V\}.$$

It is a subset of $\mathrm{M}_{m,n}(\mathbb{R})$ with a measure $\beta = \wedge_{i,j} dx_{i,j}$ ($i = 1, \ldots, m; j = 1, \ldots, n$). Finally put

$$\alpha_\infty(S,T) = c_{m-n+1} \lim_{V \to S} \frac{\int_U \beta_X}{\int_V \alpha_T}. \tag{5.3.61}$$

The product in (5.3.58) converges absolutely if $m \geq 3$ and $m - n \neq 2$.

In the special case $T = S$ we have

$$\tilde{\mathrm{N}}(S,T) = \frac{1}{\mathrm{Mass}(S)}$$

and (5.3.58) becomes the Minkowski–Siegel formula ([Se86], p. 671)

$$\mathrm{Mass}(S) = c_m \alpha_\infty(S,S)^{-1} \prod_p \alpha_p(S,S)^{-1}. \tag{5.3.62}$$

If $n = 1$, $T = (t)$ then $\mathrm{N}(S,T)$ is the number of integral representations of a positive integer t by the quadratic form q_S.

Note that for almost all primes p (i.e. for all but a finite number) each solution to the congruence (5.3.59) can be lifted to a solution of the corresponding congruence modulo any p^r (using Hensel's lemma). In this case we have

$$\alpha_p(S,T) = \frac{\mathrm{N}_p(S,T;p)}{p^d}, \tag{5.3.63}$$

which makes it possible to describe explicitly almost all the local factors in (5.3.58) and to express this product in terms of special values of certain zeta–functions (for example, values of the Riemann zeta–function at integers).

Consider for example the quadratic form $q_S = \sum_{i=1}^m x_i^2$ given by the identity matrix $S = I_m$. If m is divisible by 4, then (5.3.62) takes the following form ([Se86], p.673):

$$\mathrm{Mass}(I_m) = (1 - 2{-}k)(1 + \varepsilon 2^{1-k}) \mid B_2 B_4 \cdots B_{2k}|/4k!,$$

where $k = m/2$, $\varepsilon = (-1)^{m/2}$ and B_i is the ith Bernoulli number. For m not divisible by 4 there are exactly two classes in the genus of the form I_9, and $\mathrm{Mass}(I_9) = 17/2786918400$.

We now say a few words on how Siegel's formula is proved. The proof uses the theory of integration over the locally compact group $G = \mathcal{O}_m(\mathbb{A})$ of orthogonal matrices with respect to S with coefficients in the ring of adeles $\mathbb{A}$. The group $G_\infty = \mathcal{O}_m(\mathbb{R})$ is compact in view of the positive definiteness of S. Thus G contains the compact open subgroup $\Omega = G_\infty \times \prod_p G(S_p)$, where $G(S_p) = \mathcal{O}(\mathbb{Z}_p)$ is the orthogonal group of the p–adic lattice $\Lambda_{S,p} = \Lambda_S \otimes \mathbb{Z}_p$ (preserving the quadratic form q_S). The subgroup $\Gamma = \mathcal{O}_m(\mathbb{Q})$ of orthogonal matrices with rational coefficients is discrete in G and $\Gamma \cap \Omega = \mathrm{Aut}\,\Lambda_S$ is the finite group of automorphisms of the lattice Λ_S. For every $x = (x_v)_v \in G$ with ($v = p$ or $v = \infty$) one can define a lattice Λ_{S_x} such that $\Lambda_{S_x} \otimes \mathbb{Q}_v = x_v(\Lambda_S \otimes \mathbb{Q}_v)$. According to a version of the Hasse–Minkowski theorem, there is an isomorphism $\Lambda_{S_x} \otimes \mathbb{Q} = \Lambda_S \otimes \mathbb{Q}$ and the double cosets $\Omega x \Gamma$ of G modulo Ω and Γ can be interpreted as $\mathbb{Z}$-classes of forms S_x $(x \in I)$. The finite group $\gamma_x = \Omega \cap x\Gamma x^{-1}$ of order $w(x)$ is the group of automorphisms of the lattice Λ_{S_x}. Below a normalized Haar measure τ on the group G will be constructed. This measure is invariant under both right and left group shifts, and has the property that the volume $\mathrm{vol}(G/\Gamma)$ of the compact set $G/\Gamma = \cup_{x\in I} \Omega x \Gamma/\Gamma$ is uniquely defined (not only up to a multiplicative constant). This measure is called the Tamagawa measure on G. The following formula holds

$$\mathrm{vol}(G/\Gamma) = \sum_{x\in I} \mathrm{vol}(\Omega/\gamma_x) = \mathrm{vol}(\Omega) \sum_{x\in I} \frac{1}{w(x)}. \tag{5.3.64}$$

Let g, γ be closed subgroups of G and suppose that the volume $\mathrm{vol}(g/\gamma)$ is finite. Consider a continuous function φ with compact support on G/g, invariant under left shifts of the argument by elements of Ω. For $x \in G$ put

$$\mathrm{N}_x(\varphi) = \sum_{y\in\Gamma/\gamma} \varphi(xy).$$

This sum is finite and depends only on the double coset $\Omega x \Gamma$. Consider the *weighted average* $\tilde{\mathrm{N}}(\varphi)$ of the quantities $\mathrm{N}(\varphi)$ as x runs through I,

$$\tilde{\mathrm{N}}(\varphi) = \frac{\sum_{x\in I} \mathrm{N}_x(\varphi)/w(x)}{\sum_{x\in I} 1/w(x)}. \tag{5.3.65}$$

Standard integration techniques then show that

$$\tilde{\mathrm{N}}(\varphi) = \frac{\mathrm{vol}(g/\gamma)}{\mathrm{vol}(G/\Gamma)} \int_{G/g} \varphi(x)\, dx \tag{5.3.66}$$

assuming that the measures on the groups G and g and the homogeneous space G/g are compatible.

Siegel's formula can be deduced from equation (5.3.66) by taking for g the orthogonal adelic group with respect to the quadratic module W over $\mathbb{Q}$ defined by the condition $W \oplus (\Lambda_T \otimes \mathbb{Q}) \cong \Lambda_S \otimes \mathbb{Q}$. For the group γ we take the group of rational points in g, and the homogeneous space G/g is identified with the set of embeddings $\Lambda_T \otimes \mathbb{A} \to \Lambda_S \otimes \mathbb{A}$ preserving the quadratic forms. For φ one takes the characteristic function of the set of those embeddings $\Lambda_T \otimes \mathbb{A} \to \Lambda_S \otimes \mathbb{A}$ which take $\Lambda_T \otimes \mathbb{Z}_p$ into $\Lambda_S \otimes \mathbb{Z}_p$. The quantities c_{m-n} and c_m become the Tamagawa numbers $\tau(\mathcal{O}_{m-n})$ and $\tau(\mathcal{O}_m)$ respectively. For $x = (x_v)_v \in G$ the function $\varphi(x) = \varphi(gx)$ has the form $\prod_v \varphi_v(x_v)$, where $\varphi_\infty = 1$ on G_∞ and $\varphi_v(x_v)$ is the characteristic function of $\mathcal{O}_m(\mathbb{Z}_v)$. The integral in (5.3.66) is therefore equal to the product

$$\int_{G_\infty/g_\infty} dx_\infty \cdot \prod_p \int_{G_p/g_p} dx_p,$$

where $G_p = \mathcal{O}_m(\mathbb{Z}_p)$, $g_p = \mathcal{O}_{m-n}(\mathbb{Z}_p)$, and one easily verifies that

$$\alpha_\infty(S,T) = \int_{G_\infty/g_\infty} dx_\infty, \qquad \alpha_p(S,T) = \int_{G_p/g_p} dx_p. \tag{5.3.67}$$

Then the evaluation of

$$\tau(\mathcal{O}_m) = \mathrm{vol}(G/\Gamma)$$

can also be made using (5.3.66) putting $n = 1$ and applying some known asymptotic results for the representation numbers $\mathrm{N}(S,T)$ as $t \to \infty$. The latter are obtained for example by the *circle method* (the cases $m = 2, 3$ must be treated separately).

Now we describe the Tamagawa measure on G; formulae (5.3.67) follow from this description (cf. [CF67], chap. X).

Let V be an algebraic variety over a number field K, which is a connected *linear algebraic group*. If dim $V = n$ then there is a non-vanishing, left invariant n-form ω on V defined over K. Any two of these differ by a multiplicative constant $\lambda \in K^\times$. We now construct a measure on the group $V(\mathbb{A}_K)$ of adelic points of the variety V. For this purpose one must first fix a Haar measure μ_v on the additive group K_v^+, where v is a normalized valuation on K. In order to do this we set $\mu_v(\mathcal{O}_v) = 1$ if v is non-Archimedean, $d\mu_v = dx$ for $K_v \cong \mathbb{R}$ (Lebesgue measure) and $d\mu_v = |dz \wedge \overline{dz}|$ for $z = x + iy \in K_v \cong \mathbb{C}$. Then according to (4.3.46) one has $\mu(\mathbb{A}_K/K) = |D_K|^{1/2}$, where D_K is the discriminant of K and μ is the Haar measure on $\mathbb{A}_K$ defined as the product of local measures μ_v. Define a measure ω_v on $V(\mathbb{A}_K)$ as follows. In a neighbourhood of a point P of V the form ω is defined by the expression

$$\omega = f(x)\, dx_1 \wedge \cdots \wedge dx_n,$$

where $x_1, \ldots, x_n$ are local parameters at P which are certain rational functions $x_i \in K(V)$ and $f \in K(V)$ is a rational function regular at P. The function f can be written as a formal power series in the x_is with coefficients in K, because the variety of an algebraic group is always non-singular. If the coordinates of P belong to K_v then f is a power series in the variables $x_i - x_i^0$ with coefficients in K_v, which converges in a neighbourhood of the origin in K_v^n. Thus there exists a neighbourhood U of P in $V(K_v)$ such that $\varphi : x \to (t_1(x), \ldots, t_n(x))$ is a homeomorphism of U onto a neighbourhood U' of the origin in K_v^n, and the power series converges in U'. In U' we have the positive measure $|f(x)|_v\, dt_1 \cdot \ldots \cdot dt_n$ where $dt_1 \cdot \ldots \cdot dt_n$ is the product $\mu_v \times \cdots \times \mu_v$ on K_v^n; we lift it to U using φ and thus obtain a positive measure ω_v on U. Explicitly, if g is a continuous real valued function on $V(K_v)$ supported on U then

$$\int_U g\, \omega_v = \int_{U'} g(\varphi^{-1}(t))\, dt_1 \cdot \ldots \cdot dt_n,$$

so that ω_v is in fact dependent on a choice of local parameters. If the product

$$\prod_v \omega_v(V(\mathcal{O}_v)) \tag{5.3.68}$$

converges absolutely, then we define the Tamagawa measure by the formula

$$\tau = |D_K|^{-n/2} \prod_v \omega_v. \tag{5.3.69}$$

If the product (5.3.68) does not converge absolutely then one needs to introduce certain correcting factors $\lambda_v > 0$, which ensure the convergence in such a way that the product

$$\prod_{v \not| \infty} \lambda_v^{-1} \omega_v(V(\mathcal{O}_v))$$

will converge absolutely. The Tamagawa measure (with respect to $\{\lambda_v\}$) is then defined by the formula

$$\tau = |D_K|^{-n/2} \prod_{v \not| \infty} \lambda_v^{-1} \omega_v. \tag{5.3.70}$$

In any case, it follows from the product formula that τ is independent of the choice of ω: if we replace ω by $c\omega$ ($c \in K^\times$) then $(c\omega)_v = |c|_v^n \omega_v$ and by the product formula (4.3.31) one has $\prod_v |c|_v = 1$.

Let $k(v)$ denote the residue field with respect to a non-Archimedean place v and let $V^{(v)} = V \otimes k(v)$ be the reduction of V modulo the corresponding prime ideal $\mathfrak{p}_v \subset \mathcal{O}_v$. Then one can show, generalizing Hensel's lemma, that for almost all v

$$\omega_v(V(\mathcal{O}_v)) = \mathrm{N}v^{-n} \ \mathrm{Card}\ V^{(v)}(k(v)), \tag{5.3.71}$$

where $\mathrm{N}v$ denotes the number of elements of $k(v)$ and $V^{(v)}(k(v))$ is the group of points of $V^{(v)}$ with coefficients in $k(v)$.

Examples. If $V = \Gamma_a$ (the additive group) then

$$\omega_v(V(\mathcal{O}_v)) = \mu_v(\mathcal{O}_v) = 1;$$

if $V = \Gamma_m$ (the multiplicative group) then

$$\omega_v(V(\mathcal{O}_v)) = \frac{\mathrm{N}v - 1}{\mathrm{N}v} = 1 - \frac{1}{\mathrm{N}v};$$

if $V = \mathrm{GL}_m$ then

$$\omega_v(V(\mathcal{O}_v)) = \left(1 - \frac{1}{\mathrm{N}v}\right) \cdots \left(1 - \frac{1}{\mathrm{N}v^m}\right);$$

if $V = \mathrm{SL}_m$ then

$$\omega_v(V(\mathcal{O}_v)) = \left(1 - \frac{1}{\mathrm{N}v^2}\right) \cdots \left(1 - \frac{1}{\mathrm{N}v^m}\right).$$

The product

$$\prod_{v\nmid\infty}\left(1-\frac{1}{\mathrm{N}v^{-s}}\right)=\zeta_K(s)^{-1}$$

converges for $\mathrm{Re}(s) > 1$ but diverges at $s = 1$ (here $\zeta_K(s)$ denotes the Dedekind zeta function of K). The product $\prod_v \omega_v(V(\mathcal{O}_v))$ therefore converges for $V = \mathrm{SL}_m$ but diverges for $V = \mathrm{GL}_m$. In the latter case one could take for the correcting factors the numbers $\lambda_v = 1 - \frac{1}{\mathrm{N}v}$. More generally one can show that if $V = G$ is a semi-simple algebraic group then the product (5.3.68) converges absolutely and the correcting factors are not needed. The link between Tamagawa numbers and Siegel's research in the arithmetical theory of quadratic forms was discovered by Weil in the late 50s. He formulated during this time a conjecture later proved by Kottwitz, saying that for a connected, simply connected, semi-simple algebraic group over a number field K, which contains no factors of type E_8, one has $\tau(V) = 1$. For a connected, reductive group G over K it was proved by Sansuc and Kottwitz that

$$\tau(G)=\frac{|\mathrm{Pic}(G)|}{|\text{Ш}(G)|},$$

where Ш(G) is the Shafarevich–Tate group and $\mathrm{Pic}(G)$ is the Picard group of the affine variety (linear algebraic group) G, cf. [Kott88].

Eskin, Rudnick, Sarnak in [ERS91] gave a new proof of Siegel's famous mass formula; they used harmonic analysis to obtain an asymptotic formula for the distribution of integral points on certain affine varieties. In particular, they gave a new proof of Siegel's theorem for indefinite quadrics ($n = 1$, $m \geq 4$). From this it was deduced that the Tamagawa number of any special orthogonal group is 2, which yields the general Siegel result through a computation of adelic volumes with respect to the Tamagawa measure. Note that E. Peyre studied in [Pey95] heights and Tamagawa measures on Fano varieties.

5.4 Diophantine Equations and Galois Representations

5.4.1 The Tate Module of an Elliptic Curve

Let E be an elliptic curve defined over a number field K. Then the Galois group $G_K = G(\overline{K}/K)$ acts on the group E_n of all points of order dividing n, $E_n \cong (\mathbb{Z}/n\mathbb{Z})^2$ so we obtain a Galois representation

$$\varphi_n : G_K \to \mathrm{GL}_2(\mathbb{Z}/n\mathbb{Z}) = \mathrm{Aut}\ E_n.$$

Now let l be a prime number, $n = l^m$. Set

$$T_l(E) = \varprojlim_m E_{l^m} \cong \mathbb{Z}_l^2, \tag{5.4.1}$$

$$V_l(E) = T_l(E) \otimes \mathbb{Q}_l \cong \mathbb{Q}_l^2,$$

where $\mathbb{Z}_l$ is the ring of l-adic integers and the limit is taken over the set of homomorphisms $E_{l^m} \to E_{l^{m-1}}$ which multiply each point by l. The corresponding homomorphism

$$\rho_l : G_K \to \mathrm{Aut}\ V_l(E) \cong \mathrm{GL}_2(\mathbb{Q}_l) \tag{5.4.2}$$

is a continuous representation of the group G_K over the field $\mathbb{Q}_l$. Its image $\mathrm{Im}\ \rho_l = G_l$ is a closed subgroup of $\mathrm{GL}_2(\mathbb{Z}_l) \cong \mathrm{Aut}\ T_l(E)$, and the Weil pairing (5.3.37) determines an isomorphism of $\det\rho_l$ with the representation of G_K on the one dimensional vector space

$$V_l(\mu) = T_l(\mu) \otimes \mathbb{Q}_l, \quad T_l(\mu) = \varprojlim_m \mu_{l^m}$$

(the Tate module defined as the projective limit of roots of unity of l-power degree).

It follows from recent results of Faltings that the G_K-module $T_l(E)$ uniquely determines the curve E upto an isogeny.

Serre discovered (cf. [Se68a]) that the image $\mathrm{Im}\ \rho_l$ is as large as it could possibly be for almost all primes l. More precisely this image coincides with $\mathrm{GL}_2(\mathbb{Z}_l) \cong \mathrm{Aut}\ T_l(E)$, provided that the curve E is not special in the sense that it admits no complex multiplication, or equivalently $\mathrm{Aut}(E) = \mathbb{Z}$. Moreover the index of the subgroup $\varphi_n(G_K)$ in $\mathrm{GL}_2(\mathbb{Z}/n\mathbb{Z}) = \mathrm{Aut}\ E_n$ is bounded by a constant which is independent of n, of the curve E and of the field K (cf. [Ner76], [Silv86]). The occurrence of small images $\mathrm{Im}\ \rho_l$ is closely related to the existence of K-rational *points of finite order* (or of K-rational subgroups of such points). For example, if there exists a basis P, Q of the group E_n over $\mathbb{Z}/n\mathbb{Z}$ such that the point P is K-rational, i.e. $P \in E_n(K)$, then $P^\sigma = P$ for all $\sigma \in G_K$. Elements in the image $\varphi_n(G_K)$ are therefore represented by matrices of the form $\left(\begin{smallmatrix} 1 & * \\ 0 & * \end{smallmatrix}\right)$ in $\mathrm{GL}_2(\mathbb{Z}/n\mathbb{Z})$. If the subgroup $\langle Q \rangle$ is also K-rational, i.e. $\langle Q \rangle^\sigma = \langle Q \rangle$ then elements in the image have the form $\left(\begin{smallmatrix} 1 & 0 \\ 0 & * \end{smallmatrix}\right)$. The result of Serre is therefore closely related with *Mazur's theorem* on the universal boundedness of the torsion subgroup of an elliptic curve over $\mathbb{Q}$ (cf. [Maz77]).

Let A be an Abelian variety of dimension g defined over K. Then the Tate module is defined by

$$T_l(A) = \varprojlim_m \mathrm{Ker}(A \xrightarrow{l^m} A) \cong \mathbb{Z}_l^{2g},$$

$$V_l(A) = T_l(A) \otimes \mathbb{Q}_l \cong \mathbb{Q}_l^{2g},$$

and we again have a Galois representation (see (5.3.52))

$$\rho_l : G_K \to \mathrm{Aut}\ V_l(A) \cong \mathrm{GSp}_{2g}(\mathbb{Q}_l). \tag{5.4.3}$$

Note that certain results are known on the maximality of the image of the Galois representation ρ_l for higher dimensional Abelian varieties A with End $A = \mathbb{Z}$ (i.e. without complex multiplication).

The study of the image of ρ_l is based on an examination of the *reduction* of the *elliptic curve (or Abelian variety)* modulo $\mathfrak{p}_v$, where v is a finite place of K. The condition that E has *good reduction* $\tilde{E}_v = E \mod \mathfrak{p}_v$ is equivalent to the existence of an Abelian scheme E_v over Spec $\mathcal{O}_v$ in the sense of Mumford (cf. [Mum74]) whose generic fiber coincides with E (i.e. $E_v \otimes_{\mathcal{O}_v} K_v \cong E \otimes_K K_v$) and whose closed fiber is an elliptic curve (Abelian variety) $\tilde{E}_v = E_v \otimes_{\mathcal{O}_v} k(v)$ over the residue field $k(v) = \mathcal{O}/\mathfrak{p}_v$. The (geometric) *Frobenius endomorphism* F_v of $\tilde{E}_v$ is defined by raising the coordinates of points on $\tilde{E}_v$ to their $\mathrm{N}v = |k(v)|^{\mathrm{th}}$ powers.

Now let p_v denote the characteristic of the residue field $k(v)$ and let l be another prime number (not p_v). Denote by G_v (respectively I_v) the decomposition group (respectively, the inertia group) of some extension $\overline{v}$ of v to a fixed algebraic closure $\overline{K}$ of K (compare with (4.4.2)). If E has good reduction at v then $\overline{v}$ defines (in view of Hensel's lemma) an isomorphism from E_{l^m} to the corresponding subgroup of the curve $\tilde{E}_v$. In particular, the inertia group I_v acts trivially on E_{l^m}, $T_l(E)$ and $V_l(E)$, so the action $\rho_l(Fr_v)$ of the arithmetical Frobenius automorphism Fr_v is well-defined ($Fr_v \in G_v/I_v$) and is the same as the action of the geometric Frobenius $F_v = F_{E,v}$. One therefore has

$$\det \rho_l(Fr_v) = \det(F_v) = \mathrm{N}v = \mathrm{Card}\ k(v), \tag{5.4.4}$$

and the quantity

$$\det(1_2 - \rho_l(Fr_v)) = \det(1 - F_v) = 1 - \mathrm{Tr}\ F_v + \mathrm{N}v \tag{5.4.5}$$

is equal to the number Card $\tilde{E}_v(k(v))$ of $k(v)$-points of the reduction $\tilde{E}_v$. Conversely, one has the following

Theorem 5.11 (Criterion of Neron – Ogg – Shafarevich). *If the Galois representation ρ_l is unramified at v for some $l \neq p_v$ then E has good reduction at v.*

(cf. [Silv86], Ch. 4 of [Se68a]).

5.4.2 The Theory of Complex Multiplication

(see Chapter XIII of [CF67], [La73/87], [Shi71]). One of the central aims of algebraic number theory was formulated in 1900 by Hilbert in Paris as his twelfth problem: that of finding an explicit construction of all Abelian extensions of a given number field K. For $K = \mathbb{Q}$ it is known (by the *Kronecker–Weber theorem* (comp. with §4.1.2)) that the maximal Abelian extension $\mathbb{Q}^{\rm ab}$ of $\mathbb{Q}$ is cyclotomic, and that there is an isomorphism

$$G(\mathbb{Q}^{\rm ab}/\mathbb{Q}) \cong \prod_p \mathbb{Z}_p^\times .$$

If K is an imaginary quadratic extension of $\mathbb{Q}$ then the theory of complex multiplication makes it possible to construct $K^{\rm ab}$ using elliptic curves E with complex multiplication by K, and their *points of finite order*. By definition, one has for such curves $\operatorname{End} E \otimes \mathbb{Q} = K$. If $E(\mathbb{C}) \cong \mathbb{C}/\Gamma$ for a lattice $\Gamma \subset \mathbb{C}$, then the endomorphism ring of E has the following form

$$\operatorname{End} E = \{z \in \mathbb{C} | z\Gamma \subset \Gamma\} = \mathcal{O}_f = \mathbb{Z} + f\mathcal{O}_K \subset \mathcal{O}_K,$$

where $\mathcal{O}_K$ is the maximal order of K and f is an appropriate positive integer (in view of the fact that every subring of $\mathcal{O}_K$ has the form $\mathbb{Z} + f\mathcal{O}_K$ for some f).

Theorem 5.12. *There is a one-to-one correspondence between elliptic curves E with a given endomorphism ring $\mathcal{O}_f$ (upto isomorphism), and elements of the class group $\mathrm{Cl}(\mathcal{O}_f)$ (i.e. the group of isomorphism classes of projective modules of rank one over $\mathcal{O}_f$).*

Indeed, if a lattice Γ corresponds to E then Γ is an $\mathcal{O}_f$-module such that $\Gamma \otimes \mathbb{Q} \cong K$, i.e. a projective $\mathcal{O}_f$-module of rank one. Conversely, every $\mathcal{O}_f$-module viewed as a lattice in $\mathbb{C}$ determines an elliptic curve $\mathbb{C}/\Gamma$ with the property that $\operatorname{End}(\mathbb{C}/\Gamma)$ is the ring of multipliers of Γ, i.e. $\mathcal{O}_f$. Therefore the number h_f of curves (upto isomorphism) with a given *endomorphism ring* $\mathcal{O}_f$ is finite and its order is equal to Card $\mathrm{Cl}(\mathcal{O}_f)$.

For each curve there is a canonically defined *invariant* $j(E)$ corresponding to E; if E is written in the Weierstrass form then this is given by

$$j(E) = \frac{1728 g_2^3}{g_2^3 - 27 g_3^2}, \qquad E : y^2 = 4x^3 - g_2 x - g_3. \tag{5.4.6}$$

We now consider the case $f = 1$ in more detail.

Theorem 5.13 (Weber – Fueter). *(a) All the numbers $j(E)$ are algebraic integers. (b) If $\alpha = j(E)$ is one of these numbers then $K(\alpha)$ coincides with the maximal unramified, Abelian extension of K and $G(K(\alpha)/K) \cong \mathrm{Cl}(\mathcal{O}_f)$. The action of $G(K(\alpha)/K)$ on the set of numbers $\{j(E)\}$ is transitive.*

There are precisely nine imaginary quadratic rings $\mathcal{O}_f$ with $f = 1$ and $h_f = 1$, namely $\mathbb{Z}[\sqrt{-d}]$, where $d = 1, 2, 3, 7, 11, 19, 43, 67, 163$. The corresponding elliptic curves have rational invariants, which are also algebraic integers in view of the Weber – Fueter theorem, hence $j(E) \in \mathbb{Z}$. Moreover, the values of $j(E)$ are given respectively by:

$$\begin{aligned} j =& 2^6 \cdot 3^3, \quad 2^6 \cdot 5^3, \quad 0, \quad -3^3 \cdot 5^3, \quad -2^{15}, \quad -2^{15} \cdot 3^3, \\ & - 2^{18} \cdot 3^3 \cdot 5^3, \quad -2^{15} \cdot 3^3 \cdot 5^3 \cdot 11^3, \quad -2^{18} \cdot 3^3 \cdot 5^3 \cdot 23^3 \cdot 29^3 . \end{aligned} \tag{5.4.7}$$

In the general case $f > 1$ the numbers $j(E)$ are also algebraic integers for all E with $\mathrm{End}(E) = \mathcal{O}_f$, and for all $\sigma \in \mathrm{Gal}(\overline{K}/K)$ one can explicitly describe the action of σ on $j(E)$. This description depends only on the restriction of σ to K^{ab}, which is represented via the *Artin reciprocity law* by an idele $s \in J_K$:

$$\sigma|_{K^{\mathrm{ab}}} = \psi_K(s), \qquad (\psi_K : J_K \to G(K^{\mathrm{ab}}/K) - \text{the reciprocity map}).$$

Furthermore if Γ is the *lattice* corresponding to a curve E then one can define a lattice $s^{-1}\Gamma$: if $s = (s_v)_v$ $(s_v \in K_v^\times)$ then $s^{-1}\Gamma$ is uniquely determined by the condition $(s^{-1}\Gamma) \otimes \mathcal{O}_v = s_v^{-1}(\Gamma \otimes \mathcal{O}_v)$ for all finite v.

Theorem 5.14. *Let $j(s^{-1}\Gamma)$ denote the invariant of the elliptic curve E' defined by $E'(\mathbb{C}) = \mathbb{C}/s^{-1}\Gamma$. Then one has the following formula for the action of $\sigma \in \mathrm{Gal}(\overline{K}/K)$:*

$$j(E)^\sigma = j(s^{-1}\Gamma). \tag{5.4.8}$$

From this it follows that $j(E) \in K^{\mathrm{ab}}$. To prove these theorems, one considers the action of $\sigma \in \mathrm{Gal}(\overline{K}/K)$ on the coefficients of the Weierstrass equation (5.4.6). One obtains as a result the following new curve: $E^\sigma : y^2 = 4x^3 - g_2^\sigma x - g_3^\sigma$; therefore $j(E)^\sigma = j(E^\sigma)$. Clearly, one has $\mathrm{End}(E^\sigma) = \mathrm{End}(E) \cong \mathcal{O}_f$, and thus the set $\{j(E)^\sigma | \sigma \in \mathrm{Gal}(\overline{K}/K)\}$ is finite and the numbers $j(E)$ are all algebraic. Consequently the curve E can be defined over an algebraic number field L. If the restriction of σ to L is represented by a Frobenius automorphism for some v,

$$\sigma|L = F_{L/K}(v) = Fr_v,$$

then the above formula (5.4.8) can be established using the reduction E mod $\mathfrak{P}$, where $\mathfrak{P}$ is a divisor in L which divides $\mathfrak{p}_v$. Then this formula can be rephrased as *Hasse's theorem*:

$$j(E)^{Fr_v} = j(\mathfrak{p}_v^{-1}\Gamma), \tag{5.4.9}$$

where $\mathfrak{p} \subset \mathcal{O}_K$ is a prime ideal of $\mathcal{O}_K$ defined by the conditions $(\mathfrak{p}, f) = 1$, $\mathfrak{p}_f = \mathcal{O}_f \cap \mathfrak{p}$.

The invariants $j(E)$ therefore generate an extension $K_{(f)}/K$ satisfying the property $G(K_{(f)}/K) \cong \mathrm{Cl}(\mathcal{O}_f)$. However the field $\tilde{K} = \cup_{f \geq 1} K_{(f)}$ does not yet coincide with the whole of K^{ab}, and in order to obtain K^{ab} it is necessary to adjoin also to $K_{(1)}$ the coordinates of all points of finite order on some elliptic curve E with the property $\mathrm{End}(E) = \mathcal{O}_K$. More precisely, let E be an elliptic curve with complex multiplication, i.e. $\mathrm{End}\ E \otimes \mathbb{Q} \cong K$, defined over a number field $L \supset K$. Then the image of the Galois representation

$$\rho_l : G_L \to \mathrm{Aut}\ V_l(E) \cong \mathrm{GL}_2(\mathbb{Q}_l) \tag{5.4.10}$$

is an Abelian group which is contained in $(\mathbb{Z}_l \otimes \mathcal{O}_K)^\times$, and the index of $\mathrm{Im}\ \rho_l$ is finite and is bounded by a constant independent of l. By class field theory the representation factorizes through G_L^{ab}, and for each idele $s = (s_v)_v \in J_L$ we can define an element $\rho_l(s) = \rho_l(\sigma)$, where $\sigma \in \mathrm{Gal}(\overline{K}/K)$ is determined by the condition $\sigma | L^{\mathrm{ab}} = \psi_L(s)$. It is not difficult to see that there is a unique continuous homomorphism $\varepsilon : J_L \to K^\times$ with the condition $\varepsilon(x) = \mathrm{N}_{L/K}(x)$ for all $x \in L^\times$ and $\rho_L(s) = \varepsilon(s)\mathrm{N}_{L_l/K_l}(s_l)$ for all $s \in J_L$ and all l.

The *Abelian l-adic Representations* (5.4.10) and the action of G_L on the invariants $j(E)$ describe explicitly the class field theory of the field K. We see also that in the complex multiplication case the group $\mathrm{Im}\ \rho_l$ is Abelian and is therefore much smaller than in the general case.

An analogous theory (in a less complete form) also exists for CM-fields (totally imaginary, quadratic extensions of totally real fields) and for Abelian varieties of CM-type, i.e. Abelian varieties A whose endomorphism algebras $\mathrm{End}\ A \otimes \mathbb{Q}$ are totally imaginary, quadratic extensions of totally real fields of degree $g = \dim\ A$ (cf. [Shi71]).

5.4.3 Characters of l-adic Representations

As we have seen, one can associate to each elliptic curve E defined over a number field K a system of l-adic representations $\rho_l : G_K \to \mathrm{Aut}\ T_l(E) \cong \mathrm{GL}_2(\mathbb{Z}_l)$ on the Tate module $T_l(E)$. Together (5.4.4) and (5.4.5) give the following important formula for the *traces of Frobenius endomorphisms*:

$$\mathrm{Tr}\ \rho_l(Fr_v) = \mathrm{N}v + 1 - N_v(E),$$

where $\mathrm{N}v = \mathrm{Card}(k(v))$ is the norm of v, $N_v(E) = \mathrm{Card}\ \tilde{E}_v(k(v))$ is the number of $k(v)$ – rational points on the reduction $\tilde{E}_v$ modulo v. It turns out that the values of the character $\chi_{\rho_l} = \mathrm{Tr}\ \rho_l$ form an interesting arithmetical function of the argument v. We shall later see that the list of examples of this sort is quite rich and contains for example the Ramanujan function $\tau(p)$; the numbers of representations of positive integers by positive definite quadratic forms, etc. It is known that the character of ρ uniquely determines this representation provided that ρ is a semi-simple representation, that is, a direct sum of irreducible representations. This semi-simplicity property was established for the Tate modules of elliptic curves and Abelian varieties, cf. [Fal83], [Fal85], [Fal86], [PZ88].

A remarkable finiteness property

for the characters χ_ρ of continuous finite–dimensional l-adic representations was discovered by G.Faltings (cf. [Fal83]) in his proof of the Mordell conjecture: any such character χ_ρ is uniquely determined by a finite number of values

$$\chi_\rho(Fr_v) = \mathrm{Tr}\ \rho(Fr_v), \quad (v \in Q, \ \ Q \text{ a finite set}),$$

where Fr_v denotes the class of a Frobenius element under the assumption that ρ is unramified for all v outside a finite set S of places of K. In this situation the representation ρ factorizes through a representation of the group $G_S = G(K_S/K)$, where K_S is the maximal extension of K unramified outside S. For each $v \notin S$ the value $\chi_\rho(Fr_v)$ is therefore well-defined. We shall now construct a finite set Q of places of K, $Q \cap S \neq \emptyset$, such that ρ is is uniquely determined by the values $\chi_\rho(Fr_v)$ for $v \in Q$. Let L/K denote the composite of all Galois extensions of K unramified outside S, which are of degree less than or equal to l^{2n^2}. Then by *Hermite's theorem* (see §4.1.5), the extension L/K is finite. Now we choose an appropriate Q outside S such that the elements Fr_v fill the whole Galois group $G(L/K)$. The existence of such elements follows from the Chebotarev density theorem (Theorem 4.22). We claim that the set Q constructed in this way satisfies the conditions of the theorem.

Indeed, let ρ_1 and ρ_2 be two different representations whose characters coincide on the elements $Fr_v, v \in Q$. Consider the representation

$$\rho_1 \times \rho_2 : \mathbb{Z}_l[G] \to \mathrm{M}_n(\mathbb{Q}_l) \times \mathrm{M}_n(\mathbb{Q}_l)$$

of the group algebra $\mathbb{Z}_l[G]$. Its image M is a $\mathbb{Z}_l$-submodule of rank $\leq 2n^2$. By the construction of Q the elements $\rho_1 \times \rho_2(Fr_v)$, $\quad v \in Q$ generate M/lM as a vector space over $\mathbb{F}_l$, and consequently, the whole of M over $\mathbb{Z}_l$ (by *Nakayama's lemma* for finitely generated modules over a local ring, applied to the ring $\mathbb{Z}_l$, see [Bou62], [SZ75]). Now consider the linear form

$$f(a_1, a_2) = \mathrm{Tr}(a_1) - \mathrm{Tr}(a_2) \quad (a_1, a_2 \in \mathrm{M}_n(\mathbb{Q}_l))$$

on M. By the assumption we have that

$$\chi_{\rho_1}(Fr_v) = \chi_{\rho_2}(Fr_v), \quad v \in Q,$$

and therefore $f(a_1, a_2) = 0$ on the whole $\mathbb{Z}_l$-module M, because $f = 0$ on its generators $(\rho_1 \times \rho_2)(Fr_v)$ $\ v \in Q$. Therefore $\chi_{\rho_1}(Fr_v) = \chi_{\rho_2}(Fr_v)$, establishing the theorem, see [PZ88], [Del83], [Sz(e)81].

5.4.4 Representations in Positive Characteristic

Let E be an elliptic curve over a finite field k with $q = p^d$ elements. Consider its Tate module

$$T_p(E) = \varprojlim_m \mathrm{Ker}(E \xrightarrow{p^m} E) \cong \mathbb{Z}_p^\gamma,$$

where $\gamma = 0$ or $\gamma = 1$. In this way we obtain a representation defined by

$$\rho_p : \mathrm{Gal}(\overline{k}/k) \to \mathrm{Aut}\ T_p(E),$$

in which the group $\mathrm{Gal}(\overline{k}/k)$ is a (topologically) cyclic group.

If $T_p(E) \neq 0$ then End $E \otimes \mathbb{Q}$ is an imaginary quadratic extension of K and End $E = \mathcal{O}_f$ for some $f \geq 1$, where $\mathcal{O}_f = \mathbb{Z} + f\mathcal{O}_K$ is a subring of the maximal order $\mathcal{O}_K$ of K.

In this situation one can show that:

1) the prime p does not divide the conductor f,
2) p splits in K.

In the case $T_p(E) = 0$ we have that $D_E = \mathrm{End}\ E \otimes \mathbb{Q}$ is a division algebra of degree 4 (a quaternion algebra) over $\mathbb{Q}$ which at all primes $l \neq p$ decomposes as $D_E \otimes \mathbb{Q}_l \cong \mathrm{M}_2(\mathbb{Q}_l)$. Also, End E is a maximal order in D_E. Curves with this property are called *supersingular* curves.

In positive characteristic the endomorphism algebra becomes larger when there is a Frobenius endomorphism F_q of E, which is a *purely inseparable* isogeny; its kernel and image have only one geometric point over $\overline{k}$. In particular, if $F_q \in \mathbb{Z} \subset \mathrm{End}\ E$, then also $T_p(E) = 0$. For further information on points of finite order in positive characteristic, also in Abelian varieties, see [Man61], [La73/87], [Mum74].

5.4.5 The Tate Module of a Number Field

(cf. [Sha69], [Coa73], [Iwa72], [Iw01], [MW83]). The Tate module of the Jacobian variety J_C of a curve C gives a functor from the category of curves over a field k to the category of $\mathbb{Z}_l$-modules. If k is finite then the field $k(C)$ of rational functions on C has much in common with a number field. Iwasawa has suggested an analogue of the Tate module for a number field K. The group J_{l^m} can be interpreted as the Galois group of the étale covering $C_m \to C$ of C, where C_m is the inverse image of C embedded into J, with respect to the morphism l_J^m. One verifies that the field $\cup_m k(C_m)$ is the maximal unramified, Abelian l-extension of $k(C)$. Its Galois group coincides exactly with the Tate module $T_l(J_C)$; this gives a reasonable interpretation of the Tate module for an algebraically closed field k. However if k is not algebraically closed (for example when k is a finite field) then k need not be algebraically closed inside the field $k(J_{l^m})$. In particular the field $k(J_{l^m})$ must contain roots of unity of the degree l^m since these are values of the Weil pairing. In the case of a finite field this is almost sufficient: that is, for a finite extension k'/k we have

$$\tilde{k} = \overline{k} \cap \left(\bigcup_m k(J_{l^m}) \right) = \bigcup_m k'(\zeta_{l^m}),$$

where ζ_{l^m} denotes a primitive root of unity of degree l^m and $\overline{k}$ is the algebraic closure of k. Indeed, the image of $\mathrm{Gal}(\overline{k}/k)$ in $T_l \subset \mathrm{GL}_{2g}(\mathbb{Z}_l)$ is a topologically cyclic group, whose intersection with the l-Sylow normal subgroup $S = \{g \in \mathrm{GL}_{2g}(\mathbb{Z}_l) | g \equiv 1 \mod l\}$ is an l-subgroup of finite index. Therefore on replacing k by a finite extension k' of degree prime to the characteristic of k, the extension $\tilde{k}$ becomes an l-extension of the finite field k', i.e.

$$\tilde{k} = \bigcup_m k'(\zeta_{l^m}).$$

We see now that for a finite field of constants the Tate module T_l coincides with the Galois group of the composite Galois extension

$$k(C) \subset \tilde{k}(C) \subset A^{(l)},$$

where $\tilde{k} = \cup_m k'(\zeta_{l^m})$, $A^{(l)}$ is the maximal unramified, Abelian l-extension of $\tilde{k}(C)$.

Taking this description as a starting point, we may extend the definition of the Tate module to the number field case. Let K be a number field, $K_m = K(\zeta_{l^m})$, $\tilde{K} = \cup_m K_m$, and $A^{(l)} \supset \tilde{K}$ the maximal Abelian, unramified l-extension of $\tilde{K}$. Further let

$$T_l(K) = \mathrm{Gal}(A^{(l)}/\tilde{K}). \tag{5.4.11}$$

Then $T_l(K)$ is a projective limit of l-groups (a pro-l-group), and is in particular a $\mathbb{Z}_l$-module. Iwasawa, who introduced this module (also called the *Iwasawa module*), has shown that $V_l(K) = T_l(K) \otimes \mathbb{Q}_l$ is a *finite – dimensional* $\mathbb{Q}_l$-vector space. Using class field theory one can describe $T_l(K)$ explicitly. One knows that the Galois group of the maximal Abelian, unramified extension of a number field L is isomorphic to the class group Cl_L. Denoting by $\mathrm{Cl}_L^{(l)}$ the l-component of this group, one obtains the following description:

$$T_l(K) = \varprojlim \mathrm{Cl}_{K_m}^{(l)},$$

where the inverse limit is taken with respect to the norm maps of ideals.

On $T_l(K)$ we have an obvious action of the Galois group $\mathrm{Gal}(\tilde{K}/K)$ and its subgroup $\Gamma = G(\tilde{K}/K_1) \cong \mathbb{Z}_l$. On a class represented by an ideal $\alpha \in \mathrm{Cl}_{K_m}^{(l)}$ this action is given by $\alpha \mapsto \alpha^g$, $(g \in G(\tilde{K}/K))$, and for the corresponding $h \in \mathrm{Gal}(A^{(l)}/K)$ the ideal α^g corresponds under class field theory to $g^{-1}hg$ (in view of the equality (4.4.22)).

Iwasawa regarded $T_l(K)$ as a module over the completed group ring $\Lambda = \mathbb{Z}_l[[\Gamma]] \cong \mathbb{Z}_l[[T]]$ (the ring of formal power series over $\mathbb{Z}_l$). Just using his classification theory for such modules, he obtained the following formula for the orders of the groups $\mathrm{Cl}_{K_m}^{(l)}$, which is valid for $m \geq m_0$:

$$\log_l |\mathrm{Cl}_{K_m}^{(l)}| = \lambda m + \mu l^m + \mathrm{const}. \tag{5.4.12}$$

Under some additional assumptions he described explicitly the module $T_l(\mathbb{Q})$ for all $l \leq 4001$. This module turns out to be cyclic, and one can even find a generator of its annihilator. Essentially, this generator coincides with a product of the l-adic L-functions of Kubota and Leopoldt ([Iwa72], [KuLe64], [Sha69], [Kuz84]).

The validity of the corresponding statement in the general case (the *"Main conjecture"* of *Iwasawa theory*) was established in 1984 by B. C. Mazur and A. J. Wiles [MW83]. According to the main conjecture of Iwasawa theory a module of ideal class groups can be described as the quotient of the Iwasawa algebra by an explicitly given principal ideal. A later, more accessible proof using Kolyvagin's notion of Euler systems was found by K. Rubin, cf. his appendix to [La90].

In the works of Ferrero and Washington ([FeWa79], [Fer88], [Wash82]) another conjecture of Iwasawa was proved, which says that for each Abelian extension $K/\mathbb{Q}$ and each prime l, the invariant μ of the module $T_l(K)$ vanishes.

This result implies that $T_l(K)$ is a finitely generated $\mathbb{Z}_l$-module. Washington's conjecture, according to which the orders of the groups $\mathrm{Cl}_{K_m}^{(p)}$ stabilize in the cyclotomic $\mathbb{Z}_l$-extension of an arbitrary Abelian field for $l \neq p$, was proved in ([Wash78]).

Very recently (cf. [Barsky04]) the vanishing of the Iwasawa μ invariant was proved by D.Barsky for all totally real fields. The Iwasawa μ- invariant of p-adic Hecke L- functions was studied by H.Hida in [Hi02].

The methods of Iwasawa have been considerably extended in further research related to the study of Λ-modules of various kinds: those arising from *Selmer groups* of Abelian varieties (*Mazur modules*) see [Man71], [Man76], [Man78], [Maz79], [Maz83], [Maz86], and also those arising from elliptic units in Abelian extensions of fields of CM-type, cf. [Maz83], [Rob73], and the ones arising from Heegner points on modular curves ([Koly88], [Coa73], [Coa84], [GZ86], [Rub77]).

New approaches to proving the main conjecture and its generalizations in various situations were discovered by Kolyvagin in [Koly90], who proposed the more general concept of an "Euler system", which makes it possible to deal with all known cases from a unified point of view.

For recent developments on Euler Systems we refer to [Rub98], [Kato99], [Kato2000], [MazRub04].

Interesting Euler systems could be constructed in some cases using Beilinson elements in K_2 of modular curves and the Rankin–Selberg method, cf. [Scho98]. An analogue of the Selmer groups and the groups of Shafarevich-Tate were defined in [BK90], [FP-R94] for an arbitrary motive over a number field F, cf. also a new book by B.Mazur and K.Rubin, [MazRub04].

We only mention a GL(2) version of Iwasawa theory developed by Coates et al., cf. [Coa01], [CSS03]. The GL(2) main conjecture for elliptic curves without complex multiplication was described very recently by J. Coates, T. Fukaya, K. Kato, R. Sujata, O. Venjakob in [CFKSV].

5.5 The Theorem of Faltings and Finiteness Problems in Diophantine Geometry

5.5.1 Reduction of the Mordell Conjecture to the finiteness Conjecture

A major problem in diophantine geometry was the *Mordell conjecture*, now a

Theorem 5.15 (Faltings [Fal83]). *If X is a projective algebraic curve of genus $g \geq 2$ defined over a number field K and L/K a finite extension then $X(L)$ is finite.*

Note that prior to the work of Faltings this was not known for *any* curve X. However *Siegel's Theorem* was known, the strongest finiteness result until Faltings:

Theorem 5.16 (Siegel). *If X is an affine curve of genus $g \geq 1$ defined over the ring of integers $\mathcal{O} \subset K$ and $\mathcal{O}_S \subset \mathcal{O}$ is any subring of S-integral elements (S finite) then $X(\mathcal{O}_S)$ is finite.*

Here $\mathcal{O}_S \subset K$ denotes the subring

$$\mathcal{O}_S = \{x \in K \mid \forall v \in S, v \text{ non-Archimedean}, |x|_v \leq 1\},$$

where $S \subset \mathrm{Val}(K)$ is a finite set of valuations of K.

The starting point for research leading ultimately to the (first!) proof of Mordell's conjecture by Faltings, was the following pair of conjectures, proposed by I.R.Shafarevich [Sha62], on classification the problem for algebraic curves of genus $g \geq 1$ over an algebraic number field K, with a fixed set S of bad reduction points: *)

Let $\text{Ш}(g, K, S)$ be the set of (K-isomorphism classes of) algebraic curves X of genus $g(X) \geq 1$, defined over a number field K with bad reduction contained in a finite set $S \subset \mathrm{Val}(K)$. When $g = 1$ we assume in addition that $X(K) \neq \emptyset$.

I) *Finiteness Conjecture.* Assume that $g \geq 2$ (or that $g \geq 1$ and $X(K) \neq \emptyset$). Then for any given g, K, S these exist only finitely many such curves (up to isomorphism). This finite set will be denoted by $\text{Ш}(g, K, S)$.

For the second proof of Mordell's conjecture by Bombieri-Vojta we refer to [Bom90], [Voj91]. This entirely new proof is based on methods from Diophantine approximation and arithmetic intersection theory. Faltings in [Fal91] simplified and extended these methods to prove two longstanding conjectures of Lang concerning integral points on Abelian varieties and rational points on their subvarieties, and E. Bombieri subsequently further simplified the arguments to give a comparatively elementary proof of Mordell's conjecture.

II) *Bad Reduction Conjecture.* If $S = \emptyset$ and $K = \mathbb{Q}$ then these exist no such curves, i.e. $\text{Ш}(g, K, S) = \emptyset$.

These problems generalize theorems of Hermite and Minkowski in the theory of algebraic number fields (cf. §4.1.5). It was shown by A.N.Parshin in [Par72], [Par73], how to reduce Mordell's problem to the finiteness conjecture. His remarkable construction is given below. The *Shafarevich Conjecture* and the related *Tate conjecture* were proved by Faltings [Fal83] (see also Deligne's Bourbaki talk [Del83]).

A detailed exposition of all these questions can be found in the survey [PZ88].

The construction of A.N.Parshin consists in constructing of a map

$$\alpha : X(K) \to \text{Ш}(g', K', S') \tag{5.5.1}$$

for some other data of g', K', S' with the property that the fibers of the map (5.5.1) are finite. The image of a point $P \in X(K)$ is a certain curve $X_P \in \text{Ш}(g', K', S')$, which is constructed in several steps.

1) Let us map the curve X into its Jacobian J using the Abel map (5.3.52): $\varphi_P : X \to J$, and consider the multiplication by 2 morphism $2_J : J \to J$. We define an auxiliary curve X_1 as the inverse image of X under this map (this is an example of the *fiber product*):

$$\begin{array}{ccc} X_1 & \longrightarrow & X \\ \downarrow & & \downarrow{\scriptstyle \varphi_P} \\ J & \underset{2_J}{\longrightarrow} & J \end{array}$$

The curve X_1 is smooth, it is defined over the same field K, and its genus g_1 can be computed using the *Hurwitz formula* (5.1.1). We have $2g_1 - 2 = 2^{2g}(2g-2)$, because $X_1 \longrightarrow X$ is an unramified covering of degree $2^{2g} = \text{Card} J_2$. The inverse image of the point P is then a rational divisor $D = D_P$ on X_1 of degree 2^{2g}.
2) One constructs a covering $X_P \to X_1$ of degree 2, which is ramified only over the points in D. Such a covering exists, it has genus $g' = g(X_P)$ which is also computed by the Hurwitz formula

$$2g' - 2 = 2(2g_1 - 2) + 2^{2g} = 2^{2g+1}(2g-2) + 2^{2g},$$

that is

$$g' = 2^{2g+1}(g-1) + 2^{2g-1} + 1.$$

One checks that the curve X_P is defined over an algebraic number field $K' \supset K$, $[K' : K] < \infty$, which depends only on the data g, K, S, but not of the individual point $P \in X(K)$. Moreover, the curve X_P has a good reduction over the set S' of non-Archimedean points of K', lying over S and over the prime 2. The construction of X_P is analoguous to the construction of a 2-covering attached to a rational point on an elliptic curve (see (5.3.37). In the same way one proves that the reduction of the resulting curve is good over S' (lying over $S \cup 2$). Then one deduces the finiteness of the degree of K'/K using the *Theorem of Hermite.*

The proof of the fact that the fibers of the map (5.5.1) are finite, is purely geometric, and it belongs rather to the theory of Riemann surfaces. Indeed, the resulting map $X_P \to X$ is ramified exactly over exactly one point, namely P. If there were infinitely many points P such that the curves X_P were isomorphic, say to a fixed curve Y, we would have infinitely many maps $Y \to X$ of curves of genus > 2, ramified over different points of X. Amongst these maps there are infinitely many non–isomorphic, since the group of analytic isomorphisms of a Riemann surface of genus ≥ 2 is finite (its order is bounded by $\leq 84(g-1)$, see [Hur63], [Maz86]). This leads to a contradiction with the classical theorem of de Franchis: for any closed Riemann surface Y there exists only finitely many non-constant maps $f : Y \to Z$ into closed Riemann surfaces Z of genus $g_Z \geq 2$ (upto isomorphism), see [dFr13], [Sev14]).

Note that the theorem of de Franchis is itself a special case of an analogue of Mordell's conjecture over function fields (namely, over $\mathbb{C}(t)$). This version of Mordell's problem was solved in [Man63a], [Man63b].

5.5.2 The Theorem of Shafarevich on Finiteness for Elliptic Curves

(cf. [Sha65], [Se68a]) The finiteness conjecture was proved by I.R.Shafarevich [Sha65] for hyperelliptic curves as a corollary of the *Theorem of Siegel* on the finiteness of the number of S-integral points on an affine algebraic curve of positive genus. We shall give the proof of this result in the case of elliptic curves. Let us write the curve $X = E$ in the Weierstrass form:

$$E : y^2 = 4x^3 - g_2'x - g_3' \quad (g_2', g_3' \in K) \tag{5.5.2}$$

We next note that is the curve E has a good reduction outside of S, then its equation could be reduced to the following form

$$E : y^2 = 4x^3 - g_2x - g_3 \quad \text{with} \quad \Delta = g_3^2 - 27g_2^3 \in \mathcal{O}_S^\times$$

(it is assumed that the finite set S contains the primes over 2 and 3, and S is also chosen large enough so that $\mathcal{O}_S$ is a P.I.D.). Indeed, if $v \notin S$, then the curve E can be led over the local ring $\mathcal{O}_v$ to the form

$$E : y^2 = 4x^3 - g_{2,v}x - g_{3,v} \quad \text{with} \;\; g_{2,v}', g_{3,v}' \in \mathcal{O}_v \cap K, \; \Delta_v \in \mathcal{O}_v^\times. \tag{5.5.3}$$

By the uniqueness property of the Weierstrass form (5.5.2) one can choose an element $u_v \in K^\times$, such that

$$g_{2,v} = u_v^4 g_2', \quad g_{3,v} = u_v^6 g_3', \quad \Delta_v = u_v^{12}\Delta',$$

and we may assume that $u_v = 1$ for almost all v.

As the ring $\mathcal{O}_S$ is a P.I.D., there is an element $u \in K^\times$ the equation (5.5.2) of the curve takes the form

$$E : y^2 = 4x^3 - g_2 x - g_3$$

where

$$g_2 = u^4 g_2', \quad g_3 = u^6 g_3', \quad \Delta = u^{12}\Delta',$$

and it follows that $\Delta \in \mathcal{O}_S$.

Now we can miltiply Δ by any number $u \in (\mathcal{O}_S^\times)^{12}$ keeping fixed the isomorphism class of the curve. It follows from a version of Dirichlet's unit theorem (on S-units, see §4.1.6) that the group $(\mathcal{O}_S^\times)/(\mathcal{O}_S^\times)^{12}$ is finite. There therefore exists a finite set $M \subset \mathcal{O}_S^\times$ such that any elliptic curve of the given form can be reduced to the form (5.5.3) with $g_i \in \mathcal{O}_S$, $\Delta \in M$. On the other hand, for a given Δ the equation

$$U^3 - 27V^2 = \Delta$$

is an affine curve of genus 1, which has only finite number of solutions in $\mathcal{O}_S$ by the *Theorem of Siegel* (see [Sie29], [La60], [Mah34]).

The same idea is used in the proof of the semisimplicity of the Tate module of an elliptic curve (see [Se68a]).

5.5.3 Passage to Abelian varieties

In order to prove the Shafarevich conjecture for an arbitrary curve of genus $g \geq 1$ over a field K, whose bad reduction points belong to S, one associates to X its Jacobian variety $A = J_X$, endowed with a canonical principial polarization θ, defined over the same field K. It is known that A has good reduction outside S, and X is determined by the pair (A, θ) due to the *Theorem of Torelli*, see [Wei57]. Let us prove that the number of K-isomorphism classes of curves X having the same the couple (A, θ) is also finite over the base field K. For this one fixes a natural number $m \geq 3$, and one consideres the extension $K(A_m)/K$, obtained by adjoining to K the coordinates of all points of order m on A. The extension $K(A_m)/K$ is then unramified outside $S \cup \{\text{divisors of m}\}$, and all extensions of the form $K(A_m)/K$ have a bounded degree, they are all contained in a finite extension K' by the *Theorem of Hermite* (cf. (§4.1.5)).

Let us prove that the set K'-isomorphism classes of such curves is finite. If $\sigma \in \mathrm{Gal}(\overline{\mathbb{Q}}/\mathbb{Q})$, and φ is an isomorphism of curves with the same Jacobian, preserving the polarization, then $\alpha = \varphi^\sigma \circ \varphi^{-1}$ induces the identity on A_m. It is well–known that then the morphism α is identical (see, for example, [La62]): the matrix $T_l(A) \in \mathrm{Aut} T_l(A) \cong GSP_g(\mathbb{Z}_l)$ with coefficients in $\mathbb{Q}$ is a unitary matrix with respect to the involution $\alpha \mapsto \alpha^\rho$ determined by the polarization (the *Rosatti involution*, i.e. $\alpha\alpha^\rho = 1$. The characteristic roots ω_i of the matrix

$T_i(\alpha)$ are then algebraic integers ω_i whoses absolute value is equal to one for all Archimedean valuations, i. e. ω_i are roots of unity. By the assumption, $\alpha - 1 = m\beta$ for some $\beta \in \mathrm{End}A$, that is, $\omega_i - 1 = m\beta_i$, where β_i are algebraic integers, implying $\omega_i = 1$.

In order to pass from K' to K, one can use the fact that the number of K-forms of a curve X, isomorphic to X over K', is finite. In fact, such forms are classified up to K-isomorphism by the finite cohomology set $H^1(G(K'/K), \mathrm{Aut}_{K'}(X))$, see [Se64].

We have reduced the Shafarevich conjecture for curves to an analoguous statement for Abelian varieties, more precisely to the finiteness of the set $\text{Ш}^{(1)}_{AV}(g, K, S)$ of the K-isomorphism classes of pairs θ, where A is a g-dimensional Abelian variety over K with good reduction away from S, and θ is a polarisation of degree 1, defined over K. As we have already seen in section 5.3.5, the $\mathbb{C}$-isomorphism classes of pairs (A, θ) correspond to points on the Siegel modular variety $\mathcal{A}_g(\mathbb{C}) = \mathbb{H}_g/\mathrm{Sp}_g(\mathbb{Z})$, which is a quasiprojective normal variety, and could be defined over $\mathbb{Q}$.

Another key idea, proposed by A.N.Parshin for solving Mordell's problem, was to associate to elements of $\text{Ш}^{(1)}_{AV}(g, K, S)$ certain points of the set $\mathcal{A}_g(K)$, and then to prove that all such points have *bounded height* in some projective imbedding of the variety $\mathcal{A}_g$. Note that under this correspondence the map $\text{Ш}^{(1)}_{AV}(g, K, S) \to \mathcal{A}_g(K)$ is not injective, since $\mathbb{C}$-isomorphic pairs (A, θ) need not be K-isomorphic. However, it is easy to check that the above map has finite fibers: for a field K', considered above , the corresponding map

$$\text{Ш}^{(1)}_{AV}(g, K', S') \to \mathcal{A}_g(K')$$

is already injective, and an analoguous argument shows that the fibers of the map

$$\text{Ш}^{(1)}_{AV}(g, K, S) \to \text{Ш}^{(1)}_{AV}(g, K', S') \tag{5.5.4}$$

are also finite (the *theorem of finiteness for forms of an Abelian variety*).

Consideration of Abelian varieties $A = J_X$ rather than curves X was a very fruitful idea: to each Abelian variety A on can attach its Tate module $T_l(A)$ (l is a prime), regarded as a module over G_K. If $\lambda : A \to B$ is an isogeny over K, then one checks that the corresponding maping

$$T_l(\lambda) : T_l(A) \to T_l(B)$$

is an isomorphism of G_K-modules, so A and B have the same sets of bad reduction places. Therefore, the finiteness of $\text{Ш}^{(1)}_{AV}(g, K, S)$ follows from:

I) The finiteness of the number of isomorphism classes of G_K-modules $M \cong \mathbb{Z}_l^{2g}$ which arise as Tate modules $T_l(A)$ with given g, K, S.

II) The finiteness of the K-isomorphism classes of pairs (A, θ), for which the G_K-module $T_l(A)$ is isomorphic to a given module M.

5.5.4 Finiteness problems and Tate's conjecture

Recall the property observed first by Faltings that the character of a (continuous) representation $\rho : G_K \to \mathrm{Aut}V_l(A)$ is determined by its values on a finite set of elements $\mathrm{Fr}_v (v \in Q)$, $\mathrm{Card}(Q) < \infty$, $Q \cap S = \emptyset$, which depends only on given g, K, S (cf. §5.4.3).

By a theorem of A.Weil for Abelian varieties over finite fields (cf. §5.4.1 which generalizes the theorem of Hasse for elliptic curves (cf. §5.1.3), the number $\mid \mathrm{Tr}(\rho_l(\mathrm{Fr}_v)) \mid$ is an integer not exeeding $2g\sqrt{\mathrm{N}v}$. We conclude that for every prime l there are finitely many possibilities for characters of representations ρ_l.

Thus statement I) reduces to proving the semisimplicity of the G_K-module $V_l(A)$: for any $\mathbb{Q}_l$-subspace W in $V_l(A)$, which is a G_K-submodule, there exists an endomorphism $u \in \mathrm{End}A \otimes \mathbb{Q}$ such that $u^2 = u$ and $uV_l(A) = W$, so that $(1-u)V_l(A)$ is a G_K-invariant subspace of W in $V_l(A)$.

In turn the proof of the statement II) also splits into the following steps:

1) Let us consider the set $\text{Ш}_{AV}(g, K, S)$ of K-isomorphism classes of Abelian varieties as above (but *without polarization*) with given g, K, S. Then all the fibers of the mapping

$$\text{Ш}^{(1)}_{AV}(g, K, S) \to \text{Ш}_{AV}(g, K, S) \tag{5.5.5}$$

are finite.

2) *Tate's conjecture on isogenies.* For an arbitrary homomorphism $\lambda : A \to B$ of Abelian varieties over K, consider the corresponding mapping $V_l(\lambda) : V_l(A) \to V_l(B)$. Then
   - if the G_K-modules $V_l(A)$ and $V_l(B)$ are isomorphic then the varieties A and B are isogenous over K;
   - the natural mappings

$$\mathrm{End}A \otimes \mathbb{Z}_l \to \mathrm{End}T_l(A) \quad \text{and} \quad \mathrm{End}A \otimes \mathbb{Q}_l \to \mathrm{End}V_l(A) \tag{5.5.6}$$

   are bijective.

3) *The finiteness theorem for isogenies.* The set of K-isomorphism classes of Abelian varieties B over K, for which there exists an isogeny $A \to B$, is finite.

Statement 1) is not difficult and it reduces to showing that the number of polarizations of degree 1, defined over K (upto a K isomorphism of polarazed varieties) is finite. This is deduced as follows: let us view a principial polarization θ is an isomorphism $\theta : A \to A^\vee = \mathrm{Pic}^0(A)$. An isomorphism $\lambda : A \to A$, compatible with polarizations θ_1 and θ_2 of A, gives rise to a commutative diagram

$$\begin{array}{ccc} A & \xleftarrow{\lambda^\vee} & A^\vee \\ \theta_1 \uparrow & & \uparrow \theta_2 \\ A & \xleftarrow[\lambda]{} & A \end{array}$$

that is,

$$\theta_1 = \lambda^\vee \circ \theta_2 \circ \lambda. \tag{5.5.7}$$

Let us fix an isomorphism $\theta_0 : \mathrm{End}A \cong \mathrm{End}A^\vee$. Then the mapping $\lambda \mapsto \lambda^\vee$ becomes the Rosatti invoution ρ, and all the automorphisms of the form $\theta_0^{-1}\theta_i$ will be then invariant under $\rho : (\theta_0^{-1}\theta_i)^\rho = \theta_0^{-1}\theta_i$. Moreover, the following equality holds $\theta_0^{-1}(\lambda^\vee \circ \theta_i \circ \lambda) = \lambda^\rho \circ \theta_0^{-1} \circ \theta_i \circ \lambda$. Hence the equality (5.5.7) takes the form

$$\theta_0^{-1} \circ \theta_1 = \lambda^\rho \circ (\theta_0^{-1} \circ \theta_2) \circ \lambda \tag{5.5.8}$$

This equality shows that our statement is analogous to the assertion on the finiteness of the number of classes of integral unimodular quadratic forms up to the integral equivalence. More precisely, this fact can be stated as follows: if E is an order in a semisimple algebra $E \otimes \mathbb{Q}$ with an involution ρ, then the group $E^\times$, acting by the formula $(x, h) \mapsto x^\rho h x$ $(x \in E^\times)$ on the set of all Hermitian elements $(h^\rho = h)$ of E with a fixed norm, has only a finite number of orbits. In our case we use $E = \mathrm{End}A$, and we use the semisimplicity of the algebra $\mathrm{End}A \otimes \mathbb{Q}$.

Properties a) and b) of 2) are then reduced to the semisimplicity property of the module $V_l(A)$ applied to the varieties A^2 and $A \times B$.

5.5.5 Reduction of the conjectures of Tate to the finiteness properties for isogenies

A fruitful approach to the proof of the theorem on semisimplicity and of the conjectures of Tate was developed by Yu.G. Zarhin [Zar74] in the years 1974-77. He showed that for any ground field these properties could be deduced from the property 3) (which is sometimes known as *Conjecture T*) using a "unitary trick" for Abelian varieties discovered by himself. Using this trick Zarhin proved the Tate conjecture over global fields of positive characteristic. Note that over finite fields these conjectures were proved by Tate himself [Ta66].

It suffices to show that there is an isomorphism

$$\mathrm{End}_K(A) \otimes \mathbb{Q}_l \xrightarrow{\sim} \mathrm{End}_{G_K}(T_l(A) \otimes \mathbb{Q}_l)$$

(induced by a natural map from the left to the right). Consider a non-degenerate scew-symmetric pairing (see also in section 5.3.5):

$$e^D : T_l(A) \times T_l(A) \to \mathbb{Z}_l(1) = \varprojlim_m \mu_{l^m},$$

attached to an ample divisor D over A. We choose a maximal isotropic G_K-submodule W in $T_l(A) \otimes \mathbb{Q}_l$, and let W_m be the image of $T_l(A) \cap W$ in the quotient module

$$T_l(A)/l^m T_l(A) = A_{l^m} = \mathrm{Ker}(A \xrightarrow{l^m} A).$$

There is a commutative diagram

$$\begin{array}{ccc} A & \xrightarrow{\pi_m} & A/W_m = A_{(m)} \\ & \searrow^{l^m} & \downarrow \lambda_m \\ & & A \end{array}$$

It follows from Conjecture T that infinitely many of the varieties $A_{(m)}$ should be K-isomorphic. Let us denote by

$$\nu_m : A_{(m)} \to A_{(m_0)}$$

be a fixed isomorphism. Now consider

$$u_m = \lambda_m^{-1} \circ \nu_m \circ \lambda_{m_0} \in \mathrm{End}_K(A) \otimes \mathbb{Q}_l$$

and define

$$u = \lim_{m \to \infty} u_m \in \mathrm{End}_K(A) \otimes \mathbb{Q}_l.$$

It is easy to check that we can recover the G_K-submodule W as the image of $T_l(u)$:

$$T_l(u)(T_l(A) \otimes \mathbb{Q}_l) = W.$$

We claim that for an arbitrary G_K-submodule in $\mathrm{End}_{G_K}(T_l(A) \otimes \mathbb{Q}_l)$ there exists an idempotent $u \in \mathrm{End}_K(A) \otimes \mathbb{Q}_l$, $u^2 = u$, such that

$$u(T_l(A) \otimes \mathbb{Q}_l) = W. \tag{5.5.9}$$

This can be established by considering the variety A^8, and by constructing a certain maximal isotropic G_K-submodule $W^{(8)} \in \mathrm{End}_{G_K}(T_l(A^{(8)}) \otimes \mathbb{Q}_l)$ attached to W. Then we apply to $W^{(8)}$ the assertion already proved.

Consider the coordinate projections $p_i : A^8 \longrightarrow A$ $(i = 1, 2, \cdots, 8)$, and let $D_i = p_i^{-1}(D)$, $D^{(8)} = \sum_{i=1}^8 D_i$ be the divisors on A^8. We choose $a, b, c, d \in \mathbb{Q}_l$ satisfying $a^2 + b^2 + c^2 + d^2 = -1$, and define

$$I = \begin{pmatrix} a & -b & -c & -d \\ b & a & d & c \\ c & -d & a & b \\ d & c & -b & a \end{pmatrix}.$$

It is easy to check that ${}^t I \cdot I = 1_4$, where 1_4 is the identity matrix. Consider I as an element in $\mathrm{End}_{G_K}(T_l(A^4) \otimes \mathbb{Q}_l)$ and put

$$W_1 = \{(x, Ix) \mid x \in W^4\},$$

$$W_2 = \{(x, -Ix) \mid x \in (W^4)^{\perp}\},$$

where $(W^4)^{\perp}$ is defined by the scew–symmetric form e_4 attached to the divisor $D^{(4)}$. Then we have that $W_1 \cap W_2 = \{0\}$, W_1, W_2 are orthogonal with respect to the pairing e_8 associated to the divisor $D^{(8)}$. The G_K-submodule

$$W^{(8)} = W_1 + W_2 \subset \mathrm{End}_{G_K}(T_l(A^8) \otimes \mathbb{Q}_l)$$

is a maximal isotropic submodule with respect to the pairing defined by e_8 which satisfies all desired properties. This arguments show that there exist elements $u_1, \cdots, u_8 \in \mathrm{End}_K(A) \otimes \mathbb{Q}_l$ such that

$$\sum_{i=1}^{8} u_i(T_l(A) \otimes \mathbb{Q}_l) = W_1 + W_2.$$

The right ideal in $\mathrm{End}_K(A) \otimes \mathbb{Q}_l$ generated by $u_1, \cdots, u_8$ can be generated by a single idempotent element u because this algebra is known to be semisimple (see §5.3.5). This element exactly satisfies our requirement (5.5.9).

5.5.6 The Faltings–Arakelov Height

In the previous section we reduced the Mordell conjecture to Conjecture T on the finiteness of the number of K-isomorphism classes of Abelian varieties, which are K-isogenous to a given variety A. The proof of Conjecture T uses a certain canonical height $h(A)$ of A over K introduced by Faltings using ideas of Arakelov [Ara74a]. Its principal properties are:

Finiteness Principle. For given g, K and a real number b the number of K-isomorphism classes of Abelian varieties A over K with the condition $h(A) \leq b$ is finite.

Boundedness under isogenies: there exists a constant c such that for all K-isogenous Abelian varieties A and B one has $|h(A) - h(B)| < c$.

In order to define the height $h(A)$ consider first a one–dimensional vector space L over K endowed for all places v of K by a v-adic norm $\|\cdot\| : K \longrightarrow \mathbb{R}$, satisfying the condition $\|\lambda s\|_v = |\lambda|_v \|s\|_v$, where $\|\lambda s\|_v$ is the normalized v-valuation of an element $\lambda \in K^{\times}$. Suppose that for $s \in L\backslash\{0\}$ the equality $\|\lambda s\|_v = 1$ holds for almost all v (i.e. with possible exlusion of a finite number of them). In view of the product formula (4.3.31) we have $\prod_v |\lambda|_v = 1$ for $\lambda \in K^{\times}$, therefore the product $\prod_v \|\lambda s\|_v$ is independent on a choice of $\|s\|_v$. The degree of L is defined by the formula:

$$\deg L = -\log \prod_v \|s\|_v. \tag{5.5.10}$$

Let $\mathcal{O} \subset K$ is the maximal order. Defining norms $\|\cdot\|_v$ for all finite v is equivalent to defining of an integral structure, or an $\mathcal{O}_K$-form $L_{\mathcal{O}_K}$ of L, that is, to defining of a projective $\mathcal{O}_K$-module of rang 1 such that $L_{\mathcal{O}_K} \otimes_{\mathcal{O}_K} K \cong L$. In order to define $L_{\mathcal{O}_K}$ we put

$$L_{\mathcal{O}_K} = \{s \in L| \ \ \|\lambda s\|_v \leq 1, v \text{ a finite place of } K\}.$$

Conversely, for a given $\mathcal{O}_K$-module of rank 1 $L_{\mathcal{O}_K} \subset L$ with the property $L_{\mathcal{O}_K} \otimes_{\mathcal{O}_K} K \cong L$ we define the norm $\|\cdot\|_v$ using the isomorphism of vector spaces $L_{\mathcal{O}_K} \otimes_{\mathcal{O}_K} K_v \cong K_v$ which takes $L_{\mathcal{O}_K} \otimes_{\mathcal{O}_K} \mathcal{O}_K$ to $\mathcal{O}_K$ (for a finite v). If $s \in L_{\mathcal{O}_K} \backslash \{0\}$ then $\mathcal{O}_K \cdot s$ is a submodule of $L_{\mathcal{O}_K}$ and

$$\mathrm{Card}(L_{\mathcal{O}_K}/\mathcal{O}_K \cdot s) = \prod_{v \not| \infty} \|s\|_v^{-1}.$$

Consideration of Archimedean metrics $\|s\|_v$ is a convenient replacememt of the notion of an integral structure. Defining this metric is equivalent to giving a Hermitian form $\langle\cdot,\cdot\rangle_v$ on the one-dimensional complex vector space $L_\sigma = L \otimes_{K,\sigma} \mathbb{C}$ for all embeddings $\sigma : K \to \mathbb{C}$ associated with Archimedean places v. We have that

$$\begin{aligned} \|s\|_v &= \langle s, s\rangle_\sigma^{1/2}, \text{ if } K_v \cong \mathbb{R}; \\ \|s\|_v &= \langle s, s\rangle_\sigma, \text{ if } K_v \cong \mathbb{C}. \end{aligned} \tag{5.5.11}$$

For an Abelian variety A over K, we let $\omega(A) = \Omega_K^g[A]$ denote the one-dimensional K-vector space of regular (algebraic) differential forms of maximal degree g on A where ($g = \dim A$). For a number field K there is a natural v-adic norm $\|\cdot\|_v$ on $\omega(A)$ defined as follows.

a) For non-Archimeadean places v the norm $\|\cdot\|_v$ is defined using the theory of Néron which makes it possible to define a minimal model $A_{\mathcal{O}_v}$ of A over $\mathcal{O}_v$ and a one–dimensional $\mathcal{O}_v$ - module $\omega(A_{\mathcal{O}_v})$ endowed with a canonical isomorphism

$$\omega(A_{\mathcal{O}_v}) \otimes_{\mathcal{O}_v} K \cong \omega(A) \otimes K_v. \tag{5.5.12}$$

The norms are those corresponding to the $\mathcal{O}_K$-module $\omega(A)_{\mathcal{O}_K}$.

b) For a Archimedean place v given by an embedding $\sigma : K \to \mathbb{C}$ the norm $\|\cdot\|_v$ is defined using the Hermitian form

$$\langle\alpha, \beta\rangle_\sigma = \frac{1}{(2\pi)^g} \int_{A(\mathbb{C})} \alpha \wedge \overline{\beta}, \tag{5.5.13}$$

on $\omega(A)_\sigma = \omega \otimes_{K,\sigma} \mathbb{C}$, where $\alpha \wedge \overline{\beta}$ is a $2g$-dimensional differential form, which is integrated against the (topologically) $2g$-dimensional variety $A(\mathbb{C})$.

In terms of these norms on $\omega(A)$ the height $h(A)$ of the variety A is defined as follows (see the equality (5.5.10)):

$$h(A) = \frac{1}{[K:\mathbb{Q}]} \deg \omega(A). \qquad (5.5.14)$$

For example, if $K = \mathbb{Q}$ and α is one of the two generators $\pm\alpha$ of a $\mathbb{Z}$-module $\omega(A)_{\mathbb{Z}}$ (the *Néron differential*) then

$$h(A) = -\frac{1}{2} \log \left(\frac{1}{(2\pi)^g} \int_{A(\mathbb{C})} |\alpha \wedge \overline{\alpha}| \right).$$

The proof of the finiteness principle for the height $h(A)$ may be broken up into the following steps:

1) Reduction to a finiteness statement for Abelian varieties with a principal polarization. This can again be achieved using the "unitary trick" of Yu.G.Zarhin, who proved that for an abelian variety A over K and the dual variety $A^\vee$ there always exists a principal polatization on the variety $A^4 \times (A^\vee)^4$ (see [Zar85]).
2) The study of the *moduli spaces* $\mathcal{A}_g/\mathbb{Q}$ of $\overline{K}$-isomorphism classes of pairs (A, θ) where θ is a principal polarization of A. Recall that $\mathcal{A}_1$ is an affine line defined over $\mathbb{Q}$ parametrizing elliptic curves by means of the elliptic modular invariant (5.3.16). In general, $\mathcal{A}_g$ is a normal algebraic variety of dimension $g(g+1)/2$, which is not compact for $g \geq 1$, and the structure of its various compactifications is rather complicated [Fal85]. A pair (A, θ)defined over K produces a point $J(A, \theta) \in \mathcal{A}_g$ and one can define heights of various projective embeddings of the variety $\mathcal{A}$ defined over $\mathbb{Q}$.
3) A *canonical projective embedding* of $\mathcal{A}_g$ is constructed using Siegel modular forms; these forms may be viewed as global sections of certain line bundles (more precisely, powers of the canonical line bundle) on $\mathcal{A}_g$, cf. [MZ72], [PZ88], [FW84]. The corresponding height of a point $J(A, \theta)$ is called the *modular height.* A key observation of Faltings was that the height (5.5.14) and the modular height are essentially equal. Thus the finiteness principle follows from the basic property of heights of points in a projective space: there is only finite number of isomorphism classes of pairs (A, θ) over K with bounded modular height of the corresponding points $J(A, \theta)$ (comp. with §5.2.5).

The above three steps give only a hint of the strategy of the proof of the finiteness principle; carrying out this program in detail is a quite technical task, see also the review of B. Mazur [Maz77].

5.5.7 Heights under isogenies and Conjecture T

By the finiteness principle of §5.5.6, Conjecture T and the Mordell conjecture follows from the boundedness $|h(A) - h(B)|$ for K-isogenous varieties A and B.

The proof of this fact is deduced from theorems I and II given below.

Theorem I. *Let p be a prime number unramified in K. There exists a finite set $M = M(K, p, g)$ of primes such that: if A is an Abelian variety of dimension g over K with good reduction at all places of K dividing p, and $S(A)$ is the set of all primes divisible by the bad reduction places of A, then for each isogeny $A \to B$ of the degree not dividing any prime in M we have that $h(A) = h(B)$.*

In particular, for an Abelian variety A over K there exists a finite set M of primes such that there is only a finite set of Abelian varieties B over K (up to K-isomorphism), which admit a K-isogeny $A \to B$ of the degree not dividing primes in M.

Theorem II. *Let A be an Abelian variety over K, l is a fixed prime number. Then the set of Abelian varieties B (up to K-isomorphism) which admit a K-isogeny of degree l^m $(m \geq 1)$ is finite.*

The proofs of both theorems are based on explicit formulas for the behaviour of $h(A)$ under isogenies. In the good case (Theorem I) the height does not change because in this case one can define an isomorphism of the corresponding modules with metrics $\omega(A)$ and $\omega(B)$ of the same degree.

The proof of theorem II proceeds by *reductio ad absurdum.* Suppose that there is an infinite sequence of K - isogenous Abelian varieties

$$A \to B_{(1)} \to B_{(2)} \to \cdots \to B_{(n)} \to B_{(n+1)} \to \cdots,$$

such that the kernels $W_n = \mathrm{Ker}\,(A \to B_{(n)})$ form an *l-divisible group.* Using known results on the structure and properties of l-divisible groups, one proves that the sequence $h(B_{(n)})$ stabilizes from a sufficiently large n_0, and theorem II follows by applying the finiteness principle for the height.

The conjecture T follows from theorems I and II, since every isogeny $A \to B$ can be decomposed into a composition of isogenies

$$A \to B_{(0)} \to B_{(1)} \to B_{(2)} \to \cdots \to B_{(n)} = B,$$

satisfying the following conditions:

a) the degree of $A \to B_{(0)}$ does not divide any prime in the finite set $M \cup S(A) = \{l_1, l_2, \cdots, l_n\}$ of theorem I;
b) the degree of $B_{(i-1)} \to B_{(i)}$ is a power of a prime l_i.

According to Theorem I there are only finitely many possibilities for the variety $B_{(0)}$ (up to K-isomorphism). Applying theorem II to the variety $B_{(0)}$ and to the prime l_1 shows that there are only finitely many possibilities for the variety $B_{(1)}$. By induction, there are only finitely many possibilities for $B = B_{(n)}$ (cf. [PZ88], pp.383–384).

This completes the proof of the Mordell conjecture, as well as the conjecture T, the finiteness conjecture of Shafarevich, and the conjecture of Tate for Abelian varieties.

In §5.5.1 we have mentioned the second conjecture of Shafarevich.

This conjecture can be reformulated as a statement on the non-existence of certain smooth proper schemes over Spec$\mathbb{Z}$ of relative dimension 1 and of genus ≥ 1. Of course, the classical geometric analogue of this conjecture is well known (and discussed in [Sha62] as a motivation).

Moreover, in the 80th J.–M. Fontaine [Fon81] and independently V.A. Abrashkin (see in [PZ88]) proved that over the maximal orders in $\mathbb{Q}$, $\mathbb{Q}(\sqrt{-1})$, $\mathbb{Q}(\sqrt{-2})$, $\mathbb{Q}(\sqrt{-3})$, $\mathbb{Q}(\sqrt{-7})$, $\mathbb{Q}(\sqrt{2})$, $\mathbb{Q}(\sqrt{5})$, $\mathbb{Q}(\sqrt[5]{1})$ there exist no smooth proper Abelian schemes.

6

Zeta Functions and Modular Forms

6.1 Zeta Functions of Arithmetic Schemes

6.1.1 Zeta Functions of Arithmetic Schemes

(cf. [Sha69], [Se65]). Let X be a scheme of finite type over Spec $\mathbb{Z}$ (see §5.1). Then the closed points $x \in X$ are those which satisfy the condition that the corresponding residue field $R(x)$ is finite. The cardinality of $R(x)$ is called the norm of x and is denoted by $\mathrm{N}(x)$. The set of all closed points of X is denoted by $\overline{X}$. For the moment we shall think of this as a discrete topological space.

The zeta function of X is defined to be the Euler product

$$\zeta(X, s) = \prod_{x \in \overline{X}} (1 - \mathrm{N}(x)^{-s})^{-1}. \tag{6.1.1}$$

In the case $X = \operatorname{Spec} \mathbb{Z}$ definition (6.1.1) leads to the Riemann zeta function $\zeta(s)$ in view of Euler's identity:

$$\zeta(s) = \sum_{n=1}^{\infty} n^{-s} = \prod_{p} (1 - p^{-s})^{-1}. \tag{6.1.2}$$

For an arithmetic scheme there are only finitely many points with a given norm, so the product (6.1.1) is a formal Dirichlet series $\sum_{n=1}^{\infty} a_n n^{-s}$ with integral coefficients.

Theorem 6.1. *The product (6.1.1) is absolutely convergent for* $\mathrm{Re}(s) > \dim X$, *where* $\dim X$ *is the dimension of* X *(see §5.2.1 of Chapter 5).*

The proof of this fact can be reduced to the following special cases:

(a) $X = \operatorname{Spec} \mathbb{Z}[T_1, \cdots, T_n]$. The product then takes the form:

$$\zeta(X, s) \;=\; \prod_{p} (1 - p^{n-s})^{-1} \;=\; \zeta(s - n); \tag{6.1.3}$$

(b) $X = \mathrm{Spec}\,\mathbb{F}_q[T_1, \cdots, T_n]$. We then have

$$\zeta(X,s) = (1-p^{n-s})^{-1}\,. \tag{6.1.4}$$

Equation (6.1.3) is implied by (6.1.4), which can in turn be obtained from the following calculation of the number of closed points of an arbitrary variety X over a finite field $\mathbb{F}_q$ ($q = p^r$).

Let $n_k = \mathrm{Card}\{x \in \overline{X} \mid \mathrm{N}(x) = q^k\}$ be the number of closed points with norm q^k and $\nu_l = \mathrm{Card}\, X(\mathbb{F}_{q^l})$ the number of geometric points with values in $\mathbb{F}_{q^l}$, i.e. the number of morphisms $\mathrm{Spec}\,\mathbb{F}_{q^l} \to X$.

Lemma 6.2. *The numbers ν_l and n_k are finite and are related by the following formula*

$$\nu_l = \sum_{k|l} k n_k. \tag{6.1.5}$$

This fact is implied by the observation that for a given $x \in \overline{X}$ there are precisely l/k field embeddings $\mathbb{F}_{q^k} \to R(x)$.

We now obtain the following identities:

$$\zeta(X,s) = \prod_{k=1}^{\infty}(1-q^{-ks})^{-n_k}, \tag{6.1.6}$$

$$\log\zeta(X,s) = -\sum_{k=1}^{\infty} n_k \log(1-q^{-ks})$$

$$= -\sum_{k=1}^{\infty}\sum_{m=1}^{\infty} n_k \frac{q^{-kms}}{m} = -\sum_{l=1}^{\infty}\left(\sum_{k|l}\frac{n_k}{k}l,\right) q^{-ls} = -\sum_{l=1}^{\infty}\nu_l l^{-1} q^{-ls}.$$

(Here we replaced the variable l by km, taking (6.1.5) into account.)

If $X = \mathrm{Spec}\,\mathbb{F}_q[T_1, \cdots, T_n]$ then $\nu_l = q^{ln}$ (the number of points of the affine space $\mathbb{A}^n$ over $\mathbb{F}_{q^l}$). Hence for $q = p$ we have

$$\log\zeta(\mathbb{A}^n_p, s) = -\sum_{l=1}^{\infty}\frac{p^{ln}}{l}\cdot p^{-ls} = -\sum_{l=1}^{\infty}\frac{p^{-l(s-n)}}{l} = -\log(1-p^{n-s}),$$

establishing (6.1.4).

In both cases (6.1.3) and (6.1.4) for $\mathrm{Re}(s) > \dim X$ the product in (6.1.1) converges absolutely.

Similarly we see that (cf. [Sha69])

$$\zeta(\mathbb{P}^n_{\mathbb{F}_p}, s) = \prod_{m=0}^{n}(1-p^{-(s-m)})^{-1},$$

$$\zeta(\mathbb{P}^n_{\mathbb{Z}}, s) = \prod_{p}\prod_{m=0}^{n}(1-p^{-(s-m)})^{-1} = \prod_{m=0}^{n}\zeta(s-m). \tag{6.1.7}$$

6.1.2 Analytic Continuation of the Zeta Functions

It is thought that the functions $\zeta(X,s)$ can all be analytically continued onto the entire s–plane $\mathbb{C}$. The validity of this conjecture has been verified for many varieties. However in the general case only the following weaker result is known.

Theorem 6.3. *The function* $\zeta(X,s)$ *has a meromorphic continuation to the half plane* $\mathrm{Re}(s) > \dim\ X - \frac{1}{2}$.

The singularities of $\zeta(X,s)$ in the strip $\dim\ X - \frac{1}{2} < \mathrm{Re}(s) < \dim\ X$ are described by the following theorem:

Theorem 6.4. *Let us assume that* X *is irreducible and let* $R(X)$ *be the residue field of its generic point. Then*

1) *If* $\mathrm{Char}\ R(X) = 0$ *then the only pole of* $\zeta(X,s)$ *for* $\mathrm{Re}(s) > \dim\ X - \frac{1}{2}$ *is at the point* $s = \dim\ X$, *and this pole is simple.*
2) *If* $\mathrm{Char}\ R(X) = p > 0$ *and* q *is the highest power of* p *such that* $R(X)$ *contains* $\mathbb{F}_q$ *then the only singularities of the function* $\zeta(X,s)$ *for* $\mathrm{Re}(s) > \dim\ X - \frac{1}{2}$ *are simple poles at the points*

$$s = \dim\ X + \frac{2\pi i n}{\log q} \qquad (n \in \mathbb{Z}). \tag{6.1.8}$$

Corollary 6.5. *For each non-empty scheme* X *the point* $s = \dim\ X$ *is a pole of* $\zeta(X,s)$, *whose order is equal to the number of irreducible components of* X *of dimension* $\dim\ X$.

Corollary 6.6. *The domain of absolute convergence of the Dirichlet series* $\zeta(X,s)$ *is the right half plane* $\mathrm{Re}(s) > \dim\ X$.

Theorems 6.3 and 6.4 are deeper than Theorem 6.1. Their proof is based on the analogue of the Riemann Hypothesis for curves X over $\mathbb{F}_q$ established by Weil (cf. [Wei49]), see also [Se65], [Nis54], [LaWe].

6.1.3 Schemes over Finite Fields and Deligne's Theorem

If X is a scheme over $\mathbb{F}_q$ then for all $x \in \overline{X}$ the field $R(x)$ is a finite extension of $\mathbb{F}_q$. Hence $\mathrm{N}(x) = q^{\deg x}$ for some number $\deg x$, called the degree of x. In studying $\zeta(X,s)$ in this case it is convenient to use the new variable $t = q^{-s}$. We write

$$\zeta(X,s) = Z(X, q^{-s}), \tag{6.1.9}$$

where $Z(X, q^{-s})$ is the power series given by the product

$$Z(X,t) = \prod_{x \in \overline{X}} (1 - t^{\deg x})^{-1}.$$

If $n_k = \text{Card}\{x \in \overline{X} \mid \deg x = k\}$ and $\nu_l = \text{Card}\ X(\mathbb{F}_{q^l})$ then we have seen that

$$\log Z(X,t) = -\sum_{l=1}^{\infty} \nu_l \frac{t^l}{l}, \quad \nu_l = \sum_{k|l} k n_k, \tag{6.1.10}$$

hence

$$t\frac{Z'(X,t)}{Z(X,t)} = t\frac{d}{d}t \log Z(X,t) = \sum_{l=1}^{\infty} \nu_l t^l. \tag{6.1.11}$$

Equation (6.1.10) is often taken as the definition of the zeta function, and one writes

$$Z(X,t) = \exp\left[\sum_{l=1}^{\infty} \text{Card}\ X(\mathbb{F}_{q^l}) \frac{t^l}{l}\right]. \tag{6.1.12}$$

A remarkable property of the zeta function $Z(X,t)$ is its rationality in the variable t. This was first established by B.Dwork in [Dw59]. The rationality statement has a direct arithmetical interpretation: the numbers ν_l, i.e. the numbers of solutions of a certain system of algebraic equations in finite fields must satisfy a recurrence relation of the type:

$$\nu_{l+n} = \sum_{i=0}^{n-1} \tau_i \nu_{l+i}$$

for sufficiently large l, where n and the τ_i are certain constants. It is easy to check that the rationality of the function $Z(X,t)$ is also equivalent to the existence of finitely many complex numbers α_i, β_j such that

$$\nu_l = \sum_j \beta_j^l - \sum_i \alpha_i^l \tag{6.1.13}$$

for all $l \geq 1$. Indeed, this is obtained from the logarithmic derivative of the identity:

$$Z(X,t) = \frac{\prod_i (1 - \alpha_i t)}{\prod_j (1 - \beta_j t)},$$

taking into account (6.1.11).

A fundamental role in the study of $Z(X,t)$ is played by the fact the number ν_k can be represented as the number of fixed points of a certain map F^k acting on the set of geometric points $X(\overline{\mathbb{F}}_q)$.

Definition 6.7. *The Frobenius morphism* $F : X \to X$ *of a scheme* X *over* $\mathbb{F}_q$ *is defined on each open affine subscheme* Spec $A \subset X$ *using the ring homomorphism* $a \mapsto a^q$; *on the topological space* X *the morphism* F *acts as the identity map.*

The consequent self-map of sets of geometric points (not the same thing!) is written

$$F : X(\overline{\mathbb{F}}_q) \to X(\overline{\mathbb{F}}_q) \tag{6.1.14}$$

and the set $X(\mathbb{F}_{q^k})$ coincides with the set of fixed points of $F^k : X(\overline{\mathbb{F}}_q) \to X(\overline{\mathbb{F}}_q)$: a point $\varphi \in X(\mathbb{F}_{q^k})$ is represented by a morphism $\varphi : A \to \overline{\mathbb{F}}_q$, where Spec A is an open affine subset containing $\varphi(\text{Spec } \overline{\mathbb{F}}_q)$. The point $F^k(\varphi)$ is defined by the homomorphism

$$f \mapsto \varphi(f)^{q^k} \quad (f \in A).$$

We see that the condition $\varphi \in X(\mathbb{F}_q^k)$ is equivalent to saying that Im $\varphi \subset \mathbb{F}_{q^k} \subset \overline{\mathbb{F}}_q$ and $\varphi(f) = \varphi(f)^{q^k}$ because $\mathbb{F}_{q^k} = \{x \in \overline{\mathbb{F}}_q \mid x = x^{q^k}\}$.

The rationality of the zeta function was a part of a series of conjectures stated by Weil in 1949. Dwork's proof of the rationality was a significant step towards proving these conjectures in general. The final step was made by Deligne in 1973 who proved the so called "Riemann Hypothesis" for algebraic varieties $X/\mathbb{F}_q$, cf. [Del74].

For a smooth projective variety X over $\mathbb{F}_q$ of dimension d the Weil conjectures can be stated as follows:

W1) *Rationality*:

$$Z(X,t) = \frac{P_1(X,t) \cdot \ldots \cdot P_{2d-1}(X,t)}{P_0(X,t) \cdot \ldots \cdot P_{2d}(X,t)}, \tag{6.1.15}$$

where $d = \dim X$ and $P_r(X,t) \in \mathbb{C}[t]$ for all $r = 1, 2, \ldots, 2d$ and $P_r(X,0) = 1$.

W2) *Integrality*:

$$P_0(X,t) = 1 - t, \qquad P_{2d}(X,t) = 1 - q^d t, \tag{6.1.16}$$

and for $r = 1, 2, \ldots, 2d$ we have that $P_r(X,t) = \prod(1 - \omega_{r,i}t)$, where $\omega_{r,i}t)$ are certain algebraic integers.

W3) *The Functional Equation*:

$$Z(X, 1/q^d t) = \pm q^{d\chi/2} t^\chi Z(X,t), \tag{6.1.17}$$

where χ is the Euler characteristic of X, which can be defined purely algebraically as $\chi = (\Delta \cdot \Delta)$ (the self–intersection number of the diagonal $\Delta \subset X \times X$).

W4) *The Riemann Hypothesis*: The absolute value of each of the numbers $\omega_{r,i}t$ and their conjugates are equal to $q^{r/2}$.

W5) *Degrees of polynomials* $P_r(X,t)$: If X is the reduction of a smooth projective variety Y defined over a number field embedded in $\mathbb{C}$, then the degree of $P_r(X,t)$ is equal to the rth Betti number of the complex variety $Y(\mathbb{C})$.

In the case when X is a smooth projective curve over $\mathbb{F}_q$ these properties were established by Weil, and in particular we have that

$$Z(X,t) = \frac{L(t)}{(1-t)(1-qt)}, \tag{6.1.18}$$

where $L(t) = \prod_{i=1}^{2g}(1-\omega_i t) \in \mathbb{Z}[t]$ (g is the genus of the curve X), and $|\omega_1| = \cdots = |\omega_{2d}| = \sqrt{q}$. Using the relations (6.1.11) and (6.1.13) we have

$$\text{Card } X(\mathbb{F}_{q^k}) = q^k + 1 - \sum_{i=1}^{2g} \omega_i^k,$$

$$|\text{Card } X(\mathbb{F}_q) - q - 1| < 2g\sqrt{q}. \tag{6.1.19}$$

An elementary proof of the estimate (6.1.19) was found by Stepanov S.A. (1974) in [Step74] (cf. [Step84], [Step94], [Bom72]).

The proof of the Weil conjectures is based on an idea from the theory of compact topological varieties. If F is a morphism acting on such a variety V then for the number $\nu(F)$ of fixed points of F (appropriately defined) the famous *Lefschetz fixed–point formula* holds:

$$\nu(F) = \sum_{i=0}^{\dim\ V} (-1)^i \text{Tr } F|_{H^i(V)} \tag{6.1.20}$$

(the summands are the traces of the linear operators induced by F on the cohomology groups $H^i(V)$). In this situation one can define an analogue of (the logarithmic derivative of) the zeta function $\sum_{k=1}^{\infty} \nu(F^k)t^k$. It is not difficult to calculate the sum of this series. Let $(\alpha_{ij})_{j=1,\cdots,b_i}$ be the characteristic roots of the linear operator $H^i(F) = F|_{H^i(V)}$ on $H^i(V)$ and $b_i = \dim\ H^i(V)$ the *Betti numbers*; then

$$\text{Tr } F^k|_{H^i(V)} = \sum_{j=1}^{b_i} \alpha_{ij}^k, \qquad \nu(F^k) = \sum_{i=1}^{\dim\ V} (-1)^i \sum_{j=1}^{b_i} \alpha_{ij}^k,$$

and hence

$$\begin{aligned}\sum_{k}^{\infty}\nu(F^k)t^k &= \sum_{i=1}^{\dim V}(-1)^i\sum_{j=1}^{b_i}\left(\sum_{k}^{\infty}\alpha_{ij}^k\right)\\ &= \sum_{i=1}^{\dim V}(-1)^i\sum_{j=1}^{b_i}\frac{\alpha_{ij}t}{1-\alpha_{ij}t}\\ &= \sum_{i=1}^{\dim V}(-1)^i\sum_{j=1}^{b_i}t\frac{d}{dt}[\log(1-\alpha_{ij}t)^{-1}]. \end{aligned}\tag{6.1.21}$$

The series $Z(t)$ is determined by the conditions $Z(0) = 1$ and

$$t\frac{Z'(t)}{Z(t)} = \sum_{k=1}^{\infty}\nu(F^k)t^k,$$

and it follows from (1.21) that

$$Z(t) = \prod_{i=1}^{\dim V}\left(\prod_{j=1}^{b_i}(1-\alpha_{ij}t)^{-1}\right)^{(-1)^{i-1}}.$$

We see that in this model situation the Z–function is rational and can be calculated very explicitly. Using this fact Weil proposed conjectures W1) – W4) and proved these conjectures for curves and Abelian varieties. For these varieties he introduced an analogue of the group $H^1(X)$ and proved a Lefschetz fixed–point formula of the same type as in the topological situation. The analogue of $H^1(X)$ is provided by the Tate module $T_l(J_X)$ of the Jacobian variety of the curve X (resp. of the given Abelian variety). In the general case analogues of the (topological) cohomology groups H^i were constructed by Grothendieck (the étale cohomology groups $H^i_{ét}(X, \mathbb{Q}_l)$). In order to do this he modified the very notion of a topological space, which was replaced by a certain category (in the topological situation objects of this category are open sets, and morphisms are inclusions). The larger category used by Grothendieck was called the *étale topology.*

The use of the groups $H^i_{ét}(X, \mathbb{Q}_l)$ and more generally cohomology groups of sheaves (in the étale topology) made it possible for Deligne to prove the conjectures of Weil (cf. [Del74], [Del80b] , [Kat76]).

6.1.4 Zeta Functions and Exponential Sums

([Kat76], [Kat88], [Sha69]). A traditional method for counting solutions of congruences or systems of congruences is related to exponential sums. Formulae for the quantities ν_l = Card $X(\mathbb{F}_{q^l})$ can be obtained using Dirichlet characters $\chi : \mathbb{F}_q^\times \to \mathbb{C}^\times$ (i.e. multiplicative characters of $\mathbb{F}_q^\times$). Let ε denote the trivial character which is constant function 1 on the whole set $\mathbb{F}_q$. If m divides $q-1$ then one has the relation

$$\mathrm{Card}\{x \in \mathbb{F}_q \mid x^m = a\} = \sum_{\chi^m=\varepsilon} \chi(a). \tag{6.1.22}$$

Now consider the Gauss sum of a non–trivial character χ,

$$g(\chi) = \sum_{x\in\mathbb{F}_q^\times} \chi(x)\zeta^{\mathrm{Tr}(x)},$$

where $\mathrm{Tr} : \mathbb{F}_q \to \mathbb{F}_p$ is the trace and ζ is a primitive p^{th} root of unity.

In the work of Hasse and Davenport of 1934 an interesting relation was found between exponential sums over finite fields and zeros of zeta functions (cf. [DH34]).

Example 6.8 (The zeta function of a hypersurface.). Let

$$X : a_0T_0^m + a_1T_1^m + \cdots + a_nT_n^m = 0$$

be a hypersurface in $\mathbb{P}^n$ over $\mathbb{F}_q$, where $a_0, a_1, \ldots, a_n \in \mathbb{F}_q^\times$ and $q \equiv 1 \mod m$. Then we have that

$$Z(X,t) = \frac{P(t)^{(-1)^n}}{(1-t)(1-qt)\cdots(1-q^{n-1}t)} \tag{6.1.23}$$

where $P(t)$ denotes the polynomial

$$\prod_{\chi_0,\chi_1,\cdots,\chi_n} \left(1-(-1)^{n+1}\frac{1}{q}\chi_0(a_0^{-1})\cdots\chi_n(a_n^{-1})t\right),$$

and $\chi_0, \chi_1, \cdots, \chi_n$ run through all possible $(n+1)$-tuples of Dirichlet characters satisfying the conditions

$$\chi_i \neq \varepsilon, \quad \chi_i^m = \varepsilon, \quad \chi_0\chi_1\cdot\ldots\cdot\chi_n = \varepsilon.$$

The proof of formula (6.1.23) is based on counting the quantities $\nu_l = \mathrm{Card}\, X(\mathbb{F}_{q^l})$ using relations between Jacobi sums and Gauss sums, and the *Davenport–Hasse relations*: for a non - trivial character χ of $\mathbb{F}_q$ let us define the character $\chi' = \chi \circ \mathrm{N}$ of the field $\mathbb{F}_{q^l}$, where $\mathrm{N} : \mathbb{F}_{q^l} \to \mathbb{F}_q$ is the norm map. Then the following relation holds:

$$-g(\chi') = (-g(\chi))^l. \tag{6.1.24}$$

Relation (6.1.24) makes it possible to find explicitly the numbers α_i and β_j such that for $l \geq 1$ one has:

$$\mathrm{Card}\ X(\mathbb{F}_{q^l}) = \sum_j \beta_j^l - \sum_i \alpha_i^l,$$

and (6.1.23) is then implied by (6.1.13).

Classically estimates for exponential sums were used to obtain estimates for the number of geometric points of varieties over finite fields. Conversely, one can effectively use the estimate W4) of the Weil conjectures to study exponential sums of a rather general type. We give only a simple example ([Sha69], p.87).

Let $f(T) \in \mathbb{F}_q[T]$, $0 < \deg f = m < p$ and $\zeta^p = 1, \zeta \neq 1$. Then the following estimate holds:

$$\left|\sum_{x=0}^{p-1} \zeta^{f(x)}\right| \leq m\sqrt{q}. \tag{6.1.25}$$

In order to prove (6.1.25) let us consider an auxiliary curve $y^p - y = f(x)$. Let us denote by X the curve obtained by a desingularization of its projective closure. Consider the projection $X \xrightarrow{\pi} \mathbb{P}^1$ given by $(x, y) \mapsto x$. Then

$$\prod_{\xi \in \overline{X}} (1 - \mathrm{N}(\xi)^{-s})^{-1} = \prod_{x \in \overline{\mathbb{P}}^1} \prod_{\pi(\xi)=x} (1 - \mathrm{N}(\xi)^{-s})^{-1}. \tag{6.1.26}$$

The equality $\pi(\xi) = \infty$ is satisfied for a single point $\xi \in X$ and the corresponding factor in the product (6.1.26) is equal to $1 - p^{-s}$. If $\pi \neq \infty$ then the equation $y^p - y = f(x)$ is solvable in the field $\mathbb{F}_p(x_0)$ with $\xi = (x_0, y_0)$ so that there are exactly p solutions $y_0, y_0+1, \ldots, y_0+p-1$ with norm $\mathrm{N}(x_0)$. In this case the corresponding factor in the inner product is equal to $(1 - \mathrm{N}(x)^{-s})^{-1}$. The solvability of the equation $y^p - y = a$ in $\mathbb{F}_p(a)$ is equivalent to the condition

$$\mathrm{Tr}_{\mathbb{F}_p(a)/\mathbb{F}_p}(a) = 0, \text{ i.e. } \sum_{i=0}^{\deg a - 1} a^{p^i} = 0,$$

or

$$\sum_i f(x)^{p^i} = \sum_i f(x^{p^i}) = \sum_{P(x)=0} f(x),$$

where in the last sum x runs through all of the roots of the irreducible polynomial P associated with a closed point $\pi(\xi) \in \overline{\mathbb{P}}^1 \backslash \infty$. Hence the inner product in (6.1.26) can be transformed into the following

$$\prod_{r=0}^{p-1} (1 - \chi(P)^r \mathrm{N}(P)^{-s})^{-1},$$

where

$$\chi(P) = \zeta^{\lambda(P)}, \quad \lambda(P) = \sum_{P(x)} \zeta^{f(x)}, \quad \mathrm{N}(P) = p^{\deg P}.$$

Putting $t = p^{-s}$ we see that

$$Z(X, s) = (1 - t) \prod_P \prod_{r=0}^{p-1} (1 - \chi(P)^r \mathrm{N}(P)^{-s})^{-1},$$

where P runs through all irreducible monic polynomials in $\mathbb{F}_p[t]$. Extracting the factors with $r = 0$ we obtain:

$$Z(X,t) = \frac{1}{(1-t)(1-pt)} \prod_{r=1}^{p-1} \prod_P (1 - \chi(P)^r \mathrm{N}(P)^{-s})^{-1}.$$

For each monic polynomial G we put

$$\lambda(G) = \sum_{G(x)=0} f(x), \quad \chi(G) = \zeta^{\lambda(G)}.$$

The function χ is multiplicative, so we obtain the equation

$$L_r(t) = \prod_P (1 - \chi(P)^r t^{\deg P})^{-1} = \sum_G \chi(G)^r t^{\deg G}.$$

One verifies easily that $L_r(t)$ is a polynomial and $\deg L_r(t) \leq \deg f$, and the coefficient of t is equal to

$$\sum_{a \in \mathbb{F}_p} \chi^r(T-a) = \sum_{a \in \mathbb{F}_p} \zeta^{r\lambda(T-a)} = \sum_{a=0}^{p-1} \zeta^{rf(a)}.$$

However, each of these sums is equal to a sum of some (inverse) roots of the function $Z(X,t)$. The number of these roots is equal to $\deg L_r(t) \leq \deg f$, and the absolute value of each root is less than or equal to $\sqrt{p}$, so that estimate (6.1.25) follows.

Applications of cohomological techniques and of methods from representation theory to the study of exponential sums of general type were considerably developed in the work of N.M.Katz [Kat76], [Kat88], [KL85], of Deligne and other mathematicians in the 70s and 80s (cf. [Del74] , [Del80b], [Bry86]). In this research an exponential sum is interpreted as the trace of a certain operator (the *Frobenius operator* or the *monodromy operator*) acting on the space of global sections of a specially constructed sheaf on an algebraic variety. Thus, the general exponential sums can be constructed using cohomology groups with compact support on an appropriate Artin–Schreier covering W of an affine variety V. Then estimates on the exponential sum can be reduced to the Weil estimate W4) applied to a smooth compactification of W provided that this compactification exists (above we have considered an example for curves). In the general case it is not even known whether such compactifications exist. However this difficulty can be coped with using the technology developed in the second part of Deligne's paper on the Weil conjectures [Del80b], which contains a vast generalization of these conjectures. This generalization gives absolute values of the Frobenius elements acting on cohomology with compact support on general varieties, whereas the original formulation of the conjectures concerns essentially the constant sheaves on smooth projective varieties. Impressive examples of the use of Deligne's generalizations of the Weil conjectures were given by [Kat88] in the interesting case of the multi-dimensional Kloosterman sums of type

$$\mathrm{Kl}(p, n, a) = \sum_{\substack{x_1, \dots, x_n \bmod p \\ x_1 \cdot \dots \cdot x_n \equiv a \bmod p}} \exp\left(\frac{2\pi i}{p}(x_1 + \dots + x_n)\right),$$

by proving their important equidistribution properties with p fixed and varying a. For an l-adic sheaf on an algebraic curve the equidistribution properties of the traces of local Frobenius elements were naturally formulated in terms of a certain algebraic group G_{geom} over $\overline{\mathbb{Q}}_l$ which is defined as the Zariski closure of the image of the geometric fundamental group in the corresponding l-adic representation. Under rather general assumptions on choice of an embedding of $\overline{\mathbb{Q}}_l$ into $\mathbb{C}$ one obtains a complex algebraic group $G_{\mathrm{geom}}(\mathbb{C})$, and the Frobenius elements correspond to certain points in the space $K^\natural$ of conjugacy classes in a maximal compact subgroup K of $G_{\mathrm{geom}}(\mathbb{C})$. The equidistribution property is to be understood in the sense of a measure $\mu^\natural$ on $K^\natural$ obtained from the Haar measure on K. For the multi-dimensional Kloosterman sums this construction leads to groups

$$\begin{aligned} G_{\mathrm{geom}} &= \mathrm{Sp}(n) \text{ for even } n \text{ and arbitrary } p, \\ G_{\mathrm{geom}} &= \mathrm{SL}(n) \text{ for odd } pn, \\ G_{\mathrm{geom}} &= \mathrm{SO}(n) \text{ for } p = 2 \text{ and odd } n \geq 3, n \neq 7, \\ \text{and} \quad G_{\mathrm{geom}} &= \mathrm{G}_2(n) \text{ for } p = 2 \text{ and } \ n = 7. \end{aligned} \tag{6.1.27}$$

These methods can be used to study of the equidistribution of the arguments of Gauss sums of the type

$$\theta(a) = \frac{g(\psi, \chi^a)}{\sqrt{q}} = \frac{1}{\sqrt{q}} \sum_{x \in \mathbb{F}_q^\times} \psi(x)\chi^a(x), \quad |\theta(a)| = 1,$$

where ψ is a non–trivial additive character of $\mathbb{F}_q$, χ a generator of the cyclic group of multiplicative characters of $\mathbb{F}_q^\times$ and $1 \leq a \leq q-2$. Katz also proved that for a fixed $r \geq 1$ the r-tuple of angles $(\theta(a+1),\ \theta(a+2),\ \cdots, \theta(a+r)) \in (S^1)^r$ for $0 \leq a \leq q-2-r$ becomes equidistributed with respect to the Haar measure on $(S^1)^r$ as $q \to \infty$. An interesting related construction of the l-adic Fourier transform for sheaves on $\mathbb{A}^n$ was suggested by Brylinski and Laumon (cf. [Bry86], [KL85]).

These results are related to *Sato–Tate Conjecture* on the uniform distribution of the arguments φ_p of Frobenius automorphisms in the segment $[0, \pi]$ with respect to the measure $\dfrac{2}{\pi} \sin^2 \varphi \, d\varphi$ (for cusp forms f without complex multiplication) (cf. Chapter I in [Se68a], [Mich01] and §6.5.1). For recent developments we refer to [KS99], [KS99a], [Mich01], [Sar98].

6.2 L-Functions, the Theory of Tate and Explicite Formulae

6.2.1 L-Functions of Rational Galois Representations

Let K be a number field, Σ_K its set of places (classes of normalized valuations) and

$$\rho : G_K \longrightarrow \mathrm{GL}(V) \tag{6.2.1}$$

a representation of the Galois group $G_K = \mathrm{Gal}(\overline{K}/K)$ in a finite dimensional vector space V over a field F of characteristic zero, which we shall usually assume to be embedded in $\mathbb{C}$ (in examples and applications we shall use $F = \mathbb{Q}_l, \mathbb{C}$ or $\overline{\mathbb{Q}}$). We call ρ unramified at a non–Archimedean place v if for all places w of $\overline{K}$ dividing v one has $\rho(I^{(w)}) = \{1_V\}$, where $I^{(w)}$ is the inertia group of w over v. In this case the representation ρ can be factorized through the quotient group

$$G^{(w)}/I^{(w)} \cong G_{k(v)} = \mathrm{Gal}((\mathcal{O}_{\overline{K}}/\mathfrak{p}_w)/(\mathcal{O}_K/\mathfrak{p}_v)),$$

where $G^{(w)}$ is the decomposition group of w over v and $G_{k(v)}$ the Galois group of the algebraic closure $\mathcal{O}_{\overline{K}}/\mathfrak{p}_w$ of the residue field $k(v) = \mathcal{O}_K/\mathfrak{p}_v$. The group $G_{k(v)}$ is canonically generated by the Frobenius automorphism Fr_v, $Fr_v(x) = x^{\mathrm{N}v}$. Choosing a different place w above v will lead to the element $\rho(Fr_v)$ being replaced by a conjugate element. Hence the conjugacy class of the Frobenius element $F_{v,\rho} = \rho(Fr_v)$ is well defined, and we write

$$P_{v,\rho}(t) = \det(1_V - t \cdot F_{v,\rho}) \tag{6.2.2}$$

for the characteristic polynomial of this element. Suppose that E is a number field embedded in $\mathbb{C}$. We call the representation ρ rational (resp. entire) over E if there exists a finite number of places $S \subset \Sigma_K$ such that

a) for all $v \in \Sigma_K \backslash S$ the representation ρ is unramified at v,
b) the coefficients of $P_{v,\rho}(t)$ belong to E (resp. to the maximal order $\mathcal{O}_E$ of E).

Let s be a complex number and $v \notin S$. We have

$$P_{v,\rho}(\mathrm{N}v^{-s}) = \det(1_V - \mathrm{N}v^{-s} F_{v,\rho}) = \prod_{i=1}^{d} (1 - \lambda_{i,v} \mathrm{N}v^{-s}),$$

where $d = \dim_K V$ and $\lambda_{i,v}$ are algebraic numbers viewed as complex numbers via a fixed embedding $i : \overline{\mathbb{Q}} \hookrightarrow \mathbb{C}$.

Let us define the L-function of a rational representation:

$$L(\rho, s) = \prod_{v \notin S} P_{v,\rho}(\mathrm{N}v^{-s})^{-1}. \tag{6.2.3}$$

This is a formal Dirichlet series $\sum_{n=1}^{\infty} a_n n^{-s}$ with coefficients in E. In all known cases there exists a real constant c such that $|\lambda_{i,\rho}| \le (\mathrm{N}v)^c$, which implies the absolute convergence of the series (6.2.3) in the complex right half plane $\mathrm{Re}(s) > 1 + c$.

There are various methods of completing the product in (6.2.3) at places $v \in S$. The purpose of such a completion is to obtain an L-function satisfying a certain nice functional equation. If v is a non–Archimedean place then one considers the subspace $V(v)$ consisting of elements fixed by the inertia group $I^{(w)}$ for w above v. If ρ is ramified at v then $V(v) \neq V$ (and possibly, $V(v) = \{0\}$. The conjugacy class of $\rho(Fr_v)|_{V(v)}$ and its characteristic polynomial

$$P_{v,\rho}(t) = \det(1_{V(v)} - tF_{v,\rho}|_{V(v)})$$

are then well defined, and the degree of the latter is smaller than d. Put

$$L_v(\rho, s) = P_{v,\rho}(\mathrm{N}v^{-s})^{-1}.$$

If v is an Archimedean point then a good definition of $L_v(\rho, s)$ depends on an additional structure (e.g. a *Hodge structure*) on the vector space V. In this case the v-factors can be expressed in terms of the following Γ-factors:

$$\Gamma_{\mathbb{R}}(s) = \pi^{-s/2}\Gamma(s/2), \qquad \Gamma_{\mathbb{C}}(s) = 2(2\pi i)^{-s}\Gamma(s). \tag{6.2.4}$$

These factors play an important role in the study of the functional equations satisfied by L–functions. If we put

$$\Lambda(\rho, s) = \prod_{v} L_v(\rho, s), \tag{6.2.5}$$

then in important examples it is possible to prove that the function (6.2.5) admits an analytic continuation onto the entire s–plane and satisfies a certain functional equation relating $\Lambda(\rho, s)$ and $\Lambda(\rho^\vee, k - s)$, where $\rho^\vee$ is the representation dual to ρ and k is a real number.

Example 6.9. If χ is a primitive Dirichlet character then in view of the Kronecker – Weber theorem there is associated to χ a one dimensional representation $\rho_\chi : G_{\mathbb{Q}} \to \mathbb{C}^\times$ such that

$$L(\rho_\chi, s) = \prod_{p} (1 - \chi(p)p^{-s})^{-1} = L(s, \chi) = \sum_{n=1}^{\infty} \chi(n) n^{-s}.$$

Let δ be zero or one so as to satisfy $\chi(-1) = (-1)^\delta$. We then have that

$$\Lambda(\rho_\chi, s) = \Gamma(s + \delta) L(\rho_\chi, s) = \xi(s, \chi),$$

and the following functional equation holds (cf. Ch.3 of [Shi71] and [Wei67]):

$$\Lambda(\rho_{\overline{\chi}}, 1-s) = i^{\delta}\frac{\sqrt{C_\chi}}{g(\chi)}\Lambda(\rho_\chi, s), \tag{6.2.6}$$

where C_χ is the conductor of χ and $g(\chi)$ is the Gauss sum of χ. The function $\xi(s,\chi)$ is holomorphic on the entire s-plane for nontrivial characters χ. If χ is trivial then $\xi(s) = \pi^{-s/2}\Gamma(s/2)\zeta(s)$ has a simple pole at $s = 1$, and satisfies the functional equation

$$\xi(1-s) = \xi(s). \tag{6.2.7}$$

6.2.2 The Formalism of Artin

The definition of the L-functions $L(\rho, s)$ described above is due to E.Artin (cf. [Ar30]), who studied representations with a finite image $\mathrm{Im}(\rho)$. In this situation the reprepresentation ρ is always E–rational for some number field E, and is semi-simple (by Maschke's theorem). Hence ρ is uniquely determined (upto equivalence) by its character χ_ρ ($\chi_\rho(g) := \mathrm{Tr}\ \rho(g)$, $g \in G_K$). If ρ is an arbitrary rational representation then the function $L(\rho, s)$ is uniquely defined by its character χ_ρ. This can easily be seen by taking the logarithmic derivative of the series (6.2.3)

$$\begin{aligned}\frac{L'(\rho,s)}{L(\rho,s)} &= -\sum_{v\notin S}\sum_{i=1}^{d}\sum_{m=1}^{\infty}\lambda_{i,v}^{m}\log(\mathrm{N}v)\mathrm{N}v^{-ms} \\ &= -\sum_{v,m}\frac{\mathrm{Tr}(F_{\rho,v}^{m})\log(\mathrm{N}v)}{\mathrm{N}v^{ms}},\end{aligned} \tag{6.2.8}$$

where $F_{\rho,v}^m$ denotes the conjugacy class of the mth power of the Frobenius element $F_{\rho,v}$. In view of this fact one often uses the notation $L(\chi_\rho, s) = L(\rho, s)$.

If

$$\rho_i : G_K \longrightarrow \mathrm{GL}(V_i) \quad (i = 1, 2)$$

are two rational representations with characters $\chi_i = \mathrm{Tr}\ \rho_i$ then one can construct from them the further representations

$$\begin{aligned}\rho_1 \oplus \rho_2 : G_K &\longrightarrow \mathrm{GL}(V_1 \oplus V_2), \\ \rho_1 \otimes \rho_2 : G_K &\longrightarrow \mathrm{GL}(V_1 \otimes V_2).\end{aligned} \tag{6.2.9}$$

whose characters are equal to $\chi_1 + \chi_2$ and $\chi_1 \cdot \chi_2$ respectively, We have that

$$L(\chi_1 + \chi_2, s) = L(\chi_1, s)L(\chi_2, s). \tag{6.2.10}$$

If K'/K is a finite extension and ρ' is a Galois representation of K' with character χ', then one can define the induced representation $\rho = \mathrm{Ind}\ \rho'$, whose character is given by the formula:

$$\chi(g) = \sum_{\gamma \in G_K/G_{K'}} \chi'(\gamma^{-1} g \gamma), \tag{6.2.11}$$

where γ runs over a full set of representatives of left cosets, and it is assumed that χ' is extended by zero to the whole group G_K.

In this notation the following equation holds

$$L(\rho, s) = L(\rho', s). \tag{6.2.12}$$

If K'/K is a finite Galois extension then $G_{K'}$ is a normal subgroup of finite index in G_K. Then for any representation ρ of G_K rational over E we define its restriction to the subgroup $G_{K'}$: $\rho_1 = \mathrm{Res}\ \rho$, $\rho_1 : G_{K'} \to \mathrm{GL}(V)$. Then one has the following factorization formula due to Artin:

$$L(\rho_1, s) = \prod_{\chi \in \mathrm{Irr}\ G(K'/K)} L(\rho \otimes \rho_\chi, s)^{\deg \chi}, \tag{6.2.13}$$

where the product is being taken over the set of characters χ of all irreducible representations of the quotient group $G(K'/K) = G_K/G_{K'}$, $\deg \chi = \chi(1)$ (it is assumed that the field E contains all values of characters χ, which are certain sums of roots of unity).

In the general case the representation ρ can always be replaced by its *semi-simplification* $\tilde{\rho}$, which has the same character. In order to define this consider the composition series

$$V = V^{(0)} \supset V^{(1)} \supset \cdots \supset V^{(m)} = \{0\},$$

of ρ–invariant subspaces with irreducible factors $V^{(i)}/V^{(i+1)}$ $(0 \le i \le m-1)$. Then the representation $\tilde{\rho}$ of G_K in the space

$$\tilde{V} = V^{(0)}/V^{(1)} \oplus V^{(1)}/V^{(2)} \oplus \cdots \oplus V^{(m-1)}/V^{(m)} \tag{6.2.14}$$

is semi-simple, E–rational and has the same character as ρ. Furthermore it is uniquely determined (upto equivalence) by its character.

The representation $\rho : G_K \longrightarrow \mathrm{GL}(V)$ is called *Abelian* if $\mathrm{Im}\ \rho$ is an Abelian group. In this case we may consider ρ as a representation of the group $G_K^{\mathrm{ab}} = G_K/G_K^{\mathrm{c}}$, where G_K^{c} is the commutator subgroup of G_K (i.e. the minimal closed subgroup containing all commutators), $\rho : G_K^{\mathrm{ab}} \longrightarrow \mathrm{GL}(V)$.

We have already seen in §5.3 certain examples of such representations (on the Tate module $V_l(E)$ of an elliptic curve with complex multiplication, and on the one dimensional Tate module $V_l(\mu) = \mathbb{Q}_l(1)$ of l-primarily roots of unity).

If $\mathrm{Im}\ \rho$ is a finite group (not necessarily Abelian) then for a finite Galois extension K'/K one has $\mathrm{Ker}\rho = G_{K'}$ and one uses the notation

$$L(s, \chi, K'/K) = L(\rho, s), \tag{6.2.15}$$

where χ is the character of the corresponding representation

$$\rho : \mathrm{Gal}(K'/K) \longrightarrow \mathrm{GL}(V).$$

The functions $L(s, \chi, K'/K)$ are usually called Artin L–functions; they can be reduced to products of Abelian L–functions of extensions of K using the formalism of Artin and the famous *theorem of Brauer*: if χ is a character of a representation of a finite group G over $\mathbb{C}$ then there are cyclic subgroups $G_i \subset G$ and one dimensional characters χ_i of G_i such that

$$\chi = \sum a_i \mathrm{Ind}_{G_i}^{G} \chi \quad (a_i \in \mathbb{Z}). \tag{6.2.16}$$

If $G = G(K'/K)$ then $G_i = G(K'/K_i)$ and it follows from the relations (6.2.10) that

$$L(s, \chi, K'/K) = \prod_i L(s, \chi_i, K'/K_i)^{a_i}. \tag{6.2.17}$$

The analytic properties of Abelian L-functions are well known and they are analogous to the corresponding properties of the Dirichlet L–series: all of the functions $L(s, \chi_i, K'/K_i)$ can be meromorphically continued onto the entire complex plane, possibly with a simple pole at $s = 1$ for trivial characters χ_i. This implies the existence of a meromorphic analytic continuation of arbitrary Artin L–functions. The famous *Artin conjecture* says that for a non-trivial irreducible character χ of $G = G(K'/K)$ the function $L(s, \chi, K'/K)$ is always holomorphic. (Note that another famous *Artin's Conjecture* (on primitive roots) was discussed in §1.1.4).

However this conjecture seems to be very difficult in general, as is the generalized Riemann hypothesis which says that all zeroes of the function $L(s, \chi, K'/K)$ lying in the strip $0 < \mathrm{Re}(s) \leq 1$ are actually on the line $\mathrm{Re}(s) = \frac{1}{2}$. The difficulty with the Artin conjecture is related to the fact that one lacks the non–local definition of the Dirichlet series representing (6.2.17). In the Abelian case such a description follows from the fact that the coefficients of the Dirichlet series are "periodic" modulo a positive integer (or modulo an integral ideal of a number field). However in a number of interesting cases the Artin conjecture has been proved using the Mellin transforms of automorphic forms. A general global description of Artin L–series in terms of automorphic forms is given by the Langlands program (cf. §6.4, 6.5).

6.2.3 Example: The Dedekind Zeta Function

Let $X = \mathrm{Spec}\, \mathcal{O}_K$ where $\mathcal{O}_K$ is the maximal order of K. The Dedekind zeta function of K is the following Euler product

$$\zeta_K(s) = \zeta(X, s) = \prod_{\mathfrak{p} \subset \mathcal{O}_K} (1 - \mathrm{N}\mathfrak{p}^{-s})^{-1},$$

which is absolutely convergent for $\mathrm{Re}(s) > 1$ and admits a meromorphic continuation onto the entire complex plane. The continuation is holomorphic with the exclusion of a simple pole at $s = 1$. The residue of $\zeta_K(s)$ at $s = 1$ is known to be equal to (cf. [BS85], [Wei74a])

$$\mathrm{Res}_{s=1}\zeta_K(s) = h_K 2^{r_1}(2\pi)^{r_2}\frac{R_K}{w_K\sqrt{|D_K|}}, \tag{6.2.18}$$

where h_K is the class number of K, R_K is its regulator, D_K its discriminant, w_K the number of roots of unity in K and r_1 (resp. r_2) the number of real (resp. complex) places of K. Therefore

$$K \otimes \mathbb{R} \cong \mathbb{R}^{r_1} \oplus \mathbb{C}^{r_2}.$$

From the point of view of L–functions the function $\zeta_K(s)$ corresponds to the trivial Galois representation of the group G_K. It therefore follows from Artin's factorization formula that

$$\zeta_K(s) = \zeta(s) \prod_{\chi\in \mathrm{Irr}\ G(K/\mathbb{Q})} L(s,\chi,K/\mathbb{Q})^{\deg\chi}, \tag{6.2.19}$$

where the product is taken over all non–trivial irreducible representations of the group $G(K/\mathbb{Q})$. If the extension $K/\mathbb{Q}$ is Abelian then (6.2.19) implies the following class number formula:

$$h_K = \frac{w_K\sqrt{|D_K|}}{2^{r_1}(2\pi)^{r_2}R_K}\prod_{\chi\neq\varepsilon} L(1,\chi), \tag{6.2.20}$$

since $\mathrm{Res}\ \zeta(s) = 1$. It is not difficult to compute the values $L(1,\chi)$: let C_χ be the conductor of χ. Then

a) for $\chi(-1) = -1$ one has

$$L(1,\chi) = \frac{i\pi g(\chi)}{C_\chi^2}\sum_{\substack{(k,C_\chi)=1\\ 0<k<C_\chi}} k\chi(k); \tag{6.2.21}$$

and

$$L(0,\chi) = -\frac{1}{C_\chi}\sum_{\substack{(k,C_\chi)=1\\ 0<k<C_\chi}} k\chi(k);\ \text{in particular } L(0,\chi_3) = -\frac{1}{6} \tag{6.2.22}$$

for $\chi_3(d) = \left(\frac{d}{3}\right)$ (comp. with the functional equation given by the equality (6.2.6), and see also [Hi93], p.66); the equality (6.2.22) is used in §7.2 expressing the constant term of an Eisenstein series of weight one.

b) for $\chi(-1) = 1$, $\chi \neq \varepsilon$ one has

$$L(1,\chi) = -\frac{g(\chi)}{C_\chi} \sum_{\substack{(k,C_\chi)=1 \\ 0<k<C_\chi}} \chi(k) \log|1-\zeta^{-k}|, \tag{6.2.23}$$

where $\zeta = \exp(2\pi i/C_\chi)$ is a primitive root of unity of degree C_χ and $g(\chi)$ is the Gauss sum of χ.

Formulae (6.2.21) and (6.2.23) give essential information on the class number, the regulator, and the structure of the class group Cl_K of an Abelian field K, in particular when K is quadratic or cyclotomic (cf. [BS85]).

For a quadratic field of discriminant $D = D_K > 0$ we have

$$h_K = -\frac{1}{\log\varepsilon} \sum_{\substack{(k,D)=1 \\ 0<k<D/2}} \chi(k) \log\sin(\pi k/D), \tag{6.2.24}$$

where ε is the fundamental unit of K with $\varepsilon > 1$.

If $D = D_K < -4$ then

$$h_K = -\frac{1}{|D|} \sum_{\substack{(k,D)=1 \\ 0<k<|D|}} k\chi(k) = (2-\chi(2))^{-1} \sum_{\substack{(k,D)=1 \\ 0<k<|D|/2}} \chi(k), \tag{6.2.25}$$

and for the remaining fields $K = \mathbb{Q}(\sqrt{-1})$, $\mathbb{Q}(\sqrt{-3})$ one has $h_K = 1$ (*Dirichlet's class number formula*).

We mention that there exists a purely arithmetical proof of the formula (6.2.25) in the case $D \not\equiv 1 \mod 8$ found by B.A.Venkov (cf. [V81]).

The number

$$\varkappa_K = \mathrm{Res}_{s=1}\zeta_K(s) = h_K 2^{r_1}(2\pi)^{r_2} \frac{R_K}{w_K\sqrt{|D_K|}}$$

has a geometric interpretation as the volume of a fundamental domain for $K^\times$ in J_K^1 with respect to the measure on the group $J_K^1 = \{x \in J_K \mid \|x\| = 1\}$, which comes from the normalized Haar measure $\mu^\times$ on the group J_K (see §4.3).

6.2.4 Hecke Characters and the Theory of Tate

([La70], [CF67], [Wei74a]). The Abelian L–functions of a number field K can be described using class field theory, which states in particular that there is a one-to-one correspondence between irreducible complex representations of the group G_K^{ab} and characters of finite order of the idele class group $C_K = J_K/K^\times$. In the classical theory these characters are known as "periodic" characters of the group of fractional ideals of K. For any integral ideal $\mathfrak{m} \subset \mathcal{O}_K$ write $S = S(\mathfrak{m})$ for the finite set of places of K given by $S = S(\mathfrak{m}) = \{v \in \Sigma_K \mid v \text{ divides } \mathfrak{m}\} \cup \Sigma_K^\infty$, where Σ_K^∞ is the set of Archimedean places of K. Let $I^{(S)}$ be the free Abelian group generated by the finite places prime to $\mathfrak{m}$ and let

$$P^{(\mathfrak{m})} = \{x \in K \mid x \equiv 1(\text{mod } \mathfrak{m}\mathcal{O}_v) \text{ for } v \in S\}. \tag{6.2.26}$$

Then for each one dimensional representation $\rho : G_K^{\text{ab}} \to \mathbb{C}^\times$ there exists an integral ideal $\mathfrak{m}$ and a character $\chi : I^S \to \mathbb{C}^\times$ trivial on the subgroup $(P^{(\mathfrak{m})})$ of principal ideals of type (x), $x \in P^{(\mathfrak{m})}$ such that $\rho(Fr_v) = \chi(\mathfrak{p}_v)$. The generalized Dirichlet L-series are then defined by

$$L(s, \chi) = \prod_{v \notin S} (1 - \chi(\mathfrak{p}_v)\text{N}\mathfrak{p}_v^{-s})^{-1} = \sum_{\mathfrak{n}\ :\ \mathfrak{n}+\mathfrak{m}=\mathcal{O}_K} \chi(\mathfrak{n})\text{N}\mathfrak{n}^{-s}, \tag{6.2.27}$$

where $\mathfrak{n}$ runs through the integral ideals coprime to $\mathfrak{m}$.

Hecke has introduced a new class of characters and L–functions, which, in principle, can not be reduced to L–functions of rational Galois representations. These characters are associated to arbitrary continuous homomorphisms

$$\psi : J_K/K^\times \longrightarrow \mathbb{C}^\times \tag{6.2.28}$$

and they can be described in classical terms as follows: there exists $\mathfrak{m} \subset \mathcal{O}_K$ and a homomorphism $\chi : I^S \to \mathbb{C}^\times$ such that for all $x \in P^{(\mathfrak{m})}$ one has

$$\chi((x)) = \prod_{v|\infty} \left(\frac{\tau_v x}{|\tau_v x|}\right)^{a_v} |x|_v^{it_v} \text{N}((x))^\sigma, \tag{6.2.29}$$

where $\tau_v : K \hookrightarrow \mathbb{C}^\times$ is the complex embedding which defines v; $|x|_v = |\tau_v x|^{[K_v:\mathbb{Q}]}$ the corresponding normalized norm; t_v and $\sigma \in \mathbb{R}$; $a_v \in \mathbb{Z}$ for $K_v \cong \mathbb{C}$, $a_v = 0$ or 1 for $K_v \cong \mathbb{R}$. Since $\chi((x))$ depends only on (x) the right hand side of (6.2.29) must equal 1 for all $\varepsilon \in P^{(\mathfrak{m})} \cap E_K$. The ideal $\mathfrak{m}$ can be maximally chosen for the above condition; this ideal is called the conductor of χ and is denoted by $\mathfrak{m} = \mathfrak{f}(\chi)$. The above condition $\chi((\varepsilon)) = 1$ imposes some restriction on the choice of numbers t_v, σ and a_v. One verifies that these conditions define a subgroup of $(\mathbb{Z}/2\mathbb{Z})^{r_1} \oplus \mathbb{Z}^{r_2} \oplus \mathbb{R}^{r_1+r_2} \oplus \mathbb{R}$ which is isomorphic to $(\mathbb{Z}/2\mathbb{Z})^{r_1} \oplus \mathbb{Z}^{r_2} \oplus \mathbb{Q}^{r_1+r_2-1} \oplus \mathbb{R}$.

A correspondence between χ and ψ is defined using the homomorphism which takes an idele to its divisor (cf. [CF67], Chapter 8)

$$\text{div}_S : J_K \to I^{(S)}, \quad (x_v)_v \mapsto \prod_v \mathfrak{p}_v^{v(x_v)} = \sum_v v(x_v) \cdot v, \tag{6.2.30}$$

and using an appropriate section $\pi : I^{(S)} \to J_K$ which is defined by a choice of local uniformizing parameters $\pi_V \in \mathcal{O}_K$ ($v(\pi_v) = 1$ for $v \notin S$): by this section a prime ideal $\mathfrak{p}_v$ goes to the idele $\pi(v) = (\cdots, 1, \pi_v, 1, \cdots)$ whose v-coordinate is π_v and whose other coordinates are 1. Then the character ψ corresponds to a unique χ such that $\chi(\mathfrak{p}_v) = \psi(\pi(v))$, and this is a one-to-one correspondence.

Example 6.10. 1) If $\psi(x) = \|x\|^s = \omega_s(x)$, then $\mathfrak{m} = \mathcal{O}_K$ and

$$\chi(\mathfrak{p}_v) = \mathrm{N}\mathfrak{p}_v^s, \quad t_v = \mathrm{Im}\ s, \quad \sigma = \mathrm{Re}\ s. \tag{6.2.31}$$

2) If K is an imaginary quadratic field, $K \subset \mathbb{C}$, then for an arbitrary χ there exists $\mathfrak{m}$ such that

$$\chi((x)) = \left(\frac{x}{|x|}\right)^{a_v} |x|^s = x^a |x|^{s-a}, \tag{6.2.32}$$

(for all $x \in K$, $x \equiv 1 (\mathrm{mod}\ \mathfrak{m})$).

3) If K is a real quadratic field, $K \subset \mathbb{R}$, then

$$\chi((x)) = x^{2\pi i \rho / \log \varepsilon} |x|^s \tag{6.2.33}$$

for $x \equiv 1 (\mathrm{mod}\ \mathfrak{m})$, $x, x' > 0$, $\rho \in \mathbb{Q}$, where ε is a fundamental unit in K and $x \mapsto x'$ is the quadratic conjugation.

The interpretation of Hecke characters as certain characters of the idele class group was given by C.Chevalley (cf. [Chev40]).

Tate constructed in his thesis a general theory which makes it possible in particular to establish analytic continuation and a functional equation for all functions of type $L(s, \chi)$. We describe briefly the key points of this theory, which is based on Fourier analysis in number fields ([La70], [Ta65]).

Every continuous character $\psi : J_K/K^\times \to \mathbb{C}^\times$ may be regarded as a function on J_K and it can be decomposed into a product $\psi(x) = \prod_v \psi_v(x_v)$, where $\psi_v : K_v^\times \to \mathbb{C}^\times$ are quasicharacters (i.e. continuous homomorphisms to $\mathbb{C}^\times$) such that for almost all v the quasicharacter ψ_v is unramified: $\psi_v(\mathcal{O}_v^\times) = 1$ and in view of the continuity one has $\psi_v(x_v) = |x_v|_v^\sigma$. The number $\sigma = \mathrm{Re}\ \psi_v$ is called the real part of ψ_v.

The first stage in Tate's theory is to obtain a representation for a local factor of the Hecke L–function

$$L(s, \chi) = \prod_{v \notin S} (1 - \chi(\mathfrak{p}_v) \mathrm{N}\mathfrak{p}_v^{-s})^{-1} = \sum_{\mathfrak{n}\ :\ \mathfrak{n}+\mathfrak{m}=\mathcal{O}_K} \chi(\mathfrak{n}) \mathrm{N}\mathfrak{n}^{-s}, \tag{6.2.34}$$

as a certain integral over the locally compact group $K_v^\times$ with respect to the Haar measure $\mu^\times$ normalized by the condition $\mu^\times(\mathcal{O}_v^\times) = 1$.

Let $c : K_v^\times \to \mathbb{C}^\times$ be an unramified quasicharacter. We use the decompositions $K_v^\times = \bigcup_{n \in \mathbb{Z}} \pi_v^n \mathcal{O}_v^\times$ and $\mathcal{O}_v \backslash \{0\} = \bigcup_{n \in \mathbb{Z}, n \geq 0} \pi_v^n \mathcal{O}_v^\times$ in order to calculate the integral $\int_{\mathcal{O}_v \backslash \{0\}} c(x)\, d\mu_v^\times(x)$. Consider first the integral $\int_{\pi_v^n \mathcal{O}_v^\times} c(x)\, d\mu_v^\times(x)$ and put $x = \pi_v^n \varepsilon$ with $\varepsilon \in \mathcal{O}_v^\times$. Then

$$\int_{\pi_v^n \mathcal{O}_v^\times} c(x)\, d\mu_v^\times(x) = c(\pi_v^n) \left(\int_{\mathcal{O}_v^\times} c(\varepsilon)\, d\mu_v^\times(\varepsilon) \right) = c(\pi_v)^n$$

in view of the invariance of $d\mu_v^\times(x)$ under multiplicative shifts: $d\mu_v^\times(\pi^n x) = d\mu_v^\times(x)$.

If Re $c > 0$ then it follows that c is integrable on $\mathcal{O}_v \backslash \{0\} = \cup_{n\in\mathbb{Z},n\geq 0}\pi_v^n \mathcal{O}_v^\times$ and one has

$$\int_{\mathcal{O}_v\backslash\{0\}} c(x)\, d\mu_v^\times(x) = \sum_{n\geq 0}\int_{\pi_v^n\mathcal{O}_v^\times} c(x)\, d\mu_v^\times(x) = \sum_{n\geq 0} c(\pi_v)^n = (1-c(\pi_v))^{-1}. \tag{6.2.35}$$

If $c = \psi_v\omega_s$ then $c(\pi_v) = \chi(\mathfrak{p}_v)\mathrm{N}\mathfrak{p}_v^{-s}$ in view of (6.2.29), and the expression (6.2.35) becomes the local factor of the L series (6.2.34).

The essential fact is that the whole product (6.2.34) can also be interpreted as a certain integral over a locally compact group with respect to an invariant Haar measure. The set $\mathcal{O}_v\backslash\{0\}$ in (6.2.35) refers more to an additive theory than to a multiplicative theory. This combination of additive with multiplicative theories is a characteristic feature of Tate's theory.

Let G be a locally compact group with a Haar measure μ; denote by $L^1(G)$ the vector space of all integrable functions on G. If μ_v is the Haar measure on the additive group K_v normalized by the condition $\mu_v(\mathcal{O}_v) = 1$ then a calculation analogous to (6.2.35) shows that

$$d\mu_v^\times(x) = (1-\mathrm{N}\mathfrak{p}_v^{-1})^{-1}\frac{d\mu_v(x)}{|x|_v} \tag{6.2.36}$$

in view of the multiplicative invariance of the measure $d\mu_v(x)/|x|_v$. In particular, the condition $f \in L^1(K_v^\times)$ is equivalent to saying that $f(x)/|x|_v \in L^1(K_v)$.

Let us introduce the following notation: for a quasicharacter $c : K_v^\times \to \mathbb{C}^\times$ and $f \in L^1(K_v)$

$$\zeta_v(f,c) = \int_{K_v} f(x)c(x)\, d\mu_v^\times(x) \tag{6.2.37}$$

and suppose that $fc \in L^1(K_v^\times)$ for Re $c > 0$. Let $f = \delta_{\mathcal{O}_v\backslash\{0\}}$ be the characteristic function of the set $\mathcal{O}_v\backslash\{0\}$ and c an unramified quasicharacter. Then for $\mathrm{Re}(\psi_v\omega_s) > 0$ one has the following expression for the local ζ–factor:

$$\zeta_v(f,\psi_v\omega_s) = (1-\chi(\mathfrak{p}_v)\mathrm{N}\mathfrak{p}_v^{-s})^{-1}.$$

In order to obtain a global analogue of this expression let us consider a function $f(x) = \prod_v f_v(x_v)$ on the additive group of adeles $\mathbb{A}_K$ such that $f_v(x) \in L^1(K_v)$ and $f_v = \delta_{\mathcal{O}_v\backslash\{0\}}$ for non–Archimedean places $v \notin S$. For a quasicharacter $c : J_K/K^\times \to \mathbb{C}^\times$ we shall set

$$\zeta(f,c) = \int_{J_K} f(x)c(x)\, d\mu^\times(x) = \prod_v \int_{K_v} f(x_v)c(x_v)\, d\mu_v^\times(x_v). \tag{6.2.38}$$

Then the calculation (6.2.35) implies that

$$\zeta(f,\psi\omega_s) = L(s,\chi)\prod_{v\in S}\zeta_v(f_v,\psi_v\omega_s), \tag{6.2.39}$$

and one easily verifies that under our assumptions the integral and the product are absolutely convergent for $\mathrm{Re}(\psi_v\omega_s) > 1$. An analytic continuation for the function $L(s,\chi)$ is constructed using the integral representation (6.2.38) in which all auxiliary factors can be reduced (in our applications) to Γ–functions and to Gauss sums.

The technique of analytic continuation is based on tools from the theory of additive Fourier transforms over the group $\mathbb{A}_K$. The following are the key points:

I) *A choice of duality.* Let us fix an additive character

$$\lambda : \mathbb{A}_K/K \longrightarrow \mathbb{C}^\times, \quad \lambda(x) = \prod \lambda_v(x_v),$$

where one usually puts:

$$\lambda_v(x_v) = \begin{cases} \exp(-2\pi i x_v) & \text{if } K_v \cong \mathbb{R}, \\ \exp(-4\pi i \mathrm{Re}\ x_v) & \text{if } K_v \cong \mathbb{C}, \\ \exp(-2\pi i \{\mathrm{Tr}_{K_v/\mathbb{Q}_p} x_v\}) & \text{if } [K_v : \mathbb{Q}_p] < \infty. \end{cases} \tag{6.2.40}$$

Then the following isomorphisms of locally compact groups are defined:

$$K_v \cong \hat{K}_v, \quad \mathbb{A}_K \cong \hat{\mathbb{A}}_K,$$

(where $\hat{K}_v$, $\hat{\mathbb{A}}_K$ denote the corresponding groups of (continuous) characters). These isomorphisms are constructed as follows:

$$x \mapsto (\chi_x : y \mapsto \lambda(yx)) \quad (x, y \in \mathbb{A}_K),$$

$$x_v \mapsto (\chi_{x_v} : y \mapsto \lambda_v(yx_v)) \quad (x, y \in K_v).$$

II) *Self-dual measures.* One chooses normalized measures $\tilde{\mu}_v$ and $\tilde{\mu} = \prod_v \tilde{\mu}_v$ such that for the Fourier transforms

$$\hat{f}(x) = \int_{\mathbb{A}_K} f(y)\lambda(xy)\, d\tilde{\mu}(y), \quad \hat{f}_v(x_v) = \int_{K_v} f_v(y)\lambda_v(x_v y)\, d\tilde{\mu}_v(y) \tag{6.2.41}$$

the inversion formulae can be written in the following form:

$$\hat{\hat{f}}(-x) = f(x), \quad \hat{\hat{f}}_v(-x_v) = f(x_v). \tag{6.2.42}$$

provided $f_v \in L^1(K_v)$, $f \in L^1(\mathbb{A}_K)$.
If v is a non-Archimedean place then let $\delta_v \subset \mathcal{O}_v$ denotes the local different,

$$\delta_v^{-1} = \{x \in K_v \mid \lambda_v(xy) \in \mathbb{Z} \text{ for all } \ y \in \mathcal{O}_v\}. \tag{6.2.43}$$

Then the self-dual measure is defined as follows:

$$\tilde{\mu}_v = \begin{cases} \mathrm{N}\delta_v^{-1/2}\mu_v & \text{if } v \text{ is non-Archimedean,} \\ dx \quad \text{(Lebesgue measure)} & \text{if } K_v \cong \mathbb{R}, \\ 2dx\,dy & \text{if } z = x + iy \in K_v \cong \mathbb{C}. \end{cases} \tag{6.2.44}$$

The important property of the self-dual measure is that $\tilde{\mu}(\mathbb{A}_K/K) = 1$; also one has $\delta_v = \delta_K \mathcal{O}_v$ and $\mathrm{N}\delta_K = |D_K|$.

In concrete examples the following orthogonality relation is often used: for each character λ of a compact group G

$$\int_G \lambda(x)\,d\mu(x) = \begin{cases} \mu(G) & \text{if } \lambda = id, \\ 0 & \text{otherwise.} \end{cases} \tag{6.2.45}$$

This implies the following important formula:

$$\int_{\mathcal{O}_v} \lambda_v(xy)\,d\tilde{\mu}_v(y) = \delta_{\delta_v}(x) \cdot \tilde{\mu}_v(\mathcal{O}_v).$$

III) *The Poisson summation formula.* Let f be a continuous function on $\mathbb{A}_K$ such that both $|f|$ and $|\hat{f}|$ are summable over the subset $K \subset \mathbb{A}_K$ and the series $\sum_{\alpha\in K} f(x+\alpha)$ converges uniformly on every compact subset of $\mathbb{A}_K$. Then the following summation formula holds:

$$\sum_{\alpha\in K} f(\alpha) = \sum_{\alpha\in K} \hat{f}(\alpha). \tag{6.2.46}$$

Corollary. *Under the above assumptions for all $a \in J_K$ the following holds*

$$\sum_{\alpha\in K} f(a\alpha) = \|a\|^{-1} \sum_{\alpha\in K} \hat{f}(a^{-1}\alpha). \tag{6.2.47}$$

Now we turn to main application of the summation formula: the proof of the functional equation for the ζ–functions. We assume that for all $\sigma > 1$ the function $|f(x)|\,\|x\|^\sigma$ is integrable over the group of ideles J_K. Then for $\mathrm{Re}(\psi\omega_s) > 1$ the following integral is well defined:

$$\zeta(f, \psi\omega_s) = \int_{J_K} f(x)\psi\omega_s(x)\,d\mu^\times(x). \tag{6.2.48}$$

Theorem 6.11. *The function $\zeta(f, \psi\omega_s)$ admits an analytic continuation onto the entire complex plane and it satisfies the functional equation:*

$$\zeta(f, \psi\omega_s) = \zeta(\hat{f}, \psi^{-1}\omega_{1-s}). \tag{6.2.49}$$

To prove the theorem we decompose the integral into two parts:

$$\zeta(f,c) = \int_{\|x\|\geq 1} f(x)c(x)\,d\mu^\times(x) + \int_{\|x\|\leq 1} f(x)c(x)\,d\mu^\times(x) \text{ (with } c = \psi\omega_s).$$

In view of the assumption on f the first integral converges for all c. Let us transform the second integral using the Poisson summation formula. We have

$$\int_{\|x\|\le 1} f(x)c(x)\,d\mu^\times(x) = \int_{t\in J_K/J_K^1}\left(\int_{J_K^1} f(tx)c(tx)\,d\mu^1(x)\right) d\nu,$$

where $d\mu^1(x)$ is the measure on J_K^1 which is compatible with the Haar measure $d\nu(t) = |dt/t|$ on $\mathbb{R}_+^\times$ and the original measure $d\mu^\times$ under the isomorphism $J_K/J_K^1 \cong \mathbb{R}_+^\times$. The inner integral on the right hand side transforms into the following sum

$$\int_{J_K^1} f(tx)c(tx)\,d\mu^1(x) = \int_{C_K^1}\left(\sum_{\alpha\in K^\times} f(tx\alpha)c(tx\alpha)\right) d\mu^1(x)$$

where we denote by the same symbol $d\mu^1(x)$ the measure on C_K^1 induced by $d\mu^1(x)$ on J_K^1. Now the inner sum transforms using (6.2.46) as follows:

$$\begin{aligned} c(tx)\sum_{\alpha\in K^\times} f(tx\alpha) &= c(tx)\left(\sum_{\alpha\in K} f(tx\alpha) - f(0)\right) \\ &= c(tx)\left(\|tx\|^{-1}\sum_{\alpha\in K}\hat f(t^{-1}x^{-1}\alpha) - f(0)\right) \\ &= c(tx)\left(\|tx\|^{-1}\sum_{\alpha\in K^\times}\hat f(t^{-1}x^{-1}\alpha) + \|tx\|^{-1}\hat f(0) - f(0)\right). \end{aligned}$$

We now change variables by putting $u = t^{-1}, y = x^{-1}$; the measure in J_K^1 does not change. Putting the resulting expression into the integral, and using the notation $c_1 = \omega_1 c^{-1} = \psi^{-1}\omega_{1-s}$, the integral over $\|x\| \le 1$ becomes the following

$$\begin{aligned} \int_{\|x\|\le 1} f(x)c(x)\,d\mu^\times(x) &= \int_{\|x\|\ge 1} \hat f(x)c_1(x)\,d\mu^\times(x) \\ &\quad + \int_{|t|\le 1}\int_{C_K} c(tx)\left(\|tx\|^{-1}\hat f(0) - f(0)\right) d\mu^1\,d\nu. \end{aligned}$$

This proves the theorem.

As an example of this general calculation let us deduce the classical functional equation for the Dedekind zeta function of a number field K. Set $f_v = \delta_{\mathcal{O}_v}$ for non–Archimedean v,

$$f_v(x) = \begin{cases} \exp(-\pi x^2) & \text{for } K_v \cong \mathbb{R}, \\ \exp(-2\pi|z|^2) & \text{for } K_v \cong \mathbb{C}. \end{cases}$$

Then one has

$$\zeta_v(f_v, s) = \zeta_v(\hat{f}_v, s) = \Gamma_{K_v}(s) \quad (K_v \cong \mathbb{R}, \mathbb{C})$$

and we obtain the following expressions for the global integrals (ζ–functions):

$$\zeta(f, \omega_s) = \Gamma_{\mathbb{R}}^{r_1}(s)\Gamma_{\mathbb{C}}^{r_2}(s)\zeta_K(s),$$

$$\zeta(\hat{f}, \omega_{1-s}) = |G_K|^{(1/2)-s}\Gamma_{\mathbb{R}}^{r_1}(1-s)\Gamma_{\mathbb{C}}^{r_2}(1-s)\zeta_K(1-s),$$

which implies in view of (6.2.49) the following functional equation:

$$\Lambda_K(s) = \Lambda_K(1-s), \tag{6.2.50}$$

where

$$\Lambda_K(s) = |D_K|^{s/2}\Gamma_{\mathbb{R}}^{r_1}(s)\Gamma_{\mathbb{C}}^{r_2}(s)\zeta_K(s).$$

In the general case of arbitrary quasicharacters we may and we shall assume that $\sum_{v|\infty} t_v = 0$ (by replacing s if necessary). Put

$$L_v(s, \chi) = \begin{cases} \Gamma_{\mathbb{R}}(s + it_v - |a_v|) & \text{for } K_v \cong \mathbb{R}, \\ \Gamma_{\mathbb{C}}(s + it_v - |a_v|/2) & \text{for } K_v \cong \mathbb{C}, \end{cases}$$

$$D_\chi = |D_K|\mathrm{N}\mathfrak{f}(\chi).$$

Let $g_v(\chi) = \sum_\varepsilon (\chi_v \lambda_v)(\varepsilon\pi^{-v(\chi)})$ denote the Gauss sum, where $\{\varepsilon\}$ runs over a system of coset representatives for $\mathcal{O}_v^\times/(1 + \mathfrak{f}(\chi)\mathcal{O}_v)$ with $v(\chi) \overset{\text{def}}{=} v(\mathfrak{f}(\chi)) > 0$. Then the following functional equation holds:

$$W(\chi)\Lambda(s, \chi) = \Lambda(1-s, \overline{\chi}), \tag{6.2.51}$$

in which

$$\Lambda(s, \chi) = D_\chi^{s/2} \prod_{v|\infty} L_v(s, \chi) \cdot L(s, \chi), \tag{6.2.52}$$

and $W(\chi)$ is a complex constant with absolute value 1 given by

$$W(\chi) = i^M \mathrm{N}\mathfrak{f}(\chi)^{-1/2} \prod_{v \in S_\chi} g_v(\chi) \prod_{v \notin S_\chi} \chi(\delta_v^{-1})$$

where $S_\chi = \{v \mid \mathfrak{p}_v \text{ divides } \mathfrak{f}(\chi)\}$ and $M = \sum_{v|\infty} |a_v|$.

6.2.5 Explicit Formulae

(cf. [La70], [Wei52b], [Wei72], [More77]). We already mentioned in chapter 1 a link between the zeroes of the Riemann zeta function and the behaviour of the function $\pi(x) = \sum_{p<x} 1$, p being prime numbers. This link is expressed by an explicit formula for the function $\sum_{i=1}^{\infty} \frac{1}{i}\pi(x^{1/i})$ in terms of the non trivial

zeroes of $\zeta(s)$, i.e. those zeroes in the critical strip $0 \le \mathrm{Re}\ s \le 1$ (the *Riemann - Mangoldt formula*), see Part I, §1.1.6.

A generalization of this formula for Hecke L–series $\Lambda(s,\chi)$ (see definition (6.2.52)) was proposed by Weil, and is based on the Weierstrass product expansion of this function over its zeros.

Let us assume that Re $\chi = 0$ and the normalizing condition $\sum_{v|\infty} t_v = 0$ is satisfied (see (6.2.51)). Put $\delta_\chi = 1$ if $\chi \equiv 1$ and 0 otherwise. Then it follows from the functional equation (6.2.49) that the function

$$[s(s-1)]^{\delta_\chi}\Lambda(s,\chi)$$

is an entire function of order 1. Hence by a general theorem from the theory of functions of one complex variable one has the following Weierstrass product expansion of this function over its zeros:

$$\Lambda(s,\chi) = a_\chi e^{b_\chi s}[s(s-1)]^{-\delta_\chi}\prod_\omega\left(1-\frac{s}{\omega}\right)e^{s/\omega}, \tag{6.2.53}$$

where $\omega = \alpha + i\gamma$ runs over the set of all zeroes of the function $\Lambda(s,\chi)$ counting multiplicities, and a_χ and b_χ are certain constants. The main result on explicit formulae for the L–functions $\Lambda(s,\chi)$ can then be stated as an equation relating a linear combination of the values of a certain function over (logarithms of norms of) powers of prime ideals, to the sum of the Mellin transform of this function over the zeroes of the L–function (6.2.53).

Let us consider a complex valued function $F:\mathbb{R}\to\mathbb{C}$ with the property that there exists a constant $a>0$ such that

$$F(x)e^{((1/2)+a)|x|} \in L^1(\mathbb{R}).$$

Then the Mellin transform

$$\Phi(s) = \int_{-\infty}^{+\infty} F(x)e^{(s-\frac{1}{2})x}\,dx$$

is a function which is holomorphic in the strip $-a \le \mathrm{Re}\ s \le 1+a$. We assume that the function $F(x)$ satisfies the following conditions:

A) The function $F(x)$ has continuous derivative everywhere apart from a finite number of points α_i, at which both $F(x)$ and $F'(x)$ have only breaks of the first kind, and $F(\alpha_i) = \frac{1}{2}[F(\alpha_i+0)+F(\alpha_i-0)]$.

B) For some number $b>0$ the following estimates hold as $|x|\to\infty$:

$$F(x) = O(e^{-((1/2)+b)|x|}),$$

$$F'(x) = O(e^{-((1/2)+b)|x|}).$$

Then we have that $\Phi(s) = O(|t|^{-1})$ uniformly in the strip $-a' \le \sigma \le 1+a'$ if $0 < a' < b$ ($\sigma = \mathrm{Re}\ s$, $t = \mathrm{Im}\ s$).

The explicit formula. With the above notation and assumptions consider the sum

$$\sum_{|t|<T} \Phi(\omega)$$

extended over all zeroes $\omega = \beta + it$ of the function $L(s,\chi)$ satisfying $0 \le \beta \le 1$, $|t| < T$. Then as $T \to \infty$ this tends to a limit, which is equal to

$$\begin{aligned}\lim_{T\to\infty} \sum_{|t|<T} \Phi(\omega) =& \delta_\chi \int_{-\infty}^{+\infty} F(x)\left(e^{\frac{x}{2}} + e^{-\frac{x}{2}}\right) dx \\ &+ F(0) \log A_\chi \\ &- \sum_{\mathfrak{p},n} \frac{\log \mathrm{N}\mathfrak{p}}{\mathrm{N}\mathfrak{p}^{n/2}} [\chi(\mathfrak{p})^n F(\log \mathrm{N}\mathfrak{p}^n) + \chi(\mathfrak{p})^{-n} F(\log \mathrm{N}\mathfrak{p}^{-n})] \\ &- \sum_{v|\infty} W_v(F_v), \end{aligned} \tag{6.2.54}$$

where $A_\chi = 2^{-r_1}(2\pi)^{-r_2} D_\chi \mathrm{N}\mathfrak{f}(\chi)$,

$$F_v(x) = F(x) e^{-it_v x/n_v} \quad (n_v = [K_v : \mathbb{R}]),$$

and W_v is the functional uniquely defined by the property

$$W_v(g) = \lim_{\lambda\to\infty} \left[\int_{-\infty}^{+\infty} (1 - e^{-\lambda|x|}) K_v(x) g(x)\, dx - 2g(0) \log \lambda \right].$$

Here the function $K_v(x)$ is defined by

$$K_v(x) = \begin{cases} \dfrac{e^{-(1/2)|a_v x|}}{|e^{x/2} - e^{-x/2}|} & \text{for } K_v \cong \mathbb{R}\ (n_v = 2), \\[2ex] \dfrac{e^{-((1/2)-|a_v x|)}}{|e^{x} - e^{-x}|} & \text{for } K_v \cong \mathbb{C}\ (n_v = 1). \end{cases}$$

The explicit formula (6.2.54) is a rich source of possibilities for studying very fine points in the distribution of prime ideals in number fields and the images of the corresponding Frobenius elements in Abelian Galois representations. We mention that there is an analogue of these formulae in the case of global fields of positive characteristic (the function field case). This generalization makes it possible to obtain some very precise estimates for the numbers of points on curves over finite fields, and has some other interesting applications (cf. [Se83], [Step99], [TsVl91]).

The logical structure of the proof of (6.2.54) is quite simple and is based on studying the integral of the function $\Phi\, d \log \Lambda(s,\chi)$ along a special contour using two explicit expressions for the logarithmic derivative $\dfrac{\Lambda'(s,\chi)}{\Lambda(s,\chi)}$ arising from the Euler product defining $L(s,\chi)$ and from the Weierstrass product formula (6.2.53). One uses only the following results of an arithmetical nature:

a) The boundedness of the $|L(s,\chi)|$ in each right half plane of type $\mathrm{Re}(s) \geq 1+a$, $a>0$.
b) The functional equation (6.2.51) for $\Lambda(s,\chi)$.
c) The boundedness of the function $\Lambda(s,\chi)$ in every strip $\sigma_0 \leq \mathrm{Re}(s) \leq \sigma_1$ excluding a finite number of poles.

6.2.6 The Weil Group and its Representations

(cf. [Ta79], [Wei74a]). We wish to discuss a general construction which makes it possible to treat at the same time representations of Galois groups and quasicharacters of number fields, and their L–functions. This construction is based on the notion of the Weil group, which we briefly discuss now.

Let F be a local or global field and F^s its separable algebraic closure. For a finite extension E/F in F^s let $G_E = \mathrm{Gal}(F^s/E)$ be the corresponding open subgroup. Let

$$C_E = \begin{cases} J_E/E^\times & \text{if } E \text{ is a global field,} \\ E^\times & \text{if } E \text{ is a local field.} \end{cases}$$

The relative Weil group $W_{E/F}$ can be described as a group extension

$$0 \to C_E \to W_{E/F} \to \mathrm{Gal}(E/F) \to 0, \tag{6.2.55}$$

whose isomorphism class is defined by the canonical generator $\alpha_{E/F}$ of the cohomology group $H^2(\mathrm{Gal}(E/F), C_E) = \langle \alpha_{E/F} \rangle$ given by class field theory (see §4.4).

There is also a more invariant definition which makes it possible to treat all extensions E/F at the same time (cf. [Ta79]).

The *absolute Weil group* W_F is defined as a topological group endowed with a continuous homomorphism $\varphi : W_F \to G_F$ with dense image, which satisfies the following additional conditions ([Ta79], p. 74–75):

W1) There exist isomorphisms $r_E : C_E \xrightarrow{\sim} W_E^{\mathrm{ab}}$ for which $W_E = \varphi^{-1}(G_E)$ for all finite extensions E and $W_E^{\mathrm{ab}} = W_E/W_E^{\mathrm{c}}$, W_E^{c} being the minimal closed subgroup of W_E containing all its commutators. These isomorphisms satisfy the following condition: the composition

$$C_E \overset{r_E}{\underset{\sim}{\to}} W_E^{\mathrm{ab}} \xrightarrow{\varphi} G_E^{\mathrm{ab}}$$

coincides with the homomorphism of class field theory.

W2) Let $w \in W_F$ and $\sigma = \varphi(w) \in G_F$. Then for each E the following diagram commutes:

$$\begin{array}{ccccc} & C_E & \xrightarrow{r_E} & W_E^{\mathrm{ab}} & \\ \text{isomorphism, induced by } \sigma & \downarrow & & \downarrow & \text{conjugation by } w \\ & C_{E^\sigma} & \xrightarrow{r_{E^\sigma}} & W_{E^\sigma}^{\mathrm{ab}} & \end{array}$$

W3) For $E' \subset E$ the following diagram commutes:

$$\begin{array}{ccccc} & C_{E'} & \xrightarrow{r_{E'}} & W_{E'}^{\mathrm{ab}} & \\ \text{homomorphism, induced by the inclusion } E'\subset E & \downarrow & & \downarrow & \text{transfer, see (4.4.18)} \\ & C_E & \xrightarrow{r_E} & W_E^{\mathrm{ab}}. & \end{array}$$

W4) The natural map

$$W_F \to W_{E/F} \stackrel{\mathrm{def}}{=} W_F/W_E^c$$

defines an isomorphism

$$W_F \cong \varprojlim_E W_{E/F}.$$

It is not difficult to verify that this is equivalent to the previous definition.

The group W_F can be constructed starting from the above relative Weil groups $W_{E/F}$ using certain functorial properties of the classes $\alpha_{E/F}$. From the existence of W_F with properties W1) – W4) one can deduce all the main theorems of class field theory (in both local and global cases).

Also, there exists a homomorphism $w \mapsto \|w\|$ of W_F to $\mathbb{R}_+^\times$ which corresponds under the isomorphism $r_F : C_F \xrightarrow{\sim} W_F^{\mathrm{ab}}$ to the norm homomorphism to $\mathbb{R}_+^\times$ of the idele class group $C_F = J_F^\times/F^\times$ in the global case, and to the normalized absolute value of $C_F = F^\times$ in the local case. In view of the relation $\|\mathrm{N}_{E/F}\alpha\|_F = \|\alpha\|_E$ the restriction of this norm function $\|w\|$ on W_F to the subgroup W_E coincides with the corresponding norm function for W_E, so that we can omit the index E. One verifies that the kernel of the homomorphism $w \mapsto \|w\|$ is compact.

The relation between the local and global Weil groups. Let F be a global field and v a place of F extended to $\overline{F}$. Then there exists a natural embedding $\theta_v : W_{F_v} \to W_F$ which is compatible with the inclusions $i_v : G_{F_v} \to G_F$ and $E_v \hookrightarrow C_E$ for all E/F, $[E : F] < \infty$.

Representations of the Weil groups. Denote by $M(G)$ the set of isomorphism classes of finite dimensional complex representations $\rho : G \to \mathrm{GL}(V)$ of a topological group G. A one dimensional representation $\chi : G \to \mathbb{C}^\times$ will be called a quasicharacter of G. Using the isomorphism $r_F : C_F \xrightarrow{\sim} W_F^{\mathrm{ab}}$ we can identify quasicharacters of W_F with quasicharacters of F (or of C_F). For example, the quasicharacter corresponding to the quasicharacter $c \mapsto \|c\|_F^s$ (with $\|c\|_F$ being the idele norm of $c \in C_F$) will be denoted by the same symbol ω_s, so that one has $\omega_s(w) = \|w\|^s$.

On the other hand the image of $\varphi : W_F \to G_F$ is dense, hence the set $M(G_F)$ can be identified with a subset of $M(W_F)$. Representations in this subset are called *Galois-type representations.* A representation ρ is of Galois type iff the image $\rho(W_F)$ is finite.

Under the above identification a character χ of G_F corresponds to the character of C_F obtained from χ using class field theory.

Using the embeddings $\theta_v : W_{F_v} \to W_F$ Weil has defined L–functions $L(\rho, s)$ of representations $\rho \in M(W_F)$ which include the L-functions of Artin and Hecke as special cases (*Hecke–Weil L–functions*). For these L–functions the usual Artin formalism is valid. Also, there is an analogue of the theorem of Brauer, which makes it possible to reduce the Hecke–Weil L–functions to products of integral powers of Hecke L–functions of quasicharacters of finite extensions E of F. A precise statement of the functional equation of $L(\rho, s)$ and a definition of all its local factors are given in [Ta79].

Weil has established explicit formulae of the type (6.2.54), and proposed a generalized Riemann hypothesis, and an analogue of the Artin conjecture for the Hecke–Weil L–functions $L(\rho, s)$. A remarkable fact is that both the Riemann and Artin conjectures can be reduced to positivity properties of a certain linear functional in the right hand side of the generalized explicit formula of type (6.2.54).

Apart from complex representations one can also consider l-adic representations of W_F, and compatible systems of such representations. Tate gives in [Ta79] general conjectures which indicate that complex and l-adic representations of the Weil group play a universal role in number theory.

6.2.7 Zeta Functions, *L*-Functions and Motives

(cf. [Man68], [Del79]). As we have seen with the example of the Dedekind zeta function $\zeta_K(s)$, the zeta function $\zeta(X, s)$ of an arithmetic scheme X can often be expressed in terms of L–functions of certain Galois representations. This link seems to be universal in the following sense.

Let $X \to \operatorname{Spec} \mathcal{O}_K$ be an arithmetic scheme over the maximal order $\mathcal{O}_K$ of a number field K such that the generic fiber $X_K = X \otimes_{\mathcal{O}_K} K$ is a smooth projective variety of dimension d, and let

$$\zeta(X, s) = \prod_{\mathfrak{p}} \zeta(X(\mathfrak{p}), s)$$

be its zeta function, where $X(\mathfrak{p}) = X \otimes_{\mathcal{O}_K} (\mathcal{O}_K/\mathfrak{p})$ is the reduction of X modulo a maximal ideal $\mathfrak{p} \subset \mathcal{O}_K$. The shape of the function $\zeta(X(\mathfrak{p}), s)$ is described by the Weil conjecture (W4). If we assume that all $X(\mathfrak{p})$ are smooth projective varieties over $\mathcal{O}_K/\mathfrak{p} \cong \mathbb{F}_q$ then we obtain the following expressions for $\zeta(X, s)$:

$$\zeta(X,s) = \prod_{i=0}^{2d} L_i(X,s)^{(-1)^{i+1}}, \tag{6.2.56}$$

where

$$L_i(X,s) = \prod_{\mathfrak{p}} P_{i,\mathfrak{p}}(X, \mathrm{N}\mathfrak{p}^{-s})^{-1},$$

and $P_{i,\mathfrak{p}}(X,t) \in \overline{\mathbb{Q}}[t]$ denote polynomials from the decomposition of the zeta function

$$\zeta(X(\mathfrak{p}),s) = \prod_{i=0}^{2d} P_{i,\mathfrak{p}}(X, \mathrm{N}\mathfrak{p}^{-s})^{(-1)^{i+1}}.$$

In order to prove the conjecture (W4) ("the Riemann Hypothesis over a finite field"), Deligne identified the functions $L_i(X,s)$ with the L–functions of certain rational l–adic Galois representations

$$\rho_{X,i} : G_K \to \mathrm{Aut}\ H^i_{\acute{e}t}(X_{\overline{K}}, \mathbb{Q}_l); \quad L_i(X,s) = L(\rho_{X,i}, s)$$

defined by a natural action of the Galois group G_K on the l–adic cohomology groups $H^*_{\acute{e}t}(X_{\overline{K}}, \mathbb{Q}_l)$ using the transfer of structure

$$\begin{array}{ccc} X_{\overline{K}} & = X_K \otimes \overline{K} & \\ \downarrow & & \\ \mathrm{Spec}\ \overline{K} & \xrightarrow{\sigma} \mathrm{Spec}\ \overline{K} & (\sigma \in \mathrm{Aut}\ \overline{K}). \end{array}$$

If X_K is an algebraic curve then there are G_K–module isomorphisms

$$H^1_{\acute{e}t}(X_{\overline{K}}, \mathbb{Q}_l) \cong V_l(J) = T_l(X) \otimes_{\mathbb{Z}_l} \mathbb{Q}_l$$

(the Tate module of the Jacobian of X),

$$H^0_{\acute{e}t}(X_{\overline{K}}, \mathbb{Q}_l) = \mathbb{Q}_l, \quad H^2_{\acute{e}t}(X_{\overline{K}}, \mathbb{Q}_l) \cong V_l(\mu)$$

($V_l(\mu) = T_l(\mu) \otimes_{\mathbb{Z}_l} \mathbb{Q}_l$ the Tate module of l–power roots of unity). This implies the following explicit expressions for the L–functions

$$L_0(X,s) = \zeta_K(s), \qquad L_2(X,s) = \zeta_K(s-1),$$

and the zeta function

$$L_1(X,s) = L(X,s) = \prod_{\mathfrak{p}} P_{1,\mathfrak{p}}(X, \mathrm{N}\mathfrak{p}^{-s})^{-1},$$

(where $\deg P_{1,\mathfrak{p}}(X,t) = 2g$, g is the genus of the curve X_K) is often called the L–function of the curve X.

For topological varieties cohomology classes can be represented using cycles (by Poincaré duality), or using cells if the variety is a CW–complex. Grothendieck has conjectured that an analogue of CW–decomposition must exist for algebraic varieties over K. In view of this decomposition the factorization of the zeta function (6.2.56) should correspond to the decomposition of the variety into "generalized cells", which are no longer algebraic varieties but *motives*, elements of a certain larger category $\mathcal{M}_K$. This category is constructed in several steps, starting from the category $\mathcal{V}_K$ of smooth projective varieties over K.

Step 1). One constructs first an additive category $\mathcal{M}'_K$ in which $\mathrm{Hom}(M, N)$ are $\mathbb{Q}$–linear vector spaces, and one constructs a contravariant functor H^* from $\mathcal{V}_K$ to $\mathcal{M}'_K$, which is bijective on objects (i.e. with objects $H^*(X)$ one for each $X \in \mathcal{O}b(\mathcal{V}_K)$). This category is endowed with the following additional structures:

a) a tensor product $\otimes$ satisfying the standard commutativity, associativity and distributivity constraints;
b) the functor H^* takes disjoint unions of varieties into direct sums and products into tensor products (by means of a natural transformation compatible with the commutativity and associativity).

In this definition the group $\mathrm{Hom}(H^*(X), H^*(Y))$ is defined as a certain group of classes of correspondences between X and Y. For a smooth projective variety X over K denote by $Z^i(X)$ the vector space over $\mathbb{Q}$ whose basis is the set of all irreducible closed subschemes of codimension i, and denote by $Z^i_R(X)$ its quotient space modulo cohomological equivalence of cycles. Then in Grothendieck's definition, for fields K of characteristic zero one puts

$$\mathrm{Hom}(H^*(Y), H^*(X)) = Z_R^{\dim(Y)}(X \times Y).$$

Step 2. The category $\mathcal{M}_{\mathrm{eff},K}$ of false effective motives. This is obtained from $\mathcal{M}'_K$ by formally adjoining the images of all projections (i.e. of idempotent morphisms). In this category every projection arises from a direct sum decomposition. Categories with a tensor product and with the latter property are called *caroubien* or *pseudo–Abelian* categories; $\mathcal{M}_{\mathrm{eff},K}$ is the pseudo–Abelian envelope of $\mathcal{M}'_K$, cf. [Del79].

Step 3. The category $\overset{\circ}{\mathcal{M}}_K$ of false motives. Next we adjoin to $\mathcal{M}_{\mathrm{eff},K}$ all powers of the Tate object $\mathbb{Q}(1) = \underline{\mathrm{Hom}}(L, \mathbb{Q})$, where $L = \mathbb{Q}(-1) = H^2(\mathbb{P}^1)$ is the Lefschetz object and $\underline{\mathrm{Hom}}$ denotes the internal Hom in $\mathcal{M}_{\mathrm{eff},K}$. As a result we get the category $\overset{\circ}{\mathcal{M}}_K$ of "false motives". The category $\overset{\circ}{\mathcal{M}}_K$ can be obtained by a universal construction which converts the functor $M \to M \otimes \mathbb{Q}(-1) = M(-1)$ into an invertible functor. Each object of $\overset{\circ}{\mathcal{M}}_K$ has the form $M(n)$ with some M from $\mathcal{M}_{\mathrm{eff},K}$.
Note that for $X \in \mathcal{O}b(\mathcal{V}_K)$ the objects $H^i(X)$ are defined as the images of appropriate projections and

$$H^*(X) = \bigoplus_{i=0}^{2d} H^i(X).$$

The category $\overset{\circ}{\mathcal{M}}_K$ is a $\mathbb{Q}$–linear rigid Abelian category with the commutativity rule

$$\Psi^{r,s} : H^r(X) \otimes H^s(Y) \cong H^s(Y) \otimes H^r(X), u \otimes v \mapsto (-1)^{rs} v \otimes u,$$

which implies that the rank $\mathrm{rk}(H(X)) = \sum (-1)^r \dim H^r(X)$ could be negative (in fact it coincides with the *Euler characteristic* of X).

Step 4. The category $\mathcal{M}_K$ of true motives is obtained from $\overset{\circ}{\mathcal{M}}_K$ by a modification of the above commutativity constraint, in which the sign $(-1)^{rs}$ is dropped. This is a $\mathbb{Q}$–linear *Tannakian category*, formed by direct sums of factors of the type $M \subset H^r(X)(m)$, see [Del79].

Tannakian categories are characterized by the property that every such category (endowed with a fiber functor) can be realized as the category of finite dimensional representations of some (pro–) algebraic group.
In particular, the thus obtained category of motives can be regarded as the category of finite dimensional representations of a certain (pro–) algebraic group (the so-called *motivic Galois group*).
Each standard cohomology theory $\mathcal{H}$ on $\mathcal{V}_K$ (a functor from $\mathcal{V}_K$ to an Abelian category with the Künneth formula and with some standard functoriality properties) can be extended to the category $\mathcal{M}_K$. This extension thus defines the $\mathcal{H}$–realizations of motives.

In order to construct L–functions of motives one uses the following realizations:

a) *The Betti realization H_B*: for a field K embedded in $\mathbb{C}$ and $X \in \mathcal{O}b(\mathcal{V}_K)$ the singular cohomology groups (vector spaces over $\mathbb{Q}$) are defined

$$\mathcal{H} : X \mapsto H^*(X(\mathbb{C}), \mathbb{Q}) = H_B(X).$$

One has a Hodge decomposition of the complex vector spaces

$$H_B(M) \otimes \mathbb{C} = \oplus H_B^{p,q}(M) \quad (h^{p,q} = \dim_{\mathbb{C}} H_B^{p,q}(M)),$$

so that $\overline{H_B^{p,q}(M)} = H_B^{q,p}(M)$. If $K \subset \mathbb{R}$ then the complex conjugation on $X(\mathbb{C})$ defines a canonical involution F_∞ on $H_B(M)$, which may be viewed as the Frobenius element at infinity.

b) *The l–adic realizations H_l*: if Char $K \neq l$, $X \in \mathcal{O}b(\mathcal{V}_K)$ then the l–adic cohomology groups are defined as certain vector spaces over $\mathbb{Q}_l$

$$\mathcal{H} : X \mapsto H^*_{\acute{e}t}(X_{\bar{K}}, \mathbb{Q}_l) = H_l(X).$$

There is a natural action of the Galois group G_K on $H_l(X)$ by way of which one assigns an l–adic representation to a motive $M \in \mathcal{M}_K$

$$\rho_{M,l} : G_K \longrightarrow \text{Aut } H_l(M).$$

A non–trivial fact is that these representations are E–rational for some E, $E \subset \mathbb{C}$ in the sense of §6.2.1.

Using the general construction of 6.2.1 one defines the L–functions

$$L(M, s) = \prod_v L_v(M, s) \quad (v \text{ finite}),$$

where $L_v(M, s)^{-1} = L_{\mathfrak{p}_v}(M, \mathrm{N}\mathfrak{p}_v^{-s})^{-1}$ are certain polynomials in the variable $t = \mathrm{N}\mathfrak{p}_v^{-s}$ with coefficients in E.

For Archimedean places v one chooses a complex embedding $\tau_v : K \to \mathbb{C}$ defining v. Then the factors $L_v(M, s)$ are constructed using the Hodge decomposition $H_B(M) \otimes \mathbb{C} = \oplus H_B^{p,q}(M)$ and the action of the involution F_∞ (see the table in 5.3. of [Del79]).

According to a general conjecture the product

$$\Lambda(M, s) = \prod_v L_v(M, s) \quad (v \in \Sigma_K).$$

admits an analytic (meromorphic) continuation to the entire complex plane and satisfies a certain (conjectural) functional equation of the form

$$\Lambda(M, s) = \varepsilon(M, s)\Lambda(M^\vee, 1 - s),$$

where $M^\vee$ is the motive dual to M (its realizations are duals of those of M), and $\varepsilon(M, s)$ is a certain function of s which is a product of an exponential function and a constant.

One has the following equation

$$\Lambda(M(n), s) = \Lambda(M, s + n).$$

A motive M is called pure of weight w if $h^{p,q} = 0$ for $p + q \neq w$. In this case we put $\mathrm{Re}(M) = -\dfrac{w}{2}$. The Weil conjecture W4) (see section 6.1.3) implies that for a sufficiently large finite set S of places of K the corresponding Dirichlet series (and the Euler product)

$$L_S(M, s) = \prod_{v \notin S} L_v(M, s)$$

converges absolutely for $\mathrm{Re}(M) + \mathrm{Re}(s) > 1$.

For points s on the boundary of absolute convergence (i.e. for $\mathrm{Re}(M) + \mathrm{Re}(s) = 1$ there is the following general conjecture (generalizing the theorem of Hadamard and de la Vallée–Poussin):

a) the function $L_S(M, s)$ does not vanish for $\mathrm{Re}(M) + \mathrm{Re}(s) = 1$;
b) the function $L_S(M, s)$ is entire apart from the case when M has even weight $-2n$ and contains as a summand the motive $\mathbb{Q}(n)$; in the last case there is a pole at $s = 1 - n$.

For example, for the motive $\mathbb{Q}(-1)$ one has

$$H_B(\mathbb{Q}(-1)) = H^2(\mathbb{P}^1(\mathbb{C}), \mathbb{Q}), \quad H_l(\mathbb{Q}(-1)) \cong V_l(\mu) = T_l(\mu) \otimes_{\mathbb{Z}_l} \mathbb{Q}_l,$$

$w = 2$, $n = -1$ and the L -function

$$L(\mathbb{Q}(-1), s) = \zeta_K(s - 1)$$

has a simple pole at $s = 2$.

There are some very general conjectures on the existence of a correspondence between motives and compatible systems of l–adic representations. Nowadays these conjectures essentially determine key directions in arithmetical research ([CR01], [Tay02], [BoCa79], [Bor79], [Ta79]). We mention only a remarkable fact that in view of the proof of the theorem of G. Faltings (see §5.5) an Abelian variety is uniquely determined upto isogeny by the corresponding l–adic Galois representation on its Tate module.

This important result is cruicial also in Wiles' marvelous proof: in order to show that every semistable elliptic curve E over $\mathbb{Q}$ admits a modular parametrisation (see §7.2), it is enough (due to Faltings) to check that for some prime p the L–function of the Galois representation $\rho_{p,E}$ coinsides with the Mellin transform of a modular form of weight two (Wiles has used $p = 3$ and $p = 5$). In other words, the generating series of such a representation, defined starting from the traces of Frobenius elements, is a modular form of weight two which is proved by counting all possible deformations of the Galois representation in question taken modulo p.

6.3 Modular Forms and Euler Products

6.3.1 A Link Between Algebraic Varieties and L–Functions

There is one more method of constructing L-functions, which is connected with modular forms (or more generally with automorphic forms). These forms may be regarded as certain special functions on real reductive groups $G(\mathbb{R})$ (or on the symmetric spaces associated with them). These functions, which at first sight seem to be analytic rather than number theoretical objects, turn out to be closely related to a) Diophantine equations (arithmetic schemes, see §6.1), and b) Galois representations. A link between the three types of object is given by identifying the corresponding L-functions. A non–trivial example of the link between a) and b) is given by the proof of the theorem of Faltings: the L-function $L_1(A,s)$ attached to $H^1(A)$ of an Abelian variety A uniquely determines A upto isogeny. It is even sufficient to know a finite number of the local factors (see §5.5)

$$L_{1,v}(A,s) = \det(1 - \mathrm{N}\mathfrak{p}_v^{-s} Fr_v | T_l(A))^{-1} \quad (l \nmid \mathrm{Char}\ v).$$

Therefore the finiteness problems for Abelian varieties upto isogeny can be reread in terms of the corresponding Galois representations.

A characteristic feature of the modern theory of L-functions is the study of automorphic forms together with the (infinite dimensional) representations of the groups $G(\mathbb{R})$ and $G(\mathbb{A})$ generated by these forms, where $\mathbb{A}$ denotes the adele ring of $\mathbb{Q}$ and it is assumed that G is a reductive algebraic group defined over $\mathbb{Q}$. These representations (automorphic representations) occur in the corresponding regular representations, i.e. in vector spaces of smooth (or square integrable) functions over these groups (with respect to Haar measure). This approach makes it possible to study the L-functions using methods from the representation theory of the groups $G(\mathbb{Q}_p)$, $G(\mathbb{R})$ and $G(\mathbb{A})$ (cf. [Bor79] and §6.5).

6.3.2 Classical modular forms

are introduced as certain holomorphic functions on the upper half plane $\mathbb{H} = \{z \in \mathbb{C} \mid \mathrm{Im}\ z > 0\}$, which can be regarded as a homogeneous space for the group $G(\mathbb{R}) = \mathrm{GL}_2(\mathbb{R})$:

$$\mathbb{H} = \mathrm{GL}_2(\mathbb{R})/\mathcal{O}(2) \cdot Z, \tag{6.3.1}$$

where $Z = \{\left(\begin{smallmatrix} x & 0 \\ 0 & x \end{smallmatrix}\right) | x \in \mathbb{R}^\times\}$ is the center of $G(\mathbb{R})$ and $\mathcal{O}(2)$ is the orthogonal group. The group $\mathrm{GL}_2^+(\mathbb{R})$ of matrices $\gamma = \left(\begin{smallmatrix} a_\gamma & b_\gamma \\ c_\gamma & d_\gamma \end{smallmatrix}\right)$ with positive determinant acts on $\mathbb{H}$ by fractional linear transformations; on cosets (6.3.1) this action transforms into the natural action by group shifts.

Let Γ be a subgroup of finite index in the modular group $\mathrm{SL}_2(\mathbb{Z})$. A holomorphic function $f : \mathbb{H} \to \mathbb{C}$ is called a modular form of (integral) weight k with respect to Γ iff the conditions a) and b) are satisfied:

a) *Automorphy condition*

$$f((a_\gamma z + b_\gamma)/(c_\gamma z + d_\gamma)) = (c_\gamma z + d_\gamma)^k f(z) \tag{6.3.2}$$

for all elements $\gamma \in \Gamma$;

b) *Regularity at cusps*: f is regular at cusps $z \in \mathbb{Q} \cup i\infty$ (the cusps can be viewed as fixed points of parabolic elements of Γ); this means that for each element $\sigma = \begin{pmatrix} a & b \\ c & d \end{pmatrix} \in \mathrm{SL}_2(\mathbb{Z})$ the function $(cz+d)^{-k} f\left(\frac{az+b}{cz+d}\right)$ admits a Fourier expansion over non–negative powers of $q^{1/N} = e(z/N)$ for a natural number N. One writes traditionally

$$q = e(z) = \exp(2\pi i z).$$

A modular form

$$f(z) = \sum_{n=0}^{\infty} a(n) e(nz/N)$$

is called a cusp form if f vanishes at all cusps (i.e. if the above Fourier expansion contains only positive powers of $q^{1/N}$), see [La76], [Mi89], [Ogg65], [Fom77], [Pan81] .

The complex vector space of all modular (resp. cusp) forms of weight k with respect to Γ is denoted by $\mathcal{M}_k(\Gamma)$ (resp. $\mathcal{S}_k(\Gamma)$).

A basic fact from the theory of modular forms is that the spaces of modular forms are finite dimensional. Also, one has $\mathcal{M}_k(\Gamma)\mathcal{M}_l(\Gamma) \subset \mathcal{M}_{k+l}(\Gamma)$. The direct sum

$$\mathcal{M}(\Gamma) = \bigoplus_{k=0}^{\infty} \mathcal{M}_k(\Gamma)$$

turns out to be a graded algebra over $\mathbb{C}$ with a finite number of generators.

An example of a modular form with respect to $\mathrm{SL}_2(\mathbb{Z})$ of weight $k \geq 4$ is given by the *Eisenstein series*

$$G_k(z) = \sum_{m_1, m_2 \in \mathbb{Z}}{}' (m_1 + m_2 z)^{-k} \tag{6.3.3}$$

(prime denoting $(m_1, m_2) \neq (0,0)$). For these series the automorphy condition (6.3.2) can be deduced straight from the definition. One has $G_k(z) \equiv 0$ for odd k and

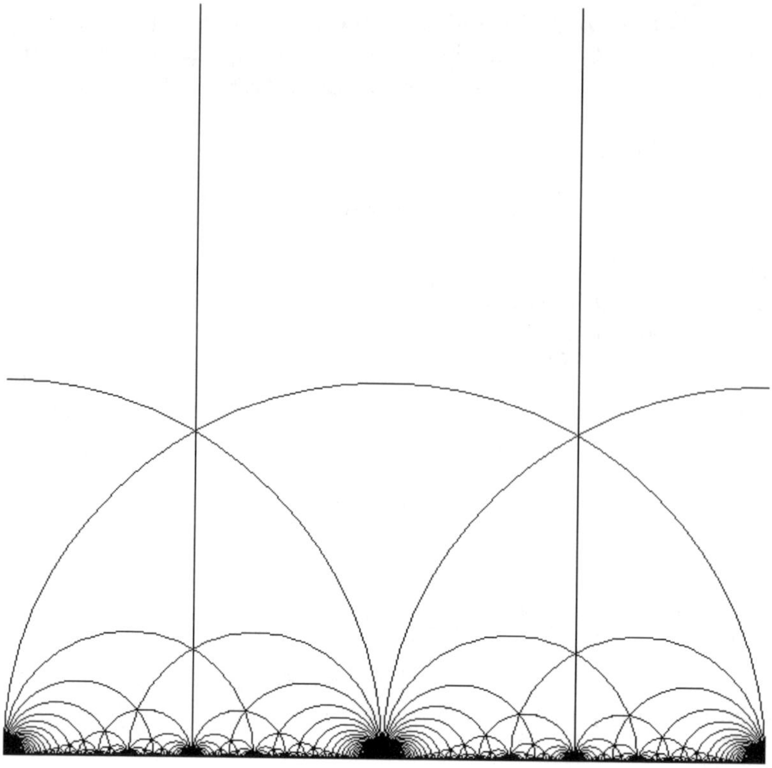

Fig. 6.1. The group $\mathrm{SL}(2,\mathbb{Z})$.
Graphics in Figure 6.1 is contributed by Curtis T. McMullen. It represents the action of the group $\mathrm{SL}(2,\mathbb{Z})$ on $\mathbb{H}$.

$$G_k(z) = \frac{2(2\pi i)^k}{(k-1)!}\left[-\frac{B_k}{2k} + \sum_{n=1}^{\infty} \sigma_{k-1}(n)e(nz)\right], \tag{6.3.4}$$

where $\sigma_{k-1}(n) = \sum_{d|n} d^{k-1}$ and B_k is the kth *Bernoulli number.*

The graded algebra $\mathcal{M}(\mathrm{SL}_2(\mathbb{Z}))$ is isomorphic to the polynomial ring of the (independent) variables G_4 and G_6.

The set $\mathbb{H}/\mathrm{SL}_2(\mathbb{Z})$ can be identified with the set of isomorphism classes of elliptic curves over $\mathbb{C}$: to each $z \in \mathbb{H}$ one associates the complex torus $\mathbb{C}/(\mathbb{Z} + z\mathbb{Z})$ which is analytically isomorphic to the Riemann surface of the elliptic curve written in Weierstrass form as follows

$$y^2 = 4x^3 - g_2(z)x - g_3(z) \tag{6.3.5}$$

where $g_2 = 60G_4(z)$, $g_3(z) = 140G_6(z)$.

If we replace z by $\gamma(z)$ for $\gamma = \left(\begin{smallmatrix} a_\gamma & b_\gamma \\ c_\gamma & d_\gamma \end{smallmatrix}\right) \in \mathrm{SL}_2(\mathbb{Z})$ then the lattice $\Lambda_z = \mathbb{Z} + z\mathbb{Z}$ is replaced by

$$\Lambda_{\gamma(z)} = \mathbb{Z} + \gamma(z)\mathbb{Z} = (cz+d)^{-1}(\mathbb{Z} + z\mathbb{Z}) = (cz+d)^{-1}\Lambda_z,$$

and the curve (6.3.5) is then replaced by the curve whose Weierstrass form has the coefficients

$$g_2(\gamma(z)) = (cz+d)^4 g_2(z), \quad g_3(\gamma(z)) = (cz+d)^6 g_3(z).$$

The discriminant of the cubic polynomial in the right hand side of (6.3.5) is a cusp form of weight 12 with respect to $\Gamma = \mathrm{SL}_2(\mathbb{Z})$:

$$2^{-4}(g_2^3 - 27g_3^2) = \tag{6.3.6}$$
$$2^{-4}(2\pi)^{12} e(z) \prod_{m=1}^{\infty} (1 - e(mz))^{24} = 2^{-4}(2\pi)^{12} \sum_{n=1}^{\infty} \tau(nz) e(nz),$$

where $\tau(n)$ is the Ramanujan function. The function

$$j(z) = 1728 \frac{g_2^3}{g_2^3 - 27g_3^2} = \frac{1}{q} + 744 + \sum_{n=1}^{\infty} c(nz) q^n \tag{6.3.7}$$

is meromorphic on $\mathbb{H}$ and at ∞, and is invariant under $\Gamma = \mathrm{SL}_2(\mathbb{Z})$. This function provides an important example of a *modular function*; it is called the *modular invariant.*

6.3.3 Application: Tate Curve and Semistable Elliptic Curves

(see [Ta74], [He97], p.343, [Se72], p.276.). Let us use the modified Weierstrass equation of §5.3.3 in the following form (with coefficients in $\mathbb{Z}[[q]]$):

$$Tate(q) : y^2 + xy = x^3 + B(q)x + C(q), \tag{6.3.8}$$

where

$$B(q) = -5\left(\frac{E_4 - 1}{240}\right) = -5\sum_{n=1}^{\infty} \sigma_3(n) q^n = -5 \sum_{n=1}^{\infty} \frac{n^3 q^n}{1 - q^n}, \tag{6.3.9}$$

$$C(q) = \frac{-5\left(\dfrac{E_4 - 1}{240}\right) - 7\left(\dfrac{E_6 - 1}{-504}\right)}{12} = -\frac{1}{12}\sum_{n=1}^{\infty} \frac{(7n^5 + 5n^3)q^n}{1 - q^n}.$$

This equation defines an elliptic curve over the ring $\mathbb{Z}((q))$ with the canonical differential ω_{can} given by

$$\frac{dx}{2y + x} = \frac{dX}{Y},$$

the variables x, y defined via the substitution

$$X = x + \frac{1}{12}, Y = x + 2y$$

as in §5.3.3.

Next let us take any p-adic number $q \in \mathbb{Q}_p^\times$ such that $|q|_p < 1$. Then for any $t \in \mathbb{Q}_p^\times/\langle q\rangle$ let us consider the following series

$$\rho(t) = \sum_{n\in\mathbb{Z}}^{\infty} \frac{q^n t}{(1-q^n t)^2} \tag{6.3.10}$$

and one obtains a Tate curve E_q over $\mathbb{Q}_p$ with the equation (6.3.8) in which

$$B(q) = -5\sum_{n=1}^{\infty} \frac{n^3 q^n}{1-q^n} \in \mathbb{Z}_p, \tag{6.3.11}$$

$$C(q) = -\frac{1}{12}\sum_{n=1}^{\infty} \frac{(7n^5+5n^3)q^n}{1-q^n} \in \mathbb{Z}_p.$$

Theorem 6.12 (Tate). *There is a $\mathbb{Q}_p$–analytic isomorphism*

$$\mathbb{Q}_p^\times/\langle q\rangle \xrightarrow{\sim} E_q(\mathbb{Q}_p), \quad t \mapsto (x(t), y(t))$$

where

$$\begin{cases} x(t) = \sum_{n\in\mathbb{Z}} \frac{q^n t}{(1-q^n t)^2} - 2\sum_{n=1}^{\infty} \frac{nq^n}{1-q^n t} \\ y(t) = \sum_{n\in\mathbb{Z}} \frac{(q^n t)^2}{(1-q^n t)^3} + \sum_{n=1}^{\infty} \frac{nq^n}{1-q^n t}. \end{cases}$$

Moreover for any semistable elliptic curve E over $\mathbb{Q}_p$ there exists a p-adic number $q \in \mathbb{Q}_p^\times$ such that $|q|_p < 1$, and E is isomorphic to E_q (over an unramified quadratic extention $\mathbb{Q}_p(\sqrt{E_4(q)})$.

Let $N \geq 1$ be a natural number. Let us define

$$Tate(q^N) : y^2 + xy = x^3 + B(q^N)x + C(q^N).$$

Next we put $t = \exp(2\pi i u)$, then the points of order N on $Tate(q^N)$ correspond to $t = \zeta_N^i q^j$, $(0 \leq i, j \leq N-1)$, $\zeta_N = \exp(2\pi i/N)$, and their coordonates are given by

$$\begin{cases} x(t) = \sum_{n\in\mathbb{Z}} \frac{q^{Nn} t}{(1-q^{Nn} t)^2} - 2\sum_{n=1}^{\infty} \frac{nq^{Nn}}{1-q^{Nn} t} \\ y(t) = \sum_{n\in\mathbb{Z}} \frac{(q^{Nn} t)^2}{(1-q^{Nn} t)^3} + \sum_{n=1}^{\infty} \frac{nq^{Nn}}{1-q^{Nn} t}. \end{cases}$$

It is important for arithmetical applications for the Tate curve that these coordonates belong to the ring $\mathbb{Z}[\zeta_N, N^{-1}][[q]]$.

Proof

uses the identity

$$\sum_{n\in\mathbb{Z}}(u+n)^{-k} = \frac{(2\pi i)^k}{(k-1)!}\sum_{n=1}^{\infty} n^{k-1}e^{2\pi inu} \quad (k \geq 2, \quad u \in \mathbb{Z}).$$

We have for the lattice $\Lambda = 2\pi i(\mathbb{Z}+\tau\mathbb{Z})$ the following equalities

$$X = \wp(2\pi iu) =$$

$$(2\pi i)^{-2}\left(u^{-2} + \sum_{m,n\in\mathbb{Z}}{}' \left((u+m\tau+n)^{-2} - (m\tau+n)^{-2}\right)\right) = \tag{6.3.12}$$

$$(2\pi i)^{-2}\left(\sum_{m\in\mathbb{Z}}\sum_{n\in\mathbb{Z}}(u+m\tau+n)^{-2} - 2\sum_{m=1}^{\infty}\sum_{n\in\mathbb{Z}}(m\tau+n)^{-2} - 2\zeta(2)\right) =$$

$$\sum_{m\in\mathbb{Z}}\sum_{n=1}^{\infty} ne^{2\pi i(u+m\tau)n} - 2\sum_{m=1}^{\infty}\sum_{n=1}^{\infty} ne^{2\pi imn\tau} + \frac{1}{12},$$

implying the above identities.

6.3.4 Analytic families of elliptic curves and congruence subgroups

For an integer N the following *congruence subgroups* are defined:

$$\Gamma_0(N) = \{\gamma \in \mathrm{SL}_2(\mathbb{Z}) \mid c_\gamma \equiv 0 \mod N\}, \tag{6.3.13}$$

$$\Gamma_1(N) = \{\gamma \in \Gamma_0(N) \mid a_\gamma \equiv d_\gamma \equiv 1 \mod N\}, \tag{6.3.14}$$

$$\Gamma(N) = \{\gamma \in \mathrm{SL}_2(\mathbb{Z}) \mid \gamma \equiv 1 \mod N\}. \tag{6.3.15}$$

More generally, a subgroup $\Gamma \subset \mathrm{SL}_2(\mathbb{Z})$ is called a congruence subgroup iff $\Gamma \supset \Gamma(N)$ for some N. Consider fundamental domains in $\mathbb{H}$ for the actions of the above congruence subgroups: (a) $\mathbb{H}/\Gamma_0(N)$, (b) $\mathbb{H}/\Gamma_1(N)$, (c) $\mathbb{H}/\Gamma(N)$. These domains can be identified respectively with the sets of isomorphism classes over $\mathbb{C}$ of the following objects:

(a) $(E, \langle P\rangle)$, an elliptic curve over $\mathbb{C}$ together with a cyclic subgroup of order N, $\langle P\rangle \subset E(\mathbb{C}), \mathrm{Card}\langle P\rangle = N$;
(b) (E, P), an elliptic curve over $\mathbb{C}$ together with a point $P \in E(\mathbb{C})$ of order N, $\mathrm{Card}\langle P\rangle = N$;

(c) $(E, \phi) = (E, P, Q)$, an elliptic curve over $\mathbb{C}$ together with an isomorphism

$$\phi : \mathbb{Z}/N\mathbb{Z} \oplus \mathbb{Z}/N\mathbb{Z} \xrightarrow{\sim} E(\mathbb{C})_N,$$

such that

$$\phi^* e_N = \exp(2\pi i \det/N),$$

where e_N is the Weil pairing (5.3.28) thus ϕ gives a basis of the points of order N:

$$P, Q \in E(\mathbb{C})_N = \langle P \rangle \oplus \langle Q \rangle \cong \mathbb{Z}/N\mathbb{Z} \oplus \mathbb{Z}/N\mathbb{Z}.$$

In order to describe this identification one associates to a point $z \in \mathbb{H}$ the following objects:

(a) $(\mathbb{C}/\Lambda_z, \langle 1/N \rangle \bmod \Lambda_z \rangle)$;
(b) $(\mathbb{C}/\Lambda_z, 1/N \bmod \Lambda_z)$;
(c) $(\mathbb{C}/\Lambda_z, 1/N \bmod \Lambda_z, z/N \bmod \Lambda_z)$,

$$1/N \bmod \Lambda_z \mapsto (1,0) \bmod N, z/N \bmod \Lambda_z \mapsto (0,1) \bmod N$$

6.3.5 Modular forms for congruence subgroups

For the study of modular forms it is convenient to use the traditional notation

$$(f|_k\gamma)(z) = \det\gamma^{k/2} f(\gamma(z))(c_\gamma z + d_\gamma)^{-k} \tag{6.3.16}$$

for the weight k action of an element $\gamma = \begin{pmatrix} a_\gamma & b_\gamma \\ c_\gamma & d_\gamma \end{pmatrix} \in \mathrm{GL}_2^+(\mathbb{R})$ with positive determinant $\det\gamma > 0$.

Let ψ be a Dirichlet character mod N. Put

$$\mathcal{M}_k(N, \psi) = \{f \in \mathcal{M}_K(\Gamma_1(N)) \mid f|_k\gamma = \psi(d_\gamma)f \text{ for all } \gamma \in \Gamma_0(N)\}, \tag{6.3.17}$$

$$\mathcal{S}_k(N, \psi) = \mathcal{M}_k(N, \psi) \cap \mathcal{S}_k(\Gamma_1(N)). \tag{6.3.18}$$

One then has the following decomposition

$$\mathcal{M}_k(\Gamma_1(N)) = \bigoplus_{\psi \bmod N} \mathcal{M}_k(N, \psi), \quad \mathcal{S}_k(\Gamma_1(N)) = \bigoplus_{\psi \bmod N} \mathcal{S}_k(N, \psi).$$

For a modular form $h \in \mathcal{M}_k(N, \psi)$ the Petersson inner product of h with $f \in \mathcal{S}_k(N, \psi)$ is defined by the formula

$$\langle f, h \rangle_N = \int_{\mathbb{H}/\Gamma_0(N)} \overline{f(z)} h(z) y^{k-2}\, dx\, dy, \tag{6.3.19}$$

where $z = x + iy$, $\mathbb{H}/\Gamma_0(N)$ is a fixed fundamental domain for $\mathbb{H}$ modulo $\Gamma_0(N)$. Then one has the following orthogonal decomposition:

$$\mathcal{M}_k(N, \psi) = \mathcal{S}_k(N, \psi) \oplus \mathcal{E}_k(N, \psi), \tag{6.3.20}$$

where $\mathcal{E}_k(N, \psi)$ is the subspace of Eisenstein series, whose basis can be explicitly described and consists of Fourier series of type (6.3.3) ([La76], [He59], [Shi71]), [Mi89].

The arithmetical significance of modular forms is well illustrated by the example of the theta series. Let

$$Q(X) = \frac{1}{2}{}^t XAX$$

be a positive definite, integral quadratic form in an even number $2k$ of variables, with an even matrix $A = (a_{ij})$, $a_{ij} \in \mathbb{Z}$, $a_{ii} \in 2\mathbb{Z}$, $k \geq 2$, where $X = {}^t(x_1, x_2, \ldots, x_{2k})$ is an integral column vector. Let us associate to $Q(X)$ the following theta series

$$\theta(z; Q) = \sum_{M \in \mathbb{Z}^{2k}} e(Q(M)z) = \sum_{n=0}^{\infty} a(n)e(nz), \tag{6.3.21}$$

where

$${}^t(m_1, m_2, \ldots, m_{2k}) = M,$$

$$a(n) = a(n; Q) = \mathrm{Card}\{M \in \mathbb{Z}^{2k} \mid Q(M) = n\} \tag{6.3.22}$$

is the *number of representations* of n by the integral quadratic form Q with matrix A.

Let N be the level of Q, i.e. the smallest positive integer N such that NA^{-1} is an even integral quadratic form. It turns out that $\theta(z; Q) \in \mathcal{M}_k(N, \varepsilon_\Delta)$, where $\varepsilon_\Delta(d) = \left(\frac{\Delta}{d}\right)$ is the quadratic character attached to the discriminant Δ of the form Q, cf. [Ogg65], [Kog71].

The theory of modular forms makes it possible to obtain good estimates, and sometimes even explicit expressions, for the numbers $a(n; Q)$. In order to do this the theta function (6.3.21) is represented as a sum of an Eisenstein series

$$E_k(z; Q) = \sum_{n=0}^{\infty} \rho_k(n; Q)e(nz) \in \mathcal{E}_k(N, \varepsilon_\Delta) \tag{6.3.23}$$

and a cusp form

$$S_k(z; Q) = \sum_{n=1}^{\infty} b_k(n; Q)e(nz) \in \mathcal{S}_k(N, \varepsilon_\Delta). \tag{6.3.24}$$

The coefficients $\rho_k(n;Q)$ are elementary arithmetical functions such as $\sigma_{k-1}(n)$. For the coefficients of cusp forms one has the famous estimate

$$b(n) = O\left(n^{\frac{k-1}{2}+\varepsilon}\right), \quad \varepsilon > 0 \tag{6.3.25}$$

which was known as the *Petersson–Ramanujan conjecture* before being proved by Deligne (cf. [Del74]), who first reduced this conjecture to the Weil conjecture (see §6.1.3) in [Del68], and then proved all these conjectures, using Grothendieck's étale l–adic cohomology.

In particular for the Ramanujan function $\tau(n)$ Deligne's estimate takes the form

$$\tau(p) < 2p^{11/2} \quad (p \text{ prime numbers}).$$

Applying the estimate (6.3.25) to the series (6.3.21) gives

$$a(n;Q) = \rho_k(n;Q) + O\left(n^{\frac{k-1}{2}+\varepsilon}\right), \quad \varepsilon > 0 \tag{6.3.26}$$

The proof of (6.3.25) is based on a geometric interpretation of a cusp form $f \in \mathcal{S}(\Gamma)$ of even weight k as a multiple differential: for $\gamma \in \Gamma$ the expression

$$f(z)\,(dz)^{k/2}$$

does not change if we replace z by $\gamma(z)$, and can therefore be defined over the modular curve $X_\Gamma = \overline{\mathbb{H}/\Gamma}$, i.e. on a projective algebraic curve, whose Riemann surface is compact and is obtained by adding to $\mathbb{H}/\Gamma$ a finite number of cusps. In particular, for $k = 2$ the expression $f(z)\,dz$ represents a holomorphic differential on X_Γ, and the estimate (6.3.25) in this case was first established by [Eich54] (the *congruence relation* of Eichler–Shimura, see a detailed exposition in [Shi71] and in Röhrlich's papers in [CSS95]).

Many interesting examples of formulae for the numbers $a(n;Q)$ can be found in the book of L.A.Kogan, [Kog71], and in [An65], [Lom78], [He59] and elsewhere.

6.3.6 Hecke Theory

In the examples of Eisenstein series and theta functions one notices an interesting fact: the *Fourier coefficients* $a(n)$ of these modular forms turn out to be either multiplicative arithmetical functions or linear combinations of such functions. For the Ramanujan function $\tau(n)$ (6.3.6) these multiplicativity properties have the following form

$$\begin{aligned}&\tau(mn) = \tau(n)\tau(m) \text{ for } (m,n) = 1,\\ &\tau(p^r) = \tau(p)\tau(p^{r-1}) - p^{11}\tau(p^{r-2}) \ (p \text{ a prime number}, r \geq 2)\end{aligned} \tag{6.3.27}$$

and it seems that not even these relations can be established using only elementary methods. They might provide an example for the theorem of Gödel (see Chapter 3) (cf. [He59], [An74], [La76]).

Let m be a positive integer, $f(z) = \sum_{n=0}^{\infty} a(n)e(nz)$ a function on $\mathbb{H}$. Then the following functions are defined

$$\begin{aligned} f|U(m)(z) &= \sum_{n=0}^{\infty} a(mn)e(nz) = m^{k/2-1} \sum_{u \bmod m} f|_k \begin{pmatrix} 1 & u \\ 0 & m \end{pmatrix}, \\ f|V(m)(z) &= \sum_{n=0}^{\infty} a(n)e(mnz) = f(mz) = m^{-k/2} f|_k \begin{pmatrix} m & 0 \\ 0 & 1 \end{pmatrix}. \end{aligned} \tag{6.3.28}$$

Imagine that the operator

$$f \mapsto f|U(m)$$

acts on the space of modular forms $\mathcal{M}_k(N, \psi)$. Then one might expect to find a basis consisting of its eigenfunctions. Assuming f to be an eigenfunction would then imply the relations

$$a(mn) = \lambda(m)a(n) \quad (n \in \mathbb{N})$$

where $\lambda(m)$ are the corresponding eigenvalues:

$$f|U(m) = \lambda(m)f.$$

The desired multiplicativity property would then follow. However, if $f \in \mathcal{M}_k(N, \psi)$ then in general one can only state that

$$f|U(m)(z), f|V(m)(z) \in \mathcal{M}_k(mN, \psi), \tag{6.3.29}$$

and

$$f|U(m)(z) \in \mathcal{M}_k(N, \psi)$$

holds only when m divides N. In order to overcome with this difficulty in the general case note that the matrices $\left(\begin{smallmatrix} 1 & u \\ 0 & m \end{smallmatrix}\right)$ in the definition (6.3.28) of $U(m)$ form part of a complete system of right coset representatives for $\Gamma_0(N)\backslash\Delta_m(N)$, where $\Delta_m(N)$ denotes the set

$$\Delta_m(N) = \left\{\gamma = \begin{pmatrix} a & b \\ c & d \end{pmatrix} \mid a, b, c, d \in \mathbb{Z}, c \equiv 0 \bmod N, \det\gamma = m\right\},$$

which is invariant under right multiplication by elements of $\Gamma_0(N)$. As a complete system of right coset representatives for $\Gamma_0(N)\backslash\Delta_m(N)$ one could take the set

$$\left\{\begin{pmatrix} a & b \\ 0 & d \end{pmatrix} \mid a, d > 0, ad = m, b = 0, \ldots, d-1\right\}. \tag{6.3.30}$$

This fact makes it possible to define instead of $U(m)$ another operator which does act on the space of modular forms $\mathcal{M}_k(N,\psi)$. This other operator is called the *Hecke operator* $T(m)$:

$$f \mapsto f|_k T(m) = m^{k/2-1} \sum_{\sigma} \psi(a_\sigma) f|_k \sigma, \tag{6.3.31}$$

where $\sigma = \begin{pmatrix} a_\sigma & b_\sigma \\ c_\sigma & d_\sigma \end{pmatrix} \in \Gamma_0(N)\backslash \Delta_m(N)$, $(m,N)=1$.

The action of $T(m)$ on the Fourier coefficients is easily calculated using the systems of representatives (6.3.30):

$$\begin{aligned} f|_k T(m) = &\sum_{m_1|m} \psi(m_1) m_1^{k-1} f|U(m/m_1)V(m_1) \\ =&a(0)\sum_{m_1|m} \psi(m_1)m_1^{k-1} \\ &+\sum_{n=1}^{\infty} \sum_{m_1|(m,n)} \psi(m_1)m_1^{k-1} a(mn/m_1^2)e(nz), \end{aligned} \tag{6.3.32}$$

where it is implicitly assumed that $a(x)=0$ for $x \notin \mathbb{Z}$.

Multiplying the systems (6.3.30) together shows that the multiplication rule for the operators $T_k(m)$ is as follows:

$$T_k(m)T_k(n) = \sum_{m_1|(m,n)} \psi(m_1)m_1^{k-1} T_k(mn/m_1^2). \tag{6.3.33}$$

In particular all the operators $T_k(n)$ commute. If $f \in \mathcal{M}_k(N,\psi)$ is an eigenfunction for all $T_k(m)$ with $(m,N)=1$, i.e. if

$$f|T_k(m) = \lambda_h(m) f \quad ((m,N)=1), \tag{6.3.34}$$

then (6.3.33) implies that

$$\lambda_f(m)\lambda_f(n) = \sum_{m_1|(m,n)} \psi(m_1)m_1^{k-1}\lambda_f(mn/m_1^2).$$

Equating the Fourier coefficients $a(n)$ in (6.3.34) one obtains the following equalities

$$\begin{aligned} a(0)\sum_{m_1|m} \psi(m_1)m_1^{k-1} &= \lambda_f(m)a(0), \\ \sum_{m_1|(m,n)} \psi(m_1)m_1^{k-1} a(mn/m_1^2) &= \lambda_f(m)a(n). \end{aligned} \tag{6.3.35}$$

In particular for $n=1$ one has

$$a(m) = \lambda_f(m)a(1), \tag{6.3.36}$$

and for $a(1) \neq 0$ the function $a(m)$ is therefore proportional to the function $\lambda(m)$ for $(m, N) = 1$.

All these properties can be especially neatly expressed in terms of Dirichlet series: for the function

$$f(z) = \sum_{n=0}^{\infty} a(n)e(nz) \in \mathcal{M}_k(N, \psi),$$

satisfying (6.3.34) put formally

$$L_N(s, f) = \sum_{\substack{n=1 \\ (N,n)=1}}^{\infty} \lambda_f(n)n^{-s}, \quad R_N(s, f) = \sum_{\substack{n=1 \\ (N,n)=1}}^{\infty} a(n)n^{-s}. \tag{6.3.37}$$

Then these (formal) Dirichlet series satisfy the following identities:

I) The Euler product expansion:

$$L_N(s, f) = \prod_{p\,:\,p \nmid N} [1 - \lambda_f(p)p^{-s} + \psi(p)p^{k-1-2s}]^{-1}. \tag{6.3.38}$$

II) $R_N(s, f) = a(1)L_N(s, f)$.

Indeed the multiplication rule (6.3.33) for distinct primes p_i, $p_i \nmid N$ implies that one has

$$L_N(s, f) = \prod_{p\,:\,p \nmid N} \left(\sum_{\delta=0}^{\infty} \lambda_f(p^\delta)p^{-\delta s} \right),$$

and each of the series can be summed over δ using the relation

$$\lambda_f(p)\lambda_f(p^\delta) = \lambda_f(p^{\delta+1}) + \psi(p)\lambda_f(p^{\delta-1}) \ (\delta \geq 1). \tag{6.3.39}$$

Equation II) follows then directly from (6.3.36).

Convergence of the series (6.3.37) for $\mathrm{Re}(s) \gg 0$ follows from the following estimates for the coefficients

a) If $f \in \mathcal{M}_k(N, \psi)$ then

$$|a(n)| = O(n^{k-1+\varepsilon}), \quad \varepsilon > 0 \tag{6.3.40}$$

and the Dirichlet series (6.3.37) converges absolutely for $\mathrm{Re}\ s > k$.

b) If $f \in \mathcal{S}_k(N, \psi)$ then

$$|a(n)| = O(n^{\frac{k-1}{2}+\varepsilon}), \quad \varepsilon > 0 \tag{6.3.41}$$

and the series $L_N(s, f)$ and $R_N(s, f)$ converge absolutely for $\mathrm{Re}\ s > \frac{(k+1)}{2}$.

The estimates (6.3.40) and especially (6.3.41) use some fine arithmetical properties of the coefficients $a(n)$. However, using only analytic properties of $f(z)$ (the fact that it is holomorphic and the automorphic condition (6.3.2)) one can easily obtain rougher estimates:

a)
$$|a(n)| = O(n^k), \ \text{for } f \in \mathcal{M}_k(N, \psi);$$

b)
$$|a(n)| = O(n^{k/2}), \ \text{for } f \in \mathcal{S}_k(N, \psi);$$

the latter estimate is implied by the estimate $|f(z)| = O(y^{k/2})$ $(y \to 0, z = x + iy)$.

A basis consisting of eigenfunctions for Hecke operators can be found using the Petersson inner product (6.3.19). One verifies that the operators $T_k(m)$ on $\mathcal{S}_k(N, \psi)$ are normal with respect to this inner product for $(m, N) = 1$. Moreover, the operators are ψ–Hermitian: for all $f, h \in \mathcal{S}_k(N, \psi)$ the following equation holds:

$$\psi(m)\langle f|T_k(m), h\rangle_N = \langle f, h|T_k(m)\rangle_N. \tag{6.3.42}$$

By a general theorem of linear algebra on families of commuting normal operators, there is an orthogonal basis of $\mathcal{S}_k(N, \psi)$ consisting of common eigenfunctions of all the $T_k(m)$, $((m, N) = 1)$. A basis with this property is called a *Hecke basis.* In the case that the number m is divisible only by prime divisors of the level N, one may use the operator $U(m)$ instead of $T_k(m)$. As was mentioned above (see (6.3.28)) these operators leave $\mathcal{M}_k(N, \psi)$ invariant. However, they are not normal and in general are not diagonalizable in $\mathcal{S}_k(N, \psi)$.

The Mellin transform of a modular form and its analytic continuation. Let

$$f(z) = \sum_{n=0}^{\infty} a(n)e(nz) \in \mathcal{M}_k(N, \psi).$$

Then the Dirichlet series

$$R(s, f) = R_1(s, f) = \sum_{n=1}^{\infty} a(n)n^{-s}$$

which converges absolutely for Re $s \gg 0$ can be analytically continued to the entire complex plane using the Mellin transform of the modular form f:

$$(2\pi)^{-s}\Gamma(s)R(s, f) = \int_0^{\infty} [f(iy) - a(0)]y^{s-1}\, dy \ \ (\mathrm{Re}(s) \gg 0). \tag{6.3.43}$$

This can be seen by integrating termwise the series

$$f(iy) - a(0) = \sum_{n=1}^{\infty} a(n) \exp(-2\pi n y)$$

and using the integral representation of the gamma function:

$$\Gamma(s) = \int_0^{\infty} e^{-y} y^{s-1}\, dy \quad (\mathrm{Re}(s) > 0).$$

The vector space of all Dirichlet series of type $R(s, f)$ for $f \in \mathcal{M}_k(N, \psi)$ can be characterized by analytic properties of these series. Following [An74], we give this characterization in the case $N = 1$.

Theorem A. *Let $f \in \mathcal{M}_k = \mathcal{M}_k(\mathrm{SL}_2(\mathbb{Z}))$. Then the Dirichlet series $R(s, f)$ admits a meromorphic continuation to the entire complex plane, and if one puts*

$$\Lambda(s, f) = (2\pi)^{-s}\Gamma(s)R(s, f),$$

then the function

$$\Lambda(s, f) + \frac{a(0)}{s} + \frac{(-1)^{k/2}a(0)}{k - s} \tag{6.3.44}$$

is entire. The following functional equation holds

$$\Lambda(k - s, f) = (-1)^{k/2}\Lambda(s, f). \tag{6.3.45}$$

Theorem B. *Every Dirichlet series $R(s) = \sum_{n=1}^{\infty} a(n)n^{-s}$ whose coefficients $a(n)$ have not more than a polynomial order of growth, and which satisfies (6.3.44) and (6.3.45), has the form $R(s, f)$ for some modular form $f \in \mathcal{M}_k = \mathcal{M}_k(\mathrm{SL}_2(\mathbb{Z}))$.*

Indeed, in order to prove theorem A we use the Mellin transform (6.3.43) and write

$$\Lambda(s, f) = \int_0^{\infty} [f(iy) - a(0)]y^{s-1}\, dy \quad (\mathrm{Re}(s) > k + 1).$$

Taking into account that $f(-1/z) = z^k f(z)$ we see that

$$\begin{aligned}
\Lambda(s, f) &= \int_1^{\infty} [f(iy) - a(0)]y^{s-1}\, dy - \frac{a(0)}{s} + \int_0^1 f(iy)y^{s-1}\, dy \\
&= \int_1^{\infty} [f(iy) - a(0)]y^{s-1}\, dy - \frac{a(0)}{s} + \int_0^1 f(-1/iy)y^{-s-1}\, dy \\
&= \int_1^{\infty} [f(iy) - a(0)](y^{s-1} + i^k y^{k-s-1})\, dy - \frac{a(0)}{s} - \frac{i^k a(0)}{k - s}.
\end{aligned}$$

The function $f(iy) - a(0)$ tends to zero exponentially as $y \to \infty$, so the last integral converges absolutely for all s and turns out to be a holomorphic function of the variable s. This proves (6.3.44). By the substitution $s \mapsto k-s$ the last written expression for $\Lambda(s,f)$ is multiplied by i^k, and both the functional equation (6.3.45) and theorem A follow.

In order to prove theorem B it suffices to use the inverse Mellin transform, and the fact that the whole modular group $\mathrm{SL}_2(\mathbb{Z})$ is generated by the matrices $\left(\begin{smallmatrix}1&1\\0&1\end{smallmatrix}\right)$ and $\left(\begin{smallmatrix}0&-1\\1&0\end{smallmatrix}\right)$.

Theorems A and B can be extended to modular forms of integral weight for congruence subgroups of $\mathrm{SL}_2(\mathbb{Z})$ (with natural technical complications). The theorem generalizing Theorem A in this situation is called the direct theorem, and the generalization of Theorem B is called the inverse theorem (or converse theorem, cf. [CoPSh02]). This inverse theorem was stated by [Wei67], [Wei71] in terms of the twisted Dirichlet series

$$\Lambda^*(s,\chi) = (2\pi)^{-s}\Gamma(s)\sum_{n=1}^{\infty}\chi(n)a(n)n^{-s}, \tag{6.3.46}$$

where χ is an arbitrary Dirichlet character. Assume that the series

$$R(s) = \sum_{n=1}^{\infty} a(n)n^{-s}$$

converges absolutely for $s = k-\delta$, $\delta > 0$, and that for all $\chi \mod r$, $(r,N)=1$ the functions (6.3.46) are entire, bounded in every vertical strip and satisfy certain functional equations relating $\Lambda^*(s,\chi)$ to $\Lambda^*(k-s,\overline{\chi})$. Then one can deduce that the Fourier series

$$f(z) = \sum_{n=0}^{\infty} a(n)e(nz)$$

represents a modular form in $\mathcal{S}_k(N,\psi)$. In other words, the automorphy properties of this Fourier series can be deduced from functional equations of the corresponding Dirichlet series twisted by Dirichlet characters; the precise form of the functional equations for these series is given in [Wei67].

6.3.7 Primitive Forms

Atkin and Lehner have made an important complement to Hecke's theory by constructing a satisfactory theory of Hecke operators for all m including the divisors of the level. We begin with a simple example (following [La76], [Fom77], [Gel75]). Consider the vector space $\mathcal{S}_{12}(\Gamma_0(2))$ containing $f_1(z) = \Delta(z)$ and $f_2(z) = \Delta(2z)$. These two functions have the same eigenvalues for all Hecke operators $T_{12}(p)$ with $p \neq 2$. However they are linearly independent. A natural question then arises: which additional conditions must be imposed on $f(z) = \sum_{n=0}^{\infty} a(n)e(nz) \in \mathcal{S}_k(N,\psi)$ so that it is uniquely determined by its eigenvalues $\lambda_f(m)$ for $(m,N)=1$.

In order to find such conditions one constructs first the subspace of old forms $\mathcal{S}_k^{\mathrm{old}}(\Gamma_1(N)) \subset \mathcal{S}_k(\Gamma_1(N))$ as the sum of images of the operators

$$V(d') : \mathcal{S}_k(\Gamma_1(N/d)) \to \mathcal{S}_k(\Gamma_1(N))$$

(see (6.3.28)) for all divisors d of the level N, and for all divisors d' of d. Set

$$\mathcal{S}_k^{\mathrm{old}}(N,\psi) = \mathcal{S}_k(N,\psi) \cap \mathcal{S}_k^{\mathrm{old}}(\Gamma_1(N)).$$

Then the vector space of new forms of level precisely N is defined to be the orthogonal complement of the old forms:

$$\mathcal{S}_k(N,\psi) = \mathcal{S}_k^{\mathrm{new}}(N,\psi) \oplus \mathcal{S}_k^{\mathrm{old}}(N,\psi). \tag{6.3.47}$$

The main result of *Atkin – Lehner* theory is that if a function $f \in \mathcal{S}_k^{\mathrm{new}}(N,\psi)$ is an eigenfunction of all Hecke operators $T_k(m)$ with $(N,m)=1$, then f is uniquely determined (upto a multiplicative constant) by the eigenvalues and one can normalize f by the condition $a(1)=1$. A primitive form of conductor N is then defined as a normalized new eigenform $f \in \mathcal{S}_k^{\mathrm{new}}(N,\psi)$. For such forms f the condition $f|U(q) = a(q)f$ for $q|N$ is automatically satisfied. One has the following Euler product expansion:

$$\begin{aligned} L(s,f) &= \sum_{n=1}^{\infty} a(n)n^{-s} \\ &= \prod_{q|N}(1-a(q)q^{-s})^{-1}\prod_{p\nmid N}(1-a(p)p^{-s}+\psi(p)p^{k-1-2s})^{-1}, \end{aligned} \tag{6.3.48}$$

in which $|a(q)| = q^{(k-1)/2}$ if the character ψ can not be defined modulo the smaller level N/q, and if ψ is defined modulo N/q then $a(q)^2 = \psi(q)q^{k-1}$ provided $q^2 \nmid N$, and $a(q)=0$ otherwise (i.e. for $q^2|N$), cf. [Li75].

Let $f(z) = \sum_{n=0}^{\infty} a(n)e(nz) \in \mathcal{S}_k(N,\psi)$ be a primitive cusp form of conductor C_f, $C_f|N$. If we put

$$W(C_f) = \begin{pmatrix} 0 & -1 \\ C_f & 0 \end{pmatrix}, \quad f^\rho(z) = \overline{f(-\overline{z})} = \sum_{n=0}^{\infty} \overline{a(n)}e(nz) \in \mathcal{S}_k(N,\overline{\psi}),$$

then there is a complex number $\lambda(f)$ with $|\lambda(f)|=1$ such that

$$f|_kW(C_f) = \lambda(f)f^\rho. \tag{6.3.49}$$

Primitive cusp forms of a given conductor are characterized by the identity (6.3.49), which is equivalent to a certain nice functional equation for the corresponding Dirichlet series (cf. [Li75])

$$\begin{aligned} L(s,f) &= \prod_p (1-a(p)p^{-s}+\psi(p)p^{k-1-2s})^{-1} \\ &= \prod_p [(1-\alpha(p)p^{-s})(1-\alpha'(p)p^{-s})]^{-1}, \end{aligned}$$

where

$$\alpha(p)\alpha'(p) = \psi(p)p^{k-1}, \quad \alpha(p) + \alpha'(p) = a(p). \tag{6.3.50}$$

If we put

$$\Lambda(s, f) = (2\pi/\sqrt{C_f})^{-s}\Gamma(s)L(s, f),$$

then this functional equation has the form

$$\Lambda(k - s, f) = i^{k/2}\lambda(f)\Lambda(s, f^{\rho}). \tag{6.3.51}$$

For a primitive Dirichlet character χ whose conductor C_χ is coprime to C_f, the twisted modular form

$$f_\chi(z) = \sum_{n=0}^{\infty} \chi(n)a(n)e(nz) \in \mathcal{S}_k(C_f C_\chi^2, \psi) \tag{6.3.52}$$

is a primitive cusp form of conductor $C_f C_\chi^2$ (comp. with (6.3.51) and (6.3.45)).

6.3.8 Weil's Inverse Theorem

A converse statement concerning analytic properties of the series (6.3.52) was found by A.Weil [Wei67] giving necessary and sufficient conditions that a Fourier series $f(z) = \sum_{n=0}^{\infty} a(n)e(nz)$ represents a modular form in $M_k(N, \psi)$ in terms of the Dirichlet series

$$\Lambda(s, f, \chi) = (2\pi)^{-s}\Gamma(s)\sum_{n=1}^{\infty} \chi(n)a(n)n^{-s},$$

where χ is a Dirichlet series.

If $f \in M_k(N, \psi)$, and χ is primitive mod$m$ let us consider the twist

$$f_\chi(z) = \sum_{n=0}^{\infty} \chi(n)a(n)e(nz) = \chi(-1)f|\frac{G(\chi)}{m} \sum_{a \bmod m} \overline{\chi}(a)\begin{pmatrix} m & a \\ 0 & m \end{pmatrix},$$

where $G(\chi) = \sum_{a \bmod m} \chi(a)e(a/m)$ is the Gauss sum and one checks

$$f_\chi(z) \in \mathcal{S}_k(Nm^2, \psi\chi^2).$$

Moreover if $f|W_N = C_1 f^{\rho}$, then

$$f_\chi|W_{Nm^2} = C_\chi f_{\overline{\chi}}^{\rho} = C_\chi(f_\chi)|K, \tag{6.3.53}$$

where

$$C_\chi = C_1\frac{G(\chi)}{G(\overline{\chi})}\chi(-N)\psi(m), \tag{6.3.54}$$

$$g|W_N(z) = g|\begin{pmatrix} 0 & -1 \\ N & 0 \end{pmatrix}(z) = N^{k/2} z^k g(-1/Nz), \tag{6.3.55}$$

$$g^\rho(z) = g|K(z) = \overline{g(-\overline{z})} = \sum_{n=0}^{\infty} \overline{a(n;g)} e(nz).$$

These statements follow from known properties of Gauss sums.

Lemma 6.13. *Let* $G_b(\chi) = \sum_{a \bmod m} \chi(a) e(ab/m)$, *and consider a primitive Dirichlet character* χ *modulo* m. *Then:*
(i) $G_b(\chi) = \overline{\chi(b)} G(\chi)$,
(ii) $G(\overline{\chi}) G(\chi) = \chi(-1) m$.

We see that (6.3.53) is equivalent to the identity

$$f|\frac{G(\chi)}{m} \sum_{a \bmod m} \overline{\chi(a)} \begin{pmatrix} m & a \\ 0 & m \end{pmatrix} = C_\chi f|\frac{G(\chi)}{m} \sum_{a \bmod m} \overline{\chi(a)} \begin{pmatrix} m & a \\ 0 & m \end{pmatrix} |KW_{Nm^2},$$

and one may commutate the terms on the right as follows:

$$\begin{aligned} C_\chi f|\frac{G(\chi)}{m} &\sum_{a \bmod m} \overline{\chi(a)} \begin{pmatrix} m & a \\ 0 & m \end{pmatrix} |KW_{Nm^2} = \\ C_\chi \frac{\overline{G(\chi)}}{m} &\sum_{a \bmod m} \chi(a) f|K \begin{pmatrix} m & -a \\ 0 & m \end{pmatrix} |W_{Nm^2} = \\ C_\chi \frac{G(\overline{\chi})}{m} &\sum_{a \bmod m} \chi(a) f|K \begin{pmatrix} m & a \\ 0 & m \end{pmatrix} |W_{Nm^2}, \end{aligned}$$

because of the equality $f|\left(\begin{smallmatrix} m & a \\ 0 & m \end{smallmatrix}\right)|KW_{Nm^2} = f|K\left(\begin{smallmatrix} m & -a \\ 0 & m \end{smallmatrix}\right) W_{Nm^2}$, and we replace a par $-a$. Notice also that

$$f|K\begin{pmatrix} m & a \\ 0 & m \end{pmatrix} W_{Nm^2} = f|KW_N\gamma(a,b)\begin{pmatrix} m & b \\ o & m \end{pmatrix} \quad \gamma(a,b) = \begin{pmatrix} m & -b \\ -Na & n \end{pmatrix} \in \Gamma_0(N),$$

because of the equality

$$W_N\gamma(a,b)\begin{pmatrix} m & b \\ o & m \end{pmatrix}\begin{pmatrix} m & 0 \\ 0 & m \end{pmatrix} = \begin{pmatrix} m & a \\ 0 & m \end{pmatrix} W_{Nm^2}, \quad -Nab \equiv 1 (\mathrm{mod} m).$$

Let us rewrite (6.3.53) in the following form (6.3.56):

$$\begin{aligned} G(\chi) &\sum_{a \bmod m} \overline{\chi(a)} f|\begin{pmatrix} m & a \\ 0 & m \end{pmatrix} \\ &= C_\chi G(\overline{\chi}) \overline{\chi(-N)} \sum_{a \bmod m} \overline{\chi(b)} f|KW_N\gamma(a,b)\begin{pmatrix} m & b \\ 0 & m \end{pmatrix}. \end{aligned} \tag{6.3.56}$$

It follows that

$$f|KW_N = (-1)^k f|W_N K = (-1)^k (C_1 f|K)K = (-1)^k \overline{C_1} f,$$

$$f|KW_N\gamma(a,b) = (-1)^k \overline{C_1 \psi n} f = (-1)^k C_1^{-1} \psi(m) f$$

hence (6.3.56) implies that C_χ is given by the above formula (6.3.54).

Let us rewrite (6.3.56) in the equivalent form:

$$\sum_{a \bmod m} \overline{\chi(a)} f| \begin{pmatrix} m & a \\ 0 & m \end{pmatrix} = \psi(m) \sum_{a \bmod m} \overline{\chi(b)} f|\gamma(a,b) \begin{pmatrix} m & b \\ 0 & m \end{pmatrix} \tag{6.3.57}$$

using $-Nab \equiv 1 (\bmod N)$.

The Mellin transform gives directly the equality:

$$\Lambda^*(s,f,\chi) = C_\chi i^k (Nm^2)^{\frac{k}{2}-s} \Lambda^*(k-s, f^\rho, \overline{\chi}).$$

A converse statement concerning analytic properties of the series (6.3.52) was found by A.Weil [Wei67].

Theorem 6.14 (A.Weil, 1967). *Suppose that a series*

$$R(s) = \sum_{n=1}^{\infty} a(n) n^{-s}$$

with complex coefficients $a(n)$ converges absolutely for $s = k - \delta$, $\delta > 0$, and that for every $\chi \bmod m$ of prime conductor m, $(m \nmid N)$ the function $\Lambda^*(s,f,\chi)$ is entier, bounded in every vertical strip and satisfies the following functional equation:

$$\Lambda^*(s,f,\chi) = C_\chi i^k (Nm^2)^{\frac{k}{2}-s} \Lambda^*(k-s, f^\rho, \overline{\chi}),$$

where $f^\rho(z) = \sum_{n=1}^{\infty} \overline{a(n)} e(nz)$,

$$C_\chi = C_1 \frac{G(\chi)}{G(\overline{\chi})} \chi(-N) \psi(m).$$

Then

$$f(z) = \sum_{n=1}^{\infty} a(n) e(nz)$$

represents a modular cusp form in $\mathcal{S}_k(N,\psi)$.

In other words, the properties of automorphy of such series can be deduced from the functional equations and the analytic properties of all the series (6.3.52)

Sketch of the proof.

Let us consider the group algebra $\mathbb{C}[[\mathrm{GL}^+(2,\mathbb{R})]]$ of $\mathrm{GL}^+(2,\mathbb{R})$, and its right ideal

$$\Omega_f \subset \mathbb{C}[[\mathrm{GL}^+(2,\mathbb{R})]]$$

consisting of all $w \in \mathbb{C}[[\mathrm{GL}^+(2,\mathbb{R})]]$ such that $f|w = 0$.

One has to show that for every $\gamma = \begin{pmatrix} a & b \\ c & d \end{pmatrix} \in \Gamma_0(N)$ the element $w = (\psi(d) - \gamma) = \psi(d)(1 - \psi(a)\gamma) \in \Omega_f$.

Let us consider first the elements $\gamma = \gamma(a,b) = \begin{pmatrix} m & -b \\ -Nb & n \end{pmatrix} \in \Gamma_0(N)$ as above and let us show

$$(1 - \psi(m)\gamma(a,b)) \in \Omega_f.$$

for all $\gamma(a,b) = \begin{pmatrix} m & -b \\ -Nb & n \end{pmatrix}$ with $m, n \in S$, where S is the set of all *primes* m for which the functional equation is satisfied for every χ mod m.

Lemma 6.15. *The equality (6.3.57) is equivalent to (6.3.58): for all* b, b' mod m, $(b,m) = (b',m) = 1$ *one has*

$$\begin{aligned}(1 - \psi(m)\gamma(a,b)) \begin{pmatrix} m & b \\ 0 & m \end{pmatrix} & \\ \equiv (1 - \psi(m)\gamma(a',b')) \begin{pmatrix} m & b' \\ 0 & m \end{pmatrix} & (\operatorname{mod} \Omega_f).\end{aligned} \tag{6.3.58}$$

Proof of Lemma 6.15 is a simple verification: for all b', b'' mod m, $(b',m) = (b'',m) = 1$ one can multiply the two parts of (6.3.57) by $\chi(b'') - \chi(b')$ and summate over χ, cf. op.cit. □

In order to finish the proof of Theorem 6.14, it sufficies to show that

$$1 - \psi(a)\gamma \in \Omega_f \text{ for all other elements } \gamma = \begin{pmatrix} a & b \\ Nc & d \end{pmatrix} \in \Gamma_0(N).$$

First, let $c = 0$, $\gamma = \begin{pmatrix} 1 & b \\ 0 & 1 \end{pmatrix}$, then $f = f|\gamma$ because f is periodic of period 1. Next, let $b = 0$, $\gamma = \begin{pmatrix} 1 & 0 \\ Nc & 1 \end{pmatrix} \in \Gamma_0(N)$. Then

$$\begin{pmatrix} 1 & 0 \\ Nc & 1 \end{pmatrix} = \begin{pmatrix} 0 & -1 \\ N & 0 \end{pmatrix} \begin{pmatrix} 1 & -c \\ 0 & 1 \end{pmatrix} \begin{pmatrix} 0 & -1 \\ N & 0 \end{pmatrix}^{-1}$$

hence

$$\begin{aligned} f| \begin{pmatrix} 1 & 0 \\ Nc & 1 \end{pmatrix} &= f|W_N \begin{pmatrix} 1 & -c \\ 0 & 1 \end{pmatrix} W_N^{-1} = (-1)^k f|W_N \begin{pmatrix} 1 & -c \\ 0 & 1 \end{pmatrix} W_N = \\ &(-1)^k f|K^2 W_N \begin{pmatrix} 1 & -c \\ 0 & 1 \end{pmatrix} W_N = f|KW_N K \begin{pmatrix} 1 & -c \\ 0 & 1 \end{pmatrix} W_N = \\ &C_1 f| \begin{pmatrix} 1 & c \\ 0 & 1 \end{pmatrix} KW_N = C_1 \overline{C}_1 f = f, \end{aligned}$$

in view of the known properties of K et W_N: $KW_N = (-1)^k W_N K$, $W_N^2 = (-1)^2$, and from the fact that $f|KW_N = C_1 f$ which is equivalent to the functional equation with $\chi = 1$, $m = 1$. Finally, if $\gamma = \left(\begin{smallmatrix} a & b \\ Nc & d \end{smallmatrix}\right) \Gamma_0(N)$, $b \neq 0$, let us choose $s, t \in \mathbb{Z}$ such that $m = a + Nbs, n = d + Nbt \in S$. Then

$$\gamma = \begin{pmatrix} 1 & 0 \\ -Nt & 1 \end{pmatrix} \begin{pmatrix} m & b \\ Nb' & n \end{pmatrix} \begin{pmatrix} 1 & 0 \\ -Ns & 1 \end{pmatrix};$$

implying

$$f|\gamma = \psi(n)f = \psi(d)f.$$

This proves the automorphy property; it remains to chack the vanishing of f at the cusps of $\Gamma_0(N)$. Let us use the absolute convergence of the series

$$R(s) = \sum_{n=1}^{\infty} a(n)n^{-s}$$

at $s = k - \delta$, $\delta > 0$. One easily deduces analytically that $f(x+iy) = \mathcal{O}(y^{-k+\delta})$ for $y \to 0$. This means that f is a cusp form in $S_k(N, \psi)$.

6.4 Modular Forms and Galois Representations

6.4.1 Ramanujan's congruence and Galois Representations

A new chapter in the theory of modular forms, and generally in arithmetic, was opened by Serre and Deligne, who discovered a link between modular forms and Galois representations. Their results have enhanced our understanding of a universal role played by modular forms in number theory, and have explained a whole series of mysterious facts concerning various arithmetical functions. Examples of these facts are the conjecture of Ramanujan–Petersson $\tau(p) < 2p^{11/2}$ for the Ramanujan function $\tau(p)$, and the congruence of Ramanujan

$$\tau(n) \equiv \sum_{d|n} d^{11} \mod 691. \tag{6.4.1}$$

The first result in this direction concerns the normalized cusp forms

$$f(z) = \sum_{n=1}^{\infty} a(n)e(nz) \in \mathcal{S}_k(\mathrm{SL}_2(\mathbb{Z})) \text{ with } k = 12, 16, 18, 20, 22, 26,$$

when $\dim\ \mathcal{S}_k(\mathrm{SL}_2(\mathbb{Z})) = 1$. Serre conjectured (in [Se68a], [Se68b])), and Deligne proved (cf. [Del68]), that for each of the above cusp forms and for every prime number l there exists a continuous Galois representation

$$\rho_l : G(K_l/\mathbb{Q}) \to \mathrm{GL}_2(\mathbb{Z}_l) \tag{6.4.2}$$

(where K_l is the maximal extension of $\mathbb{Q}$ ramified only at l) with the property that the image of the p-Frobenius element $F_{\rho,p} = \rho_l(Fr_p)$ for $p \neq l$ has characteristic polynomial $t^2 - a(p)t + p^{k-1}$, where $a(p)$ is the pth coefficient of f, and k is its weight.

One can rephrase the statement on the characteristic polynomial as saying that the representation ρ_l is $\mathbb{Z}$–integral in the sense of §6.2.1, and the following equation holds:

$$(1 - a(l)l^{-s} + l^{k-1-2s})L(s, f) = L(\rho_l, s). \tag{6.4.3}$$

This result makes it possible to study congruences modulo a prime number l for the coefficients $a(n)$. It turns out that such congruences exist only when l is exceptional for ρ_l, i.e. when the image $\mathrm{Im}\ \rho_l$ does not contain $\mathrm{SL}_2(\mathbb{Z}_l)$. In this case there are certain relations modulo l between the trace $\mathrm{Tr}\ F_{\rho,p} = a(p)$ and the determinant $\det F_{\rho,p} = p^{k-1}$ of the matrix $F_{\rho,p}$. For example, in the case of the Ramanujan function $\tau(n)$ we have $k = 12$, $l = 691$, and the image of $\rho_l \mod l$ (modulo conjugation) lies in the subgroup of upper triangular matrices $\left(\begin{smallmatrix} * & * \\ 0 & * \end{smallmatrix}\right) \mod l$. One has

$$F_{\rho_l,p} \equiv \begin{pmatrix} p^{11} & * \\ 0 & 1 \end{pmatrix} \mod l,$$

which implies $\tau(p) \equiv p^{11} + 1 \mod l$, and by multiplicativity one obtains the congruence (6.4.1). H.P.Swinnerton–Dyer in [SwD73] gave the following description of the possible exceptional primes l for the above cusp forms:

a) there exists an integer m such that $a(m) \equiv n^m \sigma_{k-1-2m} \mod l$ whenever n is a quadratic non-residue mod l; in this case

$$F_{\rho_l,p} \equiv \begin{pmatrix} p^m & * \\ 0 & p^{k-1-m} \end{pmatrix} \mod l :$$

b) $a(n) \equiv 0 \mod l$ whenever n is a quadratic residue mod l;
c) $p^{1-k}a(p)^2 \equiv 0, 1, 2, 4 \mod l$.

For the Ramanujan function $\tau(n)$ the exceptional primes are: $l = 2, 3, 5, 7, 23, 691$.

The construction of the representation ρ_l is based on methods from algebraic geometry, in particular on the study of the l–adic cohomology groups of the Kuga–Sato variety E_Γ^w which is defined as the fiber product of $w = k-2$ copies of the universal elliptic curve $E_\Gamma \to X_\Gamma$ over the *modular curve* $X_\Gamma = \overline{H/\Gamma}$, $\Gamma = \mathrm{SL}_2(\mathbb{Z})$ (cf. §6.3.2, and [Sho80]).

The variety E_Γ^w is defined over $\mathbb{Q}$ and its algebraic (and complex) dimension is equal to $w + 1 = k - 1$. Deligne has shown that the representation ρ_l of $\mathrm{Gal}(\overline{\mathbb{Q}}/\mathbb{Q})$ occurs in the vector space $H_{\acute{e}t}^{k-1}(E^w_{\Gamma,\overline{\mathbb{Q}}}, \mathbb{Q}_l)$; in other words one can associate to f a motive M_f of weight $k-1$ which occurs in the cohomology of the Kuga–Sato variety. However, the construction of M_f requires many additional cohomological techniques ([Ja90] , [Scho90]).

Ribet in [Ri77] has extended the results of Deligne to primitive modular forms of arbitrary level.

The Galois representation $\rho_l = \rho_{f,l}$ attached to a cusp form f is irreducible; if on the other hand we take for f an Eisenstein series which is an eigenfunction of all Hecke operators, $f \in \mathcal{M}_k(N, \psi)$ then it is not difficult to construct a reducible l–adic representation ρ_l whose L-function is the Mellin transform of f, i.e. such that the characteristic polynomial of $F_{\rho_l,p}$ coincides with

$$1 - \lambda_f(p)p^{-s} + \psi(p)p^{k-1-2s} \quad (f|T_k(p) = \lambda_f(p)f)$$

for $l \neq p, lp$ coprime to N. Formula (6.3.35) shows that if $a(0) \neq 0$ then $\lambda(p) = 1 + \psi(p)p^{k-1}$. Hence one may take for ρ_l the direct sum $1 \oplus (\rho_\psi \otimes \chi_l^{k-1})$, where $\chi_l : \mathrm{Gal}(\overline{\mathbb{Q}}/\mathbb{Q}) \to \mathbb{Z}_l^\times$ $(\chi_l(Fr_p) = p)$ is the cyclotomic character and ρ_ψ : $\mathrm{Gal}(\overline{\mathbb{Q}}/\mathbb{Q}) \to \overline{\mathbb{Q}}^\times \subset \overline{\mathbb{Q}}_l^\times$ is the one dimensional representation associated to ψ via the *Kronecker–Weber theorem*. For the Eisenstein series the L–function of this representation coincides with

$$\zeta(s)L(s-k+1, \psi) = \prod_p [(1-p^{-s})(1-\psi(p)p^{k-1-2s})]^{-1}.$$

6.4.2 A Link with Eichler–Shimura's Construction

The idea behind Deligne's construction dates back to Eichler's study of the zeta functions of modular curves; these functions can be characterized as the Mellin transforms of cusp forms of weight 2 (see [Eich54]).

If Γ is a congruence subgroup, then there is a one-to-one correspondence between cusp forms $f \in \mathcal{S}_2(\Gamma)$ and holomorphic differentials $f(z)\,dz$ on X_Γ. Hence $\dim\ \mathcal{S}_2(\Gamma) = g = g(X_\Gamma)$, where $g(X_\Gamma)$ denotes the genus of X_Γ. Formulae for the genera of the curves X_Γ can be found in the book of Shimura (1971). If $\Gamma = \Gamma_0(N)$ then the notation $X_\Gamma = X_0(N)$ is often used. A modular curve $X_0(N)$ is an algebraic curve defined over $\mathbb{Q}$, such that the Riemann surface $X_0(N)(\mathbb{C})$ is identified to the compact quotient $\Gamma_0(N)\backslash\overline{\mathbb{H}}$ in such a way that

$$X_0(N)(\mathbb{C}) \longleftrightarrow \Gamma_0(N)\backslash\overline{\mathbb{H}}, \overline{\mathbb{H}} = \mathbb{H} \cup \mathbb{Q} \cup \infty;$$
$$Y_0(N)(\mathbb{C}) \longleftrightarrow \Gamma_0(N)\backslash\mathbb{H}$$

for an affine algebraic curve $Y_0(N)$ which is an algebraic curve defined over $\mathbb{Q}$ (see §6.4.2).

Let us choose a *Hecke basis* $\{f_1, \dots, f_g\}$ and consider the corresponding Euler products $L(s, f_i)$ (see (6.3.37)).

Eichler discovered that the zeta function of the modular curve $X_0(N)$ has the form

$$\zeta(s)\zeta(s-1)L(X_0(N), s)^{-1}$$

where the L - function $L(X_0(N), s)$ coincides upto a finite number of Euler factors (corresponding to the divisors of the level N) with the product

$$\prod_{i=1}^{g} L(s, f_i).$$

Recall that the L–function $L(X, s)$ coincides with the L–function of the l–adic representation of $\mathrm{Gal}(\overline{\mathbb{Q}}/\mathbb{Q})$ on the Jacobian $J_X = J_0(N)$ of the curve $X_0(N)$.

Using the L–functions $L(s, f_i)$ one can also obtain a decomposition of the Jacobian into a product of simple Abelian varieties (upto isogeny):

$$J_0(N) \cong A_1 \times \cdots \times A_r. \tag{6.4.4}$$

One proves that the endomorphism algebra $\mathrm{End}\ A_j \otimes \mathbb{Q} = K_j$ is a totally real extension of $\mathbb{Q}$ generated by the Hecke eigenvalues $\lambda_f(m)$ $((m, N) = 1)$ of a cusp form $f(z) = \sum_{n=1}^{\infty} a(n)e(nz) \in \mathcal{S}_2(\Gamma_0(N))$. One has

$$L(A_j, s) = \prod_{\sigma} L(s, f^\sigma),$$

where σ runs through the embeddings $\sigma : K_j \to \mathbb{R}$, and

$$f^\sigma(z) = \sum_{n=1}^{\infty} a(n)^\sigma e(nz) \in \mathcal{S}_2(\Gamma_0(N)). \tag{6.4.5}$$

In particular, $[K_j : \mathbb{Q}] = \dim\ A_j$.

A detailed exposition of these results can be found in the book of Shimura (1971).

The especially interesting case $g(X_0(N)) = 1$ arises only for N =11, 14, 15, 17, 19, 20, 21, 24, 27, 32, 36, 49, when the vector space $\mathcal{S}_2(\Gamma_0(N))$ is generated by a single cusp form with integral Fourier coefficients. If $N = 11$ then

$$f(z) = \eta(z)^2\eta(11z)^2, \quad \text{where} \quad \eta(z) = e(z/24)\prod_{m=1}^{\infty}(1 - e(mz)) \tag{6.4.6}$$

is the Dedekind η–function. For $N = 36$ we have

$$f(z) = \eta(12z)^2\theta(z), \tag{6.4.7}$$

where $\theta(z) = \sum_{n\in\mathbb{Z}} e(n^2z)$ is the theta function (cf. [Frey86]).

6.4.3 The Shimura–Taniyama–Weil Conjecture

is discussed in more detail in Chapter 7 in connection with Fermat's Last Theorem which have been completely proved in [Wi], [Ta-Wi], together with the Shimura–Taniyama–Weil conjecture (the STW conjecture) for semi–stable elliptic curves (cf. also [CSS95] [Tan57], [Wei67], [Frey86], [Gel76]). The STW conjecture was proved in full generality in 1999 by Ch.Breuil, B.Conrad, F.Diamond and R.Taylor (cf. [Da99] and Chapter 7 for relevant techniques).

An elliptic curve E defined over $\mathbb{Q}$ is called a modular elliptic curve (a Weil curve) if there exists a non constant morphism $\varphi_N : X_0(N) \to E$. The Shimura–Taniyama–Weil conjecture says that every elliptic curve over $\mathbb{Q}$ is modular. The smallest number N with this property is called the analytic conductor of E. In this case E has good reduction modulo all primes p not dividing N, and its L–function coincides with the Mellin transform of a cusp form $f \in \mathcal{S}_2(\Gamma_0(N))$:

$$L(E, s) = L(s, f).$$

In particular, the function $L(E, s)$ admits an analytic continuation to the entire complex plane and satisfies a functional equation of the type

$$\Lambda(E, s) = (2\pi/\sqrt{N})^{-s}\Gamma(s)L(E, s) = \pm\Lambda(E, 2 - s).$$

This conjecture seems to be both very natural, and surprising since it establishes a correspondence between two quite different kinds of object: *elliptic curves* over $\mathbb{Q}$ and *primitive cusp forms of weight 2* with integral coefficients. Before Wiles' proof, the conjecture has been verified for a number of curves, in particular for all curves with complex multiplication. In the latter case the l-adic Galois representation on the Tate module turns out to be Abelian, and one proves first that its L–function coincides with the L-function of a Hecke character of the corresponding imaginary quadratic field ("Grössencharakter"). The analytic continuation and functional equation of these functions is known (see 6.2.4), so it follows from Weil's inverse theorem that $L(E,s) = L(s,f)$ for some primitive cusp form f of weight 2, which is equivalent to the Shimura–Taniyama–Weil conjecture. In the above examples the curve $X_0(36)$ admits complex multiplication by $\mathbb{Q}(\sqrt{-3})$, whereas the curve $X_0(11)$ has no complex multiplication.

The Shimura–Taniyama–Weil conjecture has a number of interesting arithmetical corollaries, in particular concerning Fermat's last theorem (cf. [Wi], [Ta-Wi], [Ri], [CSS95], [Frey86], [Se87] and Chapter 7).

There is an analogue of the Shimura–Taniyama–Weil conjecture describing all simple Abelian varieties with the property that the degree of the endomorphism algebra over $\mathbb{Q}$ coincides with the dimension of the variety. These varieties are thought to correspond to simple factors of the Jacobians of modular curves [Se87], [Wei71].

6.4.4 The Conjecture of Birch and Swinnerton–Dyer

(cf. [BSD63], [Ta65a], [Man71], [CW77], [Koly88], [Rub77], and [CR01] for a recent progress). This deep conjecture gives a relation between the most important arithmetical invariants of an elliptic curve E over a number field K, and the analytic behaviour of the L–function $L(E,s) = L(E/K,s)$ at the point $s = 1$. These arithmetical invariants are: $r_E = \mathrm{rk}\ E(K)$ (the rank of E over K), $E(K)^{\mathrm{tors}}$ the torsion subgroup, R_E the regulator of E, i.e. the determinant of the Néron–Weil pairing h_E on

$$E(K)/E(K)^{\mathrm{tors}} \subset \mathbb{R}^{r_E},$$

and the Shafarevich–Tate group $\text{Ш}(E,K)$ of E over K. By definition

$$L(E,s) = \prod_v L_v(E,s), \tag{6.4.8}$$

where

$$L_v(E,s)^{-1} = 1 - a(\mathfrak{p}_v)\mathrm{N}v^{-s} + \mathrm{N}v^{1-2s},$$
$$a(\mathfrak{p}_v) = \mathrm{N}v + 1 - \mathrm{Card}\ E(\mathcal{O}_K/\mathfrak{p}_v)$$

for all places v where E has good reduction, and

$$L_v(E,s)^{-1} = 1 - a(\mathfrak{p}_v)\mathrm{N}v^{-s}, \quad a(\mathfrak{p}_v) = \pm 1 \text{ or } 0$$

for places v with bad reduction, according to the type of bad reduction of $E \bmod \mathfrak{p}_v$. Here it is assumed that E is defined over $\mathcal{O}_K$, and that it coincides with its Néron model (minimal model) (cf. §5.2).

In view of the *"Riemann conjecture over a finite field"* (see (W4) of §6.1.3) the following estimate holds: $|a(\mathfrak{p}_v)|_v \le 2\sqrt{\mathrm{N}v}$, which implies the absolute convergence of the series $L(E,s)$ for $\mathrm{Re}(s) > 3/2$. However, in order to formulate the conjecture we need stronger analytic properties: we assume the hypothesis of Hasse–Weil on the existence of an analytic continuation of $L(E,s)$ to the entire complex plane.

The *conjecture of Birch and Swinnerton–Dyer* (BSD) consists of two parts (cf. [Bir63], [BSD63]):

a) the order of the zero $n_E = \mathrm{ord}_{s=1} L(E,s)$ coincides with the rank r_E
b) assume that the Shafarevich–Tate group of E over K is finite; then as $s \to 1$ the following asymptotic formula holds:

$$L(E,s) \sim (s-1)^{r_E} \frac{R_E |\text{Ш}(E,K)|}{|E(K)^{\mathrm{tors}}|^2} M, \tag{6.4.9}$$

where $M = \prod_{v \in S_E} m_v$ is an explicitly written product of local Tamagawa factors over the set S_E of all Archimedean places and places where E has bad reduction, $m_v = \int_{E(K_v)} |\omega|_v$, ω being the Néron differential of E.

For example, if $K = \mathbb{Q}$ an elliptic curve E can be defined by the equation

$$y^2 + a_1xy + a_3y = x^3 + a_2x^2 + a_4x + a_6 \quad (a_i \in \mathbb{Z}), \tag{6.4.10}$$

which is minimal in the sense that the absolute value of its discriminant is minimal; in this case the Néron differential has the form $dx/(2y + a_1 + a_3)$ ([Silv86]).

The BSD conjecture is closely related to the Shimura–Taniyama–Weil conjecture, because the analytic properties of the functions $L(E,s) = \sum_{n=1}^{\infty} a(n)n^{-s}$ and $L(E,\chi,s) = \sum_{n=1}^{\infty} \chi(n)a(n)n^{-s}$ imply the modular properties of the corresponding functions $f(z) = \sum_{n=1}^{\infty} \chi(n)a(n)e(nz)$ in view of the inverse theorem of Weil (see section 6.3). In all known cases the following functional equation holds:

$$\Lambda(E,s) = (2\pi/\sqrt{N})^{-s}\Gamma(s)L(E,s) = \varepsilon(E)\Lambda(E,2-s), \tag{6.4.11}$$

where the number $\varepsilon(E) = \pm 1$ is called the sign (or the "root number") of E.

These are the Weil curves for which some partial results on the validity of the BSD conjecture are known. Let $\varphi : X_0(N) \to E$ define a Weil curve E of conductor N, and let ω be the Néron differential of E. Then the pullback $\varphi^*\omega$ coincides upto a sign with the differential

$$f(z)\frac{dq}{q} = 2\pi i f(z)\, dz \quad \text{on} \quad X_0(N),$$

where $f \in \mathcal{S}_2(\Gamma_0(N))$ is a primitive cusp form of level N. One has $L(E,s) = L(f,s)$ and the following equation holds

$$L(E,1) = 2\pi \int_0^{\infty} f(iy)\, dy. \tag{6.4.12}$$

The integral is absolutely convergent in view of the exponential decay of $f(z)$ as $y \to \infty$ or $y \to 0$; it coincides essentially with the Tamagawa factor m_∞ from (6.4.9).

Theorem 6.16 (Coates J., Wiles A. (1977)). *Let E/K be an elliptic curve with complex multiplication and let K be either $\mathbb{Q}$ or the complex multiplication field. Then the condition $r_E \geq 1$ implies that $n_E \geq 1$, i.e. that $L(E,1) = 0$.*

The proof of this theorem is based on an explicit calculation of the special value (6.4.12), which is upto a rational multiplicative constant equal to the Tamagawa factor m_∞. From the existence of a point of infinite order it follows that this multiplicative constant is divisible by infinitely many primes, and is hence zero.

A generalization of this result in another direction was found by R. Greenberg (cf. [Gr83]): let $E/\mathbb{Q}$ be an elliptic curve with complex multiplication. If the number n_E is odd then either the group $\text{Ш}(E,\mathbb{Q})$ is infinite and contains a divisible group $\mathbb{Q}_p/\mathbb{Z}_p$ for every good reduction prime $p \neq 2, 3$ (i.e. for which $E \bmod p$ is an elliptic curve with a non–trivial point of order p over $\mathbb{F}_p$). Developing these ideas K. Rubin and V.Kolyvagin constructed examples of curves $E/\mathbb{Q}$ with complex multiplication and with finite Shafarevich–Tate groups $\text{Ш}(E,\mathbb{Q})$. He also proved the following deep fact: if for such a curve one has $r_E = r_E(\mathbb{Q}) > 1$ and $\varepsilon(E) = -1$, then $n_E \geq 3$. For example, the curve $E : y^2 = x^3 - 226x$ has rank 3 (generators modulo torsion are:$(-1, 15)$, $(-8, 96)$, $(121/4, 1155/8)$), and $\varepsilon(E) = -1$; hence $n_E \geq 3$ (comp. with the examples in §6.3.2). These results where extended by Kato to curves without complex multiplication, using Euler systems, cf. [Kato2000], [MazRub04].

Although these results concern L–functions of a complex variable, they use a lot of p–adic theory and properties of p–adic L–functions. Neither the domain of definition nor the set of values of these L–functions are complex; they are p–adic. These L–functions make it possible to control the p–adic behaviour of special values of the type $L(E,\chi,1)$ (where χ is a Dirichlet character) which are algebraic numbers (upto a multiplicative factor of the form m_∞) and may thus be regarded as p–adic numbers. Also, the p–adic L–functions describe the behaviour of the Selmer groups and the Shafarevich–Tate groups under Abelian extensions of the ground field K, which is either $\mathbb{Q}$ or the complex multiplication field of the given elliptic curve [Man71], [Man76], [Man78], [Iwa72], [Coa89], [MW83], [MSD74], and for recent progress, [Coa01], [Colm03], [CM98].

An important development of the BSD conjecture for modular curves (including curves with complex multiplication) was obtained by Gross and Zagier (cf. [GZ86], [GKZ87], [Coa84]). They proved for elliptic Weil curves E that if $n_E = 1$ then $r_E \geq 1$. They established furthermore the existence of elliptic curves $E/\mathbb{Q}$ for which $n_E \geq 3$. These results are based on the theory of special points on modular curves *(Heegner points)*. It was already known in the nineteenth century how to construct solutions to Pell's equation using either special values of trigonometric functions (Dirichlet), or special values of the Dedekind eta–function (Kronecker) (see Part I, §1.2). Heegner in his work [Hee52] successfully used special values of elliptic modular functions to find rational points on elliptic curves, making it possible to find effectively all imaginary quadratic fields with class number one. In the work of Birch, [Bir75], extending and clarifying the ideas of [Hee52], the existence of rational points of infinite order on certain elliptic curves was first established without explicit evaluation of the coordinates of these points, and without verification that these points indeed satisfy the equation of the given curve. Let $\varphi : X_0(E) \to E$ be a Weil parameterization of a given elliptic curve $E/\mathbb{Q}$. As was noted above (cf. §6.3.2) the set $\mathbb{H}/\Gamma_0(N) \subset X_0(N)(\mathbb{C})$ parametrizes the isomorphism classes over $\mathbb{C}$ of isogenies $E_z \to E_z/\langle P\rangle$ with cyclic kernel $\langle P\rangle$ where $E_z(\mathbb{C}) \cong \mathbb{C}/\langle 1, z\rangle$ is a varying elliptic curve associated to the point $z \in H$. Let K be an imaginary quadratic field of discriminant $D < 0$ with maximal order $\mathcal{O}$. Suppose that there exists an ideal $i \subset \mathcal{O}$ such that $\mathcal{O}/i \cong \mathbb{Z}/N\mathbb{Z}$ (this condition is satisfied for example, when $D =$ square (mod$4N$) and $(D, 2N) = 1$). Then one can associate to the isogeny $\mathbb{C}/\mathcal{O} \to \mathbb{C}/i^{-1}$ a point z on $\mathbb{H}/\Gamma_0(N)$, and it is not difficult to verify that this point is rational over the Hilbert class field H_K (the maximal unramified Abelian extension) of K. The point $\varphi(z) = y \in E(H_K)$ is called the Heegner point on E; therefore the point $y_K = \sum_{\sigma\in \mathrm{Gal}(H_K/K)} y^\sigma \in E(K)$ is defined over K. Birch and Stephens made extensive calculations of Heegner points in order to find out under which assumptions the point y_K has infinite order. They suggested a conjecture, expressing for $L(E,1) = 0$ the special value $L'(E,1)$ in terms of the product of m_∞ and the Néron–Tate height $h(y_K)$ of y_K (cf. [BS83]). This conjecture was proved by Gross and Zagier ([GZ86]).

A further significant extension of these results is contained in works of V. A. Kolyvagin in [Koly88]. He proved that if $L(E,1) \neq 0$ and y_K has finite order then the groups $E(\mathbb{Q})$ and $\text{Ш}(E,\mathbb{Q})$ are finite, proving the first part of the BSD conjecture. The methods developed by V. A. Kolyvagin make it possible to find effectively in terms of K, E and y_K the smallest positive integer annihilating the groups $E(\mathbb{Q})$ and $\text{Ш}(E,\mathbb{Q})$. Thus one also has an approach to proving the second part of the BSD conjecture. The theory of Euler systems due to V. A. Kolyvagin, cf. [Koly90], see also [MazRub04], also allows one to consider from a unified point of view Gauss sums, elliptic units, cyclotomic units and Heegner points, and it gives an approach to proving the "main conjecture" of Iwasawa theory (see §5.4.5) which describes the Iwasawa modules attached to these objects in terms of p–adic L–functions.

Heegner points, provide a lot of points on elliptic curves defined over ring class fields of imaginary quadratic fields.

Mazur formulated a number of conjectures concerning the variation of Mordell-Weil groups in towers of ring class fields with restricted ramification (cf. [Maz83], [MazRub03]) that Heegner points should account, in a suitable sense, for a majority of points rational over the fields in the tower (assuming that the elliptic curve does not have complex multiplication by the imaginary quadratic field in question).

Vatsal [Vats03] and Cornut [Cor02] proved a version of this conjecture.

A remarkable fact is that it is sometimes quite easy to verify that the special value $L(E,1)$ vanishes, but it is extremely difficult to find a point of infinite order on a curve E. In the example of Cassels–Bremner $E: y^2 = x^3 + 877x$ (cf. [Cas84], [Cas66]) the vanishing of the special value $L(E,1)$ follows from the fact that E is odd; on the other hand the generator of the group $E(\mathbb{Q})/E(\mathbb{Q})^{\mathrm{tors}} \cong \mathbb{Z}$ looks very complicated (see Part I, §1.3.2).

The results of Gross and Zagier found an unexpected application to Gauss's famous problem of finding all the imaginary quadratic fields $\mathbb{Q}(\sqrt{-d})$ with class number $h(-d)$ equal to a given number h. Previously these fields had been found explicitly in the cases $h = 1$ and $h = 2$ ([Hee52] , [Abr74], [Ba71], [Deu68], [St67], [St69]).

In 1976 Goldfeld showed in [Gol76] that if there is a cusp form $f \in \mathcal{S}_k(\Gamma_0(N))$ whose Mellin transform has a zero at 1 of essentially high order (3 or 4, depending on k and N) then for any positive integer $h \geq 1$ one can find an effective upper bound for d such that $h(-d) = h$. The desired cusp form has since been found: Mestre showed in [Me85] that the elliptic curve

$$y^2 + y = x^3 - 7x + 6 \tag{6.4.13}$$

of conductor 5077 and rank 3 ($E(\mathbb{Q}) \cong \mathbb{Z}^3$ with generators (1,0), (6,0), (0,2)) is a Weil curve, i.e. its L-function $L(E,s) = \sum_{n=1}^{\infty} a(n)n^{-s}$ is the Mellin transform of some cusp form $f(z) = \sum_{n=1}^{\infty} a(n)e(nz) \in \mathcal{S}_2(\Gamma_0(5077))$. From the results of Gross and Zagier and from the fact that E is odd (i.e. $\varepsilon(E) = -1$) one deduces that $n_E \geq 3$, so f is a cusp form with the required properties. The use of f in the theorem of Goldfeld makes it possible to prove that for a positive integer $T \geq 1$ there exists an effective constant $B(T) > 0$ such that if d possesses T different prime divisors, then the following estimate holds: $h(-d) \geq B(T) \log d$. If d is a prime then $d \log d < 55h(-d)$. Using this result all d with $h(-d) = 3$ were found, cf. [Oe83].

For recent developments concerning Gross-Zagier formulas, we refer to [BeDa97] and [Borch99].

A fruitful use of L–functions of elliptic curves and modular forms was demonstrated in the work J.Tunnell (cf. [Tun83], [Frey86]) on a classical Diophantine problem concerning *congruent numbers.* A natural number N is called a congruent number if it is the area of some right angle triangle, all of whose sides have rational lengths. For example, the number 6 is congruent as it is the area of the Egyptian triangle with the sides 3, 4, and 5. It turns out that the smallest congruent number is 5 which is equal to the area of the triangle with sides 3/2, 20/3, 41/6. The fact that $N = 1$ is not congruent provides an excellent example of use of Fermat's infinite descent argument and also proves Fermat's Last Theorem for the exponent 4.

Indeed, suppose that $Z > Y > X > 0$ are rational numbers satisfying

$$X^2 + Y^2 = Z^2, \quad \frac{1}{2}XY = N. \tag{6.4.14}$$

From these equalities we obtain

$$(X+Y)^2 = Z^2 + 4N, \quad (X-Y)^2 = Z^2 - 4N. \tag{6.4.15}$$

Multiplying the equations (6.4.15) together shows that the positive integers $u = Z/2$ and $v = (Y^2 - X^2)/4$ satisfy the equation

$$v^2 = u^4 - N^2. \tag{6.4.16}$$

Put now $N = 1$ and write the numbers $u, v > 0$, $u, v \in \mathbb{Q}$ in the form: $u = a/b$, $v = c/d$, where a and b are coprime and c and d are coprime. As a result one obtains from (6.4.16)

$$c^2b^4 = a^4b^2 - b^4d^2 = (a^4 - b^4) \cdot d^2.$$

Taking into account the fact that $\mathrm{GCD}(c,d) = 1$ and $\mathrm{GCD}(a^4 - b^4, b^4) = 1$, we see that $b^4 = d^2$, i.e.

$$a^4 - b^4 = c^2. \tag{6.4.17}$$

We now rewrite (6.4.17) in the form $(a^2 - c)(a^2 + c) = b^4$ and note that a prime number dividing both numbers $a^2 - c$ and $a^2 + c$ divides also $2a^2$ and $2c$. This implies that $\mathrm{GCD}(a^2 - c, a^2 + c) = 1$ or 2 by the coprimality of a and b. However, the product of these factors is a fourth power, so we have the following two possibilities:

$$\begin{cases} a^2 - c = 2C^4 \\ a^2 + c = 8D^4 \end{cases}, \quad \text{or} \quad \begin{cases} a^2 - c = 8D^4 \\ a^2 + c = 2C^4 \end{cases},$$

where $C > 0$, C odd, and $\mathrm{GCD}(C, D) = 1$. In both cases one has $a^2 = C^4 + 4D^4$, i.e. $D^4 = (a - C^2)(a + C^2)$. Now considering the factors $a - C^2$ and $a + C^2$, we see that $a + C^2 = 2A^4$ and $a - C^2 = 2B^4$ This in turn implies that the natural numbers A, B, C satisfy the relation $A^4 - B^4 = C^2$, and $\max\{A, B, C\} < \max\{a, b, c\}$. We have reached a contradiction.

On the other hand it is not difficult to see that the curve (6.4.16) is birationally isomorphic to a plane cubic curve E^N having Weierstrass form $y^2 = x^3 - N^2x$. In order to show this one uses the substitution

$$X = (N^2 - x^2)/y, \quad Y = 2Nx/y, \quad Z = (N^2 + x^2)/y.$$

Reducing modulo primes shows that the points of finite order on $E^N(\mathbb{Q})$ are precisely those for which $y = 0$, together with the point at infinity. Thus one obtains the remarkable fact that N is congruent if and only if the group $E^N(\mathbb{Q})$ is infinite.

J.B.Tunnell proved in 1983 that if an odd natural number N is a congruent number then

$$\mathrm{Card}\{(x,y,z) \in \mathbb{Z}^3 \mid 2x^2 + y^2 + 8z^2 = N\} =$$
$$2\mathrm{Card}\{(x,y,z) \in \mathbb{Z}^3 \mid 2x^2 + y^2 + 32z^2 = N\},$$

Assuming the Birch–Swinnerton–Dyer conjecture for the curves E^N he showed that this condition is also sufficient.

In connection with the Shimura–Taniyama–Weil conjecture we point out the result of G.V.Belyi, [Be79]: every algebraic curve defined over a number field can cover the projective line with ramification points only lying above 0, 1, and ∞, the cover being defined over a number field. From this result it follows in particular that every elliptic curve over $\mathbb{Q}$ admits a parameterization by modular forms with respect to a subgroup of finite index in $\mathrm{SL}_2(\mathbb{Z})$ (which is not necessarily a congruence subgroup). This result made it possible to solve the embedding problem over certain cyclotomic extensions of $\mathbb{Q}$, and to construct Galois extensions of such fields whose Galois groups are given finite simple groups with two generators. Previously the embedding problem over $\mathbb{Q}$ was solved by I.R.Shafarevich for all finite solvable groups in [Sha54].

6.4.5 The Artin Conjecture and Cusp Forms

(cf. [DS75], [L71b], [L80], [Tun81] [Hen76], [Hir88], [Gel95]). The correspondence between primitive cusp forms of weight 2 with respect to $\Gamma_0(N)$ and elliptic curves given above by the Shimura–Taniyama–Weil conjecture (see also Chapter 7), has a remakable analogue in the case of cusp forms of weight 1. Langlands has conjectured, and Deligne and Serre have precisely formulated a link between primitive cusp forms $f(z) = \sum_{n=1} a(n)e(nz) \in \mathcal{S}_1(N, \psi)$ and irreducible two dimensional complex Galois representations

$$\rho_f : G(\overline{\mathbb{Q}}/\mathbb{Q}) \to \mathrm{GL}_2(\mathbb{C}). \tag{6.4.18}$$

The condition that f is primitive includes the conditions

$$f|_1 T(p) = a(p)f, \quad a(1) = 1 \quad ((p, N) = 1). \tag{6.4.19}$$

Deligne and Serre proved the existence of irreducible representations ρ_f unramified outside the divisors of N such that 1) $\det\rho_f = \rho_\psi$ is a one dimensional Galois representation which corresponds via the Kronecker–Weber theorem to an odd Dirichlet character $\psi : (\mathbb{Z}/N\mathbb{Z})^\times \to \mathbb{C}^\times$, $\psi(-1) = -1$, and 2) the image Tr $F_{\rho_f,\mathfrak{p}}$ coincides with $a(p)$ for all p, $(p, N) = 1$. For Eisenstein series $f \in \mathcal{M}_1(N, \psi)$ with conditions (6.4.19) such Galois representations can be easily constructed and turn out to be direct sums of Dirichlet characters.

A remarkable consequence of the construction is the proof of the *Ramanujan–Petersson* conjecture in the case of weight $k = 1$:

$$|a(p)| < 2p^{(k-1)/2} = 2. \tag{6.4.20}$$

Indeed the number $\mathrm{Tr}\ F_{\rho_f,\mathfrak{p}} = a(p)$ is the sum of two roots of unity, and the estimate (6.4.20) therefore holds for Eisenstein series (however, for weight $k = 1$ this is also true for the cusp forms).

The Ramanujan–Petersson conjecture is related to the *Sato–Tate Conjecture* on the uniform distribution of the arguments φ_p of Frobenius automorphisms in the segment $[0, \pi]$ with respect to the measure $\frac{2}{\pi}\sin^2\varphi\, d\varphi$ (cf. Chapter I in [Se68a], [Mich01] and §6.5.1).

The construction of ρ_f uses the reduction mod l of the l-adic representations attached to modular forms of weight $k \geq 2$. First one proves that a given cusp form of weight 1 has the same Fourier coefficients modulo a prime ideal, as a cusp form g of a higher weight $k \geq 2$. Then one verifies that the l-adic representation ρ_g mod l can be lifted to characteristic zero, and one gets as a result the desired complex representation ρ_f, for which properties 1) and 2) are valid since they are satisfied modulo infinitely many prime numbers.

The conditions 1) and 2) on the representation ρ_f mean that the L-series $L(s, f)$ (the Mellin transform of f) coincides with the Artin L-series of the representation ρ_f, i.e.

$$L(s, f) = L(\rho_f, s), \tag{6.4.21}$$

and the analytic properties of $L(\rho_f, s)$ follow from those of $L(s, f)$ described in 6.3.3. For complex representations $\rho : G(\overline{\mathbb{Q}}/\mathbb{Q}) \to \mathrm{GL}_n(\mathbb{C})$ the statement that $L(\rho, s)$ is holomorphic is known as the Artin conjecture. It follows that this conjecture is true for representations of the type $\rho = \rho_f$. Conversely if one knows for a two dimensional representation ρ with odd determinant $\det\rho$ that the functions $L(\rho \otimes \chi, s)$ for all Dirichlet characters χ are holomorphic, then the existence of a cusp form f satisfying (6.4.17) can be deduced from Weil's inverse theorem (see §6.3.3). It turns out that the image $\rho(G(\overline{\mathbb{Q}}/\mathbb{Q}))$ of a two dimensional irreducible representation ρ in $\mathrm{PGL}_2(\mathbb{C})$ is always one of the following groups:

1) a *dihedral group* in which case the representation ρ is monomial (i.e. induced from a character of a cyclic subgroup);
2) A_4 (*tetrahedral case*);
3) S_4 (*octahedral case*);
4) A_5 (*icosahedral case*).

Langlands and Tunnell proved the conjecture on the existence of a cusp form f for which $\rho = \rho_f$ in cases 2) and 3). The validity of the Artin conjecture in case 4) remains unknown in general. However J. Buhler (see in [Hen76]) gave an example of a representation of type 4) for which the Artin conjecture is valid, as well as the existence of the corresponding cusp form. In this example $\mathrm{Ker}\rho = G(\overline{\mathbb{Q}}/K)$ where K is the splitting field of the polynomial

$$x^5 + 10x^3 - 10x^2 + 35x - 18,$$

and $N = 800$. An interesting program was outlined by R.Taylor to solve the Artin conjecture for icosahedral Galois representations (see [Tay02] for a recent progress).

The Artin conductor

The number N in the construction of Serre and Deligne has an interpretation as the *Artin conductor* of the representation ρ_f, which is defined for every finite dimensional Galois representation $\rho : G_{\mathbb{Q}} \to \mathrm{GL}(V)$ with finite image as follows. Let p be a prime, $\mathfrak{p}$ a prime ideal of the ring $\mathcal{O} \subset \overline{\mathbb{Q}}$ of all algebraic integers, $p \in \mathfrak{p}$. Then the image of the decomposition group

$$G^{(\mathfrak{p})} \quad = \quad \{\sigma \in G_{\mathbb{Q}} \mid \sigma\mathfrak{p} = \mathfrak{p}\},$$

is isomorphic to the Galois group of some finite extension $F/\mathbb{Q}_p$, $\rho(G^{(\mathfrak{p})}) = G(F/\mathbb{Q}_p)$. Let v_F be the normalized p-valuation of F, i.e. $v_F(F^\times) = \mathbb{Z}$. Define the ramification groups

$$G_{p,i} = \{\sigma \in G(F/\mathbb{Q}) \mid v_F(x - \sigma(x)) > i \text{ for all } x \in \mathcal{O}_F\}$$

and let $V_{p,i} = V^{G_{p,i}}$. In particular $G_{p,0}$ is the inertia subgroup, and the fact that ρ is unramified over p is equivalent to saying that $V = V_{p,0}$.

Then the Artin conductor is defined (cf. [Ar30], [Ar65], [Se63]) by

$$N = N(\rho) = \prod_p p^{n(p,\rho)}, \tag{6.4.22}$$

where the exponent $n(p, \rho)$ is defined by

$$n(p, \rho) = \sum_{i=0}^{\infty} \frac{1}{(G_{p,o} : G_{p,i})} \dim\ V/V_{p,0}. \tag{6.4.23}$$

This turns out to be an integer (at first sight it only looks like a rational number). One has $n(p, \rho) = \dim\ V/V_{\rho_{p,0}} + b_p(V)$, where the number $b_p(V)$ is called the wild invariant of the representation ρ over p. One can show that for one dimensional representations the Artin conductor coincides with the conductor of the corresponding character of the idele class group, attached to it by class field theory (cf. [Se63]).

6.4.6 Modular Representations over Finite Fields

(cf. [Se87]). Based on a deep analysis of previous constructions, Serre suggested in 1987 a universal description of all two dimensional Galois representations over finite fields in terms of cusp forms. Let p be a prime number, $\mathfrak{p}$ a prime ideal of the ring of all algebraic integers $\mathcal{O} \subset \overline{\mathbb{Q}}$ dividing p (i.e. $p \in \mathfrak{p}$). We call a representation

$$\rho : G_{\mathbb{Q}} \to \mathrm{GL}(2, \overline{\mathbb{F}}_p) \tag{6.4.24}$$

a modular representation of type (N, k, ψ), if for some modular form

$$f(z) = \sum_{n=1}^{\infty} a(n)e(nz) \in \mathcal{S}_k(N, \psi),$$

which is an eigenform of the Hecke operators normalized by $a(1) = 1$, the following condition is satisfied

$$\mathrm{Tr}(F_{\rho,l}) \equiv a(l) \mod \mathfrak{p} \tag{6.4.25}$$

for all primes $l \nmid Np$.

Serre conjectured that every irreducible representation (6.4.24) is modular for some N not divisible by p. He also described explicitly the numbers N and k and the character ψ, assuming that N and k are minimal under the condition $(N, p) = 1$. According to this conjecture the number N is determined by the ramification of ρ outside p in the same way as the Artin conductor:

$$N = N(\rho) = \prod_{l \neq p} l^{n(l,\rho)}.$$

The weight k is defined by ramification properties of ρ at p, and the character is determined by the following condition on the determinant of ρ:

$$\det\rho(\mathrm{Frob}_l) \equiv \psi(l)l^{k-1} \mod \mathfrak{p} \quad (l \not| Np).$$

Serre gave many concrete examples of representations ρ for which the corresponding cusp form f with N, k and ψ as predicted by the conjecture, can be explicitly constructed.

We point out some remarkable consequences of this conjecture. First of all it implies the validity of the Shimura–Taniyama–Weil conjecture for elliptic curves over $\mathbb{Q}$ and for simple Abelian varieties with real multiplication (see §6.4.3 and Chapter 7). Also this conjecture would imply Fermat's last theorem. This corollary can be shown using the elliptic curve of Frey $E : y^2 = x(x - A)(x + B)$, where

$$A = a^p, \ B = b^p, \ C = c^p \quad a, b, c \in \mathbb{Z}, \quad p > 5$$

are integers satisfying the condition $A+B+C=0$, $ABC \neq 0$ (a non–trivial solution of Fermat's equation), see Chapter 7.

According to this conjecture of Serre, the Galois representation

$$\rho : G_{\mathbb{Q}} \to \operatorname{Aut} E_p \cong \mathrm{GL}_2(\mathbb{F}_p)$$

on points of order p of the elliptic curve of Frey–Hellegouarch should correspond to a cusp form $f \in \mathcal{S}_2(\Gamma_0(2))$. However,

$$\dim \mathcal{S}_2(\Gamma_0(2)) = g(X_0(2)) = 0,$$

hence such cusp form cannot exist.

Another approach to proving the non–existence of the elliptic curves of Frey–Hellegouarch consists of applying to the corresponding "arithmetic surface" (a scheme over Spec $\mathbb{Z}$ of dimension 2) an analogy of a result on non-singular projective surfaces of general type over an algebraically closed field of characteristic zero (the inequality of Bogomolov–Miyaoka–Yau, [Miy77], [Par87].

6.5 Automorphic Forms and The Langlands Program

6.5.1 A Relation Between Classical Modular Forms and Representation Theory

(cf. [Bor79], [PSh79]). The domain of definition of the classical modular forms (the upper half plane) is a homogeneous space $\mathbb{H} = \{z \in \mathbb{C} \mid \operatorname{Im} z > 0\}$ of the reductive group $G(\mathbb{R}) = \mathrm{GL}_2(\mathbb{R})$:

$$\mathbb{H} = \mathrm{GL}_2(\mathbb{R})/\mathcal{O}(2) \cdot Z,$$

where $Z = \{\left(\begin{smallmatrix} x & 0 \\ 0 & x \end{smallmatrix}\right) \mid x \in \mathbb{R}^\times\}$ is the center of $G(\mathbb{R})$ and $\mathcal{O}(2)$ is the orthogonal group, see (6.3.1). Therefore each modular form

$$f(z) = \sum_{n=0}^{\infty} a(n)e(nz) \in \mathcal{M}_k(N, \psi) \subset \mathcal{M}_k(\Gamma_N) \tag{6.5.1}$$

can be lifted to a function $\tilde{f}$ on the group $\mathrm{GL}_2(\mathbb{R})$ with the invariance condition

$$\tilde{f}(\gamma g) = \tilde{f}(g) \quad \text{for all} \quad \gamma \in \Gamma_N \subset \mathrm{GL}_2(\mathbb{R}).$$

In order to do this let us consider the function

$$\tilde{f}(g) = \begin{cases} f(g(i))j(g,i)^{-k} & \text{if } \det g > 0, \\ f(g(-i))j(g,-i)^{-k} & \text{if } \det g < 0, \end{cases} \tag{6.5.2}$$

where $g = \begin{pmatrix} a & b \\ c & d \end{pmatrix} \in \mathrm{GL}_2(\mathbb{R})$ and $j(g,i) = |\det g|^{-1/2}(cz+d)$ is the factor of automorphy.

One has $\tilde{f}(xg) = \exp(-ik\theta)\tilde{f}(g)$ if $x = \begin{pmatrix} \cos\theta & -\sin\theta \\ \sin\theta & \cos\theta \end{pmatrix}$ is the rotation through the angle θ.

Consider the group $\mathrm{GL}_2(\mathbb{A})$ of non-degenerate matrices with coefficients in the adele ring $\mathbb{A}$ and its subgroup

$$U(N) = \left\{ g = 1 \times \prod_p g_p \in \mathrm{GL}_2(\mathbb{A}) \mid g_p \in \mathrm{GL}_2(\mathbb{Z}_p), g_p \equiv \begin{pmatrix} 1 & 0 \\ 0 & 1 \end{pmatrix} \bmod N\mathbb{Z}_p \right\}. \tag{6.5.3}$$

From the *chinese remainder theorem* (the *approximation theorem*) one obtains the following coset decomposition:

$$\Gamma_N \backslash \mathrm{GL}_2(\mathbb{R}) \cong \mathrm{GL}_2(\mathbb{Q}) \backslash \mathrm{GL}_2(\mathbb{A})/U(N), \tag{6.5.4}$$

using which we may consider $\tilde{f}$ as a function on the homogeneous space (6.5.4), or even on the adele group $\mathrm{GL}_2(\mathbb{A})$.

The action of $\mathrm{GL}_2(\mathbb{A})$ on $\tilde{f}$ by group shifts defines a representation $\pi = \pi_f$ of the group $\mathrm{GL}_2(\mathbb{A})$ in the space of smooth complex valued functions on $\mathrm{GL}_2(\mathbb{A})$, for which

$$\left(\pi(h)\tilde{f}\right)(g) = \tilde{f}(gh) \quad (g, h \in \mathrm{GL}_2(\mathbb{A})).$$

The condition that the representation π_f be irreducible has a remarkable arithmetical interpretation: it is equivalent to f being an eigenfunction of the Hecke operators for almost all p. If this is the case then one has an infinite tensor product decomposition

$$\pi = \bigotimes_v \pi_v, \tag{6.5.5}$$

where the π_v are representations of the local groups $\mathrm{GL}_2(\mathbb{Q}_v)$ with $v = p$ or ∞.

Jacquet and Langlands chose irreducible representations of groups such as $\mathrm{GL}_2(\mathbb{Q}_v)$ as a starting point for the construction of L–functions (cf. [JL70], [Bor79]). These representations can be classified and explicitly described. Thus for the representations π_v in (6.5.5) one can verify for almost all $v = v_p$ that the representation π_v has the form of an induced representation $\pi_v = \mathrm{Ind}(\mu_1 \otimes \mu_2)$ from a one dimensional representation of the subgroup of diagonal matrices

$$(\mu_1 \otimes \mu_2) \begin{pmatrix} x & 0 \\ 0 & y \end{pmatrix} = \mu_2(x)\mu_1(y),$$

where $\mu_i : \mathbb{Q}_p^\times \to \mathbb{C}^\times$ are unramified quasicharacters (see §6.2.4). This classification makes it possible to define for almost all p an element $h_p = \begin{pmatrix} \mu_1(p) & 0 \\ 0 & \mu_2(p) \end{pmatrix} \in \mathrm{GL}_2(\mathbb{C})$. From this one can construct the following Euler product (the L–function of the automorphic representation π)

$$L(\pi, s) = \prod_{p \notin S} L(\pi_p, s) = \prod_{p \notin S} \det(1_2 - p^{-s} h_p)^{-1} \tag{6.5.6}$$

in which the product is extended over all but a finite number of primes.

It turns out that the function $L(\pi, s)$ coincides essentially with the Mellin transform of the modular form f:

$$L(s, f) = L(\pi_f, s + (k-1)/2).$$

The notion of a primitive form f also takes on a new meaning: the corresponding function $\tilde{f}$ from the representation space of an irreducible representation π must have a maximal stabilizer. The theory of Atkin–Lehner can be reformulated as saying that the representation π_f occurs with multiplicity one in the regular representation of the group $\mathrm{GL}_2(\mathbb{A})$ (the space of all square integrable functions).

More generally, an *automorphic representation* is defined as an irreducible representation of an adele reductive group $G(\mathbb{A})$ in the space of functions on $G(\mathbb{A})$ with some growth and smoothness conditions.

Jacquet and Langlands constructed for irreducible admissible automorphic representations π of the group $\mathrm{GL}_2(\mathbb{A})$ analytic continuations of the corresponding L–functions $L(\pi, s)$, and established functional equations relating $L(\pi, s)$ to $L(\tilde{\pi}, 1-s)$, where $\tilde{\pi}$ is the dual representation. For the functions $L(\pi_f, s)$ this functional equation is exactly Hecke's functional equation (see (6.3.44)).

Note that the notion of an automorphic representation includes as special cases: 1) the classical elliptic modular forms, 2) the real analytic wave modular forms of Maass, 3) Hilbert modular forms, 4) real analytic Eisenstein series of type $\sum' \dfrac{y^s}{|cz+d|^{2s}}$, 5) Hecke L-series with Grössen–characters (or rather their inverse Melin transforms), 6) automorphic forms on *quaternion algebras* etc.

Interesting classes of Euler products are related to finite dimensional complex representations

$$r : \mathrm{GL}_2(\mathbb{C}) \to \mathrm{GL}_m(\mathbb{C}).$$

Let us consider the Euler product

$$L(\pi, r, s) = \prod_p L(\pi_p, r, s), \tag{6.5.7}$$

where

$$L(\pi_p, r, s) = \det(1_m - p^{-s} r(h_p))^{-1}.$$

These products converge absolutely for $\mathrm{Re}(s) \gg 0$, and, conjecturally, admit analytic continuations to the entire complex plane and satisfy functional equations (cf. [Bor79], [BoCa79], [L71a], [Del79], [Se68a]).

This conjecture has been proved in some special cases, for example when $r = \mathrm{Sym}^i\mathrm{St}$ is the ith symmetric power of the standard representation $\mathrm{St} : \mathrm{GL}_2(\mathbb{C}) \to \mathrm{GL}_2(\mathbb{C})$ for $i = 2, 3, 4, 5$ (cf. [Sh88]).

The Ramanujan–Petersson conjecture, proved by Deligne, can be formulated as saying that the absolute values of the eigenvalues of $h_p \in \mathrm{GL}_2(\mathbb{C})$ for a cusp form f are all equal to 1.

As a consequence of the conjectured analytic properties of the functions (6.5.7) one could deduce the following conjecture of Sato and Tate about the distribution of the arguments of the Frobenius elements: let $\alpha(p) = e^{i\varphi_p}$ $(0 \le \varphi_p \le \pi)$ be an eigenvalue of the matrix h_p defined above. Then for cusp forms f without complex multiplication (i.e. the Mellin transform of f is not the L–function of a Hecke Grössencharacter (see §6.2.4) of an imaginary quadratic field) the arguments φ_p are conjecturally uniformly distributed in the segment $[0, \pi]$ with respect to the measure $\dfrac{2}{\pi}\sin^2\varphi\, d\varphi$ (cf. [Se68a]).

In the case of complex multiplication the analytic properties of the L-functions are reduced to the corresponding properties of the L-functions of Hecke Grössencharacters (see §6.2.4), which imply the uniform distribution of the arguments φ_p with respect to the usual Lebesgue measure.

The arithmetical nature of the numbers $e^{i\varphi_p}$ is close to that of the signs of Gauss sums $\alpha(p) = g(\chi)/\sqrt{p}$ where $g(\chi) = \sum_{u=1}^{p-1} \chi(u)e(u/p)$, χ being a primitive Dirichlet character modulo p. Even if χ is a quadratic character, the precise evaluation of the sign $\alpha(p) = \pm 1$ is rather delicate (see [BS85]). If χ is a cubic character, i.e. if $\chi^3 = 1$ then $p = 6t + 1$, and the sums lie inside the 1st, the 3rd or the 5th sextant of the complex plane. Using methods from the theory of automorphic forms S.J.Patterson and D.R.Heath–Brown solved the problem of Kummer on the distribution of the arguments of cubic Gauss sums by means of a cubic analogue of the theta series, which is a certain automorphic form on the threefold covering of the group GL_2 ([Del80a], [HBP79], [Kub69]).

6.5.2 Automorphic L-Functions

The approach of Jacquet–Langlands made it possible to extend the whole series of notions and results concerning L-functions to the general case of automorphic representations of reductive groups over a global field K. Let G be a linear group over K, $G_{\mathbb{A}} = G(\mathbb{A})$ its group of points with coefficients in the adele ring of the field K. Automorphic representations are often defined as representations belonging to the regular smooth representation of the group $G_{\mathbb{A}}$, and one denotes by the symbol $\mathfrak{A}(G/K)$ the set of equivalence classes of irreducible admissible automorphic representations of $G_{\mathbb{A}}$. A representation π from this class admits a decomposition $\pi = \otimes_v \pi_v$ where $v \in \Sigma_K$ runs through the places of K and the π_v are representations of the groups $G_v = G(K_v)$. In order to construct L-functions, the L-group ${}^LG$ of G is introduced. Consider the tuple of root data (cf. [Bor79], [Spr81])

$$\psi_0(G) = (X^*(T), \Delta, X_*(T), \Delta^\vee) \tag{6.5.8}$$

of the group G; here T is a maximal torus of G (over a separable closure of the ground field K); $X^*(T)$ is the group of characters of T; $X_*(T)$ the group of one parameter subgroups of T and Δ (resp. $\Delta^\vee$) is a basis of the root system (resp. the dual basis of the system of coroots). The connected component of the Langlands L-group ${}^LG^0$ is defined to be the complex reductive group obtained by inversion $\psi_0 \mapsto \psi_0^\vee$, whose root data is isomorphic to the inverse

$$\psi_0(G)^\vee = (X_*(T), \Delta^\vee, X^*(T), \Delta). \tag{6.5.9}$$

If G is a simple group, then the group ${}^LG(\mathbb{C})$ can be characterized upto a central isogeny by one of the types A_n, B_n, ..., G_2 of the Cartan–Killing classification. It is known that the map $\psi_0 \mapsto \psi_0^\vee$ interchanges the types B_n and C_n, and leaves all other types fixed. Thus if $G = \mathrm{Sp}_n$ (respectively

GSp_n), then ${}^LG^0 = \mathrm{SO}_{2n+1}(\mathbb{C})$ (resp. ${}^LG^0 = \mathrm{Spin}_{2n+1}(\mathbb{C})$). The whole group ${}^LG$ is then defined as the semi–direct product of ${}^LG^0$ with the Galois group $\mathrm{Gal}(K^s/K)$ of an extension K^s of the ground field K over which G splits (i.e. its maximal torus T becomes isomorphic to GL_1^r). This semi-direct product is determined by the action of the Galois group $\Gamma_K = \mathrm{Gal}(K^s/K)$ on the set of maximal tori defined over K^s.

The most important classification result of the Langlands theory states that if

$$\pi = \bigotimes_v \pi_v \in \mathfrak{A}(G/K)$$

then for almost all v the local component π_v corresponds to a unique conjugacy class of an element h_v in the group ${}^LG$.

Let us consider the Euler product

$$L(\pi, r, s) = \prod_{v \notin S} L(\pi_v, r, s), \tag{6.5.10}$$

where S is a finite set of places of K,

$$L(\pi_v, r, s) = \det(1_m - \mathrm{N}v^{-s} r(h_v))^{-1}.$$

Langlands has shown that if $\pi \in \mathfrak{A}(G/K)$ then the product in (6.5.10) converges absolutely for all s with sufficiently large real part $\mathrm{Re}(s)$ (cf. [L71a]). The product (6.5.10) defines an automorphic L-function only up to a finite number of Euler factors. Although this is sufficient for certain questions related to analytic continuation of these functions, the precise form of these missing factors is very important in the study of the functional equations. A list of standard conjectures on the analytic properties of the L-functions (6.5.10) can be found in A.Borel's paper [Bor79]

We refer to recent introductory texts to the theory of automorphic L-functions and the Langlands program: [BCSGKK3], [Bum97], [Iw97],

For the group $G = \mathrm{GL}_n$ and the standard representation $r = r_n = \mathrm{St} : {}^LG^0 \xrightarrow{\sim} \mathrm{GL}_n(\mathbb{C})$ the main analytic properties of the L-functions (6.5.10) are proved in [JPShS], [GPShR87], [Sh88], [JSh] (see also [Bum97], [BCSGKK3], [CoPSh94]).

Also in the case $G = \mathrm{GL}_n$ the multiplicity one theorem (an analogue of the theorem of Atkin–Lehner) (cf. [AL70], [Mi89], [Li75]) has been extended (cf. [Gel75], [Gel76]). This is closely related to the non-vanishing theorem: for a cuspidal representation π one has $L(\pi, r_n, 1) \neq 0$.

For GL_3 an analogue of Weil's inverse theorem (see §6.3.8) has been proved: if all the L-functions of type $L(\pi \otimes \chi, r_3, s)$ (where χ is a Hecke character and π is an irreducible admissible representation) can be holomorphically continued to the entire complex plane, then the representation π can be realized in the space of cusp forms ([CoPSh94], [JPShS]). More recent results on the case of GL_n, cf. [CoPSh02].

Interesting classes of L-functions attached to Siegel modular forms were introduced and studied in [An74], [An79a], [AK78]. These modular forms and their L-functions have deep arithmetical significance and are closely related to the classical problem on the number of representations of a positive definite integral quadratic form by a given integral quadratic form (as generating functions, or theta–series). These numbers arise in Siegel's general formula considered above (5.3.71). From the point of view of the theory of automorphic representations, Siegel modular forms correspond to automorphic forms on the symplectic group $G = \mathrm{GSp}_n$. In this case the dual Langlands group coincides with the universal covering $\mathrm{Spin}_{2n+1}(\mathbb{C})$ of the orthogonal group $\mathrm{SO}_{2n+1}(\mathbb{C})$. To construct L-functions one uses the following two kinds of representation of the L-group ${}^LG = \mathrm{Spin}_{2n+1} \rtimes \mathrm{Gal}(K^s/K)$: ρ_{2n+1} and r_n, where ρ_{2n+1} is the standard representation of the orthogonal group, and r_n is the spinor representation of dimension 2^n. It is convenient to consider the following matrix realization of the orthogonal group:

$$\mathrm{SO}_{2n+1}(\mathbb{C}) = \{g \in \mathrm{SL}_{2n+1}(\mathbb{C}) \mid {}^tgG_ng = G_n\},$$

with a quadratic form defined by the matrix

$$G_n = \begin{pmatrix} 0_n & 1_n & 0 \\ \cdots & \cdots & \cdots \\ 1_n & 0_n & 0 \\ 0 & \cdots & 1 \end{pmatrix}, \quad 1_n = \begin{pmatrix} 1 & 0 & \cdots & 0 \\ 0 & 1 & \cdots & 0 \\ \cdots & \cdots & \cdots & 0 \\ 0 & 0 & \cdots & 1 \end{pmatrix}.$$

If $\pi = \otimes_v \pi_v \in \mathfrak{A}(\mathrm{GSp}_n/K)$ then for almost all v the representation π_v corresponds to a conjugacy class h_v in ${}^LG$ whose image in the standard representation is given by a diagonal matrix of the type

$$\rho_{2n+1}(h_v) = \{\alpha_{1,v}, \cdots, \alpha_{n,v}, \alpha_{1,v}^{-1}, , \alpha_{n,v}^{-1}, 1\},$$

and in the spinor representation it becomes

$$r_n(h_v) = \{\beta_{0,v}, \beta_{0,v}\alpha_{1,v}, \cdots, \beta_{0,v}\alpha_{i_1,v}\alpha_{i_2,v} \cdots \alpha_{i_m,v}, \cdots\},$$

where for every $m \leq n$ all possible products of the type

$$\beta_{0,v}\alpha_{i_1,v}\alpha_{i_2,v} \cdots \alpha_{i_m,v}, \quad 1 \leq i_1 < i_2 < \cdots < i_m \leq n$$

arise.

The element h_v is uniquely defined upto the action of the Weyl group W_n generated by the substitutions

$$\beta_{0,v} \mapsto \beta_{0,v}\alpha_{i,v}, \quad \alpha_{i,v} \mapsto \alpha_{i,v}^{-1}, \quad \alpha_{j,v} \mapsto \alpha_{j,v} \quad (j \neq i)$$

and by all possible substitutions of the coordinates

$$\alpha_{i_1,v}, \alpha_{i_2,v}, \cdots \alpha_{i_n,v}.$$

A.N.Andrianov has established meromorphic continuations and functional equations for automorphic L-functions of the type $L(\pi_f, r_n, s)$ where π_f is the automorphic representation of $\mathrm{GSp}_n(\mathbb{A})$ over $\mathbb{Q}$ attached to a Siegel modular form f with respect to $\Gamma_n = \mathrm{Sp}_n(\mathbb{Z})$, $n = 2$. He has also studied the holomorphy properties of these *spinor L-functions* for various classes of Siegel modular forms f, cf. [An74], [An79a] . Analytic properties of such functions are related to versions of the theory of new forms in the Siegel modular case for $n = 2$, cf. [AP2000]. A.N. Andrianov and V.L.Kalinin in [AK78] have studied the analytic properties of the *standard L-functions* $L(\pi_f, \rho_{2n+1}, s)$, where π_f is the automorphic representation of $\mathrm{GSp}_n(\mathbb{A})$ over $\mathbb{Q}$ attached to a Siegel modular form f with respect to the congruence subgroup $\Gamma_0^n(N) \subset \mathrm{Sp}_n(\mathbb{Z})$. For $n = 1$ these L-functions coincide with the symmetric squares of Hecke series, previously studied by Shimura.

A general *doubling method* giving explicit constructions of many automorphic L-functions, was developed in [Boe85] and [GPShR87].

Further analytic properties of automorphic L-functions

We refer to Sarnak's plenary lecture [Sar98] to ICI–1998, and to the related papers [IwSa99], [KS99], [KS99a], [LRS99], [KiSha99].

In [IwSa99], four fundamental conjectures were discussed: (A) Grand Riemann hypothesis; (B) Subconvexity problem; (C) Generalized Ramanujan conjecture; (D) Birch and Swinnerton-Dyer conjecture. Another problem which is related to (D) is a special value problem. Namely, the question as to whether an L-function vanishes at a special point on the critical line.

From the classical point of view, analytic and arithmetic properties of new classes of automorphic L-functions where studied in new Shimura's books [Shi2000], [Shi04], using a developed theory of Eisenstein series on reductive groups.

6.5.3 The Langlands Functoriality Principle

(cf. [Bor79], [BoCa79], [Gel75], [Pan84] , and for recent developments, [Lau02], [Hen01], [Car2000], [Li2000], [BCSGKK3], [CKPShSh]). This important principle establishes ties between automorphic representations of different reductive groups H and G. A homomorphism of the L-groups $u : {}^LH \to {}^LG$ attached to G and H is called an L-homomorphism if the restriction of u to ${}^LH^0(\mathbb{C})$ is a complex analytic homomorphism to ${}^LG^0(\mathbb{C})$, and u induces the identity map on the Galois group G_K. The functoriality principle is formulated in terms of the conjugacy classes of the matrices h_v corresponding to the local components π_v of an irreducible admissible representation $\pi = \otimes_v \pi_v$ of the group $H(\mathbb{A}_K)$. It includes the following statements:

1) *locally*: for almost all v there exists an irreducible admissible representation $u_*(\pi_v)$ of the group $G_v = G(K_v)$ which corresponds to the conjugacy class of the element $u(h_v)$ in ${}^LG$;

2) *globally*: there exists an irreducible admissible representation $u_*(\pi) = \pi' = \otimes_v \pi'_v \in \mathfrak{A}(G/K)$ such that $\pi'_v = u_*(\pi_v)$ for almost all v. In this situation the representation π' is also called the lifting of the representation π.

In particular, according to this principle every automorphic L-function of the type $L(\pi, r, s)$ where $r : {}^LG \to \mathrm{GL}_m(\mathbb{C})$, must coincide with the L-function $L(r_*(\pi), r_m, s)$ of the general linear group GL_m with the standard representation r_m of the L-group ${}^LG^0 \xrightarrow{\sim} \mathrm{GL}_m(\mathbb{C})$. These automorphic L-functions are called standard L-functions, and as was noted above their analytic properties have to a certain extent already been studied.

Liftings of automorphic forms can be studied using the Selberg trace formula ([BoCa79], [Sel89, Sel89], [Arth83], [ArCl89]). This powerful tool establishes a connection between characters of irreducible representations and conjugacy classes, generalizing the classical result for finite groups.

The functoriality principle for automorphic forms is closely related to the problem of parametrizing the set of equivalence classes of irreducible admissible representations over global and local fields by means of representations of the Galois group (or more precisely by means of homomorphisms from the Weil group of the ground field (6.2.6) to the L-group ${}^LG$, regarded as a group over $\mathbb{C}$ in the local case, or as a group over all completions E_λ of some number field E in the global case). It is conjectured that to an admissible homomorphism of that type must correspond a non-empty set, referred to as an L-packet, of classes of irreducible admissible representations of the group $G(K_v)$ or $G(\mathbb{A}_K)$ (this is the Langlands conjecture). In this correspondence the L–function of a representation of the Weil group (6.2.6) is identified with the L-function of the associated automorphic (irreducible, admissible) representation of the reductive group.

In the case $G = \mathrm{GL}_1$ this conjecture is the essential content of class field theory (both local and global) establishing a correspondence between characters of the group $\mathrm{Gal}(\overline{K}/K)$ and automorphic forms on GL_1, which are characters of the idele class group (in the global case) or characters of the multiplicative group (in the local case).

The task of passing from GL_1 to other reductive groups is a vast noncommutative generalization of class field theory. We have considered above special cases of this correspondence attached to classical modular forms, the group GL_2 and two–dimensional Galois representations (both complex and l–adic). These examples seem to be a promising start to a theory, which is intended to tie together algebraic varieties (motives), Galois representations and automorphic forms (automorphic representations). An excellent introduction to the Langlands program is contained in [BCSGKK3] and [CKM04].

6.5.4 Automorphic Forms and Langlands Conjectures

For some recent developments in automorphic forms and applications we also refer to [Laff02], [Lau02], [Hen01], [Car2000], [Ha98], [Li2000]. A significant progress in the area of automorphic forms and their applications has been made in the last decade.

A fundamantal problem of number theory is to classify representations of the Galois group $\mathrm{Gal}(F^s/F)$ where F^s is a separable closure of a global field F, and a fundamental problem of group theory is to give the spectral decomposition of the space $L^2(G(F)\backslash G(\mathbb{A}_F))$ of automorphic forms over a reductive group G over F.

We only mention the work on the local Langlands conjecture (for $G = \mathrm{GL}_n(K)$ over a local field K), cf. [Hen01], [Car2000], [Ha98], and [LRS93], where the general case in positive characteristic was treated.

Also, we only mention Lafforgue's work on the Langlands conjecture in positive characteristic, cf. [Lau02], [Laff02], [L02], where the Langlands correspondence was established for $G = \mathrm{GL}_r$ with arbitrary r over a function global field $F = \mathbb{F}_q(X)$ of characteristic $p > 0$ where X is a smooth projective curve over $\mathbb{F}_q$. For a geometric version of the Langlands correspondence we refer to [BCSGKK3], and to [Ngo2000], containing a proof of a Frenkel-Gaitsgory-Kazhdan-Vilonen conjecture for general linear groups.

7

Fermat's Last Theorem and Families of Modular Forms

7.1 The Shimura–Taniyama–Weil Conjecture and Higher Reciprocity Laws

7.1.1 Problem of Pierre de Fermat (1601–1665)

This chapter is based on the courses of lectures given by the second author in the Ecole Normale Supérieure de Lyon (February-May 2001), in the Moscow State University (May 2001), and in the Institut Fourier (October-December 2001). Wiles' proof of Fermat's Last Theorem and of the Shimura–Taniyama–Weil Conjecture provides a magnificent example of a synthesis of different ideas and theories from previous Chapters, such as algebraic number theory, ring theory, algebraic geometry, the theory of elliptic curves, representation theory, Iwasawa theory, and deformation theory of Galois representations.

Pierre de Fermat (1601–1665) raised his most famous problem (c.1637) in the margin of a translation of Diophantus' "Arithmetic" (see also [He97] and [KKS2000]):

> Cubum autem in duos cubos, aut quadratoquadratum in duos quadratoquadratos, et generaliter nullam in infinitum ultra quadratum potestatem in duos ejusdem nominis fas est dividere: cujus rei demonstrationem mirabilem sane detexi. Hanc marginis exiguitas non caperet
>
> (It is impossible to separate a cube into two cubes, or a fourth power into two fourth powers, or in general, any other power higher than the second, into two like powers. I have discovered a truly marvelous proof of this, which this margin is too narrow to contain).

In the modern language "Fermat's Last Theorem" says that

$$\text{for } n > 2 \quad \begin{cases} x^n + y^n = z^n \\ x, y, z \in \mathbb{Z} \end{cases} \Longrightarrow xyz = 0 \qquad (FLT(n))$$

According to Ram Murty in [Mur99], "FLT deserves a special place in the history of civilization. Because of its simplicity, it has tantalized amateurs and professionals alike .... It is as if some supermind planned it all and over centuries had been developing diverse streams of thought only to have them fuse in a spectacular synthesis to resolve FLT. No single brain can claim expertise in all of the ideas that have gone into this "marvelous proof". In the age of specialization, where each one of us knows "more and more things about less and less", it is vital for us to have an overview of the masterpiece such as the one provided by this book."

The following is a summary of early progress on FLT.

Case $n = 4$ (Fermat himself in a letter to Huygens);
Case $n = 3$ (Euler in 1753);
Case $n = 5$ (Dirichlet, Legendre, c.1825);
Case $n = 7$ (G.Lamé, 1839; $n = 14$ was already done by Dirichlet in 1832);
The "first case", $FLT_{\mathrm{I}}(p)$ for all primes p for which $q = 2p + 1$ is also prime (Sophie Germain in a letter to Gauss in 1820):

$$\begin{cases} x^p + y^p = z^p \\ x, y, z \in \mathbb{Z} \end{cases} \Longrightarrow \quad xyz \equiv 0 (\mathrm{mod} p) \qquad (FLT_{\mathrm{I}}(p)).$$

7.1.2 G.Lamé's Mistake

On March 1, 1847, a French mathematician G.Lamé informed the Academy of Sciences in Paris that he had found a complete proof of FLT based on the identity

$$x^p + y^p = (x + y)(x + \zeta y) \cdot \ldots \cdot (x + \zeta^{p-1} y), \quad \zeta = \zeta_p = \exp(2\pi i/p), p \neq 2,$$

assuming the uniqueness of factorization in the ring $\mathbb{Z}[\zeta_p]$.
Immediately J.Liouville reacted by saying: "N'y a-t-il pas là une lacune à remplir?" ("Isn't there a gap to be filled?"), and indeed few months later A.Cauchy discovered non-unique factorizations in $\mathbb{Z}[\zeta_{23}]$.

E.Kummer's Work

E.Kummer in 1847 introduced the notion of a *regular prime* p, which in modern language is:

$$\text{for every ideal } \; I \subset \mathbb{Z}[\zeta_p] \; \left(I^p \text{ principal} \Longrightarrow I \text{ principal} \right).$$

He proved FLT(p) for all regular primes p (for which he was awarded the Golden Medal of the Academy of Sciences in Paris in 1850). The smallest irregular prime is $p = 37$, see [BS85], [He97].

7.1.3 A short overview of Wiles' Marvelous Proof

(see also [Ste95], [RubSil94], [Da95]). In lectures at the Newton Institute in June 1993, Andrew Wiles announced a proof of a large part of the Shimura–Taniyama–Weil conjecture (STW) and, as a consequence, Fermat's Last Theorem (using an earlier result of K.Ribet [Ri]). A final corrected version of this proof, completed together with R. Taylor, has appeared in [Wi] and [Ta-Wi]. In this truly marvelous proof, a traditional argument of *reductio ad absurdum* is presented in the following form: if $a^p + b^p = c^p$, $abc \neq 0$, for a prime $p \geq 5$ (a primitive solution (a, b, c)), then one shows the existence of a non-zero holomorphic function $f = f_{a,b,c} : \mathbb{H} \to \mathbb{C}$ on the Poincaré upper half–plane $\mathbb{H}$, defined by a certain Fourier series with the first coefficient equal to 1. It turns out that this function has too many symmetries, forcing $f \equiv 0$, a contradiction with its construction.

In the discussion below, the following group of symmetries plays an important role:

$$\Gamma_0(N) = \left\{ \begin{pmatrix} \alpha & \beta \\ \gamma & \delta \end{pmatrix} \in \mathrm{SL}_2(\mathbb{Z}) \;\middle|\; \gamma \equiv 0 \bmod N \right\} \subset \mathrm{SL}_2(\mathbb{Z}).$$

This group is called the *congruence subgroup* of *level* N and the corresponding compact Riemann surface

$$X_0(N) = \Gamma_0(N)\backslash\overline{\mathbb{H}}, \quad \text{where } \overline{\mathbb{H}} := \mathbb{H} \cup \mathbb{Q} \cup i\infty,$$

is called a *modular curve* of *level* N. A function $f : \overline{\mathbb{H}} \to \mathbb{C}$ is called a *modular form* of weight 2 and level N if the differential $\omega_f = f(z)dz$ descends to a holomorphic differential on $X_0(N)$, that is, ω_f is holomorphic and invariant under the action of $\Gamma_0(N)$:

$$\forall \sigma = \begin{pmatrix} \alpha & \beta \\ \gamma & \delta \end{pmatrix} \in \Gamma_0(N), \;\; f\left(\frac{\alpha z + \beta}{\gamma z + \delta}\right) = (\gamma z + \delta)^2 f(z),$$

and $f(x) = 0$ for all $x \in \mathbb{Q} \cup i\infty$. We write $\mathcal{S}_2(N) \cong \Omega^1(X_0(N))$ for the $\mathbb{C}$-vector subspace of such forms; it has dimension $\dim \mathcal{S}_2(N) = \mathrm{genus}(X_0(N))$.

The main point is to show that $f = f_{a,b,c}$ is a (non–trivial) modular form of weight 2 and level 2: $f \in \mathcal{S}_2(2)$. However, $X_0(2) \cong \mathbb{S}$ (the Riemann sphere) has no non–trivial differentials, a contradiction.

To prove the existence of a non–zero function f with these properties one starts with the Frey–Hellegouarch curve

$$E = E_{a^p,b^p,c^p} : y^2 = P_3(x), \quad \text{where} \quad P_3(x) = x(x - a^p)(x + b^p), \tag{7.1.1}$$

(assuming without loss of generality that $a^p \equiv -1 \bmod 4$ and b is even). One observes first of all that the *discriminant* of the cubic polynomial P_3 is equal to $(abc)^{2p} \neq 0$. Hence the projectivization E of the affine curve with the equation (7.1.1) is smooth of genus 1, and has a rational point at infinity. Let us consider the *generating series*: for any prime l define

$$N_l(E) = \#\left\{(x,y) \in \mathbb{F}_l^2 \mid y^2 \equiv P_3(x) \bmod l\right\}, \tag{7.1.2}$$

$$b_l = b_l(E) = l - N_l(E),$$

$$g = g_{E,S} = \sum_{n\geq 1} b_n e^{2\pi i n z} \tag{7.1.3}$$

$$\text{with } \sum_{n\geq 1} b_n n^{-s} = \prod_{l\notin S} \frac{1}{1 - b_l l^{-s} + l^{1-2s}},$$

for a finite set S of primes containing all prime divisors of the discriminant of P_3. In particular, in the definition (7.1.3) one has $b_1{=}1$.

One sees that the series g converges and hence defines a holomorphic function on the complex Poincaré upper half–plane $\mathbb{H}$, and that

$$N_l(E) = l + \sum_{x \bmod l} \left(\frac{P_3(x)}{l}\right), \quad \left(\frac{x}{l}\right) \text{ being the Legendre symbol.}$$

Now one proceeds to show that:

- *Modularity:* if E is a semistable elliptic curve then the generating series $g = g_E$ is modular – this is the main ingredient of the proof;
- *Controlling the level:* there exists a modular form f of an appropriate minimal level $N_0 = N(E,p)$ such that the Fourier coefficients of f are congruent to those of g modulo an appropriate prime ideal λ_p. It turns out that for the Frey–Hellegouarch curve one has $N_0 = 2$.

Remark 7.1. According to the theorem of Faltings, the modularity of the series $g = g_{E,S}$ is equivalent to the existence of a modular parametrization

$$\varphi_N : X_0(N) \to E,$$

since E is isogenous to a factor of the Jacobian $J_0(N)$ coming from the choice of a cusp eigenform given by the generating series (7.1.3) (see §6.4.3).

7.1.4 The STW Conjecture

Conjecture 7.2 (Shimura–Taniyama–Weil). For any elliptic curve E over $\mathbb{Q}$ there exist a finite set S of primes and an $N = N(E,S) \in \mathbb{N}$ such that the generating series $g = g_{E,S}$ given by (7.1.3) is a modular form of weight 2 and level N.

If the set S of exceptional primes is *minimal*, then N is the *minimal conductor* of E.

Example 7.3. Let

$$E : y^2 + y = x^3 - x^2 \iff y^2 = x^3 - x^2 + \frac{1}{4}, \quad S = \{11\}.$$

Then the generating series is given by

$$g = q \prod_{m \geq 1} (1-q^m)^2(1-q^{11m})^2 = q-2q^2-q^3+2q^4+q^5+2q^6-2q^7+\cdots \in S_2(11)$$

l	2	3	5	7	11	13	17	19	23	29	31	$\cdots$	10007
N_l	4	4	4	9	-	9	19	19	24	29	24	$\cdots$	9989
b_l	-2	-1	1	-2	-	4	-2	0	-1	0	7	$\cdots$	18

7.1.5 A connection with the Quadratic Reciprocity Law

(cf. [Da99]). Conjecture 7.2 may be viewed as a far reaching generalization of the Quadratic Reciprocity Law of Gauss. Indeed, let us consider for any $d \in \mathbb{Z}$,

$$N_l(d) = \# \left\{ x \in \mathbb{F}_l \mid x^2 \equiv d \bmod l \right\}$$

$$\text{hence } N_l(d) = 1 + \left(\frac{d}{l}\right) = \begin{cases} 2, & d \in (\mathbb{F}_l^\times)^2, \\ 1, & d \equiv 0 \bmod l, \\ 0, & d \notin (\mathbb{F}_l)^2. \end{cases}$$

By quadratic reciprocity, $N_l(d)$ depends only on $l \bmod 4|d|$, and the generating series

$$g = \sum_{n \geq 1} b_n e^{2\pi i n z} \quad \text{with} \quad \sum_{n \geq 1} b_n n^{-s} = \prod_l \left(1 - \left(\frac{d}{l}\right) l^{-s}\right)^{-1} \tag{7.1.4}$$

belongs in fact to the finite dimensional complex vector space consisting of all formal Fourier series with coefficients d_n periodic modulo $4|d|$ (as functions of n). The generating series (7.1.4) is in fact attached to the Galois representation

$$\rho = \rho_{\chi_d} : G_{\mathbb{Q}} \to \{\pm 1\} = \mathrm{GL}_1(\mathbb{Z}), \quad (\rho_{\chi_d}(n) = \left(\frac{d}{n}\right)),$$

and therefore has the same nature as the series (7.1.3), also coming from a Galois representation (attached to E).

7.1.6 A complete proof of the STW conjecture

The STW conjecture was proved in full generality in 1999 by Ch.Breuil, B.Conrad, F.Diamond and R.Taylor (see [Da99]). In 1994 A.Wiles proved this conjecture for the important subset of *semistable* elliptic curves. This was sufficient to deduce FLT from STW, since if one assumes the existence of a Frey–Hellegouarch curve, then such a curve would necessarily be semistable. We shall explain the notion of semistability below. Following a theorem of K.Ribet (1986) (cf. [Ri]), conjectured by G.Frey in autumn 1984, a Frey–Hellegouarch curve *can not be modular*, since its generating series would have too many symmetries. Therefore the semistable STW implies FLT by the non–modularity of the Frey–Hellegouarch curves.

Note that the full STW is necessary, for example, in order to prove that

$$a^p + b^p = c^3 \Longrightarrow abc = 0 \quad \text{for} \quad p \geq 5.$$

One proves this result using a curve analogous to the Frey–Hellegouarch curve; however, this analogous curve is no longer semistable.

Controlling the level: the existence of a modular form f of minimal level $N_0 = N(p, E)$ attached to an E and p, is given as a consequence of the following theorem of Mazur-Ribet (which is only briefly discussed in this chapter, but see a detailed exposition in [Ri], [Edx95]).

The theorem of Mazur-Ribet is formulated in terms of the *Artin conductor* $N(\rho_0)$ of a Galois representation

$$\rho_0 : \mathrm{Gal}(\overline{\mathbb{Q}}/\mathbb{Q}) \to \mathrm{GL}_2(\overline{\mathbb{F}}_p),$$

and in terms of the formal power series

$$g = g_{\rho_0} = \sum_{n \geq 1} b_n q^n \in \overline{\mathbb{F}}_p[[q]],$$

which may be attached to any such representation. Assuming that the series g is the q-expansion (modulo a prime) of a modular form of weight 2 and some level N, we say that ρ_0 is modular of weight 2 and level N.

Assuming this (and some other conditions, including the irreducibility of ρ_0, see Theorem 7.4) the theorem of Mazur-Ribet states the existence of a modular form of the minimal level $N(\rho_0)$ congruent to the series $g \in \overline{\mathbb{F}}_p[[q]]$. In particular, this nice result is applicable to the Galois representation

$$\rho_0 = \rho_{p,E} : \mathrm{Gal}(\overline{\mathbb{Q}}/\mathbb{Q}) \to \mathrm{GL}_2(\mathbb{F}_p),$$

given by the Galois action on the points of order p of E. The notation $N_0 = N(p, E) = N(\rho_0)$ is used for $N(\rho_0)$. Assuming the modularity of E, one deduces the modularity of the Galois representation ρ_0, and this implies the existence of a modular form of minimal level $N(\rho_0)$ with this property. It turns out that for the Frey–Hellegouarch curve we have $N(p, E) = N(\rho_0) = 2$, and this is sufficient to deduce FLT from STW.

We shall give here only some formulations and brief comments on results about modular forms of minimal level:

Theorem 7.4 (Mazur-Ribet). *Let $p \geq 3$ be a prime. Let $\rho_0 : \mathrm{Gal}(\overline{\mathbb{Q}}/\mathbb{Q}) \to GL_2(\overline{\mathbb{F}}_p)$ be an irreducible Galois representation, which is modular of weight 2 and of a square-free level N.*

If ρ_0 is finite *at p, then it is modular of weight 2 and of level $N(\rho_0)$, where $N(\rho_0)$ is the Artin conductor of ρ_0 (see (6.4.22)). If ρ_0 is not finite at p, then it is modular of weight 2 and of level $pN(\rho_0)$.*

Remark 7.5. 1) The condition "ρ_0 is *finite* at p" is a local condition concerning the restriction $\rho_0|_{D_p}$ to the decomposition group at p. This condition means that $\rho_0|_{D_p}$ comes from a finite flat group scheme over $\mathbb{Z}_p$ (cf. Tate's paper in [CSS95]);

2) By a general property of the Galois representation

$$\rho_0 : \mathrm{Gal}(\overline{\mathbb{Q}}/\mathbb{Q}) \to \mathrm{GL}_2(\overline{\mathbb{F}}_p),$$

attached to a cusp form of level N, it follows that $N(\rho_0)$ always divides N (see §6.4.2 and [Edx95]).

The proof of Theorem 7.4 has two parts. The first part, due to Mazur, deals with primes l dividing $N/N(\rho_0)$ that are not congruent to 1 modulo p. The second part, due to Ribet, deals with an arbitrary $l \neq p$ at the cost of introducing a prime q in the level that could be then removed by the first part.

Theorem 7.6 (Mazur). *Let $p \geq 3$ be a prime. Let $\rho_0 : \mathrm{Gal}(\overline{\mathbb{Q}}/\mathbb{Q}) \to GL_2(\overline{\mathbb{F}}_p)$ be an irreducible Galois representation, which is modular with weight 2 and some level N. Suppose that l is a prime not congruent to* $1 \bmod p$*, that l divides N but l^2 does not, that ρ_0 is unramified at l if $l \neq p$ and that ρ is finite at p if $l = p$.*

Then ρ_0 is modular of weight 2 and level N/l.

Theorem 7.7 (Ribet). *Let $p \geq 3$ be a prime. Let $\rho_0 : \mathrm{Gal}(\overline{\mathbb{Q}}/\mathbb{Q}) \to GL_2(\overline{\mathbb{F}}_p)$ be an irreducible Galois representation, which is modular of weight 2 and some level N. Suppose that $l \neq p$ is a prime, that l divides N but l^2 does not, that ρ_0 is unramified at l if $l \neq p$.*

Then there exists a prime number q not dividing N and congruent to $-1 \bmod p$*, such that ρ_0 is modular of weight 2 and level qN/l.*

Corollary 7.8. *Let E be a semistable elliptic curve. Assume that the generating series*

$$g = g_{E,S} = \sum_{n\geq 1} b_n q^n \in \mathbb{Z}[[q]]$$

is modular (i.e. $g \in S_2(N)$ for some N). Then the conditions of Theorem 7.4 are satisfied for the Galois representation

$$\overline{\rho}_{p,E} : \mathrm{Gal}(\overline{\mathbb{Q}}/\mathbb{Q}) \to GL_2(\mathbb{F}_p).$$

Hence there exists a modular form of weight 2 and minimal level N_0,

$$f = \sum_{n \geq 1} a_n q^n \in \mathcal{O}_{\overline{\mathbb{Q}}}[[q]]$$

with coefficients in the ring of algebraic integers $\mathcal{O}_{\overline{\mathbb{Q}}} \subset \overline{\mathbb{Q}} \subset \mathbb{C}$ such that

$$f \in S_2(N_0) \quad \text{and} \quad \forall l \notin S, a_l \equiv b_l \bmod \lambda_p$$

for some maximal ideal $\lambda_p \subset \mathcal{O}_{\overline{\mathbb{Q}}}$ containing p (in particular, $f \not\equiv 0$).

Corollary 7.9. *If $p \geq 5$, then the Frey–Hellegouarch curve $E = E_{a^p,b^p,c^p}$: $y^2 = x(x - a^p)(x + b^p)$ with $a^p \equiv -1 \bmod 4$ and b even, can not be modular.*

In fact, using Tate's curve (see §6.3.3) one can easily calculate the Artin conductor in this case: it turns out that $N_0 = 2$. On the other hand $S_2(2)$ is zero because $S_2(2) \cong \Omega^1(X_0(2)) = 0$. Hence by theorem 7.4, f is identically zero, which is absurd since its first coefficient is 1.

7.1.7 Modularity of semistable elliptic curves

The main purpose of this chapter is to explain Wiles' proof of the modularity of all semistable elliptic curves over $\mathbb{Q}$.

Definition 7.10. *a) An elliptic curve E over $\mathbb{Q}$ is said to be semistable at a prime l if one can choose its equation in the form $\Phi(x, y) = 0$ in such a way that*

$$\Phi(x, y) \in \mathbb{Z}_l[x, y] \text{ and} \tag{7.1.5}$$

$$\text{the singular points of the reduction } \overline{\Phi}(x, y) \in \mathbb{F}_l[x, y] \text{ are simple} \tag{7.1.6}$$

(that is, the quadratic part of $\overline{\Phi}(x, y)$ at any singular point (x_0, y_0) is non-degenerate; recall that a singular point (x_0, y_0) is a point such that:

$$\overline{\Phi}(x_0, y_0) = \overline{\Phi}'_x(x_0, y_0) = \overline{\Phi}'_y(x_0, y_0) = 0.$$

b) An elliptic curve E over $\mathbb{Q}$ is called semistable if it is semistable at all primes l.

Remark 7.11. The definition 7.10 is entirely geometric. However one can give a purely algebraic definition 7.18 of the notion of semistability using Tate's uniformization, see §6.3.3: an elliptic curve E is semistable if and only if the representations $\rho_{p,E}$ on the Tate modules of E satisfy the condition:

$$\forall p, \forall l \neq p, \quad \rho_{p,E}(I_l) \text{ is conjugate to a subgroup of } \begin{pmatrix} 1 & * \\ 0 & 1 \end{pmatrix},$$

where $I_l \subset G_{\mathbb{Q}} = \mathrm{Gal}(\overline{\mathbb{Q}}/\mathbb{Q})$ denotes the inertia subgroup.

Example 7.12. a) Consider the Frey–Hellegouarch curve E given by $y^2 = x(x-a^p)(x+b^p)$, where $p \geq 5$ is prime. We shall assume that b is even and $a^p \equiv -1 \bmod 4$. After the substitution $x = 4X$, $y = 8Y + 4X$ we obtain the following equation

$$Y^2 + XY = X^3 + \gamma X^2 + \delta X, \quad \gamma = \frac{b^p - a^p - 1}{4}, \quad \delta = -\frac{a^p b^p}{16}.$$

The point $(X, Y) = (0, 0)$ is in fact the only singular point $\bmod 2$, and here the quadratic part is given by

$$\begin{cases} Y^2 + XY, & \text{if 8 divides } a^p + 1 \\ X^2 + XY + Y^2, & \text{if 8 does not divide } a^p + 1. \end{cases}$$

Thus in either case E is semistable at $l = 2$.
Now suppose $l \neq 2$, $a^p + b^p = c^p$ and $l | abc$. The equation reduces to one of the form $y^2 \equiv x^2(x - \alpha) \bmod l$ with $\alpha \not\equiv 0$. The only singular point here is $(x, y) = (0, 0)$, and the quadratic part $y^2 - \alpha x^2$ is non-degenerate.

b) The modular elliptic curve $X_0(15) : y^2 = x(x + 3^2)(x - 4^2)$ is semistable, whereas the curve $E : y^2 = x(x - 3^2)(x + 4^2)$ is not semistable.

Now the main result of the chapter says:

Theorem 7.13 (semistable STW Conjecture, A.Wiles (1994)). *Every semistable elliptic curve is modular.*

Corollary 7.14. *There exist no Frey–Hellegouarch curves $E = E_{a^p,b^p,c^p}$, and hence $FLT(p)$ is true for any prime $p \geq 5$.*

7.1.8 Structure of the proof of theorem 7.13 (Semistable STW Conjecture)

I. Modularity modulo p (with $p = 3, 5$)

Let $E : y^2 = P_3(x)$ be a semistable elliptic curve. We may assume that $P_3(x) \in \mathbb{Z}[x]$, and we let $g = g_{E,S} = \sum_{n \geq 1} b_n q^n \in \mathbb{Z}[[q]]$ be its generating series,

$$b_l = l - N_l(E) = -\sum_{x \bmod l} \left(\frac{P_3(x)}{l}\right), \quad \left(\frac{x}{l}\right) \text{ is the Legendre symbol.}$$

One constructs a modular form $h = \sum_{n \geq 1} c_n q^n \in \mathcal{O}_{\overline{\mathbb{Q}}}[[q]]$ with $c_l \equiv b_l \bmod \lambda_p$ for all $l \notin S$, the finite set of exceptional primes S, where $\lambda_p \subset \mathcal{O}_{\overline{\mathbb{Q}}}$ is a prime ideal containing p. This problem was solved only for $p = 3$ (the Tunnell-Langlands-Serre Theorem) under the assumption of *absolute irreducibility modulo 3* of the representation

$$\overline{\rho}_{3,E} : \mathrm{Gal}(\overline{\mathbb{Q}}/\mathbb{Q}) \to \mathrm{GL}_2(\mathbb{F}_3)$$

(on the points of order 3 of E). This condition is not always satisfied, and this was considered by A.Wiles as a significant difficulty in his proof. In May 1993 A.Wiles found a way to overcome this difficulty; he found a way of switching the problem from $p = 3$ to $p = 5$. He used a family E_t of elliptic curves with the property that all the representations $\overline{\rho}_{5,E_t}$ are isomorphic, and such that there exists a curve $E' = E_{t_0}$ in this family with an irreducible representation $\overline{\rho}_{3,E'}$. This made it possible to replace E by E' in the above argument, see §7.7.

II. Modular lifting

Any series $\tilde{h} = \sum_{n\geq 1} \tilde{c}_n q^n \in \mathcal{O}[[q]]$ with coefficients in a finite extension $\mathcal{O}$ of $\mathbb{Z}_p$ (with maximal ideal λ), satisfying certain *necessary conditions* of modularity, and having the property that:

$$\forall l \notin S, \tilde{c}_l \equiv c_l \bmod \lambda$$

is *automatically* a modular form (which is called a *lifting* of $h \bmod \lambda$, or an *admissible deformation* of h). One gives these necessary conditions in terms of the *absolute irreducibility* of the Galois representations attached to modular forms (these conditions are used also in Theorem 7.4 of Ribet). In other words, one shows that under these conditions any admissible lifting of $h \bmod \lambda$ is in fact modular.

III. Absolute irreducibility

One shows that the absolute irreducibility conditions are satisfied for the Galois representations

$$\overline{\rho}_{p,E} : \mathrm{Gal}(\overline{\mathbb{Q}}/\mathbb{Q}) \to \mathrm{GL}_2(\mathbb{F}_p)$$

either for $p = 3$ or for $p = 5$.

IV. End of the proof: passage from $p = 3$ to $p = 5$

Let us consider the series $\tilde{h} = g$ (the generating series of our representation). By I) this series is an admissible deformation of a modular form $h \bmod 3$ (under the conditions of absolute irreducibility for $p = 3$), and II) implies that in this case g must be a modular form. In other cases, III) says that the irreducibility condition is satisfied for $p = 5$. Moreover, II) implies that $g' = g_{E',S}$ is a modular form. By the construction of E', we know that $g' \bmod 5 = g \bmod 5 \in \mathbb{F}_5[[q]]$ again satisfies the conditions II) for modular lifting (with $p = 5$). Hence g is also modular in this case.

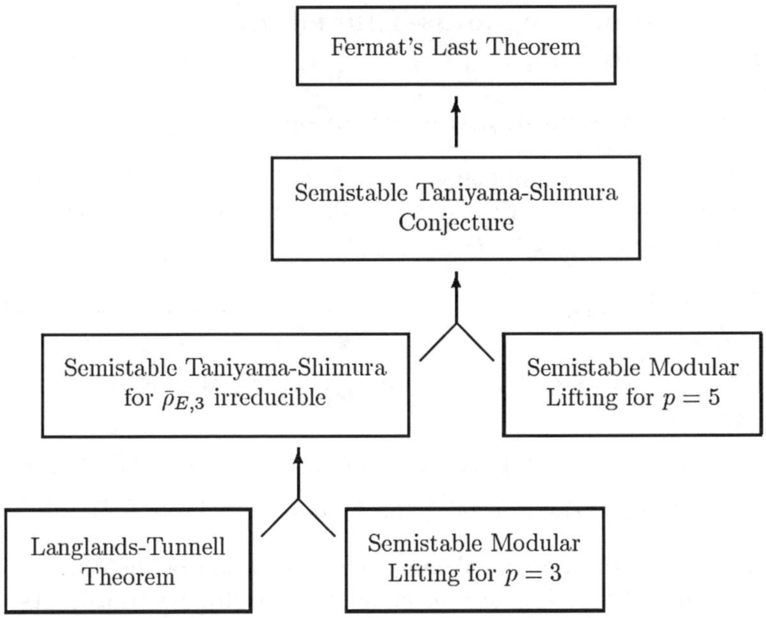

Semistable Modular Lifting Conjecture $\Rightarrow$ Fermat's Last Theorem .

Fig. 7.1.

The main stages of Wiles' Proof are presented in the Figure 7.1, which is reproduced here from [RubSil94], p. 6, with a kind permission of K.Rubin, A.Silverberg and AMS.

In the next section we describe the Langlands-Tunnell Theorem.

7.2 Theorem of Langlands-Tunnell and Modularity Modulo 3

7.2.1 Galois representations: preparation

Recall that the Galois group $\mathrm{Gal}(\overline{\mathbb{Q}}/\mathbb{Q})$ of $\overline{\mathbb{Q}} \subset \mathbb{C}$ contains for any prime l and for any maximal ideal $\lambda \subset \mathcal{O}_{\overline{\mathbb{Q}}}$ over l the decomposition subgroup D_l and its normal inertia subgroup $I_l = I_{\lambda/l}$,

$$\begin{aligned} G_{\mathbb{Q}} &= \mathrm{Gal}(\overline{\mathbb{Q}}/\mathbb{Q}) \\ &\cup \\ D_l &= D_{\lambda/l} = \{g \in \mathrm{Gal}(\overline{\mathbb{Q}}/\mathbb{Q}) \mid g(\lambda) = \lambda\} \cong \mathrm{Gal}(\overline{\mathbb{Q}}_l/\mathbb{Q}_l) \\ &\triangledown \\ I_l &= I_{\lambda/l} = \{g \in D_l \mid \forall x \in \mathcal{O}_{\overline{\mathbb{Q}}}, g(x) \equiv x (\mathrm{mod} \lambda)\} \cong \mathrm{Gal}(\overline{\mathbb{Q}}_l/\mathbb{Q}_l^{\mathrm{nr}}), \\ D_l/I_l &\cong \mathrm{Gal}(\overline{\mathbb{F}}_l/\mathbb{F}_l) = \langle \mathrm{Frob}_l \rangle, \quad \mathrm{Frob}_l(x) = x^l, \quad \mathrm{Frob}_l \in \mathrm{Gal}(\overline{\mathbb{F}}_l/\mathbb{F}_l) \end{aligned} \tag{7.2.1}$$

where $\mathbb{Q}_l^{\mathrm{nr}}$ denotes the maximal unramified extension of the l-adic field $\mathbb{Q}_l$. The subgroups $D_l = D_{\lambda/l}$ are all conjugate due to the transitivity of the action of $\mathrm{Gal}(\overline{\mathbb{Q}}/\mathbb{Q})$ on the maximal ideals λ above l. This means that for any l, the Frobenius element Frob_l lifts to a conjugacy class of cosets of I_l in $G_{\mathbb{Q}}$.

Definition 7.15. *a) Let A be a topological ring over $\mathbb{Z}_p$. We define a Galois representation over A to be a continuous homomorphism*

$$\rho : G_{\mathbb{Q}} \to \mathrm{GL}_n(A).$$

b) ρ is said to be unramified at l if $\rho(I_l) = \{1\}$; in this case the trace $tr(\rho(\mathrm{Frob}_l)) \in A$ is well defined.
c) ρ is said to be reducible if there exists a matrix $C \in \mathrm{GL}_n(A)$ such that

$$\forall g \in G_{\mathbb{Q}}, C^{-1}\rho(g)C \in \left\{ \begin{pmatrix} * & * \\ 0 & * \end{pmatrix} \right\}.$$

Arithmetical examples.

a) *The Cyclotomic character.* Let $A = \mathbb{Z}_p$, $n = 1$, $\rho = \chi_p : G_{\mathbb{Q}} \to \mathbb{Z}_p^{\times}$ the character of the group $G_{\mathbb{Q}}$ defined by the action on the roots of unity of degree p^r: $\varprojlim_r \mu_{p^r} \cong \mathbb{Z}_p$. One has $\chi_p(\mathrm{Frob}_l) = l \in \mathbb{Z}_p^{\times}$ for any $l \neq p$.

b) For an elliptic curve E over $\mathbb{Q}$ one has $E(\mathbb{C}) \cong \mathbb{C}/\langle \omega_1, \omega_2 \rangle$ (Weierstrass' theorem). Hence for any positive integer m the group of m-torsion points

$$E[m] := \mathrm{Ker}(u \mapsto mu) \cong (\mathbb{Z}/m\mathbb{Z})^2$$

of E is a $G_{\mathbb{Q}}$-module: $\overline{\rho}_{m,E} : G_{\mathbb{Q}} \to GL_2(\mathbb{Z}/m\mathbb{Z})$. Putting $m = p^r$ and passing to the limit

$$\lim_{\substack{\longleftarrow \\ r}} E[p^r] \cong \mathbb{Z}_p^2,$$

one obtains a Galois representation

$$\rho_{p,E} : G_{\mathbb{Q}} \to GL_2(\mathbb{Z}_p)$$

with the properties:

$$\det \rho_{p,E} = \chi_p \tag{7.2.2}$$

$$\rho_{p,E} \text{ is unramified at all primes } l \nmid p\Delta_E, \tag{7.2.3}$$

$$\text{and for these primes } \mathrm{tr}\rho_{p,E}(\mathrm{Frob}_l) = l + 1 - \#\tilde{E}(\mathbb{F}_l),$$

where $\tilde{E}$ denotes the reduction $E \bmod l$ (which is good in this case due to the Néron–Ogg–Shafarevich criterium, see §5.4.1).

Theorem 7.16 (Langlands–Tunnell–Serre). *Let E be an elliptic curve and let*

$$\overline{\rho}_{p,E} : \mathrm{Gal}(\overline{\mathbb{Q}}/\mathbb{Q}) \to \mathrm{GL}_2(\mathbb{F}_p)$$

be the representation on the points of order p of E. Assume that $\overline{\rho}_{3,E}$ is irreducible. Then there exists a cusp form $h = \sum_{n\geq 1} c_n q^n, c_n \in \mathcal{O}_{\overline{\mathbb{Q}}}$ of weight 2 and a maximal ideal $\lambda_3 \subset \mathcal{O}_{\overline{\mathbb{Q}}}$ above 3, such that for all l outside a finite set $S = S(E)$,

$$c_l \equiv l + 1 - \#\tilde{E}(\mathbb{F}_l) \pmod{\lambda_3}.$$

Proof is explained in §7.2. It makes use of the commutative diagram

$$\begin{array}{ccc} \Psi : \mathrm{GL}_2(\mathbb{F}_3) & \xrightarrow{\text{irreducible}} & \mathrm{GL}_2(\mathbb{Z}[\sqrt{-2}]) \subset \mathrm{GL}_2(\mathbb{C}) \\ & \searrow^{\text{id}} & \downarrow {\scriptstyle \bmod (1+\sqrt{-2})} \\ & & \mathrm{GL}_2(\mathbb{F}_3) \end{array}$$

where Ψ denotes the two-dimensional complex irreducible representation of $\mathrm{GL}_2(\mathbb{F}_3)$ given by

$$\Psi\begin{pmatrix} -1 & 1 \\ -1 & 0 \end{pmatrix} = \begin{pmatrix} -1 & 1 \\ -1 & 0 \end{pmatrix}, \Psi\begin{pmatrix} 1 & -1 \\ 1 & 1 \end{pmatrix} = \begin{pmatrix} 1 & -1 \\ -\sqrt{-2} & -1 + \sqrt{-2} \end{pmatrix}.$$

Remark 7.17. Note that the homomorphism Ψ is a section of the natural homomorphism

$$\mathrm{GL}_2(\mathbb{Z}([\sqrt{-2}]) \to \mathrm{GL}_2(\mathbb{F}_3),$$

induced by the reduction modulo $(1 + \sqrt{-2})$:

$$\mathbb{Z}[\sqrt{-2}] \to \mathbb{Z}[\sqrt{-2}]/(1 + \sqrt{-2}) \cong \mathbb{F}_3.$$

7.2.2 Modularity modulo p

The modularity of an arbitrary irreducible Galois representation modulo 3, given by the proof of Theorem 7.16, is a special case of a general conjecture of Serre [Se87] (see §6.4.6):

Let p be a prime number, $\mathfrak{p}$ a prime ideal of the ring of all algebraic integers $\mathcal{O} \subset \overline{\mathbb{Q}}$ dividing p (i.e. $p \in \mathfrak{p}$). We call a representation

$$\rho : G_{\mathbb{Q}} \to \mathrm{GL}_2(\overline{\mathbb{F}}_p) \tag{7.2.4}$$

a modular representation of type (N, k, χ), if there is a modular form (see §6.3.5)

$$h(z) = \sum_{n=1}^{\infty} c_n e(nz) \in \mathcal{S}_k(N, \chi)(e(z) = \exp(2\pi i z)),$$

which is an eigenform of all the Hecke operators, normalized by $c_1 = 1$, such that for all primes $l \nmid Np$ the representation ρ is unramified at l and we have

$$\mathrm{Tr}\ \rho(\mathrm{Frob}_l) \equiv c_l \mod \mathfrak{p}. \tag{7.2.5}$$

Serre conjectured that every irreducible representation (7.2.4) is modular for some N not divisible by p. He also described explicitly the numbers N and k and the character χ, assuming that N and k are minimal subject to the condition $(N, p) = 1$. According to Serre's conjecture, the number N is determined by the ramification of ρ outside p in the same way as the Artin conductor:

$$N = N(\rho) = \prod_{l \neq p} l^{n(l,\rho)}.$$

The weight k is given by ramification properties of ρ at p, and the character $\chi : (\mathbb{Z}/N)^\times \to \overline{\mathbb{Q}}^\times$ can be obtained from the determinant of ρ as follows:

$$\det \rho(\mathrm{Frob}_l) \equiv \chi(l) l^{k-1} \mod \mathfrak{p} \quad (l \nmid Np).$$

It was noticed by Serre [Se87] that one can easily deduce this conjecture for all representations into $\mathrm{GL}_2(\mathbb{F}_3)$ ($p = 3$), from a general result of Langlands–Tunnell, cf. [L80], [Tun81], [Gel95], and §6.5. The Langlands–Tunnell Theorem states that every two-dimensional odd complex Galois representation $\rho : G_{\mathbb{Q}} \to \mathrm{GL}_2(\mathbb{C})$ with solvable image is modular. More precisely, let $g_\rho = \sum_{n=1}^{\infty} b_n q^n$ be the generating series of ρ, i.e. the series whose coefficients are those of the Artin L-series of ρ (see §6.2.2, 6.4.5)

$$\sum_{n=1}^{\infty} b_n n^{-s} = \prod_l \frac{1}{\det(1 - l^{-s} \rho(\mathrm{Frob}_l) | V^{I_l})} = L(\rho, s),$$

where $\mathrm{GL}_2(\mathbb{C}) = \mathrm{GL}(V)$, $V = \mathbb{C}^2$ and I_l is the inertia group at l. Then the Langlands–Tunnell Theorem states that g_ρ is a weight one modular form, and is a cusp form if ρ is irreducible (see §6.4.5).

To explain how this result implies a proof of Theorem 7.16, we shall consider the complex representation $\rho = \Psi \circ \overline{\rho}_{3,E}$. The image of ρ is certainly solvable, as it is isomorphic to a subgroup of the solvable group $\mathrm{GL}_2(\mathbb{F}_3)$.

7.2.3 Passage from cusp forms of weight one to cusp forms of weight two

Next, we shall construct a cusp form $h = \sum_{n=1}^{\infty} c_n q^n$ of weight two starting from a cusp form g of weight one given by the Langlands–Tunnell Theorem. For this purpose one uses Eisenstein series of weight k and Dirichlet character χ (generalizing the series of §6.3.1):

$$F_{k,\chi}(z) = -\frac{B_{k,\chi}}{2k} + \sum_{n=1}^{\infty}\sum_{d|n} \chi(d) d^{k-1} e(nz) \in \mathcal{M}_k(N,\chi), \tag{7.2.6}$$

$$E_{k,\chi}(z) = 1 - \frac{2k}{B_{k,\chi}} \sum_{n=1}^{\infty}\sum_{d|n} \chi(d) d^{k-1} e(nz) \in \mathcal{M}_k(N,\chi), \tag{7.2.7}$$

where $k \geq 1$, and $B_{k,\chi}$ is the kth *generalized Bernoulli number* (or *Bernoulli–Leopoldt number*), defined by the equality

$$\sum_{k\geq 0} \frac{B_{k,\chi} t^k}{k!} := \sum_{a=1}^{N} \frac{\chi(a) t e^{at}}{e^{Nt}-1}.$$

For $k = 1, 2$, one requires for convergence that χ is non–trivial. The important property of these numbers is that

$$L(1-k,\chi) = -\frac{B_{k,\chi}}{k}.$$

In particular, if $N = 3$, $k = 1$, $\chi(d) = \chi_3(d) = \left(\frac{d}{3}\right)$ is an odd Dirichlet character and we have $B_{1,\chi_3} = -\frac{1}{3}$. Thus

$$E_{1,\chi_3}(z) = 1 + 6\sum_{n=1}^{\infty}\sum_{d|n} \left(\frac{d}{3}\right) e(nz) \in \mathcal{M}_k(N,\chi).$$

In order to finish the proof of Theorem 7.16 we construct a cusp form $h = \sum_{n\geq 1} c_n q^n, c_n \in \mathcal{O}_{\overline{\mathbb{Q}}}[[q]]$ as the following product:

$$h = g_\rho E_{1,\chi_3} = \left(\sum_{n\geq 1} b_n q^n\right)\left(1 + 6\sum_{n=1}^{\infty}\sum_{d|n}\left(\frac{d}{3}\right) q^n\right) =: \sum_{n\geq 1} c_n q^n, \quad c_n \in \mathcal{O}_{\overline{\mathbb{Q}}}.$$

Then h is a cusp form of weight 2 with the desired properties: we take S to be the set of all primes dividing $3N$, where N is the level of g; then for any $l \notin S$,

$$c_l \equiv b_l \equiv l + 1 - \#\tilde{E}(\mathbb{F}_l) \pmod{\mathfrak{p}}$$

for any maximal ideal $\mathfrak{p} \subset \mathcal{O}_{\overline{\mathbb{Q}}}$ containing $1 + \sqrt{-2}$.

7.2.4 Preliminary review of the stages of the proof of Theorem 7.13 on modularity

I) Any elliptic curve E over $\mathbb{Q}$ is "modular modulo 3", if its Galois representaion $\overline{\rho}_{3,E}$ is irreducible.

II) The result of I) gives only a necessary condition for modularity for $p = 3$ (together with properties (7.2.2) and (7.2.3)). One can use these conditions as a starting point for proving modularity in the semistable case. We shall reformulate the modularity statement as an assertion about an isomorphism of certain "deformation rings".

More precisely, let $\mathcal{O} \supset \mathbb{Z}_p$ be the ring of integers of a finite extension $K \supset \mathbb{Q}_p$. Then $\mathcal{O}$ is a discrete valuation ring (DVR), and we shall write λ for its maximal ideal.

Consider a cusp form $h = \sum_{n \geq 1} c_n q^n \in \mathcal{O}_{\overline{\mathbb{Q}}}[[q]]$ (for example, the one just constructed), and choose a maximal ideal $\lambda_p \subset \mathcal{O}_{\overline{\mathbb{Q}}}$ such that

$$\mathcal{O}_{\overline{\mathbb{Q}}}/\lambda_p \cong \overline{\mathbb{F}}_p \supset \mathcal{O}/\lambda \supset \mathbb{F}_p$$

Then the meaning of the stage II) is to show that if we take *any formal power series*

$$\tilde{h} = \sum_{n \geq 1} \tilde{c}_n q^n, \quad \tilde{c}_n \in \mathcal{O}$$

over the local ring $\mathcal{O}$, satisfying certain necessary conditions for modularity and irreducibility, as well as the congruences

$$c_l \equiv \tilde{c}_l \bmod \lambda_p \text{ for all } l \text{ outside a finite set } S,$$

then the series $\tilde{h}$ must be *modular*, i.e. it represents the Fourier expansion of a modular form. Recall that we have fixed an embedding $i_p : \overline{\mathbb{Q}} \hookrightarrow \overline{\mathbb{Q}}_p \supset \mathcal{O} \supset \lambda$ with the property

$$i_p^{-1}(\lambda) \subset \lambda_p \subset \mathcal{O}_{\overline{\mathbb{Q}}} \subset \overline{\mathbb{Q}},$$

so we may regard $\tilde{h}$ as the image under i_p of a series in $\mathcal{O}_{\overline{\mathbb{Q}}}[[q]]$.

Remark 7.18 (Algebraic meaning of the semistability of E). Using Tate's uniformization, see §6.3.3 one can show that an elliptic curve E is semistable if and only if the representations $\rho_{p,E}$ on the Tate modules of E satisfy the condition:

$$\forall p, \forall l \neq p, \quad \rho_{p,E}(I_l) \text{ is conjugate to a subgroup of } \begin{pmatrix} 1 & * \\ 0 & 1 \end{pmatrix},$$

where $I_l \subset G_{\mathbb{Q}} = \mathrm{Gal}(\overline{\mathbb{Q}}/\mathbb{Q})$ denotes the inertia subgroup.

7.3 Modularity of Galois representations and Universal Deformation Rings

7.3.1 Galois Representations over local Noetherian algebras

The generating series $\tilde{h}$ to be constructed at stage II), may be interpreted next as a generating series of a Galois representation. In this section we treat the problem of modularity of certain Galois representations

$$\rho : G_{\mathbb{Q}} \to \mathrm{GL}_m(A),$$

with coefficients in a local $\mathcal{O}$–algebra A with maximal ideal $\mathfrak{m}_A$. As before in section 7.2, $\mathcal{O} \supset \mathbb{Z}_p$ denotes the ring of integers of a finite extension $K \supset \mathbb{Q}_p$. Thus $\mathcal{O}$ is a discrete valuation ring (DVR), and λ denotes the maximal ideal of $\mathcal{O}$. We always assume that:

$$A/\mathfrak{m}_A \cong \mathcal{O}/\lambda = k \supset \mathbb{F}_p.$$

Definition 7.19. *Let $\mathcal{C} = \mathcal{C}_{\mathcal{O}}$ be the category of local noetherian $\mathcal{O}$–algebras equipped with an augmentation $\pi : A \to \mathcal{O}$: its objects are given by*

$$\mathcal{C} = \mathcal{C}_{\mathcal{O}} = \{(A, \pi) \mid \pi : A \to \mathcal{O} \text{ surjective}\},$$

and its morphisms given by commuting triangles:

$$\begin{array}{ccc} A & \xrightarrow{\varphi} & A' \\ & {\scriptstyle \pi_A} \searrow \quad \swarrow {\scriptstyle \pi_{A'}} & \\ & \mathcal{O} & \end{array}$$

Example 7.20. a) $A = \mathcal{O} = \mathbb{Z}_p$;

b) $A = \mathcal{O}[[X_1, \cdots, X_n]]$, $\mathfrak{m}_A = (\lambda, X_1, \cdots, X_n)$, $\pi_A(f) = f(a_1, \cdots, a_n)$ for some fixed a_i;

c)

$$A = \mathbb{Z}_p[[X]]/(X(X - p^n)) \cong \left\{(a, b) \in \mathbb{Z}_p^2 \mid a \equiv b \bmod p^n\right\},$$
$$\mathfrak{m}_A = \left\{(a, b) \in p\mathbb{Z}_p^2 \mid a \equiv b \bmod p^n\right\}, \ \pi_A(a, b) = a.$$

7.3.2 Deformations of Galois Representations

Definition 7.21. *a) We fix a representation $\rho_0 : G_{\mathbb{Q}} \to \mathrm{GL}_m(k)$ over the finite field k as above. Then a lift ρ of ρ_0 to A is a representation $\rho : G_{\mathbb{Q}} \to \mathrm{GL}_m(A)$ such that $\rho \bmod \mathfrak{m}_A = \rho_0$.*

b) Two lifts $\rho : G_{\mathbb{Q}} \to \mathrm{GL}_m(A)$ and $\rho' : G_{\mathbb{Q}} \to \mathrm{GL}_m(A)$ are called strictly equivalent if there exists a matrix $C \in \mathrm{GL}_m(A)$, such that $C \equiv I_m \bmod \mathfrak{m}_A$, and

$$\text{for all } g \in G_{\mathbb{Q}}, \ \rho'(g) = C^{-1}\rho(g)C.$$

c) A deformation of ρ_0 in A is a strict equivalence class of lifts of ρ_0 to A.

Let $\rho : G_{\mathbb{Q}} \to \mathrm{GL}_2(A)$ be (a representative in the class of) a deformation of $\rho_0 : G_{\mathbb{Q}} \to \mathrm{GL}_2(k)$:

$$\begin{array}{ccc} & & \mathrm{GL}_2(A) \\ & \overset{\rho}{\nearrow} & \downarrow \\ G_{\mathbb{Q}} & \xrightarrow[\rho_0]{} & \mathrm{GL}_2(k) \end{array}$$

For a fixed representation $\rho_0 : G_{\mathbb{Q}} \to \mathrm{GL}_m(k)$ over the finite field k let us denote by

$$S = \{l \text{ prime } \mid \rho_0(I_l) \neq I_m\}.$$

In order to deal with finite sets of deformations of ρ_0, we shall also fix the *type* $\mathcal{D} = \mathcal{D}_\Sigma$ of our deformations, where Σ denotes a finite set of primes such that $\Sigma \cap S = \emptyset$.

The main discovery in Wiles' marvelous proof is a method for counting two different types of objects:

1) *Galois representations* coming from elliptic curves over $\mathbb{Q}$ of a given type;
2) *primitive cusp eigenforms* of weight 2 and given level N, and with Fourier coefficients in $\mathbb{Q}$.

Let us fix a two-dimensional representation $\rho_0 : G_{\mathbb{Q}} \to \mathrm{GL}_2(k)$ over the finite field k, and sets S and Σ as above.

Definition 7.22. *a) A deformation ρ of ρ_0 in A is said to be of type $\mathcal{D} = \mathcal{D}_\Sigma$ if the following conditions i)–iv) are satisfied:*
i) ρ is unramified outside of set $S \cup \Sigma \cup \{p\}$;
ii) $\det \rho = \chi_p : G_{\mathbb{Q}} \to \mathbb{Z}_p^\times \hookrightarrow A^\times$, *the cyclotomic character;*
iii) for all $l \in S$ with $l \neq p$, $\rho|_{I_l} \sim \left(\begin{smallmatrix} 1 & * \\ 0 & 1 \end{smallmatrix}\right)$ (semistability at l);
iv) the restriction $\rho|_{D_p}$ satisfies a certain local condition (it is "good"). This means that $\rho|_{D_p}$ is either "flat" or "ordinary".
"Flat" means that for any ideal $\alpha \subset A$ of finite index, the reduced representation $\rho \bmod \alpha : G_{\mathbb{Q}} \to \mathrm{GL}_m(A/\alpha)$ *comes from a finite flat group scheme over $\mathbb{Z}_p$.*
"Ordinary" means that

$$\rho|_{D_p} \sim \begin{pmatrix} * & * \\ 0 & \chi \end{pmatrix} \text{ (with an unramified character } \chi).$$

b) A deformation ρ of ρ_0 is said to be admissible, if it is of type $\mathcal{D}_\Sigma$ for some finite set Σ.

For finite flat group schemes, we refer to [Ta95].

Our goal is to show that any admissible deformation ρ of a modular representation $\rho_0 : G_{\mathbb{Q}} \to \mathrm{GL}_2(k)$ is also modular (under some absolute irreducibility conditions on ρ_0).

Remark 7.23. Definition 7.22 implies that any deformation ρ of type $\mathcal{D}_\Sigma$ factorizes through the projection $G_\mathbb{Q} \to G_{\Sigma_S} = \mathrm{Gal}(\mathbb{Q}_{\Sigma_S}/\mathbb{Q})$, where $\mathbb{Q}_{\Sigma_S}$ is the maximal algebraic extension of $\mathbb{Q}$ unramified outside $\Sigma_S = S \cup \Sigma \cup \{p\}$.

7.3.3 Modular Galois representations

We have already encountered modular Galois representations over finite fields and local fields in §6.1 and §7.2. We now describe a more general notion of a modular Galois representation $\rho : G_\mathbb{Q} \to \mathrm{GL}_2(A)$ over a ring A. Rather than fixing a modular form of type (N, k, χ)

$$h(z) = \sum_{n=1}^{\infty} c_n e(nz) \in \mathcal{S}_k(N, \chi),$$

which is an eigenform of the Hecke operators, normalized by $c_1 = 1$, such that $\mathrm{Tr}\, \rho(\mathrm{Frob}_l)$ is expressed in terms of c_l for all primes $l \nmid Np$, we shall instead use a homomorphism $\pi : \mathbb{T}'(N) \to A$ from an appropriate Hecke $\mathbb{Z}$-algebra such that $\mathrm{Tr}\, \rho(\mathrm{Frob}_l)$ is expressed in terms of $\pi(T_l)$ for all "good" Hecke operators T_l (indexed by the primes $l \nmid Np$, see §6.3.6).

Recall that there is an isomorphism (see §6.3.1)

$$\Gamma_0(N)/\Gamma_1(N) \cong (\mathbb{Z}/N\mathbb{Z})^\times, \quad \sigma_d = \begin{pmatrix} a & b \\ c & d \end{pmatrix} \bmod \Gamma_1(N) \mapsto d \bmod N.$$

Definition 7.24. *(a) The "diamond" operator $\langle d \rangle$ on*

$$\mathcal{M}_k(N) = \mathcal{M}_k(\Gamma_1(N)) = \bigoplus_{\psi \bmod N} \mathcal{M}_k(N, \psi)$$

is defined by

$$\langle d \rangle f = f|_k \sigma_d = (cz+d)^{-k} f\left(\frac{az+b}{cz+d}\right).$$

In particular, for any $f \in \mathcal{M}_k(\Gamma_1(N))$ we have:

$$f \in \mathcal{M}_k(N, \psi) \iff \forall d \in (\mathbb{Z}/N\mathbb{Z})^\times, \ \langle d \rangle f = \psi(d) f.$$

b) The Hecke operators T_l (see (6.3.32)) are defined for all $f(z) = \sum_{n=0}^{\infty} a_n e(nz)$ in $\mathcal{M}_k(N)$ by

$$T_l f = U_l f + l^{k-1} V_l \langle l \rangle (f),$$

where

$$\begin{cases} U_l f = \sum_{n=0}^{\infty} a_{ln} e(nz) \\ V_l \langle l \rangle (f) = \sum_{n=0}^{\infty} a_n(\langle l \rangle (f)) e(lnz) \end{cases}$$

c) The Hecke algebra $\mathbb{T}'(N)$ *over* $\mathbb{Z}$ *is defined by*

$$\mathbb{T}'(N) = \mathbb{Z}[T_l, \langle d\rangle \mid l \nmid N, d \in (\mathbb{Z}/N\mathbb{Z})^\times].$$

Definition 7.25. *A Galois representation over a ring* A *is said modular of level* N*, if there exists a ring homomorphism* $\pi : \mathbb{T}'(N) \to A$ *such that for all primes* $l \nmid N$

$$\begin{cases} \mathrm{tr}\ \rho(\mathrm{Frob}_l) = \pi(T_l) \\ \det\ \rho(\mathrm{Frob}_l) = \pi(\langle l\rangle)l^{k-1}. \end{cases}$$

(see [Ste95] and §6.4.1). Following Hecke and Petersson (see §6.3.6), the action of $\mathbb{T}'(N)$ on the complex vector space $\mathcal{S}_k(N)$ can be orthogonally diagonalized. Suppose $\{f\}$ is an orthogonal basis and each $f(z) = \sum_{n=0}^{\infty} a_n e(nz) \in \mathcal{S}_k(N)$ is primitive. Then $a_1 = 1$ and $T_l f = a_l f$ with $a_l \in \mathcal{O}_{\overline{\mathbb{Q}}}$. As before we fix an embedding $i_p : \overline{\mathbb{Q}} \hookrightarrow \overline{\mathbb{Q}}_p$, and consider a finite extension K of $\mathbb{Q}_p$ containing all the $i_p(a_l)$ with $l \nmid Np$. Let $\mathcal{O} \supset \mathbb{Z}_p$ be the ring of integers of K, and let λ denote its maximal ideal.

In the following theorem, we identify $\overline{\mathbb{Q}}$ with its image $i_p(\overline{\mathbb{Q}}) \subset \overline{\mathbb{Q}}_p$. Thus we identify the elements $a_l \in i_p^{-1}(\mathcal{O}) \subset \overline{\mathbb{Q}}$ with $i_p(a_l)$, omitting the symbol i_p.

Theorem 7.26 (Eichler–Shimura–Deligne). *For any prime* p*, and for any primitive cusp eigenform* $f(z) = \sum_{n=0}^{\infty} a_n e(nz) \in \mathcal{S}_k(N, \chi)$ *there exists a modular representation* $\rho = \rho_{f,\lambda}$ *with coefficients in* $A = \mathcal{O}$ *such that* $\pi : T_l \mapsto a_l$.

Idea of the construction. Assume for simplicity that $k = 2$, χ is trivial and $a_n \in \mathbb{Z}$ $(n \geq 1)$. Let us consider the holomorphic differential $\omega_f = f(z)dz$, then $\mathcal{O} = \mathbb{Z}_p$, and $\lambda = p\mathbb{Z}_p$. We consider next the lattice of periods (see §5.3.5 and §6.3.2)

$$\Lambda_f = \left\langle \int_\gamma \omega_f \mid \gamma \text{ is a closed path on } X_0(N) \right\rangle \subset \mathbb{C}.$$

It turns out that $E = E_f = \mathbb{C}/\Lambda_f$ is then an elliptic curve defined over $\mathbb{Q}$. We define the Galois representation

$$\rho_{f,p} = \rho_{f,\lambda} = \rho_{p,E} : G_{\mathbb{Q}} \to \mathrm{GL}_2(\mathbb{Z}_p).$$

According to the *congruence relation of Eichler–Shimura,*

$$\mathrm{tr}(\rho_{p,E}(\mathrm{Frob}_l)) = a_l = l + 1 - \#\tilde{E}(\mathbb{F}_l), \quad \det(\rho_{p,E}(\mathrm{Frob}_l)) = l \text{ for all } l \nmid pN,$$

where $\tilde{E}$ denotes the reduction $E \bmod l$ (which is good in this case by the criterion of Néron–Ogg–Shafarevich, [Se68a]). In fact E is isogenous to a factor of Jacobian $J_0(N)$ coming from the cusp form f given by (7.1.3) (see §6.4.3).

When $\mathbb{Q}(\langle a_n \rangle_{n\geq 1}) \neq \mathbb{Q}$, but $k = 2$ and χ is trivial, one obtains $\rho_{f,p}$ via an Abelian variety with real multiplication (cf. [Shi71]). When $k > 2$ or χ is nontrivial, the construction is more complicated (cf. [Del68]), but these cases are not used in Wiles' proof.

In order to define a homomorphism $\pi : \mathbb{T}'(N) \to A$, we put $\pi(X) = \lambda_f(X)$ for any $X \in \mathbb{T}'(N)$, where $f|X = \lambda_f(X)f$, and $\lambda_f(T_l) = a_l \in \mathcal{O} \subset A$. This gives a homomorphism $\pi : \mathbb{T}'(N) \to A$.

7.3.4 Admissible Deformations and Modular Deformations

Consider again a local $\mathcal{O}$–algebra A with maximal ideal $\mathfrak{m}_A$, where $\mathcal{O} \supset \mathbb{Z}_p$ denotes (as in §7.2) the ring of integers of a finite extension $K \supset \mathbb{Q}_p$. Thus $\mathcal{O}$ is a discrete valuation ring (DVR), and λ denotes its maximal ideal. We assume always that

$$A/\mathfrak{m}_A \cong \mathcal{O}/\lambda = k \supset \mathbb{F}_p,$$

and let us fix a two-dimensional representation $\rho_0 : G_\mathbb{Q} \to \mathrm{GL}_2(k)$ over the finite field k, and sets S and Σ as above.

Let $\rho : G_\mathbb{Q} \to \mathrm{GL}_2(A)$ denote (a representative in the class of) a deformation of $\rho_0 : G_\mathbb{Q} \to \mathrm{GL}_2(k)$ of type $\mathcal{D}_\Sigma$:

$$\begin{array}{ccc} & & \mathrm{GL}_2(A) \\ & \overset{\rho}{\nearrow} & \downarrow \\ G_\mathbb{Q} & \xrightarrow[\rho_0]{} & \mathrm{GL}_2(k) \end{array}$$

Definition 7.27. *Let $DA_\Sigma(A)$ denote the set of all admissible deformations of ρ_0 of type $\mathcal{D}_\Sigma$ and $DM_\Sigma(A)$ the subset of $DA_\Sigma(A)$ consisting of all modular deformations of ρ_0 of type $\mathcal{D}_\Sigma$.*

We shall see that the set $DA_\Sigma(A)$ is finite (in fact, $A \mapsto DA_\Sigma(A)$ is a functor with values in finite sets, see [Da95]). The main theorem of Wiles' proof says that under suitable conditions on ρ_0, both sets coincide for any A as above.

Consider the subgroup of index 2

$$G_{\mathbb{Q}\left(\sqrt{(-1)^{\frac{p-1}{2}}p}\right)} \subset G_\mathbb{Q}$$

corresponding to the unique quadratic extension of $\mathbb{Q}$ unramified outside p.

Theorem 7.28 (Modularity of admissible deformations). *Suppose that $\rho_0 : G_\mathbb{Q} \to \mathrm{GL}_2(k)$ is a modular representation over a finite field k, and that the restriction*

$$\rho_0|_{G_{\mathbb{Q}\left(\sqrt{(-1)^{\frac{p-1}{2}}p}\right)}}$$

is absolutely irreducible.

Then

$$DA_\Sigma(A) = DM_\Sigma(A). \tag{7.3.1}$$

That is, every admissible deformation is modular.

Remark 7.29. i) Under the assumptions of Theorem 7.28 both representations ρ_0 and ρ are semistable at all places $l \in S$.

ii) The strong condition of absolute irreducibility of the restriction of ρ_0 implies (trivially) that ρ_0 itself is absolutely irreducible. This fact is important in Theorem 7.4 of Mazur–Ribet, which we now restate in the following form:

Theorem 7.30 (of K.Ribet on the existence of minimal deformations). *Suppose that $\rho_0 : G_\mathbb{Q} \to \mathrm{GL}_2(k)$ is a modular and absolutely irreducible representation over a finite field k. Then the set $DM_\emptyset(A)$ of minimal deformations of ρ_0 is non–empty (that is, there exists a modular deformation ρ of ρ_0 of minimal level $N(\rho_0)$, where $N(\rho_0)$ is the Artin conductor of ρ_0).*

Example 7.31 (The Frey–Hellegouarch curve). Consider again $E = E_{a^p,b^p,c^p}$ with $a^p \equiv -1 \bmod 4, 2|b, p \geq 5$, and let $\rho_0 = \overline{\rho}_{p,E} : G_\mathbb{Q} \to \mathrm{GL}_2(\mathbb{F}_p)$. Then

i) $N(\rho_0) = 2$ (see [Se87], p. 201).

ii) For Frey–Hellegouarch curves, we have $|E[2](\overline{\mathbb{Q}})| = 4$, since the points of order 2 correspond to the roots the cubic polynomial $x(x - a^p)(x + b^p)$. Using this fact we may show that ρ_0 is irreducible for $p \geq 5$ by Mazur's theorem [Maz77] and 1.3.7: the only possibilities for the torsion subgroup of $E(\mathbb{Q})$ are (up to isomorphism):

$$\begin{aligned}&\mathbb{Z}/n\mathbb{Z} \ (1 \leq n \leq 10, \text{ and } n = 12),\\ &\mathbb{Z}/2\mathbb{Z} \times \mathbb{Z}/2n\mathbb{Z} \ (1 \leq n \leq 4).\end{aligned}$$

Suppose the $G_\mathbb{Q}$–module $V = E[p]$ has a $G_\mathbb{Q}$-invariant line W over $\mathbb{F}_p$. If W is fixed pointwise by the action of $G_\mathbb{Q}$, then $E(\mathbb{Q})$ has a torsion subgroup isomorphic to $\mathbb{Z}/2\mathbb{Z} \times \mathbb{Z}/2p\mathbb{Z}$, which would directly contradict Mazur's theorem. If on the other hand W is invariant under $G_\mathbb{Q}$ only as a line over $\mathbb{F}_p$, then the quotient curve $E' = E/W$, isogenuous to E, can also be defined over $\mathbb{Q}$. One can show that E' has a rational point of order p and three points of order 2, again contradicting Mazur's Theorem (see [He97]).

iii) Assuming the modularity of ρ_0 (modulo p), one can simply apply Ribet's Theorem 7.30 to deduce that $E = E_{a^p,b^p,c^p}$ does not exist. However, it seems that we can only prove the modularity of ρ_0 by proving the modularity of $\rho_{p,E}$; this is why we require Theorem 7.28 (on modularity of admissible deformations).

7.3.5 Universal Deformation Rings

We need to show that under the notations and assumptions of Theorem 7.28 one has

$$DA_\Sigma(A) = DM_\Sigma(A),$$

for any local Noetherian $\mathcal{O}$–algebra A with maximal ideal $\mathfrak{m}_A$, and for any two-dimensional representation $\rho_0 : G_\mathbb{Q} \to \mathrm{GL}_2(k)$ as above ($A/\mathfrak{m}_A \cong \mathcal{O}/\lambda = k \supset \mathbb{F}_p$).

Reformulation of the identity (7.3.1):

We shall use the representability of the two functors

$$DA_\Sigma \supset DM_\Sigma : \mathcal{C}_\mathcal{O} \to \mathcal{S}ets_{fin}$$

(for any finite set Σ). This means that there exist "universal" objects (called universal deformation rings) $R_\Sigma, \mathbb{T}_\Sigma \in \mathcal{C}_\mathcal{O}$ such that for any $A \in \mathcal{C}_\mathcal{O}$ one has

$$DA_\Sigma(A) = \mathrm{Hom}_{\mathcal{C}_\mathcal{O}}(R_\Sigma, A) \supset DM_\Sigma(A) = \mathrm{Hom}_{\mathcal{C}_\mathcal{O}}(\mathbb{T}_\Sigma, A).$$

In particular, substituting $A = \mathbb{T}_\Sigma$ we obtain a canonical morphism

$$\varphi_\Sigma : R_\Sigma \to \mathbb{T}_\Sigma, \tag{7.3.2}$$

and the canonical universal pairs

$$(R_\Sigma, \rho_\Sigma^{univ}), \text{ and } (\mathbb{T}_\Sigma, \rho_\Sigma^{univ.mod.})$$

are related by the commutative diagram:

$$\begin{array}{ccc} & & \mathrm{GL}_2(R_\Sigma) \\ & \overset{\rho_\Sigma^{univ}}{\nearrow} & \\ G_\mathbb{Q} & & \downarrow \varphi_\Sigma \\ & \underset{\rho_\Sigma^{univ.mod.}}{\searrow} & \\ & & \mathrm{GL}_2(\mathbb{T}_\Sigma) \end{array} \tag{7.3.3}$$

In order to count the sets $DA_\Sigma(A)$ and $DM_\Sigma(A)$, a clever choice of the augmentations is used for the local algebras R_Σ and $\mathbb{T}_\Sigma$:

$$\pi_{R_\Sigma} : R_\Sigma \to \mathcal{O}$$
$$\pi_{\mathbb{T}_\Sigma} : \mathbb{T}_\Sigma \to \mathcal{O}$$

Remark 7.32. (J.–P. Serre) The universal deformation rings R_Σ and $\mathbb{T}_\Sigma$ are topologically generated by the elements $\mathrm{tr}(\rho_\Sigma^{univ}(\mathrm{Frob}_l)) \in R_\Sigma$ for all primes $l \notin \Sigma_S$.

This fact means also that both universal representations

$$\rho_\Sigma^{univ} : G_\mathbb{Q} \to \mathrm{GL}_2(R_\Sigma), \quad \rho_\Sigma^{univ.mod.} : G_\mathbb{Q} \to \mathrm{GL}_2(\mathbb{T}_\Sigma)$$

are determined by their traces (see [Da95]).

We may therefore define the augmentation maps by the following relations:

$$\begin{aligned} \pi_{R_\Sigma} &: \mathrm{tr}\ \rho_\Sigma^{univ}(\mathrm{Frob}_l) \mapsto a_l \\ \pi_{\mathbb{T}_\Sigma} &: \mathrm{tr}\ \rho_\Sigma^{univ.mod.}(\mathrm{Frob}_l) \mapsto a_l \end{aligned} \tag{7.3.4}$$

where $a_l \in \mathcal{O}$ are the Fourier coefficients of Ribet's cusp eigenform

$$f(z) = \sum_{n=0}^{\infty} a_n e(nz) \in \mathcal{S}_k(N_0),$$

of minimal level (whose Galois representation $\rho_{f,\lambda}$ corresponds to a deformation $\rho \in DM_\emptyset(\mathcal{O})$).

We shall denote by

$$\tilde{\rho} = \rho_{f,\lambda} : G_\mathbb{Q} \to \mathrm{GL}_2(\mathcal{O})$$

the corresponding modular Galois representation of the minimal level N_0.

Using the map (7.3.2), one can interpret Theorem 7.28 as an isomorphism of the local Noetherian $\mathcal{O}$–algebras R_Σ and $\mathbb{T}_\Sigma$:

Theorem 7.33 (Main Theorem on isomorphism of universal deformation rings). *Under the assumptions of Theorem 7.28, the canonical morphism (7.3.2)*

$$\varphi_\Sigma : R_\Sigma \to \mathbb{T}_\Sigma$$

of universal deformation rings is an isomorphism in the category $\mathcal{C}_\mathcal{O}$.

7.4 Wiles' Main Theorem and Isomorphism Criteria for Local Rings

7.4.1 Strategy of the proof of the Main Theorem 7.33

Let us consider again a local $\mathcal{O}$–algebra A with maximal ideal $\mathfrak{m}_A$, where $\mathcal{O} \supset \mathbb{Z}_p$ denotes (as in section 7.2) the ring of integers of a finite extension $K \supset \mathbb{Q}_p$; $\mathcal{O}$ is a dicrete valuation ring (DVR), and λ denotes the maximal ideal of $\mathcal{O}$. We always assume that

$$A/\mathfrak{m}_A \cong \mathcal{O}/\lambda = k \supset \mathbb{F}_p,$$

and we fix a two-dimensional representation $\rho_0 : G_{\mathbb{Q}} \to \mathrm{GL}_2(k)$ over the finite field k, together with sets S and Σ as described above.

Recall that Ribet's modular Galois representation

$$\tilde{\rho} = \rho_{f,\lambda} : G_{\mathbb{Q}} \to \mathrm{GL}_2(\mathcal{O})$$

of minimal level N_0 given by Theorem 7.30 belongs to the (non–empty) set $DM_{\emptyset}(\mathcal{O})$. This gives a distinguished element of each of the sets $DM_{\Sigma}(A) \subset DA_{\Sigma}(A)$. This representation $\tilde{\rho}$ is used in an explicit construction of the modular universal deformation ring $\mathbb{T}_{\Sigma}$, see [CSS95].

Surjectivity of the map $\varphi_{\Sigma} : R_{\Sigma} \to \mathbb{T}_{\Sigma}$ (7.3.2) can be easily deduced from the fact (see 7.32) that the universal deformation rings R_{Σ} and $\mathbb{T}_{\Sigma}$ are topologically generated by the elements $\mathrm{tr}(\rho_{\Sigma}^{univ}(\mathrm{Frob}_l)) \in R_{\Sigma}$ for primes $l \notin \Sigma_S$, see below §7.4.2.

Injectivity of $\varphi_{\Sigma} : R_{\Sigma} \to \mathbb{T}_{\Sigma}$ was proved by A.Wiles by an induction argument on Σ. For a prime l not in Σ_S, we let $\Sigma' = \Sigma \cup \{l\}$. Wiles deduced the bijectivity of $\varphi_{\Sigma'}$ from the bijectivity of φ_{Σ} using an isomorphism criterion for local rings. This criterion was formulated in terms of certain invariants (discovered by Wiles earlier, in spring 1991, see the introduction of his paper [Wi]). However, in order to start the induction one needed the case $\Sigma = \emptyset$ (the base of induction). This was the point which caused a problem in 1993, after the announcement of a complete proof of FLT, and which was repaired in 1994 by A.Wiles and R.Taylor using a horizontal version of Iwasawa theory together with a second isomorphism criterion for local rings. In this section we describe these criteria and give explicit constructions (due to H.Lenstra and B.Mazur) of the universal deformation ring R_{Σ}.

7.4.2 Surjectivity of φ_{Σ}

In order to prove the surjectivity, we assume the existence of the universal deformation rings $R_{\Sigma}, \mathbb{T}_{\Sigma} \in \mathcal{C}_{\mathcal{O}}$. Thus for any $A \in \mathcal{C}_{\mathcal{O}}$ we have

$$DA_{\Sigma}(A) = \mathrm{Hom}_{\mathcal{C}_{\mathcal{O}}}(R_{\Sigma}, A) \supset DM_{\Sigma}(A) = \mathrm{Hom}_{\mathcal{C}_{\mathcal{O}}}(\mathbb{T}_{\Sigma}, A),$$

implying the existence of a canonical morphism (7.3.2)

$$\varphi_\Sigma : R_\Sigma \to \mathbb{T}_\Sigma.$$

Lemma 7.34. *Let $A = R_\Sigma$ (resp. $A = \mathbb{T}_\Sigma$), and denote by A^0 subring of A which is the topological closure of the $\mathcal{O}$-subalgebra in A generated by all elements $\mathrm{tr}(\rho_\Sigma^{univ}(\mathrm{Frob}_l)) \in R_\Sigma$ (resp. $\mathrm{tr}(\rho_\Sigma^{univ.mod.}(\mathrm{Frob}_l)) \in \mathbb{T}_\Sigma$). Then $A^0 = A$.*

This lemma can be deduced from the following:

Proposition 7.35. *Let $A^0 \subset A$ be two local rings with maximal ideals satisfying*

$$\mathfrak{m}_{A^0} = \mathfrak{m}_A \cap A^0$$

and with the same finite residue field k. Suppose

$$\rho : G \to \mathrm{GL}_m(A)$$

is a representation of a group over A such that

1) $\overline{\rho} = \rho \bmod \mathfrak{m}_A$ is absolutely irreducible;
2) $\mathrm{tr}\rho(\sigma) \in A^0$ for all $\sigma \in G$.

Then ρ is conjugate over A to a representation

$$\rho^0 : G \to \mathrm{GL}_m(A^0)$$

Proof of Proposition 7.35.

Let B denote the A^0-subalgebra in $M_m(A)$ generated by $\rho(G)$. The image of B in $M_m(k)$ is a central simple algebra over the finite field k. It follows from the triviality of the Brauer group (see §5.5.5) of the finite field$k$ that the image of B in $M_m(k)$ is the whole of $M_m(k)$. Let $e_1, \cdots, e_{m^2}$ be elements of B whose reductions modulo $\mathfrak{m}_A$ form the standard basis of $M_m(k) = B \bmod \mathfrak{m}_A$. We shall show that $e_1, \cdots, e_{m^2}$ is a basis for B over A^0. By Nakayama's lemma elements of B may be expressed in the form:

$$b = \sum_{i=1}^{m^2} a_i e_i, \text{ with } a_i \in A.$$

Hence

$$\mathrm{tr}(b \cdot {}^t e_j) = \sum_{i=1}^{m^2} a_i \mathrm{tr}(e_i \cdot {}^t e_j), \text{ with } j = 1, \cdots, m^2. \tag{7.4.1}$$

Let us define

$$c_{ij} = \mathrm{tr}(e_i {}^t e_j) \in A \Rightarrow (c_{ij}) \equiv I_{m^2} \bmod \mathfrak{m}_A.$$

Hence the system (7.4.1) is solvable over the local ring A^0. One defines $V \subset A^m$ to be the submodule generated by the columns of elements in B. Thus $V \cong (A^0)^m$ is free, and we deduce that $B \xrightarrow{\sim} \mathrm{End}(V) \cong M_m(A^0)$ by Nakayama's lemma.

7.4.3 Constructions of the universal deformation ring R_Σ

We assume that ρ_0 is absolutely irreducible.

To prove the existence of R_Σ one can either appeal to a general criterion of Schlessinger (cf. Mazur's paper in [CSS95]), or instead use a more explicit method of H.Lenstra (cf. the paper of Bart de Smit and H.W.Lenstra in [CSS95]).

Consider first a finite group G, and let us define an $\mathcal{O}$-algebra $\mathcal{O}[G,m]$ with generators:

$$\{X_{ij}^g \mid i,j = 1,\cdots m; g \in G\},$$

and the following relations:

$$X_{ij}^e = \delta_{ij}, \quad X_{ij}^{gh} = \sum_{l=1}^{m} X_{il}^g X_{lj}^h \;\; i,j = 1,\cdots m; g,h \in G$$

As these relations mimic the relations satisfied by matrix coefficients of a representation of G, it follows that for any $A \in \mathcal{C}_\mathcal{O}$ there is a canonical identification

$$\mathrm{Hom}_{\mathcal{O}-alg}(\mathcal{O}[G,m], A) = \mathrm{Hom}(G, \mathrm{GL}_m(A)). \tag{7.4.2}$$

Substituting $A = \mathcal{O}/\lambda = k$ in the above formula, we obtain a homomorphism π_0 of $\mathcal{O}$–algebras corresponding to ρ_0:

$$\begin{array}{ccc} \mathrm{Hom}_{\mathcal{O}-alg}(\mathcal{O}[G,m], k) & = & \mathrm{Hom}(G, \mathrm{GL}_m(k)) \\ \Downarrow & & \Downarrow \\ \pi_0 & \longleftarrow & \rho_0. \end{array}$$

Let $\mathfrak{m}_0 = \mathrm{Ker}\,\pi_0$; we define the $\mathcal{O}$-algebra R_G to be the completion of $\mathcal{O}[G,m]$ with respect to $\mathfrak{m}_0$:

$$R_G = \varprojlim_n \mathcal{O}[G,m]/\mathfrak{m}_0^n.$$

Now suppose we have a profinite group:

$$G_\Sigma = \varprojlim_j G_j.$$

Then we put

$$R_j = R_{G_j}, \quad R_\Sigma = \varprojlim_j R_j.$$

It may be verified that

a)

$$\mathrm{Hom}_{\rho_0}(G, \mathrm{GL}_m(A)) = \mathrm{Hom}_{\mathcal{O}-alg}(R_\Sigma, A). \tag{7.4.3}$$

b) R_Σ is a local Noetherian $\mathcal{O}$-algebra (to show this, one uses a universal bound for the dimension of the tangent space of R_j, and the absolute irreducibility of ρ_0).

7.4.4 A sketch of a construction of the universal modular deformation ring $\mathbb{T}_\Sigma$

Let us again fix a two-dimensional modular representation $\rho_0 : G_\mathbb{Q} \to \mathrm{GL}_2(k)$ over the finite field k, together with sets S and Σ as above.

We shall consider a slightly different Hecke algebra than in 7.3, Definition 7.24, namely,

$$\mathbb{T}(\Sigma) = \mathcal{O}[T_l, U_q, \langle d\rangle \mid l \nmid N_\Sigma, d \in (\mathbb{Z}/N\mathbb{Z})^\times, q \in S \cup \Sigma].$$

We shall regard $\mathbb{T}(N_\Sigma)$ as a subalgebra of $\mathrm{End}_\mathcal{O} S_2(N_\Sigma, \mathcal{O})$. In the above we have

$$N_\Sigma = p \prod_{\tilde{q}\in S} \tilde{q} \prod_{\tilde{l}\in\Sigma} \tilde{l}^2.$$

Furthermore $S_2(N_\Sigma, \mathcal{O})$ denotes the $\mathcal{O}$–submodule of $\mathcal{O}[[q]]$, generated by all formal q–expansions of the form

$$\sum_{n\geq 1} i_p(a_n)q^n \in \mathcal{O}[[q]],$$

such that

$$f = \sum_{n\geq 1} a_n q^n \in S_2(N_\Sigma; \overline{\mathbb{Q}})$$

is a cusp form with coefficients $a_n \in \overline{\mathbb{Q}} \cup i_p^{-1}(\mathcal{O})$.

Let

$$\tilde{f} = f_\emptyset = \sum_{n\geq 1} \tilde{a}_n q^n$$

denote Ribet's modular form of Theorem 7.30, attached to a two-dimensional modular representation $\rho_0 : G_\mathbb{Q} \to \mathrm{GL}_2(k)$ over the finite field k.

Recall that Ribet's modular Galois representation

$$\tilde{\rho} = \rho_{f,\lambda} : G_\mathbb{Q} \to \mathrm{GL}_2(\mathcal{O})$$

of minimal level N_0 given by Theorem 7.30 belongs to the (non–empty) set $DM_\emptyset(\mathcal{O})$. For any Σ as above, we define

$$f_\Sigma = \sum_{n\geq 1} \tilde{a}_n(f_\Sigma) q^n$$

by removing from the Mellin transform of $\tilde{f}$ the Euler factors at $\tilde{l} \in \Sigma$:

$$\begin{aligned} L(f_\Sigma, s) &= \sum_{n\geq 1} \tilde{a}_n(f_\Sigma) n^{-s} \\ &= \prod_{\tilde{q}\in S} (1 - \tilde{a}_{\tilde{q}} \tilde{q}^{-s})^{-1} \prod_{l\nmid N_\Sigma} (1 - \tilde{a}_l l^{-s} + l^{1-2s})^{-1}. \end{aligned} \tag{7.4.4}$$

Now consider the following ideal of the Hecke algebra:

$$\mathcal{M}_\Sigma = (\lambda, T_l - \tilde{a}_l, U_{\tilde{q}} - \tilde{a}_{\tilde{q}}, T_{\tilde{l}})_{l \notin \Sigma \cup S \cup \{p\},\ \tilde{q} \in S, \tilde{l} \in \Sigma}. \tag{7.4.5}$$

This ideal is actually prime, since

$$\begin{aligned} \mathcal{M}_\Sigma = \mathrm{Ker}(\mathbb{T}(\Sigma) &\xrightarrow{\pi_{f_\Sigma}} k[[q]]), \\ T_l &\mapsto a_l \bmod \lambda, \\ U_{\tilde{q}} &\mapsto a_{\tilde{q}} \bmod \lambda, \\ T_{\tilde{l}} &\mapsto 0 \\ &(l \notin \Sigma \cup S \cup \{p\}, \tilde{q} \in S, \tilde{l} \in \Sigma), \end{aligned} \tag{7.4.6}$$

and the ring $k[[q]]$ is an integral domain.

We define $\mathbb{T}_\Sigma$ to be the completion of $\mathbb{T}(\Sigma)$ with respect to the ideal $\mathcal{M}_\Sigma$:

$$\mathbb{T}_\Sigma = \varprojlim_n \mathbb{T}(\Sigma)/\mathcal{M}_\Sigma^n.$$

One can check that $\mathbb{T}_\Sigma$ is a finite flat local Noetherian $\mathcal{O}$-algebra (i.e. it is a free $\mathcal{O}$-module of finite rank), and one defines an augmentation map $\mathbb{T}_\Sigma \to \mathcal{O}$ using $\tilde{f}$.

Theorem 7.36. *There exists, up to isomorphism, a unique admissible Galois representation*

$$\rho_\Sigma^{univ.mod.} : G_\mathbb{Q} \to \mathrm{GL}_2(\mathbb{T}_\Sigma), \tag{7.4.7}$$

with the following properties:

$$\begin{aligned} \mathrm{tr}(\rho_\Sigma^{univ.mod.}(\mathrm{Frob}_l)) &= T_l, \\ \det(\rho_\Sigma^{univ.mod.}(\mathrm{Frob}_l)) &= l (l \notin \Sigma \cup S \cup \{p\}). \end{aligned} \tag{7.4.8}$$

The construction by A.Wiles of the universal representation $\rho_\Sigma^{univ.mod.}$ was obtained from the Eichler–Shimura Theorem 7.26 by patching together all the modular deformations of type $\mathcal{D}_\Sigma$. To achieve this he used of the theory of pseudo–representations. The strong absolute irreducibility condition of theorem 7.28, concerning the restriction

$$\rho_0|_{G_{\mathbb{Q}\left(\sqrt{(-1)^{\frac{p-1}{2}}p}\right)}},$$

was essential in this construction.

7.4.5 Universality and the Chebotarev density theorem

Let us recall Theorem 4.22 in the following form:

Theorem 7.37 (Chebotarev density theorem). *Let L/K be a finite extension of number fields, and let X be a non–empty subset of $G(L/K)$, invariant under conjugation. Denote by P_X the set of places $v \in \Sigma_K^0$, unramified in L, such that the classes of Frobenius elements of these places belong to X: $F_{L/K}(P_X) \subset X$. Then the set P_X is infinite and has a density, which is equal to* $\mathrm{Card}\, X/\mathrm{Card}\ G(L/K)$.

Corollary 7.38. *The canonical morphism (7.3.2) is compatible with the augmentation maps π_{R_Σ} and $\pi_{\mathbb{T}_\Sigma}$*

$$\varphi_\Sigma : R_\Sigma \to \mathbb{T}_\Sigma \tag{7.4.9}$$

In fact, the traces of representations π_{R_Σ} and $\pi_{\mathbb{T}_\Sigma} \circ \varphi_\Sigma$ coincide on the subset of elements $\mathrm{Frob}_l (l \notin \Sigma_S)$ (which is dense in the group G_{Σ_S}). It follows that the corresponding universal deformations are equivalent, hence they coincide by their universal property.

7.4.6 Isomorphism Criteria for local rings

To prove that the canonical morphism (7.3.2)

$$\varphi_\Sigma : R_\Sigma \to \mathbb{T}_\Sigma$$

of universal deformation rings is an isomorphism (in the category $\mathcal{C}_\mathcal{O}$), one argues by induction on Σ. Let $\Sigma' = \Sigma \cup \{l\}$ for some prime l not in Σ_S. Wiles deduced the bijectivity of $\varphi_{\Sigma'}$ from the bijectivity of φ_Σ using an isomorphism criterion for local rings. This criterion is formulated in terms of certain invariants, which will be described next. In order to start the induction, one needs to prove the case $\Sigma = \emptyset$; this is achieved by a second isomorphism criterion for local rings.

Definition 7.39. *A local Noetherian $\mathcal{O}$–algebra A is called a complete intersection if:*

a) A is a free $\mathcal{O}$–module of finite rank;
b) $A \cong \mathcal{O}[[X_1, \cdots, X_n]]/(f_1, \cdots, f_n)$.

(cf. [Mats70]).

We shall use the following *invariants of a local $\mathcal{O}$–algebra A*:

$$I_A = \mathrm{Ker}\pi_A, \quad \Phi_A = I_A/I_A^2, \quad \eta_A = \pi_A(\mathrm{Ann} I_A) \subset \mathcal{O} \tag{7.4.10}$$

These are called respectively the *kernel of augmentation*, the *tangent space* and the *congruence module.*

Example 7.40. a) $A = \mathcal{O} = \mathbb{Z}_p$, $\Phi_A = I_A/I_A^2 = \{0\}$

b) $A = \mathbb{Z}_p[[X,Y]]/(X(X-p), Y(Y-p))$, $\Phi_A = \mathbb{Z}/p\mathbb{Z} \times \mathbb{Z}/p\mathbb{Z}$, $\eta_A = (p^2)$. The augmentation map in this case is given by

$$\pi_A(f) = f(0,0) \in \mathbb{Z}_p, \ A \text{ is a complete intersecton ring.}$$

The ring A is a complete intersection.

c) $A = \mathbb{Z}_p[[X,Y]]/(X(X-p), Y(Y-p), XY)$, $\Phi_A = \mathbb{Z}/p\mathbb{Z} \times \mathbb{Z}/p\mathbb{Z}$, $\eta_A = (p)$. The augmentation is given by

$$\pi_A(f) = f(0,0) \in \mathbb{Z}_p.$$

In this case A is not a complete intersection.

Theorem 7.41 (Criterion I). *Let $\varphi : A \to B$ be a surjective morphism in the category $\mathcal{C}_{\mathcal{O}}$. Then the following are equivalent:*

(i) φ is an isomorphism of two local complete intersection $\mathcal{O}$–algebras;
(ii) $\#\Phi_A \le \#\mathcal{O}/\eta_B < \infty$;
(iii) $\#\Phi_A = \#\mathcal{O}/\eta_B < \infty$.

Remark 7.42. In the first version of his proof A.Wiles had made the assumption that the ring B is *Gorenstein* (i.e. $\check{B} = \mathrm{Hom}(B, \mathcal{O})$ is a free B–module of rank 1). This restriction was later removed by H.Lenstra.

Corollary 7.43. *An $\mathcal{O}$-algebra $A \in \mathcal{C}_{\mathcal{O}}$ is a complete intersection ring if and only if*

$$\#\Phi_A = \#\mathcal{O}/\eta_A < \infty.$$

This is proved by applying Criterion I to the identity map $\mathrm{id}_A : A \to A$.

7.4.7 J–structures and the second criterion of isomorphism of local rings

Let us consider the distinguished ideals

$$J_m = (\omega_m(S_1), \cdots, \omega_m(S_n)) \subset \mathcal{O}[[S_1, \cdots, S_n]],$$

where

$$\omega_m(S_1) = (1+S_1)^{p^m} - 1, \quad \omega_m(S_n) = (1+S_n)^{p^m} - 1, \quad J_0 = (S_1, \cdots, S_n).$$

Definition 7.44. *Let $\varphi : A \to B$ be a surjective morphism in $\mathcal{C}_{\mathcal{O}}$. One says that φ admits a J–structure, if there is a family of commutative diagrams, indexed by $m \in \mathbb{N}$:*

$$\begin{array}{ccccc}
 & & \mathcal{O}[[S_1, \cdots, S_n]] & & \\
 & & \downarrow{\scriptstyle \sigma_m} & & \\
\mathcal{O}[[T_1, \cdots, T_n]] & \xrightarrow{\xi_m} & A_m & \xrightarrow{\varphi_m} & B_m \\
 & & \downarrow & & \downarrow \\
 & & A & \xrightarrow{\varphi} & B
\end{array}$$

with the following properties for each m:

i) ξ_m *is surjective;*
ii) φ_m *is surjective;*
iii) $A_m/J_0A_m \cong A$ *and* $B_m/J_0B_m \cong B$.
iv) B_m/J_mB_m *is a torsion free module of finite rank over the* $\mathcal{O}$*-algebra* $\mathcal{O}[[S_1, \cdots, S_n]]/J_m$.

Theorem 7.45 (Criterion II).

Let $\varphi : A \to B$ *be a surjective morphism in the category* $\mathcal{C}_{\mathcal{O}}$.

If φ *admits a* J*-structure then* φ *is an isomorphism of two local complete intersection* $\mathcal{O}$*–algebras.*

Proof of both criteria belong to commutative algebra. We refer therefore the reader to [CSS95], [Ta-Wi].

7.5 Wiles' Induction Step: Application of the Criteria and Galois Cohomology

7.5.1 Wiles' induction step in the proof of Main Theorem 7.33

In §7.4.3–7.4.4 we explained the existence of the universal deformation rings $R_\Sigma, \mathbb{T}_\Sigma \in \mathcal{C}_\mathcal{O}$. These universal rings represent the functors of admissible deformations (respectively modular deformations). This means that for any $A \in \mathcal{C}_\mathcal{O}$ one has

$$DA_\Sigma(A) = \mathrm{Hom}_{\mathcal{C}_\mathcal{O}}(R_\Sigma, A) \supset DM_\Sigma(A) = \mathrm{Hom}_{\mathcal{C}_\mathcal{O}}(\mathbb{T}_\Sigma, A).$$

In particular, substituting of $A = \mathbb{T}_\Sigma$ we obtain a canonical morphism (7.3.2):

$$\varphi_\Sigma : R_\Sigma \to \mathbb{T}_\Sigma,$$

and the canonical universal pairs

$$(R_\Sigma, \rho_\Sigma^{univ}), \text{ and } (\mathbb{T}_\Sigma, \rho_\Sigma^{univ.mod.})$$

are related by the commutative diagrams :

$$\begin{array}{ccc} R_\Sigma & \xrightarrow{\varphi_\Sigma} & \mathbb{T}_\Sigma \\ & {\scriptstyle \pi_{R_\Sigma}} \searrow \quad \swarrow {\scriptstyle \pi_{\mathbb{T}_\Sigma}} & \\ & \mathcal{O} & \end{array} \qquad \begin{array}{ccc} & & \mathrm{GL}_2(R_\Sigma) \\ & \nearrow^{\rho_\Sigma^{univ}} & \downarrow {\scriptstyle \varphi_\Sigma} \\ G_\mathbb{Q} & & \\ & \searrow_{\rho_\Sigma^{univ.mod.}} & \\ & & \mathrm{GL}_2(\mathbb{T}_\Sigma) \end{array}$$

(see §7.4.5).

Let us recall the Main Theorem 7.33 in the following form:

Theorem 7.46 (Main Theorem). *Suppose that $\rho_0 : G_\mathbb{Q} \to \mathrm{GL}_2(k)$ is a modular representation over a finite field k, such that the restriction*

$$\rho_0|_{G_{\mathbb{Q}\left(\sqrt{(-1)^{\frac{p-1}{2}}p}\right)}}$$

is absolutely irreducible.

Then the canonical morphism (7.3.2)

$$\varphi_\Sigma : R_\Sigma \to \mathbb{T}_\Sigma$$

of universal deformation rings is an isomorphism of complete intersection rings (in the category $\mathcal{C}_\mathcal{O}$).

Remark 7.47. i) We have already seen in §7.4.3 that φ_Σ is *surjective.*
ii) Recall that the proof of *injectivity* of $\varphi_\Sigma : R_\Sigma \to \mathbb{T}_\Sigma$ was given by A.Wiles using induction on Σ. We shall denote $\Sigma' = \Sigma \cup \{l\}$ for a prime l not in Σ_S. Wiles deduced the bijectivity of $\varphi_{\Sigma'}$ from the bijectivity of φ_Σ using Criterion I (see §7.4.6) for isomorphisms of local rings. This criterion is formulated in terms of the invariants (7.4.10). To begin the induction one needs the case $\Sigma = \emptyset$. This case is proved using Criterion II of §7.4.6 for isomorphisms of local rings.

The Induction Step

Let us consider

$$\Sigma' = \Sigma \cup \{l\}, \quad \mathcal{D}' = \mathcal{D}_{\Sigma'}, \quad \mathcal{D} = \mathcal{D}_\Sigma, \quad A = R_\Sigma, \quad B = \mathbb{T}_{\Sigma'}.$$

We shall assume the induction hypothesis, i.e.

$$\varphi_\Sigma : R_\Sigma \xrightarrow{\sim} \mathbb{T}_\Sigma.$$

This implies the equality of the corresponding invariants:

$$\#\Phi_{R_\Sigma} = \#\mathcal{O}/\eta_{\mathbb{T}_\Sigma} < \infty. \tag{7.5.1}$$

According to Criterion I, it suffices to prove the inequality

$$\#\Phi_{R_{\Sigma'}} \le \#\mathcal{O}/\eta_{\mathbb{T}_{\Sigma'}} < \infty, \tag{7.5.2}$$

given the equality (7.5.1). The left hand side of this inequality is controlled by a Galois cohomology group. The right hand side will be computed using a determinant representing a relative invariant $\eta_{\Sigma',\Sigma}$, which relates $\#\mathcal{O}/\eta_{\mathbb{T}_{\Sigma'}}$ and $\#\mathcal{O}/\eta_{\mathbb{T}_\Sigma}$. A fundamental inequality relating these quantities will imply the induction step.

Base of induction: the minimal case

We shall construct in the next section §7.6.3 a J-structure for the surjective morphism $\varphi_\emptyset$. This will show by Criterion II (see §7.4.6) that both rings $R_\emptyset$ and $\mathbb{T}_\emptyset$ are isomorphic complete intersection rings.

7.5.2 A formula relating $\#\Phi_{R_\Sigma}$ and $\#\Phi_{R_{\Sigma'}}$: preparation

We explain below in §7.5.5 a formula relating $\#\Phi_{R_\Sigma}$ and $\#\Phi_{R_{\Sigma'}}$ using Galois cohomology groups with coefficients certain $G_\mathbb{Q}$–modules, which we shall now describe. Let

$$\tilde{f} = f_\emptyset = \sum_{n\ge 1} \tilde{a}_n q^n$$

denote Ribet's modular form of Theorem 7.30, attached to a two-dimensional modular representation $\rho_0 : G_\mathbb{Q} \to \mathrm{GL}_2(k)$ over the finite field k.

Recall that Ribet's modular Galois representation

$$\tilde{\rho} = \rho_{f,\lambda} : G_\mathbb{Q} \to \mathrm{GL}_2(\mathcal{O})$$

of minimal level N_0 given by Theorem 7.30 belongs to the (non–empty) set $DM_\emptyset(\mathcal{O})$.

Consider the reduction

$$\tilde{\rho} \bmod \lambda^n : G_\mathbb{Q} \to \mathrm{GL}_2(\mathcal{O}/\lambda^n).$$

We shall use the following finite $G_\mathbb{Q}$-modules X_{λ^n} defined by

$$X_{\lambda^n} = \mathrm{Ad}^0_{\tilde{\rho} \bmod \lambda^n} \subset \mathrm{M}_2(\mathcal{O}/\lambda^n) \subset \mathrm{Ad}^0(\tilde{\rho} \otimes K/\mathcal{O}) = X_{\tilde{\rho}}.$$

The notation Ad will be explained below, cf. §7.5.4 (we identify the $\mathcal{O}$-modules $\mathcal{O}/\lambda^n$ and $\lambda^{-n}/\mathcal{O}$ by choosing a uniformizer).

7.5.3 The Selmer group and Φ_{R_Σ}

In order to compute the invariant $\#\Phi_{R_\Sigma}$, A.Wiles established an isomorphism of $\mathcal{O}$–modules

$$\mathrm{Hom}_{\mathcal{O}-mod}(\Phi_{R_\Sigma}, K/\mathcal{O}) \cong \mathrm{Sel}_{\mathcal{D}_\Sigma} \subset H^1(G_\mathbb{Q}, X_{\tilde{\rho}})$$

where

$$\mathrm{Sel}_{\mathcal{D}_\Sigma} = H^1_{\mathcal{D}_\Sigma}(G_\mathbb{Q}, X_{\tilde{\rho}})$$

is a *generalized Selmer group*. This is a finite $\mathcal{O}$-submodule of the (infinite) $\mathcal{O}$-module $H^1(G_\mathbb{Q}, X_{\tilde{\rho}})$. The group $\mathrm{Sel}_{\mathcal{D}_\Sigma}$ is *contained* in the (usual) *Σ-Selmer group*, which is a finite $\mathcal{O}$-submodule Sel_Σ of $H^1(G_\mathbb{Q}, X_{\tilde{\rho}})$, consisting of *all* cohomology classes *unramified* outside of $\Sigma \cup S \cup \{p\}$, compare with (5.3.40):

$$\mathrm{Sel}_\Sigma(X_{\tilde{\rho}}) := \{x \in H^1(G_\mathbb{Q}, X_{\tilde{\rho}}) \mid \forall l \notin \Sigma, \mathrm{Res}^{G_\mathbb{Q}}_{I_l} x = 0\}, \tag{7.5.3}$$

where $\mathrm{Res}^{G_\mathbb{Q}}_{I_l} x$ denotes the restriction of x to the inertia subgroup I_l.

7.5.4 Infinitesimal deformations

Consider a representation $\rho : G \to \mathrm{GL}_n(A)$ of a group G. This determines an $A[G]$–module structure on the free A-module $M = A^n$, with the action of G given by ρ. We shall also consider the following $A[G]$–modules:

$$\begin{aligned} &\mathrm{Ad}(M) = \mathrm{End}_A M, \qquad g : x \mapsto \rho(g) x \rho(g)^{-1} \\ &\cup \\ &\mathrm{Ad}(M)^{\mathrm{tr}=0} = \mathrm{End}_A^{\mathrm{tr}=0}(M) \text{ (an } A[G]-\text{submodule).} \end{aligned} \tag{7.5.4}$$

Definition 7.48. *1) An extension of M by M is a short exact sequence of $A[G]$-modules of the form*

$$0 \longrightarrow M \xrightarrow{\alpha} E \xrightarrow{\beta} M \longrightarrow 0.$$

2) Consider the ring $A[\varepsilon]$ generated over A by an element ε subject to the relation $\varepsilon^2 = 0$. There is a projection map $\pi_{\varepsilon=0} : A[\varepsilon] \to A$ which takes ε to 0. By an infinitesimal deformation of a representation $\rho : G \to \mathrm{GL}_n(A)$, we shall mean a representation $\rho' : G \to \mathrm{GL}_n(A[\varepsilon])$, such that $\rho = \pi_{\varepsilon=0} \circ \rho'$:

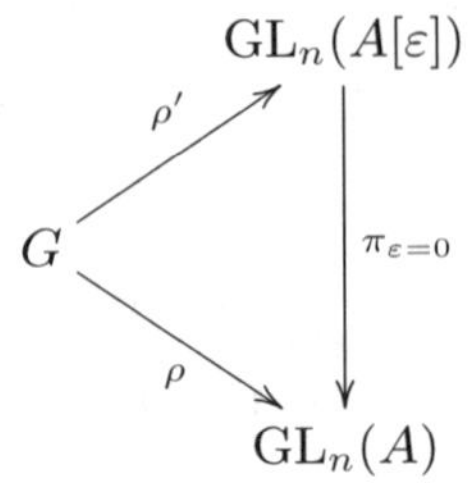

3) Two infinitesimal deformations $\rho', \rho'' : G \to \mathrm{GL}_n(A[\varepsilon])$ are said to be strictly equivalent if there exists a matrix $C \in 1_n + \varepsilon \mathrm{M}_n(A[\varepsilon])$ such that

$$\textit{for all } g \in G, \rho''(g) = C^{-1}\rho'(g)C.$$

Remark 7.49. The construction of an infinitesimal deformation may be viewed as the first step in the construction of any deformation.

Theorem 7.50. *There are canonical bijections between the following three sets a), b), c) :*

a) $H^1(G, \mathrm{Ad}\rho)$;
b) The set $\mathrm{Ext}^1(M, M)$ of equivalence classes of extensions of M by M;
c) The set of strict equivalence classes of infinitescimal deformations of ρ.

Proof of Theorem 7.50
Let us consider an extension

$$0 \longrightarrow M \xrightarrow{\alpha} E \xrightarrow{\beta} M \longrightarrow 0.$$

The module M is a free A-module. Hence we may choose a section $\phi : M \to E$ of β, which is morphism of A–modules but not of $A[G]$-modules.

This means that for all $m \in M$ and $g \in G$ we have

$$g\phi(g^{-1}m) - \phi(m) \in \mathrm{Ker}(\beta) = \mathrm{Im}(\alpha),$$

since

$$\beta(g\phi(g^{-1}m) - \phi(m)) = g\beta\phi(g^{-1}m) - \beta\phi(m) = 0.$$

From this we obtain the following 1-cocycle (representing a cohomology class in $H^1(G, \mathrm{Ad}(\rho))$):

$$T_g := \left(m \mapsto \alpha^{-1}(g\phi(g^{-1}m) - \phi(m))\right) \in \mathrm{End}_A(M) = \mathrm{Ad}\rho.$$

One rereads the cocycle condition as by saying that

$$\rho'(g) := (1_n + \varepsilon T_g)\rho(g)$$

is an infinitesimal deformation.

Conversely, any cocycle $\{T_g\} \in H^1(G, \mathrm{Ad}\rho)$, defines an extension $M \otimes_A A[\varepsilon] \cong M \oplus \varepsilon M$ with the action given by ρ'.

Remark 7.51. If $\det\rho' = \det\rho$, then there is the identity

$$\det(1_n + \varepsilon T_g)\det\rho(g) = \det\rho(g) \Rightarrow \mathrm{tr}T_g = 0$$

and it follows that $H^1(G, \mathrm{Ad}^0\rho)$ describes the deformations with fixed determinant.

7.5.5 Deformations of type $\mathcal{D}_\Sigma$

The deformations of type $\mathcal{D}_\Sigma$ are defined starting from the restriction maps

$$\mathrm{Res}_{D_l}^{G_\mathbb{Q}} : H^1(G_\mathbb{Q}, X) \longrightarrow H^1(D_l, X)$$

which are used in the definition of the *generalized Selmer groups.*

Examples of computations of Selmer groups

(comp. with §5.3, §4.5). Given a short exact sequence $0 \longrightarrow A \longrightarrow B \longrightarrow C \longrightarrow 0$ of three G-modules, there is a long exact sequence (4.5.23) of cohomology groups:

$$\begin{aligned} 0 \longrightarrow & H^0(A) \longrightarrow H^0(B) \longrightarrow H^0(C) \\ \longrightarrow & H^1(A) \longrightarrow H^1(B) \longrightarrow \cdots, H^n(A) = H^n(G, A). \end{aligned} \tag{7.5.5}$$

We shall use this sequence to deduce Kummer theory:

Example 7.52 (Kummer theory and Quadratic Characters). Let K be a field containing the group μ_m of all roots of unity of degree m in $\overline{K}$. Assume further that Char K does not divide m. For an arbitrary Galois extension L/K with Galois group $G = \mathrm{Gal}(L/K)$ the map $x \mapsto x^m$ defines a homomorphism of G–modules: $\nu : L^\times \longrightarrow L^\times$, and one has the following exact sequence

$$1 \longrightarrow \mu_m \longrightarrow \overline{K}^\times \xrightarrow{\nu} \overline{K}^\times \longrightarrow 1.$$

Passing to cohomology groups (4.5.22) we obtain the following long exact sequence

$$\begin{aligned} &H^0(G_K, \mu_m) \longrightarrow H^0(G_K, \overline{K}^\times) \xrightarrow{\nu} H^0(G_K, \overline{K}^\times) \longrightarrow \\ &H^1(G_K, \mu_m) \longrightarrow H^1(G_K, \overline{K}^\times) \xrightarrow{\nu} H^1(G_K, \overline{K}^\times) \longrightarrow \cdots . \end{aligned} \tag{7.5.6}$$

Since G acts trivially on μ_m, it follows that $H^1(G, \mu_m)$ coincides with the group $\mathrm{Hom}(G, \mu_m)$. The group $H^0(G, L^\times)$ is the subgroup of all fixed points of the Galois action, i.e. $H^0(G, L^\times) = L^{\times \mathrm{Gal}(L/K)} = K^\times$. Also, $H^0(G, \mu_m) = \mu_m$, and $H^1(G, L^\times) = \{1\}$, by Hilbert's theorem 90. We thus have the following exact sequence

$$1 \longrightarrow \mu_m \longrightarrow K^\times \xrightarrow{\nu} K^\times \longrightarrow \mathrm{Hom}(G, \mu_m) \longrightarrow 1,$$

which is equivalent to the isomorphism of Kummer:

$$K^\times / K^{\times m} \cong \mathrm{Hom}(G_K, \mu_m).$$

In particular, letting $m = 2$ and $K = \mathbb{Q}$, we have

$$H^1(G_\mathbb{Q}, \{\pm 1\}) = \mathbb{Q}^\times / \mathbb{Q}^{\times 2} \cong \mathrm{Hom}(G_\mathbb{Q}, \mu_2).$$

The right hand side here is the (infinite) set of all quadratic characters of $G_\mathbb{Q}$.

Fixing a finite set Σ of primes, we have

$$\mathrm{Sel}_\Sigma(\{\pm 1\}) \cong \left\{ \begin{array}{c} \text{the finite set of all quadratic characters of } G_\mathbb{Q} \\ \text{unramified outside } \Sigma \end{array} \right\}.$$

Example 7.53 (The inflation-restriction sequence and Local Tate duality). Let H be an open normal subgroup in G and let A be a G–module. Then one has the following "inflation - restriction" exact sequence (comp. with (4.5.24)):

$$\begin{aligned} 0 \longrightarrow H^1(G/H, A^H) \xrightarrow{\mathrm{Inf}} H^1(G, A) \xrightarrow{\mathrm{Res}} \\ H^1(H, A)^{G/H} \longrightarrow H^2(G/H, A^H). \end{aligned} \tag{7.5.7}$$

Theorem 7.54 (Local Tate duality). *Let X be a finite D_l module, where $D_l = \mathrm{Gal}(\overline{\mathbb{Q}}_l/\mathbb{Q}_l) \subset G_\mathbb{Q}$, and let $n = |X|$. We shall also consider the dual module*

$$X^* = \mathrm{Hom}(X, \mu_n) \quad ((gx^*)(x) = gx^*(g^{-1}x).$$

Then:

a) The groups $H^i(D_l, X)$ are finite for all $i \geq 0$, and are trivial for $i \geq 3$;
b) There is a non–degenerate pairing

$$H^i(D_l, X) \times H^{2-i}(D_l, X^*) \longrightarrow H^2(D_l, \overline{\mathbb{Q}}_l^\times) \cong \mathbb{Q}/\mathbb{Z};$$

c) *If $l \nmid |X|$ then the subgroups of unramified classes*

$$H^1(D_l/I_l, X^{I_l}) \quad \text{and} \quad H^1(D_l/I_l, (X^*)^{I_l})$$

are each other's annihilators with respect to the above pairing;

d) *The Euler characteristic of X is given by the following explicit formula:*

$$\frac{\#H^1(D_l, X)}{\#H^0(D_l, X) \cdot \#H^2(D_l, X)} = l^{v_l(\#X)} = \frac{\#H^1(D_l, X^*)}{\#H^0(D_l, X^*) \cdot \#H^2(D_l, X^*)}.$$

Generalized Selmer groups

Definition 7.55. *Let X be a finite $G_{\mathbb{Q}}$–module and suppose we have a family $\mathfrak{L} = \{L_l\}$ of subgroups $L_l \subset H^1(D_l, X)$, which are finite by Theorem 7.54. We shall assume that for $l \notin \Sigma$ we have*

$$L_l = \operatorname{Ker}(H^1(D_l, X) \longrightarrow H^1(I_l, X)). \tag{7.5.8}$$

The Selmer group attached to a family $\mathfrak{L} = \{L_l\}$ is defined by

$$\operatorname{Sel}_{\mathfrak{L}}(X) = \{x \in H^1(G_{\mathbb{Q}}, X) \mid \forall l,\ \operatorname{Res}^{G_{\mathbb{Q}}}_{D_l}(x) \in L_l\}. \tag{7.5.9}$$

Remark 7.56. It follows from (7.5.8) and (7.5.3) that $\operatorname{Sel}_{\mathfrak{L}}(X) \subset \operatorname{Sel}_{\Sigma}(X)$.

Interpretation of Φ_{R_Σ} as a generalized Selmer group

Consider the finite $G_{\mathbb{Q}}$–module $X_{\lambda^n} = \operatorname{Ad}^0\tilde{\rho} \bmod \lambda^n$, where

$$\tilde{\rho} : G_{\mathbb{Q}} \longrightarrow \mathrm{GL}_2(\mathcal{O})$$

denotes Ribet's modular Galois representation of minimal level (see Theorem 7.30). Let us fix a (ramification) type $\mathcal{D}_\Sigma$ and consider the tangent space $\Phi_{R_\Sigma} = I_{R_\Sigma}/I^2_{R_\Sigma}$. This group was interpreted by Wiles using the generalized Selmer group attached to the following family

$$L_l = \begin{cases} \operatorname{Ker}(H^1(D_l, X_{\lambda^n}) \longrightarrow H^1(I_l, X_{\lambda^n})), & \text{if } l \notin \Sigma_S \\ H^1(D_l, X_{\lambda^n}), & \text{if } l = p \\ \operatorname{Ker}(H^1(D_l, X_{\lambda^n}) \longrightarrow H^1(I_l, X_{\lambda^n}/X^0)), X^0 = \left\{\left(\begin{smallmatrix} 0 & * \\ 0 & 0 \end{smallmatrix}\right)\right\}, & \text{if } l \in \Sigma \cup S \end{cases} \tag{7.5.10}$$

Definition 7.57. *With the choice of $X = X_{\lambda^n}$ as above, and $\mathfrak{L}_\Sigma = \{L_l\}$ given by 7.5.10, we define*

$$\begin{aligned} \operatorname{Sel}_{\mathcal{D}_\Sigma}(X) = &\operatorname{Sel}_{\mathfrak{L}}(X) = H^1_{\mathcal{D}_\Sigma}(G_{\mathbb{Q}}, X) = \\ &\{x \in H^1(G_{\mathbb{Q}}, X) \mid \forall l, \operatorname{Res}^{G_{\mathbb{Q}}}_{D_l}(x) \in L_l\}. \end{aligned} \tag{7.5.11}$$

Theorem 7.58. *There is an isomorphism of finite $\mathcal{O}$–modules:*

$$\mathrm{Hom}_{\mathcal{O}-mod}(I_{R_\Sigma}/I_{R_\Sigma}^2, \mathcal{O}/\lambda^n) = H^1_{\mathcal{D}_\Sigma}(G_\mathbb{Q}, X_{\lambda^n}).$$

Lemma 7.59. *There is an isomorphism of finite $\mathcal{O}$–modules:*

$$\mathrm{Hom}_k(\mathfrak{m}_{R_\Sigma}/(\lambda, \mathfrak{m}_{R_\Sigma}^2), k) = H^1_{\mathcal{D}_\Sigma}(G_\mathbb{Q}, \mathrm{Ad}^0\rho_0).$$

Proof of Lemma 7.59. By Theorem 7.50, the restriction map

$$\mathrm{Hom}_{\mathcal{O}-alg}(R_\Sigma, k[\varepsilon]) \ni \phi \longmapsto \phi|_{\mathfrak{m}_{R_\Sigma}} : \mathfrak{m}_{R_\Sigma}/(\lambda, \mathfrak{m}_{R_\Sigma}^2) \to k$$

gives a canonical isomorphism

$$\mathrm{Hom}_{\mathcal{O}-alg}(R_\Sigma, k[\varepsilon]) \cong \mathrm{Hom}_k(\mathfrak{m}_{R_\Sigma}/(\lambda, \mathfrak{m}_{R_\Sigma}^2), k).$$

Indeed, any such ϕ vanishes on $(\lambda, \mathfrak{m}_{R_\Sigma}^2)$, and is determined by its restriction to $\mathfrak{m}_{R_\Sigma}$ by the formula

$$R_\Sigma \ni x \mapsto \iota(x) + \varepsilon\phi(x - \iota(x)) \in k + \varepsilon k.$$

In the above,

$$\iota : R_\Sigma \to R_\Sigma/\mathfrak{m}_{R_\Sigma} \xrightarrow{\sim} k \hookrightarrow R_\Sigma$$

is the canonical projection and ϕ is the identity on $\iota(R_\Sigma)$. By the universal property of R_Σ, the map ϕ is may be identified with an infinitesimal deformation

$$\rho' : G_\mathbb{Q} \longrightarrow \mathrm{GL}_2(k[\varepsilon])$$

of ρ_0 of type $\mathcal{D}_\Sigma$, i.e. with an element of $H^1_{\mathcal{D}_\Sigma}(G_\mathbb{Q}, \mathrm{Ad}^0\rho_0)$. This implies the lemma.

To deduce Theorem 7.58 from Lemma 7.59, one can replace $\mathfrak{m}_{R_\Sigma}$ by $(I_{R_\Sigma}, \lambda^n)$, where $I_{R_\Sigma} = \mathrm{Ker}(\pi_{R_\Sigma} : R_\Sigma \to \mathcal{O})$, using a version of Nakayama's lemma (we omit the details).

Now we explain a formula relating $\#\Phi_{R_\Sigma}$ and $\#\Phi_{R_{\Sigma'}}$. An explicite formula for $\#\Phi_{R_\Sigma}$ can be obtained from the cohomological exact sequence of Poitou–Tate.

However, a weaker result suffices for Wiles' induction, namely:

$$\#\Phi_{R_{\Sigma'}} \le \#\Phi_{R_\Sigma} \cdot \#H^1(I_l, X)^{D_l}, \text{ where } \Sigma' = \Sigma \cup \{l\}, \tag{7.5.12}$$

which follows from

$$0 \longrightarrow \mathrm{Sel}_{\mathcal{D}_\Sigma} \longrightarrow \mathrm{Sel}_{\mathcal{D}_{\Sigma'}} \longrightarrow H^1(I_l, X)^{D_l},$$

and may be deduced from the following form of the inflation-restriction sequence (7.5.7) (we use the fact that X is unramified at l, i.e. I_l acts trivially on X):

$$\begin{aligned} 0 \longrightarrow H^1(D_l/I_l, X) &\xrightarrow{\mathrm{Inf}} H^1(D_l, X) \xrightarrow{\mathrm{Res}} \\ H^1(I_l, X)^{D_l} &\longrightarrow H^2(D_l/I_l, X) \longrightarrow H^2(D_l, X). \end{aligned} \tag{7.5.13}$$

Induction step: reformulation

Let us use the induction hypothesis and assume that

$$\varphi_\Sigma : R_\Sigma \xrightarrow{\sim} \mathbb{T}_\Sigma,$$

which implies the equality of the corresponding invariants (7.5.1)

$$\#\Phi_{R_\Sigma} = \#\mathcal{O}/\eta_{\mathbb{T}_\Sigma} < \infty.$$

According to Criterion I, it suffices to prove the inequality (7.5.2)

$$\#\Phi_{R_{\Sigma'}} \le \#\mathcal{O}/\eta_{\mathbb{T}_{\Sigma'}} < \infty.$$

The left hand side is controlled by a Galois cohomology group: we know by (7.5.12) that

$$\#\Phi_{R_{\Sigma'}} \le \#\Phi_{R_\Sigma} \cdot \#H^1(I_l, X)^{D_l}, \text{ where } \Sigma' = \Sigma \cup \{l\}.$$

The right hand side will be computed below using a *relative invariant* $\eta_{\Sigma',\Sigma}$ which relates $\#\mathcal{O}/\eta_{\mathbb{T}_{\Sigma'}}$ and $\#\mathcal{O}/\eta_{\mathbb{T}_\Sigma}$:

$$\#\mathcal{O}/\eta_{\mathbb{T}_{\Sigma'}} = \#\mathcal{O}/\eta_{\mathbb{T}_\Sigma} \cdot \#\mathcal{O}/\pi_{\mathbb{T}_\Sigma}(\eta_{\Sigma',\Sigma}).$$

We explain next the main inequality between these quantities implying the induction step.

7.6 The Relative Invariant, the Main Inequality and The Minimal Case

7.6.1 The Relative invariant

Recall that we use the following invariants (7.4.10) of a local $\mathcal{O}$–algebra A:

$$I_A = \mathrm{Ker}\pi_A, \quad \Phi_A = I_A/I_A^2, \quad \eta_A = \pi_A(\mathrm{Ann}I_A) \subset \mathcal{O}$$

(the “kernel of augmentation”, resp. the “tangent space”, resp. the “congruence module”).

Under the induction hypothesis there is an isomorphism of complete intersection $\mathcal{O}$-algebras:

$$\varphi_\Sigma : R_\Sigma \xrightarrow{\sim} \mathbb{T}_\Sigma,$$

and this implies the following identity:

$$\#\Phi_{R_\Sigma} = \#\mathcal{O}/\eta_{\mathbb{T}_\Sigma} < \infty.$$

According to Criterion I, it suffices to prove the following inequality (7.5.2):

$$\#\Phi_{R_{\Sigma'}} \le \#\mathcal{O}/\eta_{\mathbb{T}_{\Sigma'}} < \infty,$$

where $\Sigma' = \Sigma \cup \{l\}$, given (7.5.1). The left hand side is controlled by a Galois cohomology group: we know by (7.5.12) that

$$\#\Phi_{R_{\Sigma'}} \le \#\Phi_{R_\Sigma} \cdot \#H^1(I_l, X)^{D_l}, \text{ where } \Sigma' = \Sigma \cup \{l\}.$$

The right hand side will be computed below using a certain *relative invariant* $\eta_{\Sigma',\Sigma}$ which relates $\#\mathcal{O}/\eta_{\mathbb{T}_{\Sigma'}}$ and $\#\mathcal{O}/\eta_{\mathbb{T}_\Sigma}$:

$$\#\mathcal{O}/\eta_{\mathbb{T}_{\Sigma'}} = \#\mathcal{O}/\eta_{\mathbb{T}_\Sigma} \cdot \#\mathcal{O}/\pi_\Sigma(\eta_{\Sigma',\Sigma}).$$

We have by definition

$$\eta_{\mathbb{T}_\Sigma} = \pi_{\mathbb{T}_\Sigma}(\mathrm{Ann}I_{\mathbb{T}_\Sigma}) \subset \mathcal{O}, \eta_{\mathbb{T}_{\Sigma'}} = \pi_{\mathbb{T}_{\Sigma'}}(\mathrm{Ann}I_{\mathbb{T}_{\Sigma'}}) \subset \mathcal{O}. \tag{7.6.1}$$

Notice that by the universal property of $\mathbb{T}_\Sigma$, we have a commutative diagram:

$$\begin{array}{ccc} \mathbb{T}_{\Sigma'} & \xrightarrow{\pi_{\Sigma',\Sigma}} & \mathbb{T}_\Sigma \\ & {\scriptstyle \pi_{\mathbb{T}_{\Sigma'}}} \searrow \quad \swarrow {\scriptstyle \pi_{\mathbb{T}_\Sigma}} & \\ & \mathcal{O} & \end{array} \tag{7.6.2}$$

From this we obtain

$$\mathcal{O} \supset \eta_{\mathbb{T}_{\Sigma'}} = \eta_{\mathbb{T}_\Sigma} \cdot \pi_{\mathbb{T}_\Sigma}(\eta_{\Sigma',\Sigma}), \tag{7.6.3}$$

where

$$\eta_{\Sigma',\Sigma} = \pi_{\Sigma',\Sigma}(\mathrm{Ann} I_{\mathbb{T}_{\Sigma'}})$$

is called the *relative invariant* of the morphism $\pi_{\Sigma',\Sigma}$.

An explicit computation of the invariant using the construction of the universal modular deformation ring $\mathbb{T}_\Sigma$ in §7.4.4 leads to

$$\eta_{\Sigma',\Sigma} = \pi_{\Sigma',\Sigma}(\mathrm{Ann} I_{\mathbb{T}_{\Sigma'}}) = (l-1)(T_l^2 - (l+1)^2) \in \mathbb{T}_\Sigma. \tag{7.6.4}$$

This calculation involve computations with the reduction of modular curves (see [Da95]). We omit the details.

Below we shall explain the main inequality between these quantities:

$$\# H^1(I_l, X)^{D_l} \le \# \mathcal{O}/\pi_{\mathbb{T}_\Sigma}(\eta_{\Sigma',\Sigma}), \text{ where } \Sigma' = \Sigma \cup \{l\}. \tag{7.6.5}$$

This inequality impies the induction step. The use of the relative invariant in (7.6.5)

$$\# \mathcal{O}/\pi_{\mathbb{T}_\Sigma}(\eta_{\Sigma',\Sigma})$$

require the choice of augmentations $\pi_{\mathbb{T}_\Sigma}$ and π_{R_Σ} given above by (7.3.4):

$$\pi_{R_\Sigma} : \mathrm{tr}\rho_\Sigma^{univ}(\mathrm{Frob}_l) \mapsto a_l$$
$$\pi_{\mathbb{T}_\Sigma} : \mathrm{tr}\rho_\Sigma^{univ.mod.}(\mathrm{Frob}_l) \mapsto a_l$$

where $a_l \in \mathcal{O}$ are the Fourier coefficients of Ribet's cusp eigenform

$$f(z) = \sum_{n=0}^{\infty} a_n e(nz) \in \mathcal{S}_k(N_0)$$

of minimal level (whose Galois representation $\rho_{f,\lambda}$ corresponds to a deformation $\rho \in DM_\emptyset(\mathcal{O})$).

7.6.2 The Main Inequality

Applying the augmentation $\pi_{\mathbb{T}_\Sigma}$ to $\eta_{\Sigma',\Sigma}$ given above by (7.3.4) to the relative invariant $\eta_{\Sigma',\Sigma}$ introduced above, we obtain:

$$\begin{aligned} \pi_{\mathbb{T}_\Sigma}(\eta_{\Sigma',\Sigma}) &= \pi_{\mathbb{T}_\Sigma}((l-1)(T_l^2 - (l+1)^2)) \\ &= (l-1)(a_l^2 - (l+1)^2) \in \mathcal{O}. \end{aligned} \tag{7.6.6}$$

Notice that the quantity (7.6.6) does not vanish due to Deligne's bound: $|a_l| < 2\sqrt{l}$:

$$\pi_{\mathbb{T}_\Sigma}(\eta_{\Sigma',\Sigma}) = (l-1)(a_l^2 - (l+1)^2) \ne 0.$$

We shall deduce the main inequality (7.6.5) in the following more explicit form:

Theorem 7.60 (Main Inequality). *Consider for any $n \geq 1$ the finite $G_{\mathbb{Q}}$–module $X = \mathrm{Ad}^0 \tilde{\rho} \bmod \lambda^n$, where*

$$\tilde{\rho} : G_{\mathbb{Q}} \longrightarrow \mathrm{GL}_2(\mathcal{O})$$

denotes Ribet's modular Galois representation of minimal level (see Theorem 7.30). Then the following inequality holds:

$$\#H^1(I_l, X)^{D_l} \leq \#\mathcal{O}/(l-1)(a_l^2 - (l+1)^2), \text{ where } \Sigma' = \Sigma \cup \{l\}. \quad (7.6.7)$$

Proof follows from a computation of a determinant. Taking into account that

$$H^1(I_l, X)^{D_l} = \mathrm{Hom}_{D_l}(I_l, X). \quad (7.6.8)$$

Since the action of I_l on $X = \mathrm{Ad}^0 \tilde{\rho} \bmod \lambda^n$ is trivial for $l \nmid N(\tilde{\rho})$, it follows that

$$H^1(I_l, X)^{D_l} = \mathrm{Hom}_{D_l}(I_l, X). \quad (7.6.9)$$

Notice that $I_l \cong \mathrm{Gal}(\mathbb{Q}_l/\mathbb{Q}_l^{nr})$, where $\mathbb{Q}_l^{nr} \supset \mu_{p^\infty}$ is the maximal non–ramified extension of $\mathbb{Q}_l$ (which contains all p–power roots of unity because $p \neq l$).

This means that there exists a canonical surjection

$$I_l \longrightarrow \mathbb{Z}_p(1) = \varprojlim_n \mu_{p^n}$$

(the right hand side here is the Galois group of the maximal Kummer p-extension of $\mathbb{Q}_l^{nr}$).

Next, notice that the order of the finite module $X = \mathrm{Ad}^0 \tilde{\rho} \bmod \lambda^n$ is a power of p. Hence any homomorphism in (7.6.9) factorizes through $\mathbb{Z}_p(1)$. We therefore have

$$\mathrm{Hom}_{D_l}(I_l, X) = \mathrm{Hom}_{D_l}(\mathbb{Z}_p(1), X) = X(-1)^{D_l} \quad (7.6.10)$$

where

$$X(-1) = X \otimes \mathbb{Z}_p(-1), \mathbb{Z}_p(-1) = \mathrm{Hom}(\mathbb{Z}_p(1), \mathbb{Z}_p).$$

Since the action of I_l on $X = \mathrm{Ad}^0 \tilde{\rho} \bmod \lambda^n$ is trivial, we now see that

$$X(-1)^{D_l} = X(-1)^{D_l} = X(-1)^{\mathrm{Frob}_l} = \mathrm{Ker}(\mathrm{Frob}_l - 1)|_{X(-1)}. \quad (7.6.11)$$

We may calculate further:

$$\begin{aligned} &\#\mathrm{Ker}(\mathrm{Frob}_l - 1)|_{X(-1)} = \#\mathrm{Coker}(\mathrm{Frob}_l - 1)|_{X(-1)} \\ &= \#\mathcal{O}/\det(\mathrm{Frob}_l - 1)|_{X(-1)} \end{aligned} \quad (7.6.12)$$

and one only needs to verify that that

$$\#\mathcal{O}/\det(\mathrm{Frob}_l - 1)|_{X(-1)} = \#\mathcal{O}/(l-1)(a_l^2 - (l+1)^2).$$

To prove this we shall calculate explicitely the eigenvalues of $\mathrm{Frob}_l - 1$ on $X(-1)$. Let $\alpha = \alpha_l$ and $\beta = \beta_l$ be the eigenvalues of of Frob_l acting on the $\mathcal{O}$-module $M = \mathcal{O}^2$. Thus we have:

$$\alpha + \beta = a_l, \quad \alpha\beta = l$$

We shall now determine the eigenvalues of $Frob_l$ acting on the following $\mathcal{O}$-modules:

- on $\check{M} = \mathrm{Hom}(M, \mathbb{Q}/Z)$ the eigenvalues of Frob_l are $\{\alpha^{-1}, \beta^{-1}\}$;
- on $\mathrm{End}M \cong M \otimes \check{M}$ the eigenvalues of Frob_l are: $\{1, 1, \alpha\beta^{-1}, \beta\alpha^{-1}\}$. To see this, note that

$$\begin{pmatrix} \alpha & 0 \\ 0 & \beta \end{pmatrix} \otimes \begin{pmatrix} \alpha^{-1} & 0 \\ 0 & \beta^{-1} \end{pmatrix} = \begin{pmatrix} 1 & 0 & 0 & 0 \\ 0 & 1 & 0 & 0 \\ 0 & 0 & \alpha\beta^{-1} & 0 \\ 0 & 0 & 0 & \beta\alpha^{-1} \end{pmatrix}$$

- on $\mathrm{Ad}^0\tilde{\rho} \subset \mathrm{End}M$ there are only three eigenvalues of Frob_l:

$$\left\{1, \alpha\beta^{-1} = \frac{\alpha^2}{\alpha\beta} = \frac{\alpha^2}{l}, \beta\alpha^{-1} = \frac{\beta^2}{\alpha\beta} = \frac{\beta^2}{l}\right\}.$$

- on $X(-1)$ there are the following eigenvalues of Frob_l:

$$\left\{\frac{1}{l}, \frac{\alpha^2}{l^2}, \frac{\beta^2}{l^2}\right\}.$$

This is because $X(-1) = \mathbb{Z}_p(-1) \otimes X$, and Frob_l acts on $\mathbb{Z}_p(-1)$ as $\frac{1}{l}$;
- finally the eigenvalues of $\mathrm{Frob}_l - 1$ on $X(-1)$ are the following:

$$\left\{\frac{1}{l} - 1, \frac{\alpha^2}{l^2} - 1, \frac{\beta^2}{l^2} - 1\right\}.$$

It follows that:

$$\begin{aligned} \det(\mathrm{Frob}_l - 1)|_{X(-1)} &= (\frac{1}{l} - 1)(\frac{\alpha^2}{l^2} - 1)(\frac{\beta^2}{l^2} - 1) = \\ -\frac{(l-1)(l^2-\alpha^2)(l^2-\beta^2)}{l^5} &= \\ -\frac{(l-1)(l^4 - l^2(\alpha^2+\beta^2) + \alpha^2\beta^2)}{l^5} &= \\ -\frac{(l-1)(l^2+1-(\alpha^2+\beta^2))}{l^3} &= -\frac{(l-1)(a_l^2-(l+1)^2)}{l^3}. \end{aligned}$$

This proves the Main inequality (7.6.7), because $l \in \mathcal{O}^\times$ is a unit. $\square$

7.6.3 The Minimal Case

We keep the notations and assumptions of Main theorem 7.33. To complete the proof of Main theorem 7.33, we now treat the minimal case $\Sigma = \emptyset$, and establish the following

Theorem 7.61. *The surjective morphism*

$$\varphi_\emptyset : R_\emptyset \longrightarrow \mathbb{T}_\emptyset$$

in the category $\mathcal{C}_\mathcal{O}$ is an isomorphism of complete intersection $\mathcal{O}$-algebras.

Proof of Theorem 7.61. We notice first that the $\mathcal{O}$–algebra $R_\emptyset$ admits n (topological) generators, where

$$n = \dim_k I_{R_\emptyset}/I^2_{R_\emptyset} = \dim_k \mathrm{Sel}_{\mathcal{D}_\emptyset}(X).$$

Here $X = \mathrm{M}_2(k)^{\mathrm{tr}=0}$ is a finite $G_\mathbb{Q}$-module with the action given by the representation ρ_0.

We shall construct a J-structure on the surjective morphism $\varphi_\emptyset$. By Criterion II of §7.4.6, this will show that $R_\emptyset$ and $\mathbb{T}_\emptyset$ are isomorphic complete intersection rings.

The construction of the J–structure uses Nakayama's lemma and the Chebotarev density theorem (see Theorem 4.22).

For any $m \in \mathbb{N}$, we choose a finite set of primes:

$$Q_m = \{q_{m1}, \cdots, q_{mn}\} \tag{7.6.13}$$

with the properties

- $q_{mj} \equiv 1 \bmod p^m$ ($\Longleftrightarrow p^m$ divides $\#(\mathbb{Z}/q_{mj}\mathbb{Z})^\times$);
- the eigenvalues of $\rho_0(\mathrm{Frob}_{q_{mj}})$ in $\overline{k}$ are distinct.
- $Q_m \cap S = \emptyset$.

We begin defining the J-structure using using these finite sets of primes $\Sigma = Q_m$ as follows:

$$A_m = R_{Q_m}, B_m = \mathbb{T}_{Q_m}, A = R_\emptyset, B = \mathbb{T}_\emptyset.$$

We have:

$$\begin{aligned} \mathcal{O}[[S_1, \cdots, S_n]]/J_m &= \mathcal{O}[[S_1, \cdots, S_n]]/(\omega_m(S_1), \cdots, \omega_m(S_n)) \\ &= \mathcal{O}[\Delta_1 \times \cdots \times \Delta_n], \end{aligned} \tag{7.6.14}$$

where Δ_j denotes the cyclic subgroup of order p^m in $(\mathbb{Z}/q_{mj}\mathbb{Z})^\times$.

We shall consider deformations ρ of type $\mathcal{D}_m = \mathcal{D}_{Q_m}$.

Proposition 7.62. *Under the notations and assumptions of the Main Theorem 7.33, there is an isomorphism*

$$\mathrm{Sel}_{\emptyset} \cong \mathrm{Sel}_{\mathcal{D}_m};$$

the $\mathcal{O}$-algebra $A_m = R_{Q_m}$ admits n generators.

Proof of Proposition 7.62 uses an explicit formula for the order of the finite group $\#\mathrm{Sel}_{\mathcal{D}_m}$ obtained from the cohomological exact sequence of Poitou–Tate. We omit the details here, but see §7.5.5.

In order to have a J–structure notice that for any $q = q_{mj} \in Q_m$ the action of the inertia group I_q on X factorizes through the quotient Δ_q of $(\mathbb{Z}/q\mathbb{Z})^\times$ of order p^m, by a result of Faltings:

$$I_q \to \mathrm{Gal}(\mathbb{Q}(\mu_q)/\mathbb{Q}) \cong (\mathbb{Z}/q\mathbb{Z})^\times \to \Delta_q.$$

(see Appendix by G. Faltings to [Wi]). Next, notice that

$$\mathbb{Z}_p[\Delta_q] \cong \mathbb{Z}_p[[S]]/\omega_m(S), \omega_m(S) = (1+S)^{p^m} - 1,$$

and there is an isomorphism

$$\mathcal{O}[\Delta_{q_{m1}} \times \cdots \times \Delta_{q_{mn}}] := C_m \cong \mathcal{O}[[S_1, \cdots, S_n]]/J_m, \tag{7.6.15}$$

where $J_m = (\omega_m(S_1), \cdots, \omega_m(S_n)) \subset \mathcal{O}[[S_1, \cdots, S_n]]$. According to a result of de Shalit on the structure of certain Hecke algebras, $B_m/J_m B_m$ is a torsion free module of finite rank over C_m (see in [CSS95]; the proof of this result uses diamond operators in the Hecke algebras $\mathbb{T}(N_\Sigma)$).

Consider the surjective morphism $\varphi : A \to B$ of Theorem 7.61. Then the isomorphism shows that φ admits a J–structure: there exists a family of commutative diagrams, for all $m \in \mathbb{N}$:

$$\begin{array}{ccccc}
 & & \mathcal{O}[[S_1, \cdots, S_n]] & & \\
 & & \downarrow \sigma_m & & \\
\mathcal{O}[[X_1, \cdots, X_n]] & \xrightarrow{\xi_m} & A_m & \xrightarrow{\varphi_m} & B_m \\
 & & \downarrow & & \downarrow \\
 & & A & \xrightarrow{\varphi} & B
\end{array}$$

with the following properties:

i) ξ_m is surjective;

ii) φ_m is surjective;

iii) $A_m/J_0 A_m \cong A, \quad B_m/J_0 B_m \cong B.$

iv) $B_m/J_m B_m$ is a torsion free module of finite rank over the $\mathcal{O}$-algebra $\mathcal{O}[[S_1, \cdots, S_n]]/J_m$.

In order to conclude the proof of Theorem 7.61, we apply directly Criterion II of §7.4.7: let $\varphi : A \to B$ be a surjective morphisme in the category $\mathcal{C}_\mathcal{O}$, which admits a J-structure. Then φ is an isomorphism of two local complete intersection $\mathcal{O}$–algebras (see also [Ta-Wi]).

7.7 End of Wiles' Proof and Theorem on Absolute Irreducibility

7.7.1 Theorem on Absolute Irreducibility

In this section we explain Wiles' deduction of the Shimura–Taniyama–Weil conjecture (Theorem 7.13) from the theorem on modularity of admissible deformations 7.33.

In order to use Theorem 7.33 one needs to have an absolutely irreducible Galois representation ρ_0 over a finite field k, starting from an elliptic curve.

Theorem 7.63 (Irreducibility). *Let*

$$E : y^2 = 4x^3 - g_2x - g_3$$

be the Weierstrass form of a semistable elliptic curve over $\mathbb{Q}$. Suppose that the Galois representations $\overline{\rho}_{3,E}$ and $\overline{\rho}_{5,E}$ are both reducible.

Then we have the only four possibilities for j_E:

$$j_E \in \left\{ -\frac{25}{2}, -\frac{5^2 \cdot 241^3}{2^3}, -\frac{5 \cdot 29^3}{2^5}, \frac{5 \cdot 211^3}{2^{15}} \right\},$$

and E is modular.

Proof of Theorem 7.63. The modularity of E in the exceptional cases above is checked directly, using for example the Cremona tables [Cre97], (see also [Rub95], Lemma 9).

We shall use modular parameterization of the set of equivalence classes of elliptic curves, see §6.3.2, (6.3.13)):

$$\Gamma_0(N)\backslash\mathbb{H} \longleftrightarrow$$

$$\left\{(E, C_N) \;\middle|\; \begin{array}{l}\text{an elliptic curve over } \mathbb{C} \\ \text{together with a cyclic subgroup of order } N\end{array}\right\} \Big/ \underset{(isomorphism)}{\sim}$$

This set can be identified with the set of $\mathbb{C}$-points of an affine algebraic curve $Y_0(N)$. The $\mathbb{C}$-points of its Zariski closure $X_0(N)$ (called a *modular curve*) can be identified with the compact quotient $\Gamma_0(N)\backslash\overline{\mathbb{H}}$. Both curves are defined over $\mathbb{Q}$. The boundary

$$X_0(N)(\mathbb{C})\backslash Y_0(N)(\mathbb{C}) \longleftrightarrow \Gamma_0(N)\backslash(\mathbb{Q} \cup \infty)$$

is the set of *cusps* ($\Gamma_0(N)$-equivalent classes of parabolic points, defined over $\mathbb{Q}$ by a theorem of Manin–Drinfeld). Under this identification, rational points $Y_0(N)(\mathbb{Q})$ correspond to the set

$$\left\{(E, C_N) \;\middle|\; \begin{array}{l}\text{an elliptic curve over } \mathbb{Q} \text{ together with a} \\ G_{\mathbb{Q}} - \text{ invariant cyclic subgroup of order } N\end{array}\right\} \Big/ \underset{(isom)}{\sim},$$

Therefore:

$$X_0(N)(\mathbb{Q}) \longleftrightarrow$$

$$\{cusps\} \bigcup \left\{ (E, C_N) \;\middle|\; \begin{array}{l} \text{an elliptic curve } E \text{ over } \mathbb{Q} \text{ with a} \\ G_{\mathbb{Q}}\text{-invariant cyclic subgroup of order } N \end{array} \right\} \Big/ \underset{(isom)}{\sim},$$

Example 7.64. The modular curve $X_0(15)$ has 4 cusps and it is in fact an elliptic curve which can be defined by the equation $y^2 = x(x+3^2)(x-4^2)$, with the following rational points:

$$X_0(15)(\mathbb{Q}) \longleftrightarrow \left\{ \begin{array}{l} (0,0), (-9,0), (16,0), \infty \\ (-4, \pm 20)), (-36, \pm 180) \end{array} \right\}.$$

Now let $\overline{\rho}_{3,E}$ be a reducible representation, then there exists a $G_{\mathbb{Q}}$-invariant cyclic subgroup $C_3 \subset E(\mathbb{Q})$; and if $\overline{\rho}_{5,E}$ is also a reducible representation, then there exists a $G_{\mathbb{Q}}$-invariant cyclic subgroup $C_5 \subset E(\mathbb{Q})$ of order 5.

We obtain therefore a $G_{\mathbb{Q}}$-invariant cyclic subgroup $C_{15} = C_3 + C_5 \subset E(\mathbb{Q})$, which gives one of the four points of type (E, C_{15}) in the set

$$X_0(15)(\mathbb{Q}) \backslash \{cusps\}$$

and an explicit evaluation of the invariants j_E finishes the proof of Theorem 7.63.

Proposition 7.65 ([Se95], Proposition 1). *Let E be a semistable elliptic curve over $\mathbb{Q}$. Then either*

- $\rho_0 = \overline{\rho}_{3,E}$ *is surjective or*
- *the image $\rho_0(G_{\mathbb{Q}})$ is conjugate to a subgroup of $\left\{ \left(\begin{smallmatrix} * & * \\ 0 & * \end{smallmatrix} \right) \right\}$, i.e. ρ_0 is reducible.*

Proof. Let $\mathfrak{G}$ be the image of ρ_0 under the projection to $\mathrm{PGL}_2(\mathbb{F}_3) \cong \mathfrak{S}_4$. Each $g \in G_{\mathbb{Q}}$ maps to a permutation $\sigma_g \in \mathfrak{S}_4$ on the set of 4 lines in $\mathbb{F}_3^2$.

Suppose that neither i) nor ii) holds. Then $\mathfrak{G} \neq \mathfrak{S}_4$ and there are no fixed points. But ρ_0 is odd and

$$\det(\rho_0(g)) = \mathrm{sgn}(\sigma_g) \in \{\pm 1\} \in \mathbb{F}_3^\times : \mathrm{GL}_2(\mathbb{F}_3) \longrightarrow \mathbb{F}_3^\times.$$

Hence $\mathfrak{G}$ is not contained in $\mathfrak{A}_4$. It follows that either $\mathfrak{G} \cong \mathfrak{D}_4 \cong \langle (1234), (14)(23) \rangle$ (the dihedral group), or $\mathfrak{G} \subset \mathfrak{D}_4$ is a subgroup of index 2 in $\mathfrak{D}_4$. In both cases, the group $\mathfrak{G}^{ab} = \mathfrak{G}/[\mathfrak{G}, \mathfrak{G}]$ has order 4. However ρ_0 is *semistable* so for all $l \neq 3$, $\rho_0(I_3) \sim \left\{ \left(\begin{smallmatrix} 1 & * \\ 0 & 1 \end{smallmatrix} \right) \right\} = 1$. Since the order of $\mathfrak{G}$ is not divisible by 3, this implies that $\rho_0(I_l) = 1$. On the other hand, the Galois group of the maximal abelian extension of $\mathbb{Q}$ unramified outside 3 is isomorphic to $\mathbb{Z}_3^\times$. Since it has no factor groups of order 4, contradicting the fact that $|\mathfrak{G}^{\mathrm{ab}}| = 4$. □

Note that a similar result holds for all $\overline{\rho}_{l,E}$, including $l = 5$.

Lemma 7.66 (J.-P. Serre, [Se95]). *Let E be a semistable elliptic curve over $\mathbb{Q}$ such that $\overline{\rho}_{3,E}$ is irreducible, then the restriction*

$$\rho_0|G_{\mathbb{Q}(\sqrt{-3})}$$

is absolutely irreducible.

Proof of Lemma 7.66: we use the fact that $\overline{\rho}_{3,E}$ is odd, and one may assume that $\mathrm{Im}(\overline{\rho}_{3,E}) \ni \begin{pmatrix} 1 & 0 \\ 0 & -1 \end{pmatrix}$. Suppose that $\rho_0|G_{\mathbb{Q}(\sqrt{-3})}$ is not absolutely irreducible, and that there exists a subspace $W \subset E[3] \otimes \overline{\mathbb{F}}_3$ invariant under $G_{\mathbb{Q}(\sqrt{-3})} \lhd G_{\mathbb{Q}}$. Then W^{τ} is also $G_{\mathbb{Q}(\sqrt{-3})}$-invariant, where τ is the image of the complex conjugation, implying that

$$\rho_0|G_{\mathbb{Q}(\sqrt{-3})} \sim \left\{ \begin{pmatrix} a & 0 \\ 0 & \overline{a} \end{pmatrix} \right\} \subset \mathrm{GL}_2(\overline{\mathbb{F}}_3) \Rightarrow \rho_0(G_{\mathbb{Q}(\sqrt{-3})}) \text{ is commutative.}$$

But under the assumption of Lemma 7.66, ρ_0 is surjective due to Proposition 7.65.

Therefore, $\mathrm{Im}(G_{\mathbb{Q}(\sqrt{-3})})$ in $\mathrm{PGL}_2(\mathbb{F}_3) \cong \mathfrak{S}_4$ is *not commutative* (since it is a subgroup of $\mathfrak{S}_4$ of index ≤ 2). This contradiction proves Lemma 7.66.

Theorem 7.63 becomes in this case the "Theorem on Absolute Irreducibility". In particular, the assumptions of Theorem 7.28 on modularity of admissible deformations are satisfied for $\overline{\rho}_{3,E}$.

7.7.2 From $p = 3$ to $p = 5$

Let E be a semistable elliptic curve over $\mathbb{Q}$. According to §7.7.1, only the following three cases are possible:

(1) $\rho_0 = \overline{\rho}_{3,E}$ is irreducible;
(2) $\rho_0 = \overline{\rho}_{5,E}$ is irreducible;
(3) $j_E \in \left\{ -\frac{25}{2}, -\frac{5^2 \cdot 241^3}{2^3}, -\frac{5 \cdot 29^3}{2^5}, \frac{5 \cdot 211^3}{2^{15}} \right\}$.

It was noticed in Theorem 7.63 that in the exceptional case (3) E is modular. As mentioned after Lemma 7.66, in Case (1) $\overline{\rho}_{3,E}$ safisfies the assumptions of Theorem 7.28 on modularity of admissible deformations. Hence $\rho_{3,E}$, being an admissible deformation of $\overline{\rho}_{3,E}$, is modular. This implies the modularity of E. Case (2) is covered by the following

Theorem 7.67 ([Se95]). *Let E be a semistable elliptic curve over $\mathbb{Q}$ such that $\overline{\rho}_{5,E}$ is irreducible. Then there exists an elliptic curve E' over $\mathbb{Q}$ such that $\overline{\rho}_{5,E'} \sim \overline{\rho}_{5,E}$ and $\overline{\rho}_{3,E'}$ is absolutely irreducible.*

Proof of the theorem is carried out in §7.7.3.

Let us show how this implies Theorem 7.13: The same argument as in Case (1) shows that the curve E' is modular. Hence the representation $\overline{\rho}_{5,E'}$ is modular. However $\overline{\rho}_{5,E'}$ is equivalent to $\overline{\rho}_{5,E}$, hence $\overline{\rho}_{5,E}$ is also modular.

Moreover, let us check that $\overline{\rho}_{5,E}$ satisfies the assumptions of Theorem 7.28. Knowing that $\overline{\rho}_{5,E}$ is irreducible, we show that the restriction

$$\overline{\rho}_{5,E}|_{G_{\mathbb{Q}(\sqrt{5})}}$$

is absolutely irreducible.

Indeed, the complex conjugation $\tau \in G_{\mathbb{Q}(\sqrt{5})}$ satisfies

$$\det\overline{\rho}_{5,E}(\tau) = -1 \Rightarrow \overline{\rho}_{5,E}(\tau) \sim \begin{pmatrix} 1 & 0 \\ 0 & -1 \end{pmatrix}.$$

Hence both eigenspaces of $\overline{\rho}_{5,E}(\tau)$ are defined over $\mathbb{F}_5$, and the irreductibility over $\mathbb{F}_5$ implies the absolute irreducibility (over $\overline{\mathbb{F}}_5$).

It follows that the assumptions of Theorem 7.28 on modularity of admissible deformations are satisfied for the representation $\overline{\rho}_{5,E}$.

This implies that the representation $\rho_{5,E}$ is modular because it is an admissible deformation of $\overline{\rho}_{5,E}$. This means that the curve E is modular, proving STW Conjecture also in this case (Theorem 7.13).

It remains to explain the proof of Theorem 7.67.

7.7.3 Families of elliptic curves with fixed $\overline{\rho}_{5,E}$

We explain a construction of $E' \in \{E_t\}$ as a fiber of an elliptic fibration $E_t \to \mathbb{P}^1_t$ over $\mathbb{Q}$ with $\overline{\rho}_{5,E_t} \sim \overline{\rho}_{5,E}$ for all t and such that $E = E_{t_0}$ for some t_0.

We again use a modular curve, but this time it is attached to the congruence subgroup $\Gamma(5)$ (see (6.3.15)). Let us use the modular parameterization of the set of equivalence classes of elliptic curves, see §6.3.2, (6.3.13):

$$\Gamma(N)\backslash\mathbb{H} \longleftrightarrow$$

$$\left\{(E,\phi) \;\middle|\; \begin{array}{l}\text{an elliptic curve over } \mathbb{C} \text{ together with}\\ \text{an isomorphism } \phi : E[N] \xrightarrow{\sim} (\mathbb{Z}/N\mathbb{Z})^2, \phi^*\det = e_{N,E'}\end{array}\right\} \Big/ \underset{(isom)}{\sim}$$

where

$$e_{N,E} : E[N] \wedge E[N] \to \mu_N$$

is the Weil pairing which is algebraically defined by (6.3.31). Recall that over $\mathbb{C}$ for $E = \mathbb{C}/\langle 1, z\rangle$, one has

$$E[N] = \langle \frac{1}{N}, \frac{z}{N}\rangle/\langle 1, z\rangle \cong (\mathbb{Z}/N\mathbb{Z})^2; e_{N,E} = \exp(\pm 2\pi i(ad - bc)/N).$$

Moreover, the modular curve $X(N)$ is an algebraic curve defined over the cyclotomic field $\mathbb{Q}(\zeta_N)$, $\zeta_N = \exp(2\pi i/N)$ such that the Riemann surface $X(N)(\mathbb{C})$ is identified to the compact quotient $\Gamma(N)\backslash\overline{\mathbb{H}}$ in such a way that

$$X(N)(\mathbb{C}) \longleftrightarrow \Gamma(N)\backslash\overline{\mathbb{H}}, \overline{\mathbb{H}} = \mathbb{H} \cup \mathbb{Q} \cup \infty;$$
$$Y(N)(\mathbb{C}) \longleftrightarrow \Gamma(N)\backslash\mathbb{H}$$

for an affine algebraic curve $Y(N)$ defined over $\mathbb{Q}(\zeta_N)$ (see §6.4.2).

In order to work with curves defined over $\mathbb{Q}$ let us fix a curve E over $\mathbb{Q}$ and consider the twisted curve X_E over $\mathbb{Q}$ in such a way that it has an affine model Y_E over $\mathbb{Q}$ defined via the following modular description:

$$Y_E(\mathbb{C}) \longleftrightarrow$$
$$\left\{(E', \phi) \;\middle|\; \begin{array}{l}\text{an elliptic curve over } \mathbb{C} \text{ together with}\\ \text{an isomorphism } \phi : E'[N] \xrightarrow{\sim} E[N], \phi^* e_{N,E} = e_{N,E'}\end{array}\right\} \Big/ \underset{(isom)}{\sim}.$$

This description gives the set of $\mathbb{Q}$-rational points of Y_E:

$$Y_E(\mathbb{Q}) \longleftrightarrow$$
$$\left\{(E', \phi) \;\middle|\; \begin{array}{l}\text{an elliptic curve over } \mathbb{Q} \text{ with an isomorphism}\\ \text{of } G_{\mathbb{Q}}\text{-modules } \phi : E'[N] \xrightarrow{\sim} E[N], \phi^* e_{N,E} = e_{N,E'}\end{array}\right\} \Big/ \underset{(isom)}{\sim}.$$

Note that X_E is isomorphic to $X(N)$ over $\mathbb{C}$ (and even over $\mathbb{Q}(\zeta_N)$).

We are interested in the case $N = 5$. In this case we have the following explicit description of the curve X_E by K.Rubin and A. Silverberg (see in [RubSil94] and in [CSS95]):

Proposition 7.68 ([Rub95], Proposition 11). *Let E be an elliptic curve over $\mathbb{Q}$, and X_E the curve over $\mathbb{Q}$, obtained from $X(5)$ by twist as above.*

Then the genus of X_E is equal to 0, and there is an explicit parameterization:

$$\psi : \mathbb{P}^1 \longrightarrow X_E \text{ over } \mathbb{Q} \text{ with } t \mapsto (E_t, \phi_t) \tag{7.7.1}$$
$$E_t : y^2 = x^3 + f_E(t)x + g_E(t), f_E, g_E \in \mathbb{Q}[t], \tag{7.7.2}$$
$$\text{and } \deg(f_E) = 30, \deg(g_E) = 20.$$

Under this parameterization elements $t \in \mathbb{Q}$ are in bijection with elements of $Y_E(\mathbb{Q})$.

Note that by this construction there is an isomorphism of $G_{\mathbb{Q}}$-modules ϕ : $E'[5] \xrightarrow{\sim} E[5]$, in other words, for any $t \in \mathbb{Q}$, $\overline{\rho}_{5,E_t} \sim \overline{\rho}_{5,E_t}$.

In order to finish the proof of Theorem 7.67, we need to choose an element $E' = E_t$ in this family ($t \in \mathbb{Q}$) such that i) $\overline{\rho}_{3,E_t}$ is irreducible and ii) E_t is semistable.

7.7.4 The end of the proof

In order to obtain an irreducible $\overline{\rho}_{3,E_t}$ let us consider an auxiliary twisted modular curve X'_E over $\mathbb{Q}$, in such a way that it has an affine model Y'_E over $\mathbb{Q}$ defined via the following modular description:

$$Y'_E(\mathbb{C}) \longleftrightarrow \frac{\left\{(E',\phi,C_3) \;\middle|\; \begin{array}{l}\text{an elliptic curve over } \mathbb{C} \text{ together with}\\ \text{an isomorphism } \phi: E'[5] \xrightarrow{\sim} E[5], \phi^* e_{5,E} = e_{5,E'}, |C_3| = 3\end{array}\right\}}{(isomorphism \sim)}$$

This description gives the set of $\mathbb{Q}$-rational points of Y'_E:

$$Y'_E(\mathbb{Q}) \longleftrightarrow \frac{\left\{(E',\phi,C_3) \;\middle|\; \begin{array}{l}E_{/\mathbb{Q}}, C_3 \subset E(\mathbb{Q})(G_{\mathbb{Q}}\text{-submodule}, |C_3| = 3) \text{ with an isomorphism}\\ \text{of } G_{\mathbb{Q}}\text{-modules } \phi: E'[N] \xrightarrow{\sim} E[N], \phi^* e_{N,E} = e_{N,E'}\end{array}\right\}}{(isomorphism \sim)}$$

Note that X'_E is isomorphic over $\mathbb{C}$ to the compact quotient

$$(\Gamma(5) \cap \Gamma_0(3))\backslash\overline{\mathbb{H}}.$$

The genus of X'_E can be computed with the classical techniques described in Chapter I of Shimura's book [Shi71]; the answer is $g(X'_E) = 9$.

By the theorem of Faltings (see §5.5) $Y'_E(\mathbb{Q})$ is a *finite* set. Thus, for all but finitely many $t \in \mathbb{Q}$ the representation $\overline{\rho}_{3,E_t}$ is irreducible.

Concerning semistability, let us observe that:

- for any $l \neq 5$ we have

$$\overline{\rho}_{5,E_t}|_{I_l} \sim \overline{\rho}_{5,E}|_{I_l} \sim \left\{\begin{pmatrix}1 & * \\ 0 & 1\end{pmatrix}\right\}.$$

  This directly implies the semistability for such l, using the algebraic definition of semistability.
- for $l = 5$, the semistability at 5 follows from the explicit equation (7.7.1) of the curves, via the geometric definition 7.10 of semistability: if t tends to t_0 in the 5-adic topology, then the coefficients of the minimal (5-adic) equations of E_t converge in the 5-adic topology to the coefficients of the minimal (5-adic) equation of $E = E_{t_0}$. Thus E_t becomes semistabe at 5 for all $t \in \mathbb{Q}$ sufficiently close to t_0.

The most important insights.

Here is how they were described by Wiles himself in the introduction to his paper [Wi]:

- the discovery in the spring 1991 of the invariants η_A (known at that time as "congruence modules" of Hida in the case of p-adic Hecke algebras);
- the switch from $p = 3$ to $p = 5$ (found in May 1993);
- the use of the horizontal Iwasawa theory (J–structures) in September 1994.

An interesting account on the history of this subject is given in [Mozz].

Part III

Analogies and Visions

III-0

Introductory survey to part III: motivations and description

III.1 Analogies and differences between numbers and functions: ∞-point, Archimedean properties etc.

This part was conceived as an explanation of some basic intuitive ideas that underlie modern number–theoretical thinking. One subject could have been called *Analogies between numbers and functions.*

III.1.1 Cauchy residue formula and the product formula

Let us explain the analogy between a number field, i.e., a finite extension of $\mathbb{Q}$, and the field $\mathbb{C}(S)$ of meromorphic functions on a smooth complete curve S, following [SABK94], p.3.

An example of this analogy is given by the Cauchy residue formula,

$$\sum_{x\in S} v_x(f) = \sum_{x\in S} \mathrm{Res}_x\left(\frac{df}{f}\right) = 0, \tag{III.1.1}$$

where Res_x denotes the residue at x of differential forms.

When $f \in \mathbb{Q}^*$ is a rational number we have the product formula

$$|f| = \prod_p p^{v_p(f)} \tag{III.1.2}$$

where p runs over all integral primes and $v_p(f) \in \mathbb{Z}$ is the p-adic valuation of f.

If we define

$$v_\infty(f) = -\log|f| \in \mathbb{R},$$

we may rewrite the equality (III.1.2) as

$$\sum_p v_p(f)\log(p) + v_\infty(f) = 0, \tag{III.1.3}$$

an analog for $\mathbb{Q}$ of equation (III.1.1) for the field $\mathbb{C}(S)$. One can see from this example that, in this analogy, the complete curve in S is analogous to the affine scheme Spec($\mathbb{Z}$) to which is added a point at infinity (at this point the Archimedean norm is used instead of discrete valuations).

III.1.2 Arithmetic varieties

In general, let X be an arithmetic variety. By this we mean a regular scheme, projective and flat over $\mathbb{Z}$.

In other words, we consider a system of polynomial equations

$$f_1(T_0, \cdots, T_N) = f_2(T_0, \cdots, T_N) = \cdots = f_m(T_0, \cdots, T_N) = 0 \qquad \text{(III.1.4)}$$

where $f_1, \cdots, f_m \in \mathbb{Z}[T_0, \cdots, T_N]$ are homogeneous polynomials with integral coefficients. These define the projective scheme $X = \mathrm{Proj}(S)$, where S is the quotient of $\mathbb{Z}[T_0, \cdots, T_N]$ by the ideal, generated by $f_1, \cdots, f_m$. The points of X are those homogeneous prime ideal $\mathcal{P}$ in S which do not contain the augmentation ideal. The structure morphism $f : X \to \mathrm{Spec}\mathbb{Z}$ maps $\mathcal{P}$ to $\mathcal{P} \cap \mathbb{Z}$. The fiber of f over a prime integer (special fiber) is the variety $f^{-1}(p\mathbb{Z}) = X/p = \mathrm{Proj}(S/pS)$ over the field with p elements. The generic fiber is $f^{-1}((0)) = X_{\mathbb{Q}} = \mathrm{Proj}(S \otimes_{\mathbb{Z}} \mathbb{Q})$. We assume that X is regular and that f is flat, i.e. S is torsion free. It follows that X/p is smooth, except for finitely many values of p, like q in Fig.III-0.1, where it may not even be reduced.

In the same way that we completed Spec$\mathbb{Z}$ by adding a point ∞ to it, we "complete" the family X of varieties over Spec$\mathbb{Z}$ by adding to it the complex variety $X_\infty = X(\mathbb{C})$, i.e. the set of complex solutions of (III.1.4), viewed as the fiber at infinity. We think of the whole familly as analogous to a complex manifold Y fibered over a smooth complete curve S via a flat proper map $f : Y \to S$. If the fibers of f have dimension one, X has Krull dimension two (we call it an arithmetic surface). Notice that an integral solution of (III.1.4) is a rational point on $X(\mathbb{Z}) = X(\mathbb{Q}) \subset X(\mathbb{C})$, i.e. a section of f.

III.1.3 Infinitesimal neighborhoods of fibers

One can also consider reductions of X (defined over $\mathbb{Z}$) modulo p^n for some prime p. The limit as $n \to \infty$ defines a p-adic completion of $X_{\mathbb{Z}}$. This can be thought as an "infinitesimal neiborhood" of the fiber at p.

The picture is more complicated at arithmetic infinity, since one does not have a suitable notion of "reduction modulo ∞" available to define the closed fiber. On the other hand, one does not have an analogue of the p-adic completion at hand. This is given by the Riemann surface (smooth projective algebraic curve over $\mathbb{C}$) determined by the equation of the algebraic curve over $\mathbb{Q}$, under the embedding $\mathbb{Q} \subset \mathbb{C}$,

$$X(\mathbb{C}) = X \otimes_{\mathbb{Q}} \mathrm{Spec}(\mathbb{C}),$$

with the absolute value $|\cdot|$ at the infinite prime replacing the p-adic valuation.

III.2 Arakelov geometry, fiber over ∞, cycles, Green functions (d'après Gillet-Soulé)

In order to control the size of the rational points P , S.Yu.Arakelov (cf. [Ara74a]) had the brilliant idea of considering *Hermitian metrizations* of various linear objects related to algebraic varieties (invertible sheaves, tangent bundles etc.). Let us take an algebraic vector bundle E on X, endowed with a smooth Hermitian metric h on the corresponding holomorphic vector bundle E_∞ on X_∞). The pair $\overline{E} = (E, h)$ will be called a *Hermitian vector bundle* on X. This gives a method to compactify arithmetic schemes over number fields at the arithmetical infinity.

According to Arakelov's program [Ara74a], [Ara74b], to an algebraic variety X over a number field K one can attach a completed arithmetical variety $\overline{X}$ of the dimension $\dim(X) + 1$,

$$\overline{X} \to \overline{\mathrm{Spec}\mathcal{O}_K} = \mathrm{Spec}\mathcal{O}_K \cup \{\infty_1, \cdots, \infty_r\},$$

where $\mathcal{O}_K$ is the maximal order of K, $\{\infty_1, \cdots, \infty_r\}$ the set of all places at ∞ of K (see [GS92], [SABK94], [La88]).

For an algebraic number field $\mathbb{K}$ of degree $n = [\mathbb{K} : \mathbb{Q}]$ admitting $n = r_1 + 2r_2$ embeddings

$$\alpha : \mathbb{K} \to \mathbb{C}, \tag{III.2.1}$$

there is $r = r_1 + r_2$ *Archimedean primes*, with r_1 real embeddings and r_2 paires of complex conjugate embeddings.

When $\mathrm{Spec}\mathcal{O}_K$ is compactified by adding Archimedean primes $\{\infty_1, \cdots, \infty_r\}$, one also obtains n complex varieties $X_\alpha(\mathbb{C})$, obtained from the embeddings $\alpha : \mathbb{K} \to \mathbb{C}$. Of these complex varieties, r_1 carry a real involution.

In particular, each curve has a well defined minimal model over $\mathcal{O}$ which is called an arithmetical surface (since we added an arithmetical dimension to the geometric one). Adding metrics at infinity to this, Arakelov developed the intersection theory of arithmetical divisors. Heights in this picture become the (exponentiated) intersection index, see [Ara74b], [La88]

This theory was vastly generalized by H.Gillet and C.Soulé [GS91], [GS92], [SABK94], following some suggestions in [Man84]. The Figure III-0.1 which is reproduced here from [SABK94] with a kind permission of Ch.Soulé, is a visualization of an arithmetical surface.

Its fibers over the closed points of Spec($\mathcal{O}$) can be non–singular ("non–degenerate", or with "good reduction") or singular (having "bad reduction"). Rational points of the generic fiber correspond to the horizontal arithmetical divisors; there are also vertical divisors (components of closed fibers) and "vertical divisors at infinity" added formally, together with an *ad hoc* definition of their intersection indices with other divisors defined via Green's function, see [SABK94].

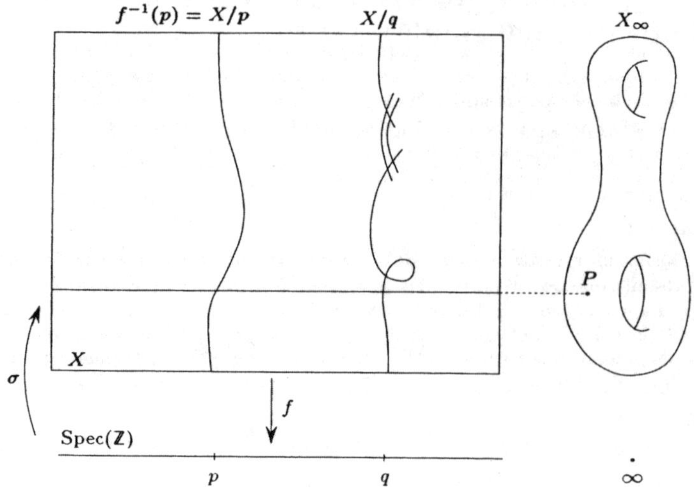

Fig. III-0.1. An arithmetic surface.

Arakelov's picture played a prominent role in Faltings' proof and the subsequent development of his work.

The Figure III-0.1 which is reproduced here from [SABK94] with a kind permission of Ch.Soulé and Cambridge University Press, is a visualization of an arithmetic surface.

III.2.1 Arithmetic Chow groups

Let X be an arithmetic variety and $\overline{E}$ a Hermitian vector bundle on X. One can attach to $\overline{E}$ *characteristic classes* with values in *arithmetic Chow groups.*

More specifically, an arithmetic cycle is a pair (Z, g) consisting of an algebraic cycle over X, i.e. a finite sum $\sum_\alpha n_\alpha Z_\alpha$, $n_\alpha \in \mathbb{Z}$, where Z_α is a closed irreducible subscheme of X, of fixed codimension p, and a Green current g for Z. By this we mean that g is a real current on X_∞ which satisfies $F_\infty^*(g) = (-1)^{p-1} g$ and

$$d\, d^c g + \delta_Z = \omega, \tag{III.2.2}$$

where ω is the current attached to a smooth form on X_∞, and δ_Z is the current given by integration on Z_∞:

$$\delta_Z(\eta) = \sum_\alpha n_\alpha \int_{Z_\alpha(\mathbb{C})} \eta, \tag{III.2.3}$$

for any smooth form η of appropriate degree (here F_∞^* denotes the morphism, induced by the complex conjugation F_∞ on X_∞).

The *arithmetic Chow group* $\widehat{CH}^p(X)$ is the Abelian group of arithmetic cycles, modulo the subgroup, generated by pares $(0, \partial u + \overline{\partial} v)$ and $(\mathrm{div} f, -\log|f|^2)$, where u and v are arbitrary currents of the appropriate degree and $\mathrm{div} f$ is the divisor of a non-zero rational function f on some irreducible closed subscheme of codimension $p-1$ in X.

III.2.2 Arithmetic intersection theory and arithmetic Riemann-Roch theorem

The important fact that the arithmetic Chow groups $\widehat{CH}^p(X)$ have fonctorial properties, is studied in [SABK94].

Given two arithmetic cycles (Z, g) and (Z', g'), we need a Green current for their intersection. The formula

$$g'' = \omega g' + g\delta_{Z'}$$

where ω is defined as in (III.2.2), involves a product of two currents $g\delta_{Z'}$. To make sense of it in general we need to show that one can take for g a smooth form on $X_\infty - Z_\infty$ of logarithmic type along Z_∞.

Then one can define characteristic classes for Hermitian vector bundles $\overline{E}$ on X, in particular, a Chern character class

$$\widehat{ch}(\overline{E}) \in \bigoplus_{\geq 0} \widehat{CH}^p(X) \otimes \mathbb{Q}. \tag{III.2.4}$$

This class satisfies the usual axiomatic properties of a Chern character, but it depends on the choice of a metric on E. It is additive only on orthogonal direct sums, but in general its failure to be additive on exact sequences is given by the secondary characteristic class of Bott–Chern. Similar results hold for the Todd class $\widehat{Td}(\overline{E})$ (cf. Chapter IV of [SABK94]). The arithmetic Riemann-Roch theorem is formulated in terms of these classes, for a proper flat map $f : X \to Y$, as a formula for the first arithmetic Chern class for the *direct image map* on Hermitian vector bundles.

Main application of this result gives an asymptotic behaviour of the type

$$h^0(X, \overline{E} \otimes \overline{L}^n) \geq \frac{rk(E)}{(d+1)!} \overline{L}^{d+1} n^{d+1} + \mathcal{O}(n^d \log n) \tag{III.2.5}$$

for an ample L with a positive metric, where $h^0(X, \overline{E} \otimes \overline{L}^n)$ is the logarithm of the number of global sections $s \in H^0(X, E \otimes L^n)$ such that $\|x\| \leq 1$ for every $x \in X_\infty$, and $\overline{L}^{d+1} \in \mathbb{R}$ denotes the arithmetic self–intersection.

The arithmetic Riemann-Roch theorem was discussed in [Fal92]; it was explained that the Hirzebruch-Riemann-Roch theorem (HRR) gives a formula for $\chi(E)$, $E \to X$ an algebraic vector bundle over an algebraic manifold X, in terms of characteristic classes of X and E: $\chi(E) = [\mathrm{Td}(X) \cdot \mathrm{ch}(E)]_{\dim X}$. The Grothendieck-Riemann-Roch (GRR) theorem, a relative version, states that for an algebraic morphism $f\colon X \to Y$ and an algebraic vector bundle E on X, one has an equality of mixed cohomology classes: $\mathrm{ch}(f_*E) = f_*(\mathrm{Td}(X|Y)\cdot\mathrm{ch}(E))$, and contains HRR as a special case where Y = point. The proof of HRR is by means of elliptic differential operators and analytic methods or cobordism. The proof of GRR, on the other hand, is easier and more algebraic.

This proof was carried in [Fal92] over to the arithmetic case.

The data for the arithmetic case are as follows: $f\colon X \to Y$ is a morphism of arithmetic varieties, both of which are defined over a common arithmetic variety B (such as $\mathrm{Spec}(\mathcal{O}_k)$), $\widehat{E}$ an arithmetic vector bundle on X. One needs the following objects: (1) $\widehat{K}(X)$, equivalence classes of $\widehat{E}$'s, (2) $\widehat{A}^*(X)$, the arithmetic Chow ring, (3) $\widehat{\mathrm{ch}}\colon \widehat{K}(X) \to \widehat{A}^*(X)$, (4) $f_*\colon \widehat{K}(X) \to \widehat{K}(Y)$, the arithmetic push forward, (5) a class $\widehat{\mathrm{Td}}^R(X|Y) \in \widehat{A}^*(X)$, and (6) an intersection product in $\widehat{A}^*(X)$. Then the arithmetic Riemann-Roch theorem (ARR) states: $\widehat{\mathrm{ch}}(f_*\widehat{E}) = f_*(\widehat{\mathrm{Td}}^R(X|Y)\cdot\widehat{\mathrm{ch}}(\widehat{E}))$ in $\widehat{A}(Y)$ for $\widehat{E} \in \widehat{K}(X)$. The HRR case of this is now $Y = B = \mathrm{Spec}(\mathcal{O}_K)$, and for surfaces ($X_K$ a curve), the component in degree 2 yields the RR for arithmetic surfaces by G. Faltings in [Fal84] the first case for which ARR was proven. Here the method was construction of volume forms on the cohomology. Following Faltings' introduction, this led to new interest in this topic, and soon there was rapid progress: Deligne generalised the volume form to more cases, and Gillet and Soulé developed an arithmetic intersection theory for general varieties, as well as Hermitian K-theory. Then they joined efforts with Bismut and managed to define the determinant of cohomology of an arithmetic variety, cf. [BiGS88]. One could then ask for a RR result on this determinant. It turned out that even for the projective space the immediate generalisation of the classical RR was false. To remedy this they introduced a secondary class $R(x)$ so that a modified version (using R) remains true. Bismut and Lebeau could prove a RR result for closed immersions, from which the RR for determinant bundles follows with some more work, see in [SABK94].

The theory of intersections and arithmetic Riemann-Roch theorem were discussed also in a Bourbaki talk [Bo90] by J.-B. Bost (see also [BoCo95]).

III.2.3 Geometric description of the closed fibers at infinity

A general picture of an arithmetic surface over $\overline{\mathrm{Spec}(\mathcal{O}_{\mathbb{K}})}$ is as follows:

$$\begin{array}{ccccccc} X_{\mathfrak{p}} & \hookrightarrow & X_{\mathrm{Spec}(\mathcal{O}_{\mathbb{K}})} & \hookrightarrow & X_{\overline{\mathrm{Spec}(\mathcal{O}_{\mathbb{K}})}} & \hookleftarrow & ??? \\ \downarrow & & \downarrow & & \downarrow & & \downarrow \\ \mathfrak{p} & \hookrightarrow & \mathrm{Spec}(\mathcal{O}_{\mathbb{K}}) & \hookrightarrow & \overline{\mathrm{Spec}(\mathcal{O}_{\mathbb{K}})} & \hookleftarrow & \alpha \end{array}$$

where we do not have an explicit geometric description of the closed fibers over the Archimedean primes.

Formally one can enlarge the group of divisors on the arithmetic surface by adding formal real linear combinations of irreducible "closed vertical fibers at infinity" $\sum_\alpha \lambda_\alpha F_\alpha$. Here the fibers F_α are only treated as formal symbols, and no geometric model of such fibers is provided. The remarkable fact is that Hermitian geometry on the complex varieties $X_\alpha(\mathbb{C})$ is sufficient to specify the contribution of such divisors to intersection theory on the arithmetical surface, even without an explicit knowledge of the closed fiber.

The main idea of the Arakelov geometry is that it is *sufficient* to work with the "infinitesimal neighborhood" $X_\alpha(\mathbb{C})$ of the fibers F_α, to have well defined intersection indices.

From the point of view of the classical *geometry of generations of algebraic curves* over a disk Δ with a special fibrer over 0, the analogous statement would say that the geometry of the special fiber is completely determined by the generic fiber (cf. [Mar04], §3 of Chapter 3). This would be a very strong statement on the form of the degeneration: for instance blowing up points on the special fiber is not seen by just looking at the generic fiber. Investigating this analogy leads one to *expect* that the fiber at infinity should behave like in the *totally degenerate case*, cf. a discussion in §8.2.3. This is the case where one has maximal degeneration, where all the components of the closed fiber are $\mathbb{P}^1$'s and the geometry of the degeneration is completely encoded by the *dual graph*, which describes in a purely combinatorial way how these $\mathbb{P}^1$'s are joined. The dual graph has a vertex for each component of the closed fiber and an edge for each double point.

For an arithmetic surface ($\dim X = 1$), the local intersection multiplicities of two finite, horizontal, irreducible divisors D_1 and D_2 on $X_{\mathcal{O}_\mathbb{K}}$ is given by

$$[D_1, D_2] = [D_1, D_2]_{fin} + [D_1, D_2]_{inf},$$

where the first term counts the contribution from the finite places (i.e. what happens over $Spec(\mathcal{O}_\mathbb{K})$), and the second term is the contribution of the Archimedean primes, i.e. the part of the intersection that happens over arithmetic infinity.

While the first term is computed in algebro geometric terms, from the local equations for the divisors D_i at P, the second term is defined as a sum of values of Green functions g_α on the Riemann surfaces $X_\alpha(\mathbb{C})$,

$$[D_1, D_2]_{inf} = -\sum_\alpha \varepsilon_\alpha \left(\sum_{\beta,\gamma} g_\alpha(P^\alpha_{1,\beta}, P^\alpha_{2,\gamma}) \right),$$

at points

$$\{P^\alpha_{i,\beta} \mid \beta = 1, \ldots, [\mathbb{K}(D_i) : \mathbb{K}]\} \subset X_\alpha(\mathbb{C}),$$

for a finite extension $\mathbb{K}(D_i)$ of $\mathbb{K}$, determined by D_i. Here $\varepsilon_\alpha = 1$ for real embeddings and $=2$ for complex embeddings (see [CS86], [GS92], [SABK94], [La88] for a detailed account of these notions of Arkelov geometry).

Further evidence for the similarity between the Archimedean and the totally degenerate fibers came from an explicite computation of the Green function at the Archmedean places derived in [Man91] in terms of a Schottky uniformization of the Riemann surface $X_\alpha(\mathbb{C})$, which is discussed in more detail in §8.1. Such uniformization has an analogue at a finite prime, in terms of p-adic Schottky groups, only in the totally degenerate case. Another sourse of evidence comes from a cohomological theory of the local factors at Archimedean primes, developed by K.Consani in [Cons98], valid for any arithmetic variety, see also §8.2 for more details. This general construction shows that the resulting description of the local factor as regularized determinant at the Archimedean primes resembles mostly the case of the totally degenerate reduction at a finite prime.

One can present both results in the light of the noncommutative space given by a spectral triple $(\mathcal{O}_A, \mathcal{H}, \mathcal{D})$ discussed in §8.3, where the data $(\mathcal{O}_A, \mathcal{H}, \mathcal{D})$ consist of a C^*-algebra $\mathcal{O}_A$ (or more generally of a smooth subalgebra of a C^*-algebra) with a representation as bounded operators on a Hilbert space $\mathcal{H}$ and an operator $\mathcal{D}$ on $\mathcal{H}$ that verifies the main properties of a classical Dirac operator D (a square root of the Laplacian) on a smooth spin manifold.

III.3 Theory of ζ-functions, local factors at ∞, Serre's Γ-factors; and generally an interpretation of zeta functions as determinants of the arithmetical Frobenius: Deninger's program

(cf. [Se70a], [Den94], [Den01], [Lei03], and §7, Ch.3 of [Mar04]).

An important invariant of arithmetic varieties is the L-function. This is written as a product of contributions from the finite primes and the Archimedean primes,

$$\prod_{\mathfrak{p}\in \mathrm{Spec}\mathcal{O}_{\mathbb{K}}} L_{\mathfrak{p}}(H^m(X), s), \qquad \text{(III.3.1)}$$

see §6.2.7. The reason why one needs to include the contribution of the Archimedean primes can be seen in the case of the "affine line" $\mathrm{Spec}(\mathbb{Z})$, where one has the Riemann zeta-function, which is written as the Euler product

$$\zeta(s) = \prod_p (1 - p^{-s})^{-1}. \qquad \text{(III.3.2)}$$

However to have a nice functional equation, one needs to consider the product

$$\hat{\zeta}(s) = \zeta(s)\Gamma(s/2)\pi^{-s/2}, \qquad \text{(III.3.3)}$$

which includes a contribution of the Archimedean prime, expressed in terms of the Gamma function.

An analogy with algebraic geometry and *Weil's conjectures* (see §6.1.3) suggest to think of the functional equation as a sort of "Poincaré duality" which holds for a compact manifold, hence the need to "compactify" by adding the Archimedean primes (and the Archimedean fibers of arithmetic varieties).

When one looks at an arithmetic variety over a finite prime $\mathfrak{p} \in \mathrm{Spec}(\mathcal{O}_{\mathbb{K}})$, the fact that the reduction lives over a residue field of positive characteristic implies that there is a special operator, the *geometric Frobenius* $\mathrm{Fr}_{\mathfrak{p}}^*$ acting on a suitable cohomology theory (étale cohomology), induced by the Frobenius automorphism $\phi_{\mathfrak{p}}$ of $\mathrm{Gal}(\overline{\mathbb{F}}_p/\mathbb{F}_p)$.

The local L-factors of (III.3.1) at finite primes encodes the action of the geometric Frobenius in the form

$$L_{\mathfrak{p}}(H^m(X),s) = \det\left(1 - \mathrm{Fr}_{\mathfrak{p}}^* N(\mathfrak{p})^{-s} | H^m(\overline{X},\mathbb{Q}_l)^{I_{\mathfrak{p}}}\right)^{-1}. \qquad \text{(III.3.4)}$$

Here we are considering the action of the geometric Frobenius $\mathrm{Fr}_{\mathfrak{p}}^*$ on the *inertia invariants* $H^m(\overline{X},\mathbb{Q}_l)^{I_{\mathfrak{p}}}$ of the étale cohomology and a precise definition of these arithmetic structures is beyond our purpose. We wish to describe the contribution of the Archimedean primes to the product (III.3.1) by giving some quick heuristic explanation of (III.3.4).

For X a smooth projective variety (in any dimension) defined over $\mathbb{Q}$, the notation $\overline{X} := X \otimes \mathrm{Spec}(\overline{\mathbb{Q}})$ is used. One can write the local L factor (III.3.4) equivalently as

$$L_{\mathfrak{p}}(H^m(X),s) = \prod_{\lambda \in \mathrm{Spec}(\mathrm{Fr}_{\mathfrak{p}}^*)} \left(1 - \lambda q^{-s}\right)^{-\dim H^m(\overline{X},\mathbb{Q}_l)_{\lambda}^{I_{\mathfrak{p}}}}, \qquad \text{(III.3.5)}$$

where $H^m(\overline{X},\mathbb{Q}_l)_{\lambda}^{I_{\mathfrak{p}}}$ is the (generalized) eigenspace of the Frobenius with eigenvalue λ.

An important conclusion is that the local L-factor in (III.3.5) depends upon the data

$$(H^*(\overline{X},\mathbb{Q}_l)^{I_{\mathfrak{p}}}, \mathrm{Fr}_{\mathfrak{p}}^*) \qquad \text{(III.3.6)}$$

of a vector space, which has a cohomological interpretation, together with a linear operator acting upon it.

III.3.1 Archimedean L-factors

Since the étale cohomology satisfies the compatibility isomorphism with the singular (Betti) cohomology:

$$H^i(\overline{X},\mathbb{Q}_l) \cong H^i(X(\mathbb{C}),\mathbb{Q}) \otimes \mathbb{Q}_l, \qquad \text{(III.3.7)}$$

this indicates that one can work with the smooth complex manifold $X(\mathbb{C})$ and gain information on the "closed fiber" at arithmetic infinity, and expect that the contribution of the Archimedean primes to the L-function may be expressed in terms of the cohomology $H^i(X(\mathbb{C}),\mathbb{Q})$ (or in terms of the de Rham cohomology with complex coefficients).

Let us recall that the expected contribution of the Archimedean primes (cf. [Se70a]) is determined by Γ-factors attached to the Hodge structure

$$H^m(X(\mathbb{C})) = \bigoplus_{p+q=m} H^{p,q}(X(\mathbb{C})).$$

Namely, one has the following product of Gamma functions

$$L(H,s) = \begin{cases} \prod_{p,q} \Gamma_{\mathbb{C}}(s-\min(p,q))^{h^{p,q}} \\ \prod_{p<q} \Gamma_{\mathbb{C}}(s-p)^{h^{p,q}} \prod_{p} \Gamma_{\mathbb{R}}(s-p)^{h^{p,+}} \Gamma_{\mathbb{R}}(s-p+1)^{h^{p,-}} \end{cases} \tag{III.3.8}$$

where $H = H^m(X(\mathbb{C}))$, $s \in \mathbb{C}$, $h^{p,q} = \dim_{\mathbb{C}} H^{p,q}$ and $h^{p,\pm}$ is the dimension of the $\pm(-1)^p$-eigenspace of the $\mathbb{C}$-linear involution F_∞ on H. Recall that by definition 6.2.4,

$$\Gamma_{\mathbb{C}}(s) = (2\pi)^{-s}\Gamma(s), \quad \Gamma_{\mathbb{R}}(s) = 2^{-\frac{1}{2}}\pi^{-\frac{s}{2}}\Gamma(\frac{s}{2}), \quad \text{where} \quad \Gamma(s) = \int_0^\infty e^{-t}t^s\frac{dt}{t}.$$

Let us try to seek a unified picture of what happens at the finite and at the infinite primes. In particular, these should be a suitable reformulation of the local factors (III.3.4) and (III.3.8) where both formulae can be expressed in the same way.

III.3.2 Deninger's formulae

Deninger in [Den91], [Den92] and [Den94] expressed both local factors (III.3.4) and (III.3.8) as certain infinite determinants.

Recall that the Ray-Singer determinant of an operator T with pure point spectrum with finite multiplicities $\{m_\lambda\}_{\lambda\in \mathrm{Spec}(T)}$ is defined as

$$\det_\infty(s-T) := \exp\left(-\frac{d}{dz}\zeta_T(s,z)|_{z=0}\right) \tag{III.3.9}$$

where the zeta function of T is defined as

$$\zeta_T(s,z) = \sum_{\lambda\in \mathrm{Spec}(T)} m_\lambda(s-\lambda)^{-z}. \tag{III.3.10}$$

Suitable conditions for the convergence of this expressions in the case of the local factors are described in [Man95]. Ch. Deninger showed that (III.3.4) can be written equivalently in the form

$$L_{\mathfrak{p}}(H^m(X), s)^{-1} = \det_{\infty}(s - \Theta_q), \tag{III.3.11}$$

for an operator with spectrum

$$\begin{aligned} &\mathrm{Spec}(s - \Theta_q) = \\ &\left\{ \frac{2\pi i}{\log q} \left(\frac{\log q}{2\pi i}(s - \alpha_\lambda) + n \right) : n \in \mathbb{Z}, \lambda \in \mathrm{Spec}(T) \right\}, \end{aligned} \tag{III.3.12}$$

with multiplicities d_λ and with $q^{\alpha_\lambda} = \lambda$.

Moreover, the local factor (III.3.8) at infinity can be written similarly in the form

$$L(H^m(X), s)^{-1} = \det_{\infty}\left(\frac{1}{2\pi}(s - \Phi)|_{\mathcal{H}^m} \right), \tag{III.3.13}$$

where $\mathcal{H}^m$ is an infinite dimensional vector space and Φ is a linear operator with spectrum $\mathrm{Spec}(\Phi) = \mathbb{Z}$ and finite multiplicities. This operator is regarded as a "logarithm of Frobenius" at arithmetic infinity.

Given Deninger's formulae (III.3.12) and (III.3.13), it is natural to ask for a cohomological interpretation of the data

$$(\mathcal{H}^m, \Phi), \tag{III.3.14}$$

somewhat analogous to the non-Archimedean data (III.3.6).

III.4 A guess that the missing geometric objects are noncommutative spaces

We shall see that cohomological interpretation of the above data $(\mathcal{H}^m, \Phi)$ in (III.3.14) leads to the notions of a spectral triple and a noncommutative space, see §8.2.

III.4.1 Types and examples of noncommutative spaces, and how to work with them. Noncommutative geometry and arithmetic

Let us follow Chapter 1 of [Mar04], and recall some basic notions of Noncommutative geometry, developed by Connes, cf. [Co94], [Co2000], which extends the tools of ordinary geometry to treat spaces that are quotients, for which the usual "ring of functions" (i.e. functions invariant with respect to the equivalence relation) is too small to capture the information on the "inner structure" of points in the quotient space. Typically, for such spaces functions on the quotients are just constants, while a non-trivial ring of functions, which remembers the structure of the equivalence relation, can be defined using a noncommutative algebra of coordinates, analogous to the noncommutative variables of quantum mechanics.

Following A.Connes (cf. [Co94], p.85) let us give a simplest example of a noncommutative quotient space, and consider the set $Y = \{a, b\}$ consisting of two elements a and b, so that the algebra $C(Y)$ of complex-valued functions on Y is the commutative algebra $\mathbb{C} \oplus \mathbb{C}$ of 2×2 diagonal matrices. There are two ways of declaring that the two points a and b of Y are identical under an equivalence relation $a \sim b$:

1) The first method is to consider the *subalgebra* $A \subset C(Y)$ of functions on Y which take the same value at a and b: $f(a) = f(b)$.
2) The second method is to consider a *larger algebra* $B \supset C(Y)$ of all 2×2 matrices:

$$\begin{pmatrix} f_{aa} & f_{ab} \\ f_{ba} & f_{bb} \end{pmatrix}$$

(these are functions on the *graph* of the equivalence relation).

The relation between the two algebras is given by the notion of *strong Morita equivalence* of C^*-algebras (cf. [Rief76]). This relation which resembles the Brauer equivalence (see §4.5.5) preserves many invariants of C^*-algebras, such as K-theory and the topology of the space of irreducible representations. One can interpret then

$$\omega_a = f_{aa}, \omega_b = f_{bb}$$

as *pure states* of B in the sense of quantum mechanics, which yield equivalent irreducible representations of $B = \mathrm{M}_2(\mathbb{C})$.

Note that if a and b are not equivalent, one obtains by the second method only the algebra of diagonal matrices, because the graph of the equivalence relation consists then just of (a, a) and (b, b).

Let us describe another simple example (cf. [Co94], p.87, and [Mar04], §2 of Ch.1), in which the above two algebraic operations of quotient 1) and 2) yield obviously different (even not strongly Morita equivalent) algebras.

Take the topological space $Y = [0, 1] \times \{0, 1\}$ with the equivalence relation $\mathcal{R} : (x, 0) \sim (x, 1)$ for $x \in (0, 1)$, and let $X = Y/\mathcal{R}$ be the (non-Hausdorff) quotient space. That is, $Y = I_1 \cup I_2$ is the disjoint union of two copies I_1 and I_2 of the interval $[0, 1]$, and the quotient space $X = Y/\mathcal{R}$ is obtained by gluing the two interiors of the intervals I_1 and I_2 but not the end points.

By the first method take continuos functions f on Y, invariant with respect to the equivalence relation, this is the algebra $A = C([0, 1])$, which is homotopic to $\mathbb{C}$, and $K_0(A) = \mathbb{Z}$. By the second method let us consider functions on the graph of the equivalence relation, then one obtains the algebra

$$B = \{f \in C([0, 1]) \otimes \mathrm{M}_2(\mathbb{C}) \mid f(0) \text{ and } f(1) \text{ are diagonal }\}$$

which is an interesting nontrivial algebra (we view a generic element $x \in \mathrm{M}_2(C([0, 1]))$ as a continous map $t \mapsto x(t) \in \mathrm{M}_2(\mathbb{C})$). Note that the space of irreducible representations of B is $X = Y/\mathcal{R}$, and the K theory of B is much less trivial than that of $A = C([0, 1])$.

In general, such "quantum spaces" are defined by extending the Gelfand–Naimark correspondence

$$X \text{ loc. compact Haussdorf space } \Longleftrightarrow C_0(X) \text{ Abelian } C^*\text{-algebra}$$

by dropping the commutativity hypothesis in the right hand side. The correspondence then becomes a definition of what's on the left hand side: a *noncommutative space.*

The idea of preserving the information on the structure of the equivalence relation in the description of quotient spaces has analogues in Grothendieck's theory of stalks in algebraic geometry. Such quotients arise from foliations, and, more recently, in number theory and in arithmetic geometry, starting from [BoCo95], where a noncommutative space related to class field theory is constructed. This space, viewed as the space of 1-dimensional $\mathbb{Q}$-lattices up to commensurability, relates the phenomena of spontaneous symmetry breaking in quantum statistical mechanics to the mathematics of Galois theory. A similar noncommutative space was used in [Co2000a] and [Co99] to obtain a spectral realization of the zeroes of the Riemann zeta function. Some other recent examples (cf. [CoMo04] , [CoMo04a]) interpret differential operators on modular forms (Rankin–Cohen brackets) in terms of the Hopf algebra of a noncommutative space of codimension one foliations. It turns out that the modular Hecke algebra appears as the "holomorphic part" of the algebra of a certain noncommutative space (cf. [CoMar04]).

Let us only mention the noncommutative elliptic curves, and noncommutative modular curves cf. op.cit.

We add to this list other interesting examples coming from the construction in [Man91], giving a description of the totally degenerate fibers at "arithmetic infinity" of arithmetic varieties over number fields, cf. [CM]. We describe in Chapter 8, how Connes theory gives a link of this construction with Deninger's approach [Den91], who suggested to reinterpret Serre's gamma-factors of zeta-functions as infinite regularized determinants of certain ∞-adic Frobenius maps acting upon new cohomological spaces. These spaces are described in §8.2.

Isomorphism of noncommutative spaces and Morita equivalence

In noncommutative geometry, isomorphisms of C^*-algebras are too resrictive to provide a good notion of isomorphism of noncommutative spaces, and the correct notion is provided by Morita equivalence of C^*-algebras (cf. [Man02a], §1.3):

Definition III-0.1 (Morita category). *Let A, B be two associative rings. A Morita morphism $A \to B$ by definition, is the isomorphism class of a bimodule ${}_AM_B$, which is projective and finitely generated separately as module over A and B.*

The composition of morphisms is given by the tensor product ${}_AM_B \otimes {}_BM'_C$, or ${}_AM \otimes {}_BM'_C$ for short.

If we associate to ${}_AM_B$ the functor

$$\mathrm{Mod}_A \to \mathrm{Mod}_B : \ N_A \mapsto N \otimes_A M_B,$$

the composition of functors will be given by the tensor product, and isomorphisms of functors will correspond to the isomorphisms of bimodules.

We imagine an object A of the (opposite) Morita category as a noncommutative space, right A–modules as sheaves on this space, and the tensor multiplication by ${}_AM_B$ as the pull–back functor. We have chosen to work with right modules, but passing to the opposite rings allows one to reverse left and right in all our statements.

Two bimodules ${}_AM_B$ and ${}_BN_A$ supplied with two bimodule isomorphisms ${}_AM \otimes_B N_A \to {}_AA_A$ and ${}_BN \otimes_A M_B \to {}_BB_B$ define mutually inverse Morita isomorphisms (equivalences) between A and B. The basic example of this kind is furnished by $B = \mathrm{Mat}\,(n, A)$, $M = {}_AA^n{}_B$ and $N = {}_BA^n{}_A$.

We will now briefly summarize Morita's theory.

(A) *Characterization of functors $S : \mathrm{Mod}_A \to \mathrm{Mod}_B$ of the form $N_A \mapsto N \otimes_A M_B$.* They are precisely functors satisfying any of the two equivalent conditions:

(i) *S is right exact and preserves direct sums.*

(ii) *S admits a right adjoint functor $T : \mathrm{Mod}_B \to \mathrm{Mod}_A$ (which is then naturally isomorphic to $\mathrm{Hom}_B(M_B, *)$).*
We will call such functors *continuous.*

(B) *Characterization of continuous functors S such that T is also continuous and $ST \cong 1$.* Let S be given by ${}_AM_B$ and T by ${}_BN_A$. Then $M \otimes_B N \cong {}_AA_A$. Moreover, in this case

(iii) *M_B and ${}_BN$ are projective.*

(iv) *${}_AM$ and N_A are generators.*

In particular, equivalences $\mathrm{Mod}_A \to \mathrm{Mod}_B$ are automatically continuous. Hence any pair of mutually quasi–inverse equivalences must be given by a couple of biprojective bigenerators as above.

(C) *Finite generation and balance.* Any right module M_B can be considered as a bimodule ${}_AM_B$ where $A = B' := \mathrm{End}_B(M_B)$. We can then similarly produce the ring $B'' = A' := \mathrm{End}_A({}_AM)$. Module M_B is called *balanced* if $B'' = B$. Similarly, one can start with a left module. With this notation, we have:

(v) *M_B is a generator iff ${}_{B'}M$ is balanced and finitely generated projective.*

Properties (i)–(v) can serve as a motivation for our definition of the Morita category above.

The tools of noncommutative geometry

The following is a list of some techniques used in order to compute invariants and extract essential information from the geometry.

- Topological invariants: K-theory
- Hochshild and cyclic cohomology
- Homotopy quotients, assembly map (Baum–Connes)
- Metric structure: Dirac operator, spectral triples
- Characteristic classes, zeta functions

We recall some of these notions in Part III, starting with Dirac operator and spectral triples.

In noncommutative geometry, the notion of a spectral triple provides the correct generalization of the classical structure of a Riemannian manifold. The two notions agree on a commutative space. In the usual context of Riemannian geometry, the definition of the infinitesimal element ds on a smooth spin manifold can be expressed in terms of the inverse of the classical Dirac operator D. This is the key remark that motivates the theory of spectral triples. In particular, the geodesic distance between two points on the manifold is defined in terms of D^{-1} (cf. [Co94] §VI). The spectral triple that describes a classical Riemannian spin manifold is (A, H, D), where A is the algebra of complex valued smooth functions on the manifold, H is the Hilbert space of square integrable spinor sections, and D is the classical Dirac operator (a square root of the Laplacian). These data determine completely and uniquely the Riemannian geometry on the manifold.

The notion of spectral triple extends to more general *noncommutative spaces*, where the data $(\mathcal{A}, \mathcal{H}, \mathcal{D})$ consist of a C^*-algebra $\mathcal{A}$ (or more generally of a smooth subalgebra of a C^*-algebra) with a representation as bounded operators on a Hilbert space $\mathcal{H}$, and an operator $\mathcal{D}$ on $\mathcal{H}$ that verifies the main properties of a Dirac operator.

III.4.2 Generalities on spectral triples

Let us recall the basic setting of Connes theory of spectral triples. For a more complete treatment we refer to [Co95], [Co94], [CoMo95].

Definition III-0.2. *A spectral triple $(\mathcal{A}, \mathcal{H}, D)$ consists of an involutive algebra $\mathcal{A}$ with a representation*

$$\rho : \mathcal{A} \to \mathcal{B}(\mathcal{H})$$

as bounded operators on a Hilbert space $\mathcal{H}$, and an operator D (called the Dirac operator) on $\mathcal{H}$, which satisfies the following properties:

1. *D is self–adjoint.*
2. *For all $\lambda \notin \mathbb{R}$, the resolvent $(D - \lambda)^{-1}$ is a compact operator on $\mathcal{H}$.*
3. *For all $a \in \mathcal{A}$, the commutator $[D, a]$ is a bounded operator on $\mathcal{H}$.*

Remark III-0.3. The property *2.* of Definition III.1.1 generalizes ellipticity of the standard Dirac operator on a compact manifold. Usually, the involutive algebra $\mathcal{A}$ satisfying property *3.* can be chosen to be a dense subalgebra of a C^*–algebra. This is the case, for instance, when we consider smooth functions on a manifold as a subalgebra of the commutative C^*-algebra of continuous functions. In the classical case of Riemannian manifolds, property *3.* is equivalent the Lipschitz condition, hence it is satisfied by a larger class than that of smooth functions.

III.4.3 Contents of Part III: description of parts of this program

We included under the heading of this part various recent topics related to Arakelov's geometry, and Noncommutative geometry. All these topics use a lot of algebraic tools such as cohomology groups and non-commutative rings.

We start Chapter 8 with §8.1 on Schottky uniformization and Arakelov geometry, based on geometric constructions in [Man91]. We give an analytic construction of degenerating curves over complete local fields, following [Mum72], in §8.1.2. Then in §8.1.5 we describe the result in [Man91] on the relation between the Arakelov Green function on a Riemann surface $X(\mathbb{C})$ with Schottky uniformization and geodesics in the 3-dimensional hyperbolic handlebody $\mathfrak{X}_\Gamma$.

Following [CM], we describe in this chapter, how Connes theory gives a link of theses constructions with Deninger's approach [Den91], who suggested to reinterpret Serre's gamma-factors of zeta-functions as infinite regularized determinants of certain ∞-adic Frobenius maps acting upon two types of cohomology spaces. *Archimedean cohomology* spaces are described in §8.2, and *dynamical cohomology* spaces are described in §8.3.

In §8.2 we describe a cohomological theory for the Archimedean fiber of an Arakelov surface. This results are based on the general theory, valid for any arithmetic variety, developed in [Cons98] (we follow an interesting discussion of the Archimedean cohomology in [CM04b] and [Mar04], Chapter 3). This construction provides a refinement for the *Archimedean cohomology* H^*_{ar} introduced by Deninger in [Den91].

Based on this construction and following [CM], we describe in §8.2.2 a cohomological spectral data $(A, H^{\cdot}(X^*), \Phi)$, where the algebra ($A$ is obtained from the $\mathrm{SL}(2,\mathbb{R})$ action on certain cohomology groups.

In Theorem 8.3 we describe how to recover the alternating product of the Archimedean factors from a zeta function of a spectral triple. In §8.3 a different construction is described, which is related to description in [Man91] of the *dual graph of the fiber at infinity.* A geometric model is given here for the dual graph as the *mapping torus* of a dynamical system T on a Cantor set.

We consider a noncommutative space which describes the action of the Schottky group on its limit set and parameterizes the "components of the closed fiber at infinity". This space is represented by a *Cuntz–Krieger algebra* $\mathcal{O}_A$, described in §8.3.5.

Next, we describe a spectral triple for this noncommutative space, via a representation on the cochains of a "dynamical cohomology", defined in terms of the tangle of bounded geodesics in the handlebody. In both constructions presented in Chapter 8, the Dirac operator agrees with the grading operator Φ, that represents the "logarithm of a Frobenius–type operator" on the Archimedean cohomology. In fact, the Archimedean cohomology embeds in the dynamical cohomology, compatibly with the action of a real Frobenius $\bar{F}_\infty$, so that the local factor can again be recovered from these data. Moreover, the "reduction mod infinity" is presented in §8.4 in terms of the homotopy quotient associated to the noncommutative space $\mathcal{O}_A$ and the μ-map of Baum–Connes, cf. [BaCo].

Suggestions for further reading to Part III

Works of A.Connes on Trace formula in noncommutative geometry and the zeros of the Riemann zeta function, [Co99], [Co2000a], work of J.–B. Bost [Bo01] on algebraic leaves of algebraic foliations over number fields, and [BoCo95], [CoMar04], [CoMo95], [CoMo04], [Ber86] sheading new light on relations of physics with number theory using noncommutative geometry (see also nice papers of P.Cartier [Car95], [Car01], [Car02] on related subjects).

We refer also to Number Theory and Physics archive at
`http://www.maths.ex.ac.uk/~mwatkins/zeta/physics.htm`
where one can find very useful material and bibliography on relations with quantum mechanics, statistical mechanics, p-adic and adelic physics, Selberg trace formula, string theory and quantum cosmology, scattering theory, dynamical and spectral zeta functions, trace formulae and explicit formulae, $1/f$ noise and signal processing, supersymmetry, QCD, renormalisation, symmetry breaking and phase transitions, quantum fields, integer partitions, time, biologically-inspired and similarly unconventional methods for finding primes, dynamical systems, entropy, specific zeta values, logic, languages, information, etc., probability and statistics, noncommutative geometry, random matrices, Fourier theory, fractal geometry, Bernoulli numbers, Farey sequences, Beurling g-primes, golden mean, zeta functions and L-functions.

8

Arakelov Geometry and Noncommutative Geometry (d'après C. Consani and M. Marcolli, [CM])

8.1 Schottky Uniformization and Arakelov Geometry

8.1.1 Motivations and the context of the work of Consani-Marcolli

Our primary motivation in this chapter is a desire to enrich the somewhat formal picture of Arakelov's geometry at arithmetical infinity (see [Ara74b], [La88], [GS92], [Man84], and §5.2.6). We try to describe geometric and algebraic objets which play the roles, respectively, of the "∞–adic completion" of a completed arithmetical Arakelov variety, of its "closed fiber at infinity", and the "reduction modulo ∞", in the spirit of the work of Mumford [Mum72]. We follow the paper [Man91], and recent works [CM], [CM03], [CM04a], [Cons98], which clarified much at the arithmetical infinity.

We try to explain then how to use Connes' theory of spectral triples (see [Co95], [Co99], [Co94], and §8.3) in order to relate the hyperbolic geometry to Deninger's Archimedean cohomology.

We recall the beginning of a dictionary which translates classical notions into the laguage of operators in the Hilbert space $\mathcal{H}$:

Complex variable	Operator in $\mathcal{H}$	
Real variable	Self-adjoint operator	(8.1.1)
Infinitesimal	Compact operator	

From the arithmetic point of view, algebraic numbers appear in commutative geometry as values of algebraic functions, whereas in noncommutative geometry they appear as values of traces of projections, or more generally values of appropriate states on observables. In both cases, a control of the action of the Galois group is gained, if this action commutes with an action of certain "geometric" endomorphisms, or correspondences, whenever the latter are defined over the ground field.

Interesting examples of noncommutative spaces come from geometric constructions in [Man91].

The choice of structures used in this section is related to Mumford's idea (see [Mum72], and §8.1.2) that p-adic curves with maximally degenerate reduction admit an (essentialy unique) p-adic Schottky uniformization. On can use his geometric picture in an Archimedean setting.

In the paper [Man91] a filling $\mathfrak{X}$ of X was constructed, which was an auxiliary three dimensional manifold endowed with a metric of constant negative curvature, such that X is its boundary, and which can be defined by a *Schottky uniformization*, described in §8.1.3.

It was also proven in [Man91] that Green's function on X can be expressed in terms of the geometry of geodesics on $\mathfrak{X}$.

An interesting analogy was suggested, in which the set of all closed geodesics in $\mathfrak{X}$ and the set of geodesics with one end at X play the roles, respectively, of "∞–adic completion" of $\overline{X}$, its "closed fiber at infinity", and the "reduction modulo ∞", in the spirit of the work of Mumford ([Mum72]).

Preliminary notions and notation

Throughout this section let us denote by K one among the following fields: (a) the complex numbers $\mathbb{C}$, (b) a finite extension of $\mathbb{Q}_p$. When (b) occurs, we write $\mathcal{O}_K$ for the ring of integers of K, $\mathfrak{m} \subset \mathcal{O}_K$ for the maximal ideal and $\pi \in \mathfrak{m}$ for a uniformizer (i.e. $\mathfrak{m} = (\pi)$). We also denote by k the residue classes field $k = \mathcal{O}/\mathfrak{m}$.

We denote by $\mathbb{H}'$ (or simply by $\mathbb{H}^3 = \mathbb{H}'$) the *three-dimensional real hyperbolic space* i.e. the quotient

$$\mathbb{H}' = \mathrm{SU}(2)\backslash\mathrm{PGL}(2,\mathbb{C}).$$

This space can also be described as the upper half space $\mathbb{H}' \simeq \mathbb{C}\times\mathbb{R}^+$ endowed with the hyperbolic metric. The group $\mathrm{PSL}(2,\mathbb{C})$ acts on $\mathbb{H}'$ by isometries. The complex projective line $\mathbb{P}^1(\mathbb{C})$ is identified with the conformal boundary at infinity of $\mathbb{H}'$ and the action of $\mathrm{PSL}(2,\mathbb{C})$ on $\mathbb{H}'$ extends to an action on $\overline{\mathbb{H}'} := \mathbb{H}' \cup \mathbb{P}^1(\mathbb{C})$. The group $\mathrm{PSL}(2,\mathbb{C})$ acts on $\mathbb{P}^1(\mathbb{C})$ by fractional linear transformations.

8.1.2 Analytic construction of degenerating curves over complete local fields and Arakelov geometry (d'après Mumford [Mum72])

The idea of investigating the p-adic analogues of classical and abelian varieties is due to John Tate [Ta74], who showed that if K is a complete non-Archimedean local field, and E is an elliptic curve over K *whose j-invariant is not an integer*, then E can be analytically uniformized. This uniformization is *not* a holomorphic map:

$$\pi : \mathbb{A}_K^1 \to E$$

generalizing the universal covering space

$$\pi : \mathbb{C} \to E(= \text{ closed points of an elliptic curve over } \mathbb{C}),$$

but instead is a holomorphic map:

$$\pi_2 : \mathbb{A}_K^1 \backslash \{0\} \to E$$

generalizing an infinite cyclic covering π_2 over $\mathbb{C}$:

$$\begin{array}{ccccc} & & \mathbb{C}^\times & & \\ & \overset{\pi_1}{\nearrow} & & \overset{\pi_2}{\searrow} & \\ \mathbb{C} & & \xrightarrow{\quad\pi\quad} & & E \end{array}$$

$$\pi_1(z) = e^{2\pi i (z/\omega_1)},$$

ω_1 is one of the two periods of E. Here one can take holomorphic map to mean holomorphic in the sense of the non-Archimedean function theory of Grauert and Remmert [GR71]. But the uniformization π_2 is more simply expressed by embedding E in $\mathbb{P}_K^2$ and defining the tree homogeneous coordinates of $\pi(z)$ by three everywere convergent Laurent series.

In order to explain what happens for curves of higher genus, let us present a bit further the interesting analogies between the real, complex and p-adic structures $PGL(2)$ (as developped by Bruhat, Tits and Serre, see [Se71]):

(A) *real case*: $PSL(2, \mathbb{R})$ acts isometrically and transitively on the upper half plane $\mathbb{H}$ and the boundary can be identified with $\mathbb{RP}^1$ (the real line plus ∞):

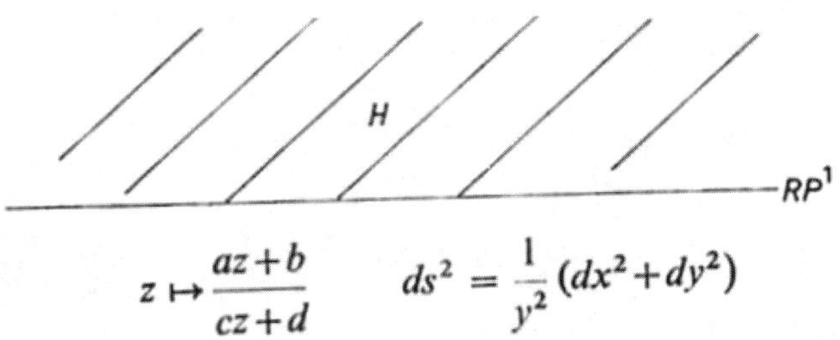

Fig. 8.1.

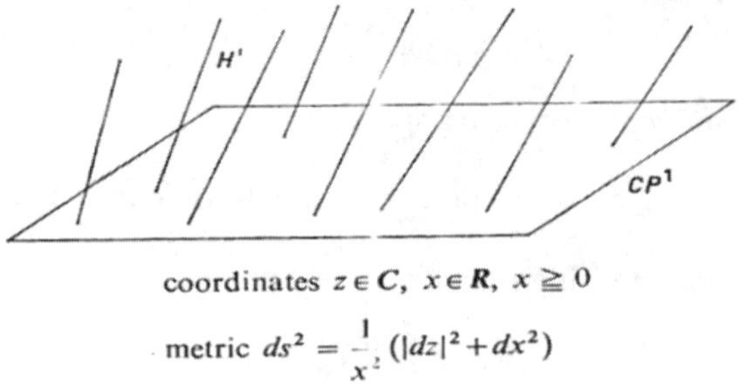

Fig. 8.2.

(B) *complex case*: $PGL(2,\mathbb{C})$ acts isometrically and transitively on the upper half space $\mathbb{H}'$ and the boundary can be identified with $\mathbb{CP}^1$: The action of $PSL(2,\mathbb{C})$ on $\mathbb{H}'$ is given by

$$(z,x) \mapsto \left(\frac{\overline{(cz+d)}(az+b)+acx^2}{|cz+d|^2+|c|^2x^2}, \frac{x}{|cz+d|^2+|c|^2x^2} \right).$$

(C) *p-adic case*: $PGL(2,K)$ acts isometrically and transitively on the *Bruhat-Tits tree* Δ, (whose vertices correspond to the subgroups $gPGL(2,\mathcal{O}_K)g^{-1}$, and whose edges have length 1 and correspond to the subgroups $gPGL(2,\mathcal{O}_K)g^{-1}$, $\left\{ \left(\begin{smallmatrix} a & b \\ c & d \end{smallmatrix} \right) \middle| a,b,c,d \in \mathcal{O}_K, c \in \mathfrak{m}, ad \notin \mathfrak{m} \right\}$ modulo $\mathcal{O}_K^*$) and the set of whose ends can be identified with $K\mathbb{P}^1$ (see [Se77], and [Mum72], p. 131).

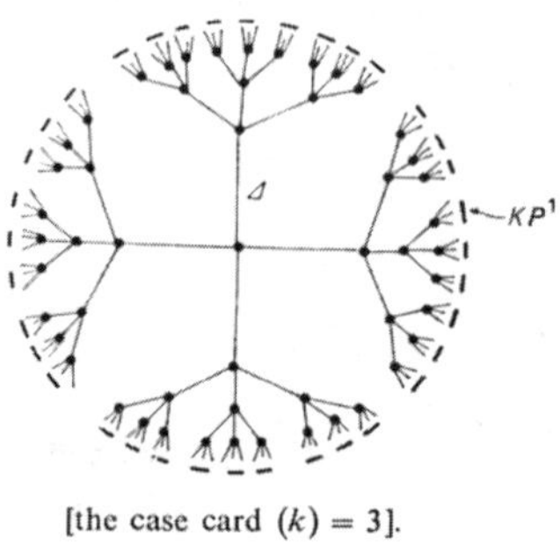

Fig. 8.3.

In Δ, for any vertex v the set of edges meeting v is naturaly isomorphic to $k\mathbb{P}^1$, the isomorphism being canonical up to an element of $PGL(2,k)$ (where $k = \mathcal{O}_K/\mathfrak{m}$).

In the first case, if $\Gamma \subset PSL(2,\mathbb{R})$ is a discrete subgroup with no elements of finite order such that $PSL(2,\mathbb{R})/\Gamma$ is compact, we obtain Koebe's uniformization

$$\mathbb{H} \to \mathbb{H}/\Gamma = X$$

of an arbitrary compact Riemann surface X of genus $g \geq 2$.

In the second, if $\Gamma \subset PSL(2,\mathbb{C})$ is a discrete subgroup which act discontinuously at at least one point of $\mathbb{CP}^1$ (a Kleinian group) and which moreever is free with n generators and has no unipotent elements in it, then according to a theorem of Maskit (see in [Mum72]), Γ is a so-called *Schottky group*, i.e. if we denote by Ω_Γ the *domain of discontinuity* of Γ acting on $\mathbb{P}^1(\mathbb{C})$ then Ω_Γ is connected and up to a homeomorphism we get a uniformization

$$\begin{array}{ccccc} (\mathbb{H}' \cup \Omega_\Gamma) & \to & (\mathbb{H}' \cup \Omega_\Gamma)/\Gamma & \underset{\text{homeo}}{\cong} & \text{solid torus with } n \text{ handles} \\ \cup & & \cup & & \cup \\ \Omega_\Gamma & \xrightarrow{\pi} & \Omega_\Gamma/\Gamma & \underset{\text{homeo}}{\cong} & \{\text{ boundary, a surface of genus } n\}. \end{array}$$

The quotient

$$X_{/\mathbb{C}} = \Omega_\Gamma/\Gamma \tag{8.1.2}$$

is a Riemann surface of genus g and the covering $\Omega_\Gamma \to X_{/\mathbb{C}}$ is called a *Schottky uniformization* of $X_{/\mathbb{C}}$. Every complex Riemann surface $X_{/\mathbb{C}}$ admits a Schottky uniformization. In particular, Γ/Ω_Γ is a compact Riemann surface of genus n and for a covering corresponding to the subgroup

$$N \subset \pi_1(\Omega_\Gamma)/\Gamma \tag{8.1.3}$$

$$N = \text{ least normal subgroup containing } a_1, \cdots, a_n. \tag{8.1.4}$$

The Schottky uniformization π admits a p-adic analog.

In the third case, let $\Gamma \subset PGL(2,K)$ be any discrete subgroup consisting entirely of hyperbolic elements. Then according to Ihara, the group Γ is free: let Γ have n generators. Again, let Ω_Γ be the set of closed points of $\mathbb{P}^1_K$ where Γ acts discontinuously (equivalently, Ω_Γ is the set of points which are not limits of fixed points of elements of Γ).

Then it was proved in [Mum72] that there is a curve C of genus n and a holomorphic isomorphism:

$$\pi : \Omega_\Gamma/\Gamma \xrightarrow{\approx} C.$$

Moreover, Δ/Γ has a very nice interpretation as a graph of the specialization of C over the ring $\mathcal{O}_K$. In fact

a) there will be a smallest subgraph

$$(\Delta)_0/\Gamma \subset \Delta/\Gamma$$

such that

$$\pi_1((\Delta)_0/\Gamma) \xrightarrow{\sim} \pi_1(\Delta/\Gamma) \text{ and } (\Delta)_0/\Gamma \text{ will be finite.}$$

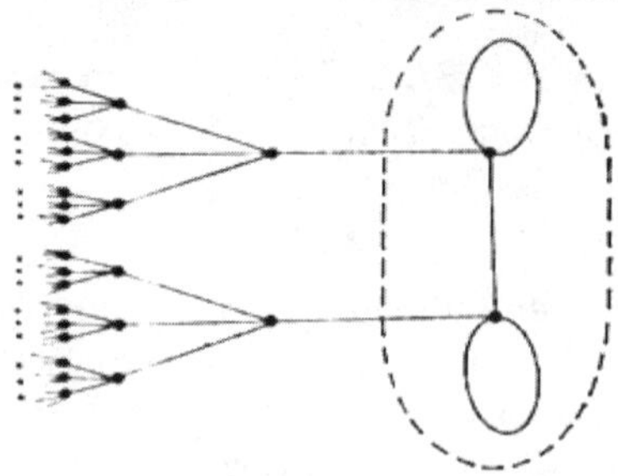

Fig. 8.4. Ends of Δ/Γ $(\Delta/\Gamma)_0$

b) C will have a canonical specialization $\overline{C}$ over $\mathcal{O}_K$ where $\overline{C}$ is a singular curve of arithmetic genus n made up from copies of $\mathbb{P}^1_k$ with a finite number of distinct pairs of k-rational points identified to form ordinary double points points. Such a curve $\overline{C}$ will be called a *k-split degenerate curve of genus n.*
c) $C(K)$, the set of K-rational points of C, will be naturally isomorphic to the set of ends of Δ/Γ; $\overline{C}(k)$, the set of k-rational points of $\overline{C}$, will be naturally isomorphic to the set of edges of Δ/Γ that meet vertices of $(\Delta/\Gamma)_0$ (so that the components of $\overline{C}$ correspond to the edges of Δ/Γ meeting a fixed vertex of $(\Delta/\Gamma)_0$ and the double points of $\overline{C}$ correspoind to the edges of $(\Delta/\Gamma)_0$; and finally the specialization map

$$C(K) \longrightarrow C(k)$$

is equal, under the above identification, to the map

$$\text{Ends of } (\Delta/\Gamma)_0 \to \begin{pmatrix} \text{edges of } \Delta/\Gamma \\ \text{meeting } (\Delta/\Gamma)_0 \end{pmatrix}$$

which takes an end to the last edge in the shortest path from that end to $(\Delta/\Gamma)_0$.

Example 8.1. The following figure 8.5 illustrates a case when the genus is 2, $\overline{C}$ has 2 components, each with one double point and meeting each other once: Because all the curves C which were constructed in this way have property (b), they are refered to as *degenerating curves.* The main theorem in [Mum72] implies that every such degenerating curve C has a unique analytic uniformization $\pi : \Gamma\backslash\Omega_\Gamma \xrightarrow{\sim} C$.

8.1.3 Schottky groups and new perspectives in Arakelov geometry

A unified description of the Archimedean and the totally split degenerate fibers of an arithmetic surface was given in [CM], using operator algebras

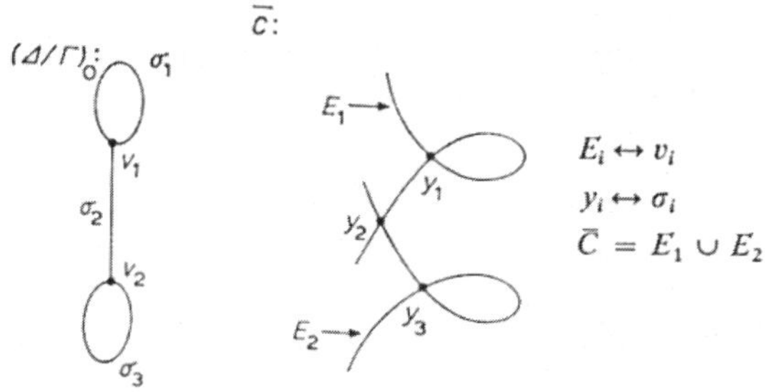

Fig. 8.5.

and Connes' theory of *spectral triples* in noncommutative geometry, [Co94]. Some of more recent results were reported in [CM04a] on a non-commutative interpretation of the totally split degenerate fibers of an arithmetic surface.

Let $\mathfrak{X}$ be an arithmetic surface defined over Spec($\mathbb{Z}$) (or over Spec($\mathcal{O}_K$), for a number field K), having the smooth algebraic curve $X_{/\mathbb{Q}}$ as its generic fiber. Then, as a *Riemann surface*, $X_{/\mathbb{C}}$ admits always a uniformization by means of a Schottky group Γ. In analogy to Mumford's p-adic uniformization of algebraic curves (cf. [Mum72]), the Riemann surface $X_{/\mathbb{C}}$ can be interpreted as the boundary at infinity of a 3-manifold $\mathfrak{X}_\Gamma$ defined as the quotient of the real hyperbolic 3-space $\mathbb{H}'$ by the action of the *Schottky group* Γ. The space $\mathfrak{X}_\Gamma$ contains in its interior an *infinite link of bounded geodesics.*

In [Man91] an expression was given for the *Arakelov Green function* on $X_{/\mathbb{C}}$ in terms of configurations of geodesics in $\mathfrak{X}_\Gamma$, thus interpreting this tangle as the dual graph $\mathcal{G}$ of the "closed fiber at infinity" of $\mathfrak{X}$.

Schottky uniformization and Schottky groups

Topologically a compact Riemann surface X of genus g is obtained by gluing the sides of a $4g$-gon. Correspondingly, the fundamental group has a presentation

$$\pi_1(X) = \langle a_1, \cdots, a_g, b_1, \cdots, b_g \mid \prod_i [a_i, b_i] = 1 \rangle,$$

where the generators a_i and b_i label the sides of the polygon.

In the genus $g = 1$ case, the parallelogram is the fundamental domain of the $\pi_1(X) = \mathbb{Z}^2$ action on the plane $\mathbb{C}$, so that $X = \mathbb{C}/(\mathbb{Z} + \mathbb{Z}\tau)$ is an elliptic curve.

For the genus at least $g \geq 2$ the hyperbolic plane $\mathbb{H}$ admits a tesselation by regular $4g$-gons, and the action of the fundamental group by deck transformation is realized by the action of a subgroup $\pi_1(X) \cong G \subset \mathrm{PSL}(2, \mathbb{R})$ by isometries of $\mathbb{H}$. This endows the compact Riemann surface X with a hyperbolic metric and a Fuchsian uniformization

$$X = G\backslash H.$$

Another, less known, type of uniformization of compact Riemann surfaces is Schottky uniformization.

Let us recall briefly some general facts on Schottky groups.

A *Schottky group* of rank $g \geq 1$ is a discrete subgroup $\Gamma \subset \mathrm{PSL}(2, \mathbb{C})$, which is purely loxodromic and isomorphic to a free group of rank g. The group $\mathrm{PSL}(2, \mathbb{C})$ acts on $\mathbb{P}^1(\mathbb{C})$ by fractional linear transformation

$$\gamma : z \mapsto \frac{az + b}{cz + d}.$$

Thus, Γ also acts on $\mathbb{P}^1(\mathbb{C})$.

Let us denote by Λ_Γ the *limit set of the action* of Γ. One sees that Λ_Γ is contained in $\mathbb{P}^1(\mathbb{C})$. This set can also be described as the closure of the set of the attractive and repelling fixed points $z^{\pm}(g)$ of the loxodromic elements $g \in \Gamma$. In the case $g = 1$ the limit set consists of two points, but for $g \geq 2$ the limit set is usually a *fractal* of some *Hausdorff dimension* $0 \leq \delta = \dim_H(\Lambda_\Gamma) < 2$ (cf. e.g. Fig.8.6 reproduced here from [MSW02] and `http://klein.math.okstate.edu/IndrasPearls/gallery/GeneralSchottky.gif` with a kind permission of David J. Wright and Cambridge UP).

Fig. 8.6. Limit set of a Schottky group

Recall that the notion of the *Hausdorff dimension* can be used for quite general subsets of $A \subset \mathbb{R}^d$, and it gives for the curves and surfaces the usual notion of dimension. The Hausdorff dimension (cf. e.g. [Arne04]*)) could be non integer for fractal subsets.

We denote by $\Omega_\Gamma = \Omega_\Gamma(\mathbb{C})$ the *domain of discontinuity* of Γ, that is, the complement of Λ_Γ in $\mathbb{P}^1(\mathbb{C})$.

The quotient space $\mathfrak{X}_\Gamma := \mathbb{H}'/\Gamma$ is topologically a *handlebody of genus g*, and the quotient

$$X_\mathbb{C} = \Omega_\Gamma/\Gamma$$

is a *Riemann surface of genus g*. The covering $\Omega_\Gamma \to X_{/\mathbb{C}}$ is called a *Schottky uniformization* of $X_\mathbb{C}$. Every complex Riemann surface $X_\mathbb{C}$ admits a Schottky uniformization. The handlebody $\mathfrak{X}_\Gamma$ can be compactified by adding the conformal boundary at infinity $X_\mathbb{C}$ to obtain $\overline{\mathfrak{X}}_\Gamma := \mathfrak{X}_\Gamma \cup X_\mathbb{C} = (\mathbb{H}' \cup \Omega_\Gamma)/\Gamma$.

Let $\{\gamma_i\}_{i=1}^\infty$ be a set of generators of the Schottky group Γ. Let us use the notation $\gamma_{i+g} := \gamma_i^{-1}$, for $i = 1, \ldots, g$. There are $2g$ Jordan curves C_k on the sphere $\mathbb{P}^1(\mathbb{C})$, with pairwise disjoint interiors D_k, such that the elements γ_k are given by fractional linear transformations that map the interior of C_k to the exterior of C_j with $|i-j| = g$. The curves C_k give a marking of the Schottky group. The markings are circles in the case of classical Schottky groups. A fundamental domain for the action of a classical Schottky group Γ on $\mathbb{P}^1(\mathbb{C})$ is the region exterior to $2g$-circles. (cf. Fig.8.7 reproduced here from the lectures [Mar04]):

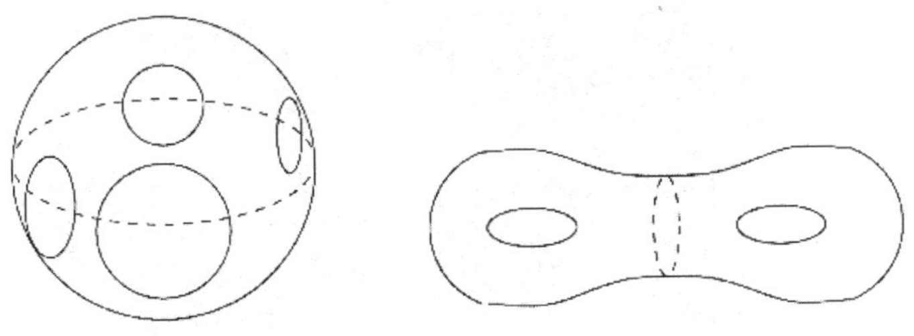

Fig. 8.7. Schottky uniformization for $g = 2$

* A family of $R = \{B_i\}_{i\in\mathbb{N}}$ of subsets $B_i \subset \mathbb{R}^d$ is an ε-covering of A if $A \subset \cup B_i$ and $\forall i, diam(B_i) \le \varepsilon$. First define $M_\varepsilon^\delta = \inf_{R=\{B_i\}} \sum (diam B_i)^\delta$, where the infimum is taken over all ε-coverings. The Hausdorff dimension is then defined by $\dim_H(A) = \sup\{\delta : \lim_{\varepsilon\to 0} M_\varepsilon^\delta = +\infty\}$. One verfies easily that the dimension of a regular curve is 1, and the dimension of a regular surface is 2. An example of a non-integer dimension is given by the three-adic Cantor set consisting of real numbers of the form $\sum_{n=1}^\infty \frac{\varepsilon_n}{3^n}$, with $\varepsilon_n \in \{0, 2\}$, whose Hausdorff dimension is equal to $\log 2/\log 3$.

Fuchsian and Schottky uniformization.

Notice that, unlike Fuchsian uniformization, where the covering $\mathbb{H}$ is the universal cover, in the case of Schottky uniformization Ω_Γ is very far from being simply connected, in fact it is the complement of a Cantor set.

A relation between Schottky and Fuchsian uniformizations is given by passing to the covering that corresponds to the normal subgroup $N\langle a_1, \cdots, a_g\rangle$ of $\pi_1(X)$ generated by half the generators $\{a_1, \cdots, a_g\}$:

$$\Gamma \cong \pi_1(X)/N\langle a_1, \cdots, a_g\rangle.$$

Surface with boundary: simultaneous uniformization

In order to see better the Schottky uniformization, one can relate it to a simultaneous uniformization of the upper and lower half planes that yelds to Riemann surfaces with boundary, joint at the boundary.

A Schottky group that is specified by real parameters so that it lies in $\mathrm{PSL}(2, \mathbb{R})$, is called *Fuchsian Schottky group* (cf. Fig.8.8 reproduced from the lectures [Mar04], Fig.3. Viewed as a group of isometries of the hyperbolic plane

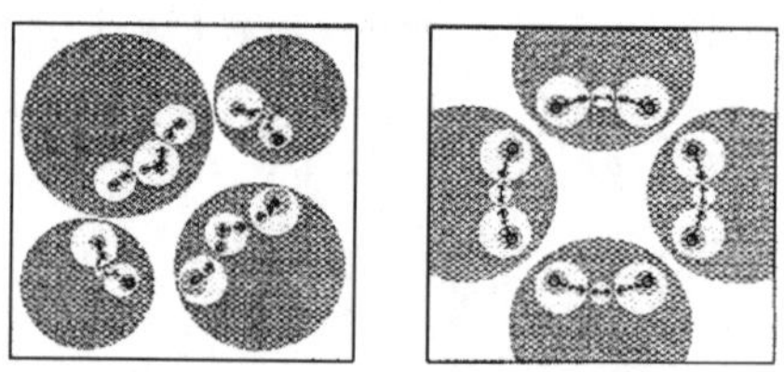

Fig. 8.8. Classical and Fuchsian Schottky groups

$\mathbb{H}$, or equivalently of the Poincaré disk, a Fuchsian Schottky group G produces a quotient $G\backslash\mathbb{H}$ which is topoligally a Riemann surface with a boundary.

A *quasi-circle* for Γ is a Jordan curve C in $\mathbb{P}^1(\mathbb{C})$ which is invariant under the action of Γ. In particular, such curve contains the limit set Λ_Γ. The existence of a quasi-circle for a Riemann surface $X(\mathbb{C})$ of genus $g \geq 2$ is known due to Bowen, cf. [Bo], [Mar04].

We have that $\mathbb{P}^1(\mathbb{C})\backslash C = \Omega_1 \cup \Omega_2$, and, for $\pi_\Gamma : \Omega_\Gamma \to \mathbb{P}^1(\mathbb{C})$, the covering map

$$\hat{C} = \pi_\Gamma(C \cap \Omega_\Gamma) \subset \mathbb{P}^1(\mathbb{C})$$

is a set of curves on $X(\mathbb{C})$ that disconnect the Riemann surface in the union of two surfaces with boundary, uniformized respectively by Ω_1 and Ω_2.

There exist conformal maps

$$\alpha_i : \Omega_i \xrightarrow{\cong} U_i, U_1 \cup U_2 = \mathbb{P}^1(\mathbb{C})\backslash\mathbb{P}^1(\mathbb{R}) \tag{8.1.5}$$

with $U_i \cong \mathbb{H}$=upper half planes in $\mathbb{P}^1(\mathbb{C})$, and with

$$G_i = \left\{ \alpha_i \gamma \alpha_i^{-1} \mid \gamma \in \hat{\Gamma} \right\} \cong \Gamma$$

Fuchsian Schottky groups $G_i \subset \mathrm{PSL}(2,\mathbb{R})$. Here $\tilde{\Gamma}$ is the Γ-stabilizer of each of two connected components in $\mathbb{P}^1(\mathbb{C}) \backslash C$.

The compact Riemann surface is then obtained as

$$X(\mathbb{C}) = X_1 \cup_{\partial X_1 = \hat{C} = \partial X_2} X_2,$$

with $X_i = U_i/G_i$ (Riemann surfaces with boundary $\hat{C}$ (cf. Fig.8.9 reproduced here from the lectures [Mar04], Fig.4).

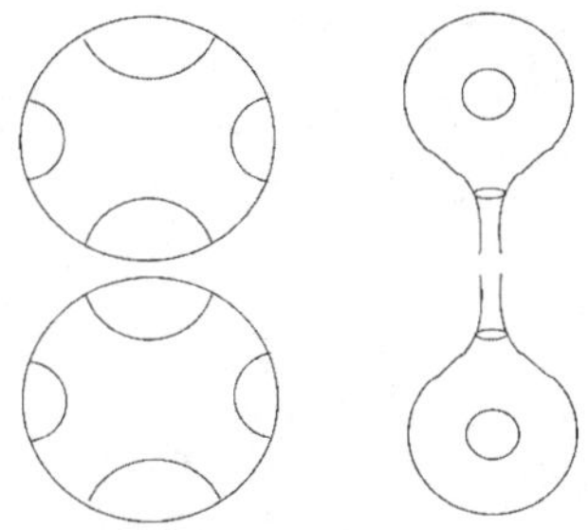

Fig. 8.9. Fuchsian Schottky groups: Riemann surfaces with boundary

In the case where $X(\mathbb{C})$ has a real structure $\iota : X \to X$, and the fixed point set $Fix(\iota) = X(\mathbb{R})$ of the involution is nonempty, we have in fact $\hat{C} = X(\mathbb{R})$, and the quasi-circle is given by $\mathbb{P}^1(\mathbb{R})$.

8.1.4 Hyperbolic handlebodies

The action of a rank g Schottky group $\Gamma \subset \mathrm{PSL}(2,\mathbb{C})$ on $\mathbb{P}^1(\mathbb{C})$, by fractional linear transformations, extends to an action by isometries on $\mathbb{H}^3$. For a classical Schottky group, a fundamental domain in $\mathbb{H}^3$ is given by the region external to $2g$ half spheres over the circles $C_k \subset \mathbb{P}^1(\mathbb{C})$ (cf. Fig.8.10 reproduced here from the lectures [Mar04], Fig.5).

The quotient

$$\mathfrak{X}_\Gamma \tag{8.1.6}$$

is topologically a handlebody of genus g filling the Riemann surface $X(\mathbb{C})$ (cf. Fig.8.11 reproduced here from the lectures [Mar04], Fig.6).

Metrically, $\mathfrak{X}_\Gamma$ is a real hyperbolic 3-manifold of infinite volume, having $X(\mathbb{C})$ as its conformal boundary at infinity $X(\mathbb{C}) = \partial\mathfrak{X}_\Gamma$.

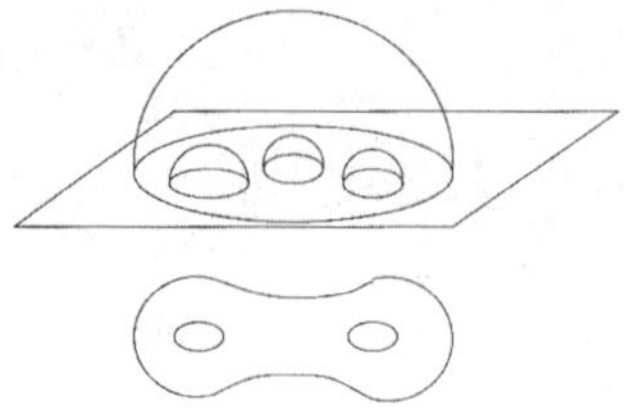

Fig. 8.10. Genus two: fundamental domain in $\mathbb{H}^3$

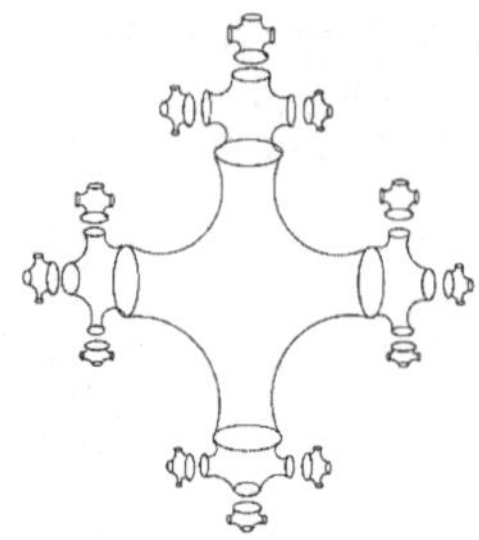

Fig. 8.11. Handlebody of genus two: fundamental domains in $\mathbb{H}^3$

We denote by $\overline{\mathfrak{X}}_\Gamma$ the compactification obtained by adding the conformal boundary at infinity,

$$\overline{\mathfrak{X}}_\Gamma = (\mathbb{H}^3 \cup \Omega_\Gamma)/\Gamma. \tag{8.1.7}$$

In the genus zero case, we just have the sphere $\mathbb{P}^1(\mathbb{C})$ as the conformal boundary at infinity of $\mathbb{H}^3$, thought of as the unit ball in the Poincaré model.

In the genus one case we have a solid torus $\mathbb{H}^3/q^{\mathbb{Z}}$, for $q \in \mathbb{C}^*$ acting as

$$q(z, y) = (qz, |q|y)$$

in the upper half space model, with conformal boundary at infinity the Jacobi uniformized elliptic curve $\mathbb{C}^*/q^{\mathbb{Z}}$.

In this case, the limit set consists of the point $\{0, \infty\}$, the domain of discontinuity is $\mathbb{C}^*$ and a fundamental domain is the annulus $\{|q| < |z| \leq 1\}$ (exterior of two circles).

The relation of Schottky uniformization to the usual Euclidean uniformization of complex tori $X = \mathbb{C}/(\mathbb{Z} + \tau\mathbb{Z})$ is is given by $q = \exp(2\pi i \tau)$.

In the case $g \geq 2$, the limit set Λ_Γ is a Cantor set with an interesting dynamics of the action of Γ. It is the dynamics of the Schottky group on its limit set that generates an interesting noncommutative space.

Geodesics in $\mathfrak{X}_\Gamma$

The hyperbolic handlebody $\mathfrak{X}_\Gamma$ has infinite volume, but it contains a region of finite volume, which is a deformation retract of $\mathfrak{X}_\Gamma$. This is called the "convex core" of $\mathfrak{X}_\Gamma$ and is obtained by taking geodesic hull of the limit set Λ_Γ in $\mathbb{H}^3$ and then the quotient by Γ.

We identify different classes of infinite geodesics in $\mathfrak{X}_\Gamma$:

Closed geodesics: since Γ is purely loxodromic, for all $\gamma \in \Gamma$ there exist two fixed points $\{z^\pm\} \in \mathbb{P}^1(\mathbb{C})$. The geodesics in $\mathbb{H}^3 \cup \mathbb{P}^1(\mathbb{C})$ with ends in at two such points $\{z^\pm\}$, for some $\gamma \in \Gamma$ correspond to closed geodesics in the quotient $\mathfrak{X}_\Gamma$.

Bounded geodesics: the images in $\mathfrak{X}_\Gamma$ of geodesics in $\mathbb{H}^3 \cup \mathbb{P}^1(\mathbb{C})$ having both ends on the limit set Λ_Γ are geodesics which remain confined within the convex core of $\mathfrak{X}_\Gamma$.

Unbounded geodesics: these are the geodesics in $\mathfrak{X}_\Gamma$ that eventually wander off the convex core towards the conformal boundary $X(\mathbb{C})$ at infinity. They correspond to geodesics in $\mathbb{H}^3 \cup \mathbb{P}^1(\mathbb{C})$ with at least one end at a point of Ω_Γ.

In the genus one case, there is a unique primitive closed geodesic, namely the image in the quotient of the geodesic in $\mathbb{H}^3$ connecting 0 and ∞. The bounded geodesics are corresponding to geodesics in $\mathbb{H}^3$ originating at 0 or ∞.

The most interesting case is that of genus $g \geq 2$, where the bounded geodesics form a complicated tangle inside $\mathfrak{X}_\Gamma$. Topologically, this is a generalized solenoid, namely it is locally the product of a line and a Cantor set.

8.1.5 Arakelov geometry and hyperbolic geometry

In this section we describe the result in [Man91] on the relation between the Arakelov Green function on a Riemann surface $X(\mathbb{C})$ with Schottky uniformization and geodesics in the 3-dimensional hyperbolic handlebody $\mathfrak{X}_\Gamma$.

Arakelov Green function

Given a divisor $A = \sum_x m_x(x)$ with support $|A|$ on a compact smooth Riemann surface $X(\mathbb{C})$, and a choice of a positive real-analytic 2-form $d\mu$ on $X(\mathbb{C})$, the Green function $g_{\mu,A} = g_A$ is a real analytic function on $X(\mathbb{C})\backslash|A|$, uniquely determined by the following conditions

Laplace equation: g_A satisfies $\partial\bar{\partial} g_A = \pi i(\deg(A)d\mu - \delta_A)$, with δ_A the δ-current $\varphi \mapsto \sum_x m_x \varphi(x)$.

Singularities: if z is a local coordinate in a neighbourhood of x, then $g_A - m_x \log|z|$ is locally real analytic.

Normalization: g_A satisfies $\int_X g_A d\mu = 0$.

If $B = \sum_y n_y(y)$ is another divisor, such that $|A| \cap |B| = \emptyset$, then the expression $g_\mu(A, B) := \sum_y n_y g_{\mu,A} = (y)$ is symmetric and biadditive in A, B. Generally, such expressions g_μ depends on μ, where the choice of μ is equivalent to the choice of a real analytic Riemann metric on $X(\mathbb{C})$, compatible with the complex structure.

However, in the special case of degree zero divisors, $\deg A = \deg B = 0$, the $g_\mu(A, B) = g(A, B)$ are conformal invariants.

In the case of the Riemann sphere $\mathbb{P}^1(\mathbb{C})$, if w_A is a meromorphic function with $Div(w_A) = A$, we have

$$g(A, B) = \log \prod_{y \in |B|} |w_B(y)|^{n_y} = \mathrm{Re} \int_{\gamma_B} \frac{dw_A}{w_A} \tag{8.1.8}$$

where γ_B is a 1-chain with boundary B.

In the case of degree zero divisors A, B on a Riemann surface of higher genus, the formula (8.1.8) can be generalized replacing the logarithmic differential $\frac{dw_A}{w_A}$ with a differential of the third kind (meromorphic differential with nonvanishing residues) ω_A with purely imagimary periods and residues m_x at x. This gives

$$g(A, B) = \mathrm{Re} \int_{\gamma_B} \omega_A. \tag{8.1.9}$$

Thus one can explicitely compute $g(A, B)$ from a basis of differentials of the third kind with purely imaginary periods.

Cross ratio and geodesics

The basic step in expressing the Arakelov Green function in terms of geodesics in the hyperbolic handlebody $\mathfrak{X}_\Gamma$, is a very simple classical fact of hyperbolic geometry, namely the fact that the cross ratio of four points on $\mathbb{P}^1(\mathbb{C})$ can be expressed in terms of geodesics in the interior $\mathbb{H}^3$:

$$\log |\langle a, b, c, d \rangle| = -\mathrm{ordist}(a * \{c, d\}, b * \{c, d\}). \tag{8.1.10}$$

Here, ordist denotes the oriented distance, and we use the notation $a * \{c, d\}$ to indicate the point on the geodesic $\{c, d\}$ in $\mathbb{H}^3$ with endpoints $c, d \in \mathbb{P}^1(\mathbb{C})$, obtained as the intersection of $\{c, d\}$ with the unic geodesic from a that cuts $\{c, d\}$ at a right angle (cf. Fig.8.12 reproduced here from the lectures [Mar04], Fig.8).

Differentials and Schottky uniformization

The next important step is an explicit construction of a basic of differentials of the third kind with purely imaginary periods for a Riemann surface $X(\mathbb{C}) = \Gamma \backslash \Omega_\Gamma$ with a Schottky uniformization. This construction uses averages over the group Γ of expressions involving the cross ratio

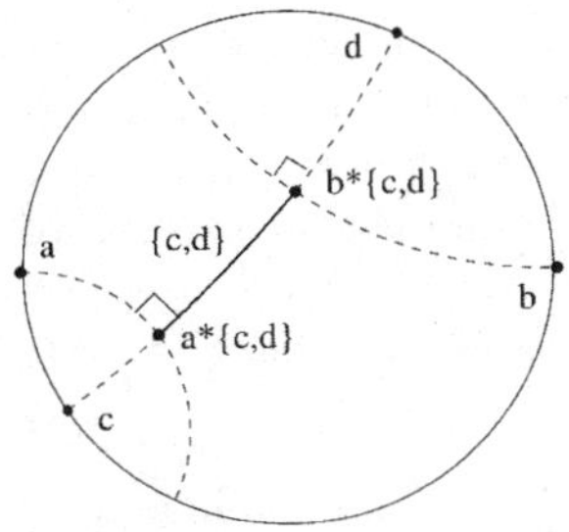

Fig. 8.12. Cross ratio and geodesic length

$$\langle a,b,c,d\rangle := \frac{(a-b)(c-d)}{(a-d)(c-b)}. \tag{8.1.11}$$

Let us denote by $C(|\gamma)$ a set of representatives for $(\rho^{\mathbb{Z}})\backslash\Gamma/(\gamma^{\mathbb{Z}})$, and by $S(\gamma)$ the conjugacy class of γ in Γ.

Let w_A be a meromorphic function on $\mathbb{P}^1(\mathbb{C})$ with divisor $A = (a) - (b)$, such that the support $|A|$ is contained in the complement of an open neighbourhood of Λ_Γ.

For a fixed choice of a base point $z_0 \in \Omega_\Gamma$, the series

$$\nu_{(a)-(b)} = \sum_{\gamma\in\Gamma} d\log\langle a,b,\gamma z,\gamma z_0\rangle \tag{8.1.12}$$

gives the lift to Ω_Γ of a differential of the third kind on the Riemann surface $X(\mathbb{C})$, endowed with the choice of Schottky uniformization. These differentials have residues ± 1 at the images of a and b in $X(\mathbb{C})$, and they have vanishing a_k periods, where $\{a_k, b_k\}_{k=1,\cdots,g}$ are the generators of the homology $X(\mathbb{C})$.

Similarly, we obtain lifts of differentials of the first kind on $X(\mathbb{C})$ by considering the series

$$\omega_\gamma = \sum_{h\in C(|\gamma)} d\log\big\langle hz^+(\gamma), hz^-(\gamma), z, z_0\big\rangle, \tag{8.1.13}$$

where we denote by $\{z^+(\gamma), z^-(\gamma)\} \subset \Lambda_\Gamma$ the pair of the attractive and repelling fixed points of $\gamma \in \Gamma$.

The series (8.1.12) and (8.1.13) converge on compact sets $K \subset \Omega_\Gamma$ whenever $\dim_H \Lambda_\Gamma < 1$. Moreover, they do not depend on the choice of the base point $z_0 \in \Omega_\Gamma$.

In particular, given a choice $\{\gamma_k\}_{k=1}^g$ of generators of the Schottky group Γ, we obtain by the series (8.1.13) a basis of holomorphic differentials that satisfy

$$\int_{a_k} \omega_{\gamma_l} = 2\pi\sqrt{-1}\delta_{kl}. \tag{8.1.14}$$

One can then use a linear combination of the holomorphic differentials ω_{γ_k} to correct the meromorphic differentials $\nu_{(a)-(b)}$ in such a way that the resulting meromorphic differentials have purely imaginary b_k-periods. Let $X_l(a,b)$ be coefficients such that the differentials of the third kind

$$\omega_{(a)-(b)} := \nu_{(a)-(b)} - \sum_l X_l(a,b)\omega_{\gamma_l} \qquad (8.1.15)$$

have purely imaginary periods. The coefficients $X_l(a,b)$ satisfy the system of equations

$$\sum_l X_l(a,b)\omega_{\gamma_l} = \mathrm{Re}\int_{b_k} \nu_{(a)-(b)} = \sum_{h\in S(g_k)} \log|\langle a,b,z^+(h),z^-(h)\rangle|. \quad (8.1.16)$$

Thus one obtains that the Arakelov Green function for $X(\mathbb{C})$ with Schottky uniformization can be computed as

$$g((a)-(b),(c)-(d)) = \sum_{h\in\Gamma} \log|\langle a,b,hc,hd\rangle| -$$

$$\sum_{l=1}^{g} X_l(a,b) \sum_{h\in S(g_l)} \log|\langle z^+(h),z^-(h),c,d\rangle|. \qquad (8.1.17)$$

Notice that this result seems to indicate that there is a choice of Schottky uniformization involved as additional data for Arakelov geometry at arithmetic infinity. However, it was already noticed that the Schottky uniformization is determined by the real structure at least for real Archimedean primes.

Green function and geodesics

One can explicitely express the Green function in terms of geodesics using the formula (8.1.10) together with the obtained expression (8.1.17):

$$g((a)-(b),(c)-(d)) = \sum_{h\in\Gamma} \mathrm{ordist}(a*\{hc,hd\}, b*\{hc,hd\})$$

$$+\sum_{l=1}^{g} X_l(a,b) \sum_{h\in S(g_l)} \mathrm{ordist}(z^+(h)*\{c,d\}, z^-(h),*\{c,d\}). \qquad (8.1.18)$$

The coefficient $X_l(a,b)$ can also be expressed in terms of geodesics, using the equation (8.1.16).

8.2 Cohomological Constructions, Archimedean Frobenius and Regularized Determinants

8.2.1 Archimedean cohomology

Given Deninger's formulae (III.3.12) and (III.3.13) as above, it is natural to ask for a cohomological interpretation of the data $(\mathcal{H}^m, \Phi)$ (see (III.3.14)). A general answer was found by C.Consani in [Cons98], for general arithmetic varieties (in any dimension), giving a cohomological interpretation of the pair $(\mathcal{H}^m, \Phi)$ on Deninger's calculation of the Archimedean L-factors as regularized determinants.

Her construction was motivated by the analogy between geometry at arithmetic infinity and the classical geometry of a degeneration over a disk. She introduced a double complex of differential forms with an endomorphism N representong the "logarithm of the monodromy" around the special fiber at arithmetic infinity, which is modelled on (a resolution of) the complex of *nearby cycles* in the geometric case. The definition of the complex of nearby cycles and of its resolution, on which the following construction is modelled is rather technical. What is easier to visualize geometrically is the related complex of *vanishing cycles* of a geometric degeneration (see Fig.8.13 reproduced here from the lectures [Mar04], Fig.15).

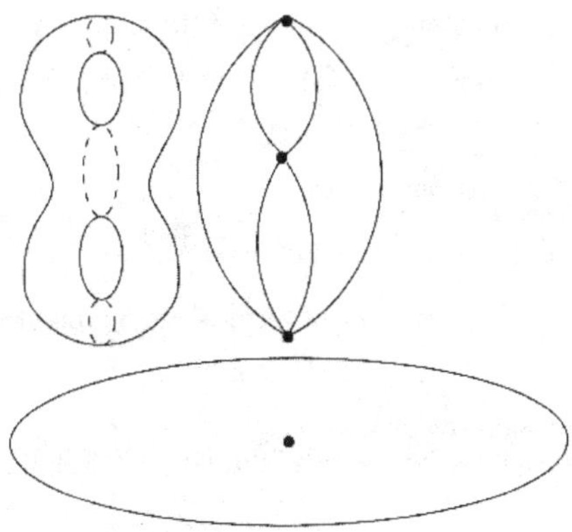

Fig. 8.13. Vanishing cycles

Let us describe here the construction of [Cons98] (see also [CM04b]). One constructs the cohomology theory underlying the data (III.3.14) in several steps.

Let $X = X(\mathbb{C})$ be a complex Kähler manifold.

Step 1: Consider first a doubly infinite graded complex

$$C^{\cdot} = \Omega^{\cdot}(X) \otimes \mathbb{C}[U, U^{-1}] \otimes \mathbb{C}[\hbar, \hbar^{-1}], \tag{8.2.1}$$

where $\Omega^{\cdot}$ is the de Rham complex of differential forms on X, while U and $\hbar$ are formal variables, with U of degree two and $\hbar$ of degree zero.
Let us consider on this complex differentials

$$d'_C := \hbar d, d''_C := \sqrt{-1}(\bar{\partial} - \partial), \tag{8.2.2}$$

with total differential $\delta_C = d'_C + d''_C$.
We also have an inner product

$$\left\langle \alpha \otimes U^r \otimes \hbar^k, \beta \otimes U^s \otimes \hbar^t \right\rangle := \langle \alpha, \eta \rangle \delta_{r,s} \delta_{k,t}, \tag{8.2.3}$$

where $\langle \alpha, \eta \rangle$ is the usual Hodge inner product of forms,

$$\langle \alpha, \eta \rangle = \int_X \alpha \wedge^* C(\bar{\eta}), \tag{8.2.4}$$

with $C(\eta) = (\sqrt{-1})^{p-q}$, for $\eta \in \Omega^{p,q}$.
Step 2: Let us use the Hodge filtration

$$F^p \Omega^m := \oplus_{p'+q=m,\, p' \geq p} \Omega^{p',q}(X) \tag{8.2.5}$$

to define linear subspaces of the complex (8.2.1) of the form

$$\mathfrak{C}^{m,2r} = \bigoplus_{\substack{p+q=m \\ k \geq \max\{0, 2r+m\}}} F^{m+r-k} \Omega^m(X) \otimes U^r \otimes \hbar^k \tag{8.2.6}$$

and the $\mathbb{Z}$-graded vector space

$$\mathfrak{C}^{\cdot} = \bigoplus_{\cdot = 2r+m} \mathfrak{C}^{m,2r}. \tag{8.2.7}$$

Step 3: Let us pass to a real vector space by considering

$$\mathfrak{T}^{\cdot} = (\mathfrak{C}^{\cdot})^{c=id} \tag{8.2.8}$$

where c denotes complex conjugation.
In terms of the intersection of the Hodge filtrations

$$\gamma^{\cdot} = F^{\cdot} \cap \bar{F}^{\cdot} \tag{8.2.9}$$

hence

$$\mathfrak{T}^{\cdot} = \bigoplus_{\cdot = 2r+m} \mathfrak{T}^{m,2r}, \tag{8.2.10}$$

where

$$\mathfrak{T}^{m,2r} = \bigoplus_{\substack{p+q=m \\ k\geq \max\{0,2r+m\}}} \gamma^{m+r-k}\Omega^m(X)\otimes U^r \otimes \hbar^k. \tag{8.2.11}$$

The $\mathbb{Z}$-graded complex vector space $\mathfrak{C}^{\cdot}$ is a subcomplex of $C^{\cdot}$ with respect to the differential d'_C and for $P^{\perp}$ the orthogonal projection onto $\mathfrak{C}^{\cdot}$ in the inner product (8.2.8), one obtains a second differential $d'' = P^{\perp}d''_C$. Similarly, $d' = d'_C$ and $d'' = P^{\perp}d''_C$ define differentials on the $\mathbb{Z}$-graded real vector space $\mathfrak{T}^{\cdot}$ in terms of the corresponding cutoffs on the indices of the complex $C^{\cdot}$. For

$$\Lambda_{p,q} = \{(r,k)\in\mathbb{Z}^2 \mid k\geq \kappa(p,q,r)\} \tag{8.2.12}$$

with

$$\kappa(p,q,r) := \max\left\{0, 2r+m, \frac{|p-q|+2r+m}{2}\right\} \tag{8.2.13}$$

(see Fig.8.14 reproduced here from the lectures [Mar04], Fig.16), we iden-

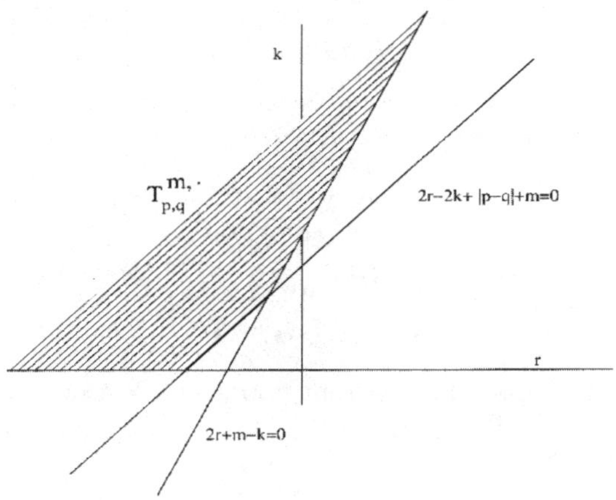

Fig. 8.14. Cutoffs defining the complex at arithmetic infinity

tify $\mathfrak{T}^{\cdot}$ as a real vector space with the span

$$\mathfrak{T}^{\cdot} = \mathbb{R}\langle \alpha\otimes U^r\otimes\hbar^k\rangle \tag{8.2.14}$$

where $(r,k)\in\Lambda_{p,q}$ for $\alpha=\xi+\bar{\xi}$ with $\xi\in\Omega^{p,q}$.

Operators

The complex $(\mathfrak{T}^{\cdot},\delta)$ has interesting structures given by the action of certain linear operators.

We have the operators N and Φ that correspond to the "logarithm of the monodromy" and the "logarithm of Frobenius". These are of the form

$$N = U\hbar, \Phi = -U\frac{\partial}{\partial U} \tag{8.2.15}$$

and they satisfy $[N, d'] = [N, d''] = 0$ and $[\Phi, d'] = [\Phi, d''] = 0$, hence they induce operators in cohomology.

Moreover, there is another important operator, which corresponds to the Lefschetz operator on forms,

$$\mathbb{L} : \eta \otimes U^r \otimes \hbar^k \mapsto \eta \wedge \omega \otimes U^{r-1} \otimes \hbar^k, \tag{8.2.16}$$

where ω is the Kähler form on the manifold X. This satisfies $[\mathbb{L}, d'] = [\mathbb{L}, d''] = 0$, and also descends on the cohomology.

The pairs of operators N and Φ or $\mathbb{L}$ and Φ satisfy interesting commutation relations

$$[\Phi, N] = -N, [\Phi, \mathbb{L}] =, \mathbb{L}$$

that can be viewed as an action of the ring of differential operators

$$\mathbb{C}[P, Q]/(PQ - QP = Q).$$

SL(2, $\mathbb{R}$) representations

Another important ingredient of the structure of the complex $(\mathfrak{T}^{\cdot}, \delta)$ are two involutions

$$S : \alpha \otimes U^r \otimes \hbar^{2r+m+l} \mapsto \alpha \otimes U^{-(r+m)} \otimes \hbar^l \tag{8.2.17}$$
$$\tilde{S} : \alpha \otimes U^r \otimes \hbar^k \mapsto C(*\alpha) \otimes U^{r-(n-m)} \otimes \hbar^k \tag{8.2.18}$$

These maps, together with the nilpotent operators N and $\mathbb{L}$ define two representations, σ^L and σ^R, of the group SL(2, $\mathbb{R}$), given explicitely on the following generators

$$\begin{aligned} \nu(s) &= \begin{pmatrix} s & 0 \\ 0 & s^{-1} \end{pmatrix}, s \in \mathbb{R}^* \\ u(t) &= \begin{pmatrix} 1 & t \\ 0 & 1 \end{pmatrix}, t \in \mathbb{R} \\ w &= \begin{pmatrix} 0 & -1 \\ 1 & 0 \end{pmatrix}, \end{aligned} \tag{8.2.19}$$

namely

$$\begin{aligned} \sigma^L(\nu(s)) &= s^{-n+m}, \sigma^R(\nu(s)) = s^{2r+m}, \\ \sigma^L(u(t)) &= \exp(t\mathbb{L}), \sigma^R(u(t)) = \exp(tN), \\ \sigma^L(w) &= (\sqrt{-1})^n C\tilde{S}, \sigma^R(w) = CS. \end{aligned} \tag{8.2.20}$$

Of these representations, σ^L extends to an action by bounded operators on the Hilbert completion of $\mathfrak{T}^{\cdot}$ in the inner product (8.2.4), while the action of the subgroup $\nu(s), s \in \mathbb{R}^*$ of $\mathrm{SL}(2,\mathbb{R})$ via the representation σ^R on the Hilbert space is by unbounded densly defined operators.

8.2.2 Local factor and Archimedean cohomology

C.Consani showed in [Cons98] that the data $(\mathcal{H}^m, \Phi)$ of (III.3.14) can be identified with

$$(\mathbb{H}(\mathfrak{T}^{\cdot}, \delta)^{N=0}, \Phi),$$

where $(\mathbb{H}(\mathfrak{T}^{\cdot}, \delta)$ is the hypercohomology (the cohomology with respect to the total differential) of the complex $\mathfrak{T}^{\cdot}$, and $\mathbb{H}(\mathfrak{T}^{\cdot}, \delta)^{N=0}$ is the kernel of the map induced by N on cohomology. The operator Φ is the one induced on cohomology by that of (8.2.15). She called $\mathcal{H}^m \cong \mathbb{H}(\mathfrak{T}^{\cdot}, \delta)^{N=0}$ the *Archimedean cohomology*. It was also shown in [Cons98], that this cohomology groups can be obtained as a piece of the cohomology of the cone of the monodromy N. This is the complex

$$\mathrm{Cone}(N) = \mathfrak{T}^{\cdot} \oplus \mathfrak{T}^{\cdot}[+1] \tag{8.2.21}$$

with differential

$$D = \begin{pmatrix} \delta & -N \\ 0 & \delta \end{pmatrix}.$$

The complex (8.2.21) inherits a positive definite inner product from $\mathfrak{T}^{\cdot}$, which descends to cohomology. The representation σ^L of $\mathrm{SL}(2,\mathbb{R})$ on $\mathfrak{T}^{\cdot}$ induces a representation on $\mathrm{Cone}(N)^{\cdot}$. The corresponding infinitesimal representation $d\sigma^L : \mathfrak{g} \to \mathrm{End}(\mathfrak{T}^{\cdot})$ of the Lie algebra $\mathfrak{g} = \mathfrak{sl}(2,\mathbb{R})$ extends to a representation of the universal enveloping algebra $U(\mathfrak{g})$ on $\mathfrak{T}^{\cdot}$ and to a representation on $\mathrm{Cone}(N)^{\cdot}$. This gives a representation in the algebra of bounded operators on the Hilbert completion of $\mathrm{Cone}(N)^{\cdot}$ under the inner product.

Theorem 8.2. *The triple*

$$(\mathcal{A}, \mathcal{H}, \mathcal{D}) = (U(\mathfrak{g}), \mathbb{H}^{\cdot}(\mathrm{Cone}(N)^{\cdot}), \Phi)$$

has the properties that $\mathcal{D} = \mathcal{D}^*$,and that $(1 + \mathcal{D}^2)^{-1/2}$ is a compact operator. The commutators $[D, a]$ are bounded operators for all $a \in U(\mathfrak{g})$, and the triple is 1^+-summable.

Thus, $(\mathcal{A}, \mathcal{H}, \mathcal{D})$ has most of the properties of a spectral triple confirming the fact that the logarithm of Frobenius Φ should be thought of as a Dirac operator. However, we are not dealing here with an involutive subalgebra of a C^*-algebra.

In any case, the structure is sufficient to consider zeta functions for this "spectral triple". In particular, we can recover the alternating products of the local L-factors at infinity from a zeta function of the spectral triple.

Theorem 8.3. *Consider the zeta-function*

$$\zeta_{a,\Phi}(z) = \mathrm{Tr}(a|\Phi|^{-s})$$

with $a = \sigma^L(w)$. This gives

$$\det_{\infty,\sigma_L(w),\Phi}(s) = \prod_{m=0}^{2n} L(H^m(X), s)^{(-1)^m}$$

8.2.3 Cohomological constructions

In the rest of the section we give, following [CM], some explanantions of a cohomological theory for the Archimedean fiber of an Arakelov surface. However, there is no need to restrict to the case $\dim X = 1$, because the general theory, valid for any arithmetic variety, was developed in [Cons98] (see also an interesting discussion of the Archimedean cohomology in [Mar04], Chapter 3). This construction provides an alternative definition and a refinement for the *Archimedean cohomology* H^*_{ar} introduced by Deninger in [Den91].

The cohomology spaces $H^{\cdot}(\tilde{X}^*)$ are *infinite dimensional* real vector spaces endowed with a monodromy operator N and an endomorphism Φ. The cohomology group $H^{\cdot}(\tilde{X}^*)$ is in fact the same as $\mathbb{H}(\mathbb{T}^{\cdot}, \delta)|_{U=(2\pi i)^{-1}}$ in the previous sections. The groups H^*_{ar} can be identified with the subspace of the N-invariants (i.e. Ker(N)) over which (the restriction of) Φ acts in the following way. The monodromy operator determines an integer, even graduation on

$$H^{\cdot}(\tilde{X}^*) = \bigoplus_{p\in\mathbb{Z}} gr^{\omega}_{2p} H^{\cdot}(\tilde{X}^*),$$

where each graded piece is still *infinite dimensional.* We will refer to it as to the *weight graduation.* This graduation induces a corresponding one on the subspace

$$H^{\cdot}(\tilde{X}^*)^{N=0} := \bigoplus_{\cdot\geq 2p} gr^{\omega}_{2p} H^{\cdot}(X^*).$$

The summands $gr^{\omega}_{2p} H^{\cdot}(X^*)$ are *finite dimensional* real vector spaces on which Φ acts as a multiplication by the weight p.

When $X_{/\kappa}$ is a non-singular, projective curve defined over $\kappa = \mathbb{C}$ or $\mathbb{R}$, the description of $gr^{\omega}_{2p} H^{\cdot}(X^*)$ $(\cdot \geq 2p)$ is particularly easy. For $\kappa = \mathbb{C}$, this space coincides with the de Rham cohomology $H^*_{DR}(X_{/\mathbb{C}}, \mathbb{R})$ of the Riemann surface $X_{/\mathbb{C}}$.

A motivation for the definition of these complexes comes from the classical theory of mixed Hodge structures for an *algebraic degeneration over a disk* (and its arithmetical counterpart, the theory of Frobenius weights). The notation: $H^{\cdot}(\tilde{X}^*)$, $H^{\cdot}(X^*)$, $H^{\cdot}(Y)$ followed in this section is purely formal. Namely, $\tilde{X}^*$, X^* and Y are only symbols although this choice is motivated by the analogy with Steenbrink's construction in [Stee76], in which $\tilde{X}^*$, X^* and Y describe resp. the geometric generic fiber and the complement of the special fiber Y in the model. The space $H^{\cdot}(\tilde{X}^*)$ is the hypercohomology group of a double complex $K^{\cdot,\cdot}$ of real, differential twisted forms on which one defines an additional structure of polarized *Lefschetz module*.

The whole theory is inspired by the expectation that the fibers at infinity of an arithmetic variety should be thought to be semi-stable and more specifically to be "maximally degenerate or totally split", cf. a discussion in §III.2.3. It would natural to think that the construction of the complex $K^{\cdot,\cdot}$ on the Riemann surface $X_{/\kappa}$, whose structure and behavior gives the arithmetical information related to the "mysterious" fibers at infinity of an arithmetic surface, fits in with Arakelov's intuition that Hermitian geometry on $X_{/\kappa}$ is enough to recover the intersection geometry on the fibers at infinity.

8.2.4 Zeta function of the special fiber and Reidemeister torsion

In this paragraph we explain, following §3.5 of [CM] that in the case $\dim X = 1$, the expression of Theorem 8.3 can be interpreted as a *Reidemeister torsion*, and it is related to a zeta function for the fiber at arithmetic infinity.

We begin by giving the definition of a zeta function of the special fiber of a semistable fibration, which motivates the analogous notion at arithmetic infinity.

Let X be a regular, proper and flat scheme over $\mathrm{Spec}(\Lambda)$, for Λ a discrete valuation ring with quotient field K and finite residue field k. Assume that X has geometrically reduced, connected and one-dimensional fibers. Let us denote by η and v resp. the generic and the closed point of $\mathrm{Spec}(\Lambda)$ and by $\bar{\eta}$ and $\bar{v}$ the corresponding geometric points. Assume that the special fiber X_v of X is a connected, effective Cartier divisor with reduced normal crossings defined over $k = k(v)$. This degeneration is sometime referred to as a *semistable* fibration over $\mathrm{Spec}(\Lambda)$.

Let N_v denote the cardinality of k. Then, define the zeta-function of the special fiber X_v as follows (u is an indeterminate)

$$Z_{X_v}(u) = \frac{P_1(u)}{P_0(u)P_2(u)}, \qquad P_i(u) = \det(1 - f^*u \mid H^i(X_{\bar{\eta}}, \mathbb{Q}_\ell)^{I_{\bar{v}}}), \quad (8.2.22)$$

where f^* is the geometric Frobenius i.e. the map induced by the Frobenius morphism $f : X_{\bar{v}} \to X_{\bar{v}}$ on the cohomological inertia-invariants at $\bar{v}$.

The polynomials $P_i(u)$ are closely related to the characteristic polynomials of the Frobenius $F_i(u) = \det(u \cdot 1 - f^* \mid H^i(X_{\bar{\eta}}, \mathbb{Q}_\ell)^{I_{\bar{v}}})$ through the formula

$$P_i(u) = u^{b_i} F_i(u^{-1}), \qquad b_i = \mathrm{degree}(\mathrm{F_i}). \quad (8.2.23)$$

The zeta function $Z_{X_v}(u)$ generalizes on a semistable fiber the description of the *Hasse-Weil zeta function* of a smooth, projective curve over a finite field.

Based on this construction we make the following definition for the fiber at an Archimedean prime of an arithmetic surface:

$$Z_{\Phi}(u) := \frac{P_1(u)}{P_0(u)P_2(u)}, \tag{8.2.24}$$

where we set

$$P_q(u) := \det_{\infty}\left(\frac{1}{2\pi} - u\frac{\Phi_q}{2\pi}\right), \tag{8.2.25}$$

with $\Phi_q = \Phi|_{H^q(\tilde{X}^*)^{N=0}}$.

In order to see how this is related to the result of Theorem 8.3, we recall briefly a simple observation of Milnor (cf. §3 [Mil68]). Suppose given a finite complex L and an infinite cyclic covering $\tilde{L}$, with $H_*(\tilde{L},\kappa)$ finitely generated over the coefficient field κ. Let $h : \pi_1 L \to \kappa(s)$ be the composition of the homomorphism $\pi_1 L \to \Pi$ associated to the cover with the inclusion $\Pi \subset \mathrm{Units}(\kappa(s))$. The Reidemeister torsion for this covering is given (up to multiplication by a unit of $\kappa\Pi$) by the alternating product of the characteristic polynomials $F_q(s)$ of the κ–linear map

$$s_* : H_q(\tilde{L},\kappa) \to H_q(\tilde{L},\kappa),$$

$$\tau(s) \simeq F_0(s)F_1(s)^{-1}F_2(s)\cdots F_n(s)^{\pm 1}. \tag{8.2.26}$$

Moreover, for a map $T : L \to L$, let $\zeta_T(u)$ be the Weil zeta

$$\zeta_T(u) = P_0(u)^{-1}P_1(u)P_2(u)^{-1}\cdots P_n(u)^{\mp 1},$$

where the polynomials $P_q(u)$ of the map T_* are related to the characteristic polynomials $F_q(s)$ by (8.2.23) and b_q are the q-the Betti number of the complex L. By analogy with (8.2.26), Milnor writes the Reidemeister torsion $\tau_T(s)$ (up to multiplication by a unit) as

$$\tau_T(s) := F_0(s)F_1(s)^{-1}F_2(s)\cdots F_n(s)^{\pm 1},$$

where $F_q(s)$ are the characteristic polynomials of the map T_*. Then the relation between zeta function and Reidemeister torsion is given by:

$$\zeta_T(s^{-1})\tau_T(s) = s^{\chi(L)}, \tag{8.2.27}$$

where $\chi(L)$ is the Euler characteristic of L.

Similarly, we can derive the relation between the zeta function of the fiber at infinity defined as in (8.2.24) and the alternating product of Gamma factors. Namely, we write

$$\frac{L_{\mathbb{C}}(H^1(X_{/\mathbb{C}},\mathbb{C}),s)}{L_{\mathbb{C}}(H^0(X_{/\mathbb{C}},\mathbb{C}),s)\cdot L_{\mathbb{C}}(H^2(X_{/\mathbb{C}},\mathbb{C}),s)} = \frac{F_0(s)\cdot F_2(s)}{F_1(s)}, \tag{8.2.28}$$

where we set

$$F_q(s) := \det_\infty \left(\frac{s}{2\pi} - \frac{\Phi_q}{2\pi} \right), \tag{8.2.29}$$

with $\Phi_q = \Phi|_{H^q(\tilde{X}^*)^{N=0}}$. For this reason we may regard (8.2.28) as the Reidemeister torsion of the fiber at arithmetic infinity:

$$\tau_\Phi(s) := \frac{F_0(s) \cdot F_2(s)}{F_1(s)}. \tag{8.2.30}$$

The relation between zeta function and Reidemeister torsion is then given as follows.

Proposition 8.4. *The zeta function Z_Φ of (8.2.24) and the Reidemeister torsion τ_Φ of (8.2.30) are related by*

$$Z_\Phi(s^{-1})\tau_\Phi(s) = s^{g-2} e^{\chi\, s \log s},$$

with g is the genus of the Riemann surface $X_{/\mathbb{C}}$ and $\chi = 2 - 2g$ its Euler characteristic.

Indeed, the result follows by a simple direct calculation of the regularized determinants. Namely, we compute (in the case $q = 0, 1$)

$$\begin{aligned}
P_q(u) &= \det_\infty \left(\frac{1}{2\pi} - u\frac{\Phi_q}{2\pi} \right) = \exp\left(-b_q \frac{d}{dz}((2\pi)^z \sum_{n \geq 0} (1+un)^{-z})|_{z=0} \right) \\
&= \exp\left(b_q \left(\log \Gamma\left(\frac{1}{u}\right) + \frac{\log 2\pi}{u} + \frac{\log u}{2} + \frac{\log u}{u} \right)\right) \\
&= u^{b_q/2} e^{-b_q \frac{\log u}{u}} (2\pi)^{-1/u} \Gamma(1/u),
\end{aligned}$$

where b_q are the Betti numbers of $X_{/\mathbb{C}}$. The case $q = 2$ is analogous, but for the presence of the $+1$ eigenvalue in the spectrum of Φ_2, hence we obtain

$$P_2(u) = \exp\left(-b_2 \frac{d}{dz} \left((2\pi)^z u^{-z} \zeta(1/u, z) - \left(\frac{1}{u} - 1\right)^{-z} \right)_{z=0} \right)$$

$$= \Gamma_{\mathbb{C}} \left(\frac{1}{u} - 1 \right)^{-1} u^{-3/2} e^{\frac{\log u}{u}}.$$

Thus, we obtain

$$Z_\Phi(s^{-1}) = \frac{L_{\mathbb{C}}(H^0(X_{/\mathbb{C}}, \mathbb{C}), s) \cdot L_{\mathbb{C}}(H^2(X_{/\mathbb{C}}, \mathbb{C}), s)}{L_{\mathbb{C}}(H^1(X_{/\mathbb{C}}, \mathbb{C}), s)} s^{g-2} e^{\chi\, s \log s}.$$

8.3 Spectral Triples, Dynamics and Zeta Functions

We have seen that in the Arakelov theory a completion of an arithmetic surface is achieved by enlarging the group of divisors by formal linear combinations of the "closed fibers at infinity". For an arithmetic surface these fibers were described in [Man91] as follows: the dual graph of any such closed fiber has the form of an infinite tangle of bounded geodesics in a hyperbolic handlebody endowed with a Schottky uniformization.

This Section is based on Sections 4, 5 and 6 of [CM], and on [CM04a]. We describe an alternative construction of cohomological spectral data (A, H, Φ), which is related to description in [Man91] of the dual graph of the fiber at infinity. We use a geometric model for the dual graph as the mapping torus of a dynamical system T on a Cantor set. We consider a noncommutative space which describes the action of the Schottky group on its limit set and parameterizes the "components of the closed fiber at infinity". This can be identified with a *Cuntz–Krieger algebra* $\mathcal{O}_A$. We describe a spectral triple for this noncommutative space, via a representation on the cochains of a "dynamical cohomology", defined in terms of the tangle of bounded geodesics in the handlebody. In the same way as for the Archimedean cohomology of the previous section (§8.2), the Dirac operator agrees with the grading operator Φ, that represents the "logarithm of a Frobenius–type operator" on the Archimedean cohomology. In fact, the Archimedean cohomology embeds in the dynamical cohomology, compatibly with the action of a real Frobenius $\bar{F}_\infty$, so that the local factor can again be recovered from these data. The duality isomorphism on the cohomology of the cone of N corresponds to the pairing of dynamical homology and cohomology. This suggests the existence of a duality between the monodromy N and the dynamical map $1 - T$.

In noncommutative geometry, the notion of a spectral triple provides the correct generalization of the classical structure of a Riemannian manifold. The two notions agree on a commutative space. In the usual context of Riemannian geometry, the definition of the infinitesimal element ds on a smooth spin manifold can be expressed in terms of the inverse of the classical Dirac operator D. This is the key remark that motivates the theory of spectral triples. In particular, the geodesic distance between two points on the manifold is defined in terms of D^{-1} (cf. [Co94] §VI). The spectral triple that describes a classical Riemannian spin manifold is (A, H, D), where A is the algebra of complex valued smooth functions on the manifold, H is the Hilbert space of square integrable spinor sections, and D is the classical Dirac operator (a square root of the Laplacian). These data determine completely and uniquely the Riemannian geometry on the manifold. It turns out that, when expressed in this form, the notion of spectral triple extends to more general non-commutative spaces, where the data (A, H, D) consist of a C^*-algebra A (or more generally of a smooth subalgebra of a C^*-algebra) with a representation as bounded operators on a Hilbert space H, and an operator D on H that verifies the main properties of a Dirac operator. The notion of smoothness is determined by D: the smooth elements of A are defined by the intersection of domains of powers of the derivation given by commutator with $|D|$. The basic geometric structure encoded by the theory of spectral triples is Riemannian geometry, but in more refined cases, such as Kähler geometry, the additional structure can be easily encoded as additional symmetries.

In the constructions of this chapter, the Dirac operator D is obtained from the *grading operator* associated to a filtration on the cochains of the complex that computes the dynamical cohomology. The induced operator on the subspace identified with the Archimedean cohomology agrees with the "logarithm of Frobenius" of [Cons98] and [Den91].

This structure further enriches the geometric interpretation of the Archimedean cohomology, giving it the meaning of spinors on a noncommutative manifold, with the logarithm of Frobenius introduced in [Den91] in the role of the Dirac operator.

An advantage of this construction is that a completely analogous formulation exists in the case of Mumford curves. This provides a unified description of the Archimedean and totally split degenerate fibers of an arithmetic surface.

Let $\mathfrak{X}$ be an arithmetic surface defined over $\mathrm{Spec}(\mathbb{Z})$ (or over $\mathrm{Spec}(\mathcal{O}_{\mathbb{K}})$, for a number field $\mathbb{K}$), having the smooth algebraic curve $X_{/\mathbb{Q}}$ as its generic fiber. Let p be a finite prime where $\mathfrak{X}$ has totally split degenerate reduction. Then, the completion $\hat{X}_p$ at p of the generic fiber of $\mathfrak{X}$ is a split-degenerate stable curve over $\mathbb{Q}_p$ (also called a Mumford curve) uniformized by the action of a p-adic Schottky group Γ. The dual graph of the reduction of $\hat{X}_p$ coincides with a finite graph obtained as the quotient of a tree Δ_Γ by the action of Γ.

The curve $\hat{X}_p$ is holomorphically isomorphic to a quotient of a subset of the ends of the Bruhat-Tits tree Δ of $\mathbb{Q}_p$ by the action of Γ. Thus, in this setting, the Bruhat-Tits tree at p replaces the hyperbolic space $\mathbb{H}'$ "at infinity", and the analog of the tangle of bounded geodesics in $\mathfrak{X}_\Gamma$ is played by doubly infinite walks in Δ_Γ/Γ.

In analogy with the Archimedean construction, the dynamical system $(\mathcal{W}(\Delta_\Gamma/\Gamma), T)$ is described in Section 8.3.8, where T is an invertible shift map on the set $\mathcal{W}(\Delta_\Gamma/\Gamma)$ of doubly-infinite walks on the graph Δ_Γ/Γ. The first cohomology group $H^1(\mathcal{W}(\Delta_\Gamma/\Gamma)_T, \mathbb{Z})$ of the mapping torus $\mathcal{W}(\Delta_\Gamma/\Gamma)_T$ of T inherits a natural filtration using which a dynamical cohomology group was introduced. One has in then a *Cuntz-Krieger graph algebra* $C^*(\Delta_\Gamma/\Gamma)$ and we can construct a spectral triple as in the case at infinity. The *Dirac operator* is related to the grading operator Φ that computes the local factor as a regularized determinant, as in [Den01], [Den94]. In [CM03], a possible way was suggested of extending such construction to places that are not of split degenerate reduction, inspired by the "foam space" construction of [Man72b] and [CM04a]. Notice however, unlike the local factor at infinity, the factor at the non-Archimedean places involves the full spectrum of D and not just its positive or negative part. It is believed that this difference should correspond to the presence of an underlying geometric space based on loop geometry, which manifests itself as loops at the non-Archimedean places and as "half loops" (holomorphic disks) at arithmetic infinity.

There is another important difference between the Archimedean and non-Archimedean cases. At the Archimedean prime the local factor is described in terms of zeta functions for a Dirac operator D (cf. [CM], [Den91]). On the other hand, at the non-Archimedean places, in order to get the correct normalization as in [Den94], we need to introduce a rotation of the Dirac operator by the imaginary unit, $D \mapsto iD$. This rotation corresponds to the so called Wick rotation that moves poles on the real line to poles on the imaginary line (zeroes for the local factor) and appears to be a manifestation of a rotation from Minkowskian to Euclidean signature $it \mapsto t$, as already remarked in ([Man95], p.135): *"imaginary time motion" may be held responsible for the fact that zeroes of $\Gamma(s)^{-1}$ are purely real whereas the zeroes of all non-Archimedean Euler factors are purely imaginary.* It is expected, therefore, that a more refined construction would involve a version of spectral triples for Minkowskian signature*).

8.3.1 A dynamical theory at infinity

Let us describe some dynamical theory tools used in constructions of Deninger-style cohomology spaces at arithmetical infinity. We explain that these spaces can be used in order to describe *Gamma*-factors of an arithmetic surface as certain zeta-reguarized determinants and a "reduction modulo ∞" by means of Noncommutative geometry (the theory of spectral triples) in Section 8.4.

Since the uniformizing group Γ is a free group, there is a simple way of obtaining a coding of the bounded geodesics in the handlebody $\mathfrak{X}_\Gamma$ (defined by (8.1.6)) The set of such geodesic can be identified with $\Lambda_\Gamma \times_\Gamma \Lambda_\Gamma$, by specifying the endpoints in $\mathbb{H}^3 \cup \mathbb{P}^1(\mathbb{C})$ modulo the action of Γ.

Given a choice of a generators $\{\gamma_i\}_{i=1}^g$ for Γ, there is a bijection between the elements of Γ and the set of all admissible walks in the *Cayley graph* of Γ, namely reduced words in the $\{\gamma_i\}_{i=1}^{2g}$, where we use the notation $\gamma_{i+g} := \gamma_i^{-1}$, for $i = 1, \dots, g$.

In the following we consider the sets $\mathcal{S}^+$ and $\mathcal{S}$ of resp. right-infinite, doubly infinite admissible sequences in the $\{\gamma_i\}_{i=1}^{2g}$:

$$\mathcal{S}^+ = \{a_0 a_1 \dots a_\ell \dots \; |a_i \in \{\gamma_i\}_{i=1}^{2g}, \; a_{i+1} \neq a_i^{-1}, \forall i \in \mathbb{N}\}, \tag{8.3.1}$$

$$\mathcal{S} = \{\dots a_{-m} \dots a_{-1} a_0 a_1 \dots a_\ell \dots \; |a_i \in \{\gamma_i\}_{i=1}^{2g}, \; a_{i+1} \neq a_i^{-1}, \forall i \in \mathbb{Z}\}. \tag{8.3.2}$$

According to B.Mazur [Maz2000], "Archimedean and non-Archimedean phenomena remain, at bottom, puzzingly different. Perhaps the next century will see a more profound understanding of the relation between these..."

The admissibility condition simply means that we only allow "reduced" words in the generators, without cancellations.

On the space $\mathcal{S}$ we consider the topology generated by the sets $W^s(x,\ell) = \{y \in \mathcal{S} | x_k = y_k, k \geq \ell\}$, and $W^u(x,\ell) = \{y \in \mathcal{S} | x_k = y_k, k \leq \ell\}$ for $x \in \mathcal{S}$ and $\ell \in \mathbb{Z}$.

There is a two-sided *shift operator* T acting on $\mathcal{S}$ as the map

$$\begin{aligned} T(\dots \; a_{-m} \; \dots a_{-1}\, a_0\, a_1 \dots \; a_\ell \; \dots) &= \\ \dots a_{-m+1} \dots \; a_0 \;\; a_1 \; a_2 \dots a_{\ell+1} \dots \end{aligned} \tag{8.3.3}$$

Then we can pass from the descrete dynamical system $(\mathcal{S}, T)$ to its suspension flow and obtain the mapping torus.

The following topological space is defined in terms of the *Smale space* $(\mathcal{S}, T)$ and will be considered as a geometric realization of the "dual graph" associated to the fiber at arithmetic infinity of the arithmetic surface $\mathfrak{X}$.

Definition 8.5. *The mapping torus (suspension flow) of the dynamical system $(\mathcal{S}, T)$ is defined as*

$$\mathcal{S}_T := \mathcal{S} \times [0,1]/(x,0) \sim (Tx, 1) \tag{8.3.4}$$

Topologically, this space is a solenoid, that is, a bundle over S^1 with fiber a Cantor set.

8.3.2 Homotopy quotion

The space $\mathcal{S}_T$ is a very natural space associated to the *noncommutative space*

$$\Lambda_\Gamma \times_\Gamma \Lambda_\Gamma \simeq \mathcal{S}/\mathbb{Z}, \tag{8.3.5}$$

where Λ_Γ is the *limit set of the action* of Γ, and $\mathbb{Z}$ is acting in (8.3.5) via the invertible shift T of (8.3.3). This space is given by given by the C^*-algebra

$$A = C(\mathcal{S}) \rtimes_T \mathbb{Z} \tag{8.3.6}$$

describing the action of the shift T on the totally disconnected space $\mathcal{S}$. One obtains the *homotopy quotient* (cf. [BaCo], [Co83]),

$$\mathcal{S}_T = \mathcal{S} \times_{\mathbb{Z}} \mathbb{R}. \tag{8.3.7}$$

This is a *commutative space* that provides, up to homotopy, a geometric model for the *noncommutative space* (8.3.5), where the noncommutative space (8.3.5) can be identified as in §8.4.1 with the *quotient space of a foliation* (8.4.3) whose generic leaf is contractible (a copy of $\mathbb{R}$).

In order to study such spaces, K–theory tools are used.

The C^*–algebra $C(\mathcal{S})$ is a commutative *AF–algebra* (approximately finite dimensional), obtained as the direct limit of the finite dimensional commutative C^*–algebras generated by characteristic functions of a covering of $\mathcal{S}$.

There is the *Pimsner–Voiculescu exact sequence* (cf. [PV80]) which has the form

$$\begin{array}{ccccc} K_1(A) & \longrightarrow K_0(C(\mathcal{S})) & \stackrel{\delta=1-T_*}{\longrightarrow} & K_0(C(\mathcal{S})) \\ \uparrow & & & \downarrow \\ K_1(C(\mathcal{S})) & \longleftarrow K_1(C(\mathcal{S})) & \longleftarrow & K_0(A) \end{array} \tag{8.3.8}$$

where $A = C(\mathcal{S}) \rtimes_T \mathbb{Z}$. Here since the space is totally disconnected, $K_0(C(\mathcal{S})) \cong C(\mathcal{S}, \mathbb{Z})$, being the direct limit of the K_0-groups of the finite dimensional commutative C^*–algebras, and $K_1(C(\mathcal{S})) = 0$ for the same reason (locally constant integer valued functions). Then the exact sequence becomes

$$\begin{aligned} 0 \to K_1(C(\mathcal{S}) \rtimes_T \mathbb{Z}) \to C(\mathcal{S}, \mathbb{Z}) \stackrel{\delta=1-T_*}{\to} C(\mathcal{S}, \mathbb{Z}) \\ \to K_0(C(\mathcal{S}) \rtimes_T \mathbb{Z}) \to 0, \end{aligned} \tag{8.3.9}$$

with $K_0(C(\mathcal{S}) \rtimes_T \mathbb{Z}) \cong C(\mathcal{S}, \mathbb{Z})_T$. Since the shift T is *topologically transitive*, i.e. it has a dense orbit, we also have $K_1(C(\mathcal{S}) \rtimes_T \mathbb{Z}) \cong C(\mathcal{S}, \mathbb{Z})^T \cong \mathbb{Z}$.

In dynamical system language, these are respectively the *invariants* and *coinvariants* of the invertible shift T (cf. [BoHa], [PaTu82]).

In terms of the homotopy quotient, one can describe this exact sequence more geometrically in terms of the *Thom isomorphism* and the μ-map

$$\mu : K^{*+1}(\mathcal{S}_T) \cong H^{*+1}(\mathcal{S}_T, \mathbb{Z}) \to K_*(C(\mathcal{S}) \rtimes_T \mathbb{Z}). \tag{8.3.10}$$

Thus we obtain

$$\begin{aligned} K_1(A) &\cong H^0(\mathcal{S}_T) = \mathbb{Z} \\ K_0(A) &\cong H^1(\mathcal{S}_T) \end{aligned}$$

The cohomology group $H^1(\mathcal{S}_T)$ can be identified with the Čech cocomology group given by the homototy classes $[\mathcal{S}_T, U(1)]$: there is the isomorphism

$$C(\mathcal{S}, \mathbb{Z})_T \cong H^1(\mathcal{S}_T, \mathbb{Z}) \tag{8.3.11}$$

given explicitly by mapping

$$f \mapsto [\exp(2\pi i t f(x))], \tag{8.3.12}$$

for $f \in C(\mathcal{S}, \mathbb{Z})$ and with $[\cdot]$ the homotopy class.

8.3.3 Filtration

Now let us consider the cohomology group $H^1(\mathcal{S}_T, \mathbb{Z})$ and recall the following combinatorial explicit description of the cohomology of the mapping torus $H^1(\mathcal{S}_T, \mathbb{Z})$.

There is an identification of $H^1(\mathcal{S}_T, \mathbb{Z})$ with the K_0-group of the crossed product C^*-algebra for the action of T on $\mathcal{S}$,

$$H^1(\mathcal{S}_T, \mathbb{Z}) \cong K_0(C(\mathcal{S}) \rtimes_T \mathbb{Z}). \tag{8.3.13}$$

Theorem 8.6. *The cohomology* $H^1(\mathcal{S}_T, \mathbb{Z})$ *satisfies the following properties. The identification* (8.3.13) *endows* $H^1(\mathcal{S}_T, \mathbb{Z})$ *with a filtration by free abelian groups* $F_0 \hookrightarrow F_1 \hookrightarrow \cdots \hookrightarrow F_n \hookrightarrow \cdots$, *with* $\mathrm{rank} F_0 = 2g$ *and* $\mathrm{rank} F_n = 2g(2g-1)^{n-1}(2g-2)+1$, *for* $n \geq 1$, *so that*

$$H^1(\mathcal{S}_T, \mathbb{Z}) = \varinjlim_n F_n.$$

In fact, by the Pimsner-Voiculescu six term exact sequence, the group $K_0(C(\mathcal{S}) \rtimes_T \mathbb{Z})$ can be identified with the kernel of he map $1-T$, acting as $f \mapsto f - f \circ T$ on the $\mathbb{Z}$-module $C(\mathcal{S}, \mathbb{Z}) \cong K_0(C(\mathcal{S})$:

$$C(\mathcal{S}, \mathbb{Z})_T = C(\mathcal{S}, \mathbb{Z})/B(\mathcal{S}, \mathbb{Z}) \cong \mathcal{P}/\delta\mathcal{P}, \tag{8.3.14}$$

where $\mathcal{P} \subset C(\mathcal{S}, \mathbb{Z})$ is the set of functions that depend only on "future coordinates", and δ is the operator $\delta(f) = f - f \circ T$.

This can be identified with functions on the limit set Λ_Γ, since each point in Λ_Γ is described by an infinite admissible sequence in the generators γ_i and their inverses.

Then the set of functions $\mathcal{P}$ can be identified with

$$C(\mathcal{S}^+, \mathbb{Z}) = C(\Lambda_\Gamma, \mathbb{Z})$$

viewed as the submodule of the $\mathbb{Z}$-module $C(\mathcal{S}, \mathbb{Z})$ of functions that only depend on future coordinates. Thus, $\mathcal{P}$ has a filtration $\mathcal{P} = \cup_{n=0}^{\infty} \mathcal{P}_n$, where $\mathcal{P}_n$ is generated by the characteristic functions of $\mathcal{S}^+(w)$ with w of length at most $n+1$. Taking into account the relations between these, we obtain that $\mathcal{P}_n$ is a free abelian group generated by the characteristic functions of $\mathcal{S}^+(w)$ with w of length exactly $n+1$. The number of such words is $2g(2g-1)^n$, hence $\mathrm{rank}\mathcal{P}_n = 2g(2g-1)^n$. The map δ satisfies $\delta : \mathcal{P}_n \to \mathcal{P}_{n+1}$, with a 1-dimensional kernel given by the constant functions. The resulting quotients

$$F_n = \mathcal{P}_n / \delta \mathcal{P}_{n-1}$$

are torsion free (cf. Theorem 19 §4 of [PaTu82]) and have ranks

$$\mathrm{rank} F_n = 2g(2g-1)^{n-1}(2g-2)+1$$

for $n \geq 1$, while $F_0 \cong \mathcal{P}_0$ is of rank $2g$. There is an injection $F_n \hookrightarrow F_{n+1}$ induced by the inclusion $\mathcal{P}_n \subset \mathcal{P}_{n+1}$, and $\mathcal{P}/\delta\mathcal{P}$ is the direct limit of the F_n under these inclusions. Thus we obtain the filtration on $H^1(\mathcal{S}_T, \mathbb{Z})$:

$$H^1(\mathcal{S}_T, \mathbb{Z}) = \varinjlim_n F_n.$$

8.3.4 Hilbert space and grading

It is convenient to consider the complex vecor space

$$\mathcal{P}_{\mathbb{C}} = C(\Lambda_\Gamma, \mathbb{Z}) \otimes \mathbb{C}$$

and the corresponding exact sequence computing the cohomology with complex coefficients:

$$0 \to \mathbb{C} \to \mathcal{P}_{\mathbb{C}} \xrightarrow{\delta} \mathcal{P}_{\mathbb{C}} \to H^1(\mathcal{S}_T, \mathbb{C}) \to 0. \tag{8.3.15}$$

The complex vector space $\mathcal{P}_{\mathbb{C}}$ sits in the Hilbert space

$$\mathcal{P}_{\mathbb{C}} \subset \mathcal{L} = L^2(\Lambda_\Gamma, d\mu),$$

where μ is the Patterson-Sullivan measure on the limit set, satisfying

$$\gamma^* \delta\mu = |\gamma'|^{\dim_H(\Lambda_\Gamma)} d\mu,$$

with $\dim_H(\Lambda_\Gamma)$ the Hausdorff dimension.

This gives a Hilbert space $\mathcal{L}$, together with a filtration $\mathcal{P}_n$ by finite dimensional subspaces. In this setting, it is natural to consider a corresponding grading operator,

$$D = \sum_n n\hat{\Pi}_n, \tag{8.3.16}$$

where Π_n denotes the orthogonal projection onto $\mathcal{P}_n$ and $\hat{\Pi}_n = \Pi_n \ominus \Pi_{n-1}$

8.3.5 Cuntz–Krieger algebra

There is a noncommutative space, that encodes nicely the dynamics of the Schottky group Γ on its limit set Λ_Γ. This space is given by the *Cuntz–Krieger algebra*, which carries a refined information on the action of the Schottky group on its limit set. In order to define this algebra, consider the $2g \times 2g$ matrix A that gives the admissibility condition for the sequences in $\mathcal{S}$: this is the matrix with $\{0,1\}$ entries satisfying $A_{ij} = 1$ for $|i-j| \neq g$ and $A_{ij} = 0$ otherwise.

Recall that a *partial isometry* is a linear operator S satisfying the relation $S = SS^*S$.

The *Cuntz–Krieger algebra* $\mathcal{O}_A$ (cf. [Cu], [CuKrie]) is defined as the universal C^*–algebra generated by partial isometries $S_1, \ldots, S_{2g}$, satisfying the relations

$$\sum_j S_j S_j^* = I \tag{8.3.17}$$

$$S_i^* S_i = \sum_j A_{ij}\, S_j S_j^*. \tag{8.3.18}$$

This algebra is related to the Schottky group by the following result.

Proposition 8.7. *There is an isomorphism*

$$\mathcal{O}_A \cong C(\Lambda_\Gamma) \rtimes \Gamma. \tag{8.3.19}$$

Up to stabilization (tensoring with compact operators), the algebra has another crossed product description as

$$\mathcal{O}_A \cong \mathcal{F}_A \rtimes \Gamma. \tag{8.3.20}$$

with $\mathcal{F}_A$ an *AF–algebra* (approximately finite dimensional) algebra, a direct limit of finite dimensional C^*-algebras).

Let us consider the cochain complex of Hilbert spaces

$$0 \to \mathbb{C} \to \mathcal{L} \xrightarrow{\delta} \mathcal{L} \to \mathcal{H} \to 0$$

Proposition 8.8. *The C^*-algebra $\mathcal{O}_A$ admits a faithful representation on the Hilbert space $L^2(\Lambda_\Gamma, d\mu)$.*

This is obtained as follows.

For $d_H = \dim_H(\Lambda_\Gamma)$ the Hausdorff dimension, consider the operators

$$(T_i f)(x) := |(\gamma_i^{-1})'|^{\delta_H/2} f(\gamma_i^{-1} x), \quad \text{and} \quad (P_i f)(x) := \chi_{\gamma_i}(x) f(x), \tag{8.3.21}$$

where γ_i are the generators of Γ, and

$$(T_{\gamma^{-1}} f)(x) := |\gamma'|^{\delta_H/2} f(\gamma x), \tag{8.3.22}$$

for all $\gamma \in \Gamma$. The

Proposition 8.9. *The operators*

$$S_i := \sum_j A_{ij} T_i^* P_j \tag{8.3.23}$$

are partial isometries on $\mathcal{L}$, satisfying the Cuntz-Krieger algebra relations for the matrix A of the subshift of finite type (8.3.1). Thus, the Cuntz–Krieger algebra $\mathcal{O}_A$ can be identified with the subalgebra of bounded operators on the Hilbert space $L^2(\Lambda_\Gamma, \mu)$ generated by the S_i as in the equality (8.3.23).

Indeed, the operators P_i are orthogonal projectors, i.e. $P_i P_j = \delta_{ij} P_j$. The composite $T_i^* P_j T_i$ satisfies

$$\sum_j A_{ij} (T_i^* P_j T_i\, f)(x) = \begin{cases} f(x), & \text{if } P_i(x) = x \\ 0, & \text{otherwise.} \end{cases}$$

In fact, for $x = g_i y$, we have

$$\sum_j A_{ij} (T_i^* P_j T_i f)(x) = \sum_j A_{ij} T_i^* P_j f(g_i^{-1} g_i y) =$$
$$T_i^* (\sum_j A_{ij} P_j) f(y) = T_i^* (I - P_{i+g}) f(y) = f(g_i y) = f(x).$$

In the remaining cases with $x \neq g_i y$, we have

$$\sum_j A_{ij}(T_i^* P_j T_i f)(x) = \sum_j A_{ij} T_i^* P_j f(g_i^{-1}x) = T_i^* \sum_j A_{i\,i+g} P_{i+g} f(g_{i+g}x) = 0,$$

where the last equality follows from the fact that the transition matrix A satisfies $A_{i\ i+g} = 0$. This implies that the S_i satisfy

$$S_i S_i^* = \sum_j A_{ij} T_i^* P_j T_i = P_i.$$

Since the projectors P_i satisfy $\sum_i P_i = I$, we obtain the relation (8.3.17). Moreover, since $T_i T_i^* = 1$, and the entries A_{ij} are all zeroes and ones, we also obtain

$$S_i^* S_i = \sum_{j,k} A_{ij} A_{ik} P_k T_i T_i^* P_j = \sum_j (A_{ij})^2 P_j = \sum_j A_{ij} P_j.$$

Replacing $P_j = S_j S_j^*$ from (8.3.17) we then obtain (8.3.18).

The Cuntz–Krieger algebra $\mathcal{O}_A$ can be described in terms of the action of the free group Γ on its limit set Λ_Γ (cf. [Rob01], [Spi91]), so that we can regard $\mathcal{O}_A$ as a *noncommutative space* replacing the classical quotient Λ_Γ/Γ.

Spectral triples for Schottky groups

Let us consider the diagonal action of the algebra $\mathcal{O}_A$ on the Hilbert space $\mathcal{H} = \mathcal{L} \oplus \mathcal{L}$, and define the *Dirac operator* as

$$\mathcal{D}|_{\mathcal{L}\oplus 0} = \sum_n (n+1)(\hat{\Pi}_n \oplus 0)$$
$$\mathcal{D}|_{0\oplus\mathcal{L}} = -\sum_n n(0 \oplus \hat{\Pi}_n). \tag{8.3.24}$$

Theorem 8.10. *For a Schottky group Γ with $\dim_H(\Lambda_\Gamma) < 1$, the data $(\mathcal{O}, \mathcal{H}, \mathcal{D})$, for $\mathcal{H} = \mathcal{L} \oplus \mathcal{L}$ with the diagonal action of $\mathcal{O}_A$ through the representation (8.3.23) and the Dirac operator (8.3.24), define a non-finitely summable, θ-summable spectral triple.*

The key point of this result is the compatibility relation between the algebra and the Dirac operator, namely the fact that the commutators $[D, a]$ are bounded operators, for all $a \in \mathcal{O}_A^{alg}$, the involutive algebra generated algebraically by the S_i, subject to Cuntz-Krieger relations.

This follows by an estimate on the norm of the commutators $\|[\mathcal{D}, S_i]\|$ and $\|[\mathcal{D}, S_i^*]\|$, in terms of the Poincaré series of the Schottky group (using $d_H < 1$):

$$\sum_{\gamma \in \Gamma} |\gamma'|^s, s = 1 > d_H,$$

where the Hausdorff dimension d_H is the exponent of convergence of the Poincaré series.

The dimension of the n-th eigenspace of $\mathcal{D}$ is $2g(2g-2)^{n-1}(2g-2)$ for $n \geq 1$, $2g$ for $n = 0$, and $2g(2g-2)^{-n-1}(2g-2)$ for $n \leq -1$, so the spectral triple is not finitely summable, since $|\mathcal{D}|^z$ is not of trace class.

It is θ-summable, since the operator $\exp(-t\mathcal{D}^2)$ is of trace class, for all $t > 0$.

Using the description (8.3.20) of the noncommutative space as crossed product of an AF-algebra by the action of the shift, $\mathcal{F}_A \rtimes_T \mathbb{Z}$, one may be able to find a 1-summable spectral triple. Here the dense subalgebra should not contain any of the group elements.

8.3.6 Arithmetic surfaces: homology and cohomology

In particular case of arithmetic surfaces, there is an identification (found in [Cons98], [CM])

$$\mathbb{H}^{\cdot}(\mathrm{Cone}, \Phi) \cong \mathcal{H} \oplus \check{\mathcal{H}} \tag{8.3.25}$$

where $\mathcal{H}$ is the Archimedean cohomology, and $\check{\mathcal{H}}$ its dual under the involution S of (8.2.17).

We can extend then the identification

$$U : \mathcal{H}^1 \stackrel{\cong}{\to} \mathcal{V} \subset \mathcal{L}$$

by considering a subspace $\mathcal{W}$ of the homology $H_1(\mathcal{S}_T)$ of a certain topolgical Smale space, attached to the Schottky uniformization of $X(\mathbb{C})$, in such a way that $\mathcal{W} \cong \check{\mathcal{H}}$. This homology group $H_1(\mathcal{S}_T)$ admit a very explicit combinatorial description, and can be computed as a direct limit

$$H_1(\mathcal{S}_T, \mathbb{Z}) = \varinjlim_N \mathcal{K}_N,$$

where the groups $\mathcal{K}_N$ are free Abelian of rank $(2g-1)^N + 1$ for N even, and $(2g-1)^N + (2g-1)$ for N odd. The $\mathbb{Z}$-module $\mathcal{K}_N$ is generated by the closed geodesics represented by periodic sequences in $\mathcal{S}$ of period $N+1$. These need not be primitive closed geodesics.

In terms of primitive closed geodesics one can write equivalently

$$H_1(\mathcal{S}_T, \mathbb{Z}) = \bigoplus_{N=0}^{\infty} \mathcal{R}_N$$

where $\mathcal{R}_N$ are free Abelian groups with

$$\mathrm{rk}\mathcal{R}_N = \frac{1}{N} \sum_{d|N} \mu(d) \mathrm{rk}\mathcal{K}_{N/d}$$

($\mu(d)$ is the Möbius function).

There is a natural pairing of homology and cohomology given by

$$\langle\cdot,\cdot\rangle : F_n \times \mathcal{K}_N \to \mathbb{Z} \quad \langle [f], x\rangle = Nf(\bar{x}) \tag{8.3.26}$$

This determines a graded subspace $\mathcal{W} \subset H_1(\mathcal{S}_T, \mathbb{Z})$ dual to $\mathcal{V} \subset H^1(\mathcal{S}_T)$. With the identification

$$\begin{array}{ccl} \mathcal{H}^1 & \xrightarrow{U} & \mathcal{V} \subset H^1(\mathcal{S}_T) \\ \downarrow_S & & \quad\downarrow_{\langle,\rangle} \\ \tilde{\mathcal{H}}^1 & \xrightarrow{U} & \mathcal{W} \subset H_1(\mathcal{S}_T, \mathbb{Z}) \end{array} \tag{8.3.27}$$

one can identify the Dirac operator of (8.3.24) with the logarithm of Frobenius

$$\mathcal{D}|_{\mathcal{V}\oplus\mathcal{V}} = \Phi_{\mathcal{H}^1\oplus\mathcal{H}^1} \tag{8.3.28}$$

8.3.7 Archimedean factors from dynamics

The dynamical spectral triple described in Theorem 8.10 is not finitely summable. However, it is still possible to recover from these data the local factor at arithmetic infinity.

As in the previous sections, we consider a fixed Archimedean prime given by a real embedding $\alpha : \mathbb{K} \hookrightarrow \mathbb{R}$, such that the corresponding Riemann surface $X_{/\mathbb{R}}$ is an orthosymmetric smooth real algebraic curve of genus $g \geq 2$. The dynamical spectral triple provides another interpretation of the Archimedean factor $L_{\mathbb{R}}(H^1(X_{/\mathbb{R}}, \mathbb{R}), s) = \Gamma_{\mathbb{C}}(s)^g$.

Proposition 8.11. *Consider the zeta functions*

$$\zeta_{\pi(\mathcal{V}),D}(s,z) := \sum_{\lambda\in \mathrm{Spec}(D)} \mathrm{Tr}\,(\pi(\mathcal{V})\Pi(\lambda, D))\,(s-\lambda)^{-z}, \tag{8.3.29}$$

for $\pi(\mathcal{V})$ the orthogonal projection on the norm closure of $0 \oplus \mathcal{V}$ in $\mathcal{H}$.

The corresponding regularized determinants satisfy

$$\exp\left(-\frac{d}{dz}\zeta_{\pi(\mathcal{V}),D/2\pi}(s/2\pi, z)|_{z=0}\right)^{-1} = L_{\mathbb{C}}(H^1(X), s), \tag{8.3.30}$$

(cf. Proposition 6.8 of [CM]).

8.3.8 A Dynamical theory for Mumford curves

Let K denote a finite extension of $\mathbb{Q}_p$ and Δ_K the Bruhat-Tits tree associated to $G = \mathrm{PGL}(2, K)$. Let us recall few results about the action of a Schottky group on a *Bruhat-Tits tree* and on C^*-algebras of graphs. Detailed explanations are contained in [Man76a], [Mum72], and [CM04a].

Recall that the Bruhat–Tits tree is constructed as follows. One considers the set of free $\mathcal{O}$-modules of rank 2: $M \subset V$. Two such modules are *equivalent* $M_1 \sim M_2$ if there exists an element $\lambda \in K^*$, such that $M_1 = \lambda M_2$. The group GL(V) of linear automorphisms of V operates on the set of such modules *on the left*: $gM = \{gm \mid m \in M\}$, $g \in \mathrm{GL(V)}$. Notice that the relation $M_1 \sim M_2$ is equivalent to the condition that M_1 and M_2 belong to the same orbit of the center $K^* \subset \mathrm{GL(V)}$. Hence, the group $G = \mathrm{GL}(V)/K^*$ operates (on the left) on the set of classes of equivalent modules.

We denote by Δ_K^0 the set of such classes and by $\{M\}$ the class of the module M. Because $\mathcal{O}$ is a principal ideals domain and every module M has two generators, it follows that

$$\{M_1\}, \{M_2\} \in \Delta_K^0, M_1 \supset M_2 \quad \Rightarrow \quad M_1/M_2 \simeq \mathcal{O}/\mathfrak{m}^l \oplus \mathcal{O}/\mathfrak{m}^k, \quad l, k \in \mathbb{N}.$$

The multiplication of M_1 and M_2 by elements of K preserves the inclusion $M_1 \supset M_2$, hence the natural number

$$d(\{M_1\}, \{M_2\}) = |l - k| \tag{8.3.31}$$

is well defined.

The graph Δ_K of the group PGL$(2, K)$ is the infinite graph with set of vertices Δ_K^0, in which two vertices $\{M_1\}, \{M_2\}$ are adjacent and hence connected by an edge if and only if $d(\{M_1\}, \{M_2\}) = 1$. (cf. [Man76a] and [Mum72].)

For a Schottky group $\Gamma \subset \mathrm{PGL}(2, K)$ there is a smallest subtree $\Delta'_\Gamma \subset \Delta_K$ containing the axes of all elements of Γ. The set of ends of Δ'_Γ in $\mathbb{P}^1(K)$ is Λ_Γ, the limit set of Γ. The group Γ carries Δ'_Γ into itself so that the quotient Δ'_Γ/Γ is a finite graph that coincides with the dual graph of the closed fibre of the minimal smooth model of the algebraic curve C/K holomorphically isomorphic to $X_\Gamma := \Omega_\Gamma/\Gamma$ (cf. [Mum72] p. 163). There is a smallest tree Δ_Γ on which Γ acts and such that Δ_Γ/Γ is the (finite) graph of the specialization of C. The curve C is a k-split degenerate, stable curve. When the genus of the fibers is at least 2 - i.e. when the Schottky group has at least $g \geq 2$ generators - the curve X_Γ is called a *Schottky–Mumford curve.*

The possible graphs Δ_Γ/Γ and the corresponding fiber for the case of genus 2 are illustrated in Figure 8.15.

Let us now describe a *dynamical system* associated to the space $\mathcal{W}(\Delta/\Gamma)$ of walks on the directed tree Δ on which Γ acts. In particular, we are interested in the cases when $\Delta = \Delta_K,\ \Delta_\Gamma$.

For $\Delta = \Delta_\Gamma$, one obtains a subshift of finite type associated to the action of the Schottky group Γ on the limit set Λ_Γ, of the type that was considered in [CM].

Let $\bar{V} \subset \Delta_\Gamma$ be a finite subtree whose set of edges consists of one representative for each Γ-class. This is a fundamental domain for Γ in the weak sense (following the notation of [Man76a]), since some vertices may be identified

Fig. 8.15. The graphs Δ_Γ/Γ for genus $g=2$, and the corresponding fibers.

under the action of Γ. Correspondingly, $V \subset \mathbb{P}^1(K)$ is the set of ends of all infinite paths starting at points in $\bar{V}$.

Consider the set $\mathcal{W}(\Delta_\Gamma/\Gamma)$ of doubly infinite walks on the finite graph Δ_Γ/Γ. These are doubly infinite admissible sequences in the finite alphabet given by the edges of $\bar{V}$ with both possible orientations. On $\mathcal{W}(\Delta_\Gamma/\Gamma)$ we consider the topology generated by the sets $\mathcal{W}^s(\omega,\ell) = \{\tilde{\omega} \in \mathcal{W}(\Delta_\Gamma/\Gamma) : \tilde{\omega}_k = \omega_k, k \geq \ell\}$ and $\mathcal{W}^u(\omega,\ell) = \{\tilde{\omega} \in \mathcal{W}(\Delta_\Gamma/\Gamma) : \tilde{\omega}_k = \omega_k, k \leq \ell\}$, for $\omega \in \mathcal{W}(\Delta_\Gamma/\Gamma)$ and $\ell \in \mathbb{Z}$. With this topology, the space $\mathcal{W}(\Delta_\Gamma/\Gamma)$ is a totally disconnected compact Hausdorff space.

The invertible shift map T, given by $(T\omega)_k = \omega_{k+1}$, is a homeomorphism of $\mathcal{W}(\Delta_\Gamma/\Gamma)$. One can describe again the dynamical system $(\mathcal{W}(\Delta_\Gamma/\Gamma), T)$ in terms of subshifts of finite type.

Lemma 8.12. *The space $\mathcal{W}(\Delta_\Gamma/\Gamma)$ with the action of the invertible shift T is a subshift of finite type, where $\mathcal{W}(\Delta_\Gamma/\Gamma) = \mathcal{S}_A$ with A the directed edge matrix of the finite graph Δ_Γ/Γ.*

Genus two example

In the example of Mumford–Schottky curves of genus $g=2$, the tree Δ_Γ is illustrated in Figure 8.16.

In the first case in the Figure 8.16, the tree Δ_Γ is just a copy of the *Cayley graph* of the free group Γ on two generators, hence we can identify doubly infinite walks in Δ_Γ with doubly infinite reduced words in the generators of Γ and their inverses. The *directed edge matrix* is given by

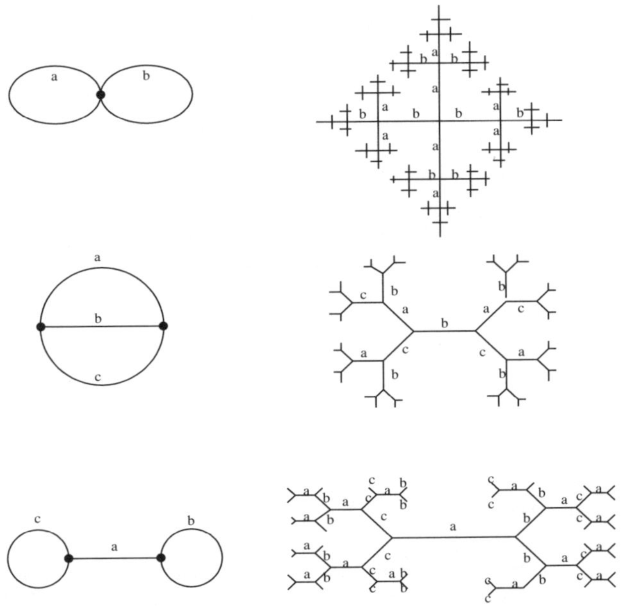

Fig. 8.16. The graphs Δ_Γ/Γ for genus $g = 2$, and the corresponding trees Δ_Γ.

$$A = \begin{pmatrix} 1\,1\,0\,1 \\ 1\,1\,1\,0 \\ 0\,1\,1\,1 \\ 1\,0\,1\,1 \end{pmatrix}.$$

In the second case in Figure 8.16, we label by $a = e_1$, $b = e_2$ and $c = e_3$ the oriented edges in the graph Δ_Γ/Γ, so that we have a corresponding set of labels $E = \{a, b, c, \bar{a}, \bar{b}, \bar{c}\}$ for the edges in the covering Δ_Γ. A choice of generators for the group $\Gamma \simeq \mathbb{Z} * \mathbb{Z}$ acting on Δ_Γ is obtained by identifying the generators g_1 and g_2 of Γ with the chains of edges $a\bar{b}$ and $a\bar{c}$. Doubly infinite walks in the tree Δ_Γ are admissible doubly infinite sequences of such labels, where admissibility is determined by the directed edge matrix

$$A = \begin{pmatrix} 0\,1\,0\,0\,0\,1 \\ 1\,0\,1\,0\,0\,0 \\ 0\,1\,0\,1\,0\,0 \\ 0\,0\,1\,0\,1\,0 \\ 0\,0\,0\,1\,0\,1 \\ 1\,0\,0\,0\,1\,0 \end{pmatrix}.$$

The third case in Figure 8.16 is analogous. A choice of generators for the group $\Gamma \simeq \mathbb{Z} * \mathbb{Z}$ acting on Δ_Γ is given by $ab\bar{a}$ and c. Doubly infinite walks in the tree Δ_Γ are admissible doubly infinite sequences in the alphabet $E = \{a, b, c, \bar{a}, \bar{b}, \bar{c}\}$, with admissibility determined by the directed edge matrix

$$A = \begin{pmatrix} 0&0&1&0&0&1 \\ 1&1&0&0&0&0 \\ 0&0&1&1&0&0 \\ 0&1&0&0&1&0 \\ 1&0&0&0&1&0 \\ 0&0&0&1&0&1 \end{pmatrix}.$$

The construction is analogous for genus $g > 2$, for the various possible finite graphs Δ_Γ/Γ. The directed edge matrix can then be written in block form as

$$A = \begin{pmatrix} \alpha_{11} & \alpha_{12} \\ \alpha_{21} & \alpha_{22} \end{pmatrix},$$

where each block α_{ij} is a $\#(\Delta_\Gamma/\Gamma)^1_+ \times \#(\Delta_\Gamma/\Gamma)^1_+$–matrix with $\alpha_{12} = \alpha_{12}^t$, $\alpha_{21} = \alpha_{21}^t$, and $\alpha_{11} = \alpha_{22}^t$.

8.3.9 Cohomology of $\mathcal{W}(\Delta/\Gamma)_T$

Let $\Delta = \Delta_\Gamma$. We identify the first cohomology group $H^1(\mathcal{W}(\Delta_\Gamma/\Gamma)_T, \mathbb{Z})$ with the group of homotopy classes of continuous maps of $\mathcal{W}(\Delta_\Gamma/\Gamma)_T$ to the circle. Let $C(\mathcal{W}(\Delta_\Gamma/\Gamma), \mathbb{Z})$ be the $\mathbb{Z}$-module of integer valued continuous functions on $\mathcal{W}(\Delta_\Gamma/\Gamma)$, and let

$$C(\mathcal{W}(\Delta_\Gamma/\Gamma), \mathbb{Z})_T := Coker(\delta),$$

for $\delta(f) = f - f \circ T$. The analog of Theorem 8.6 holds:

Proposition 8.13. *The map $f \mapsto [\exp(2\pi i t f(x))]$, which associates to an element $f \in C(\mathcal{W}(\Delta_\Gamma/\Gamma), \mathbb{Z})$ a homotopy class of maps from $\mathcal{W}(\Delta_\Gamma/\Gamma)_T$ to the circle, gives an isomorphism $C(\mathcal{W}(\Delta_\Gamma/\Gamma), \mathbb{Z})_T \simeq H^1(\mathcal{W}(\Delta_\Gamma/\Gamma)_T, \mathbb{Z})$. Moreover, there is a filtration of $C(\mathcal{W}(\Delta_\Gamma/\Gamma), \mathbb{Z})_T$ by free $\mathbb{Z}$-modules*

$$F_0 \subset F_1 \subset \cdots \subset F_n \cdots ,$$

of rank $\theta_n - \theta_{n-1} + 1$, where θ_n is the number of admissible words of length $n+1$ in the alphabet, so that we have

$$H^1(\mathcal{W}(\Delta_\Gamma/\Gamma)_T, \mathbb{Z}) = \varinjlim_n F_n.$$

The quotients F_{n+1}/F_n are also torsion free.

(cf. [CM04a]).

The space $\mathcal{W}(\Delta_\Gamma/\Gamma)_T$ corresponds to a space of "bounded geodesics" on the graph Δ_K/Γ, where geodesics, in this setting, are just *doubly infinite walks* in Δ_K/Γ. In particular, a closed geodesic is the image under the quotient map $\pi_\Gamma : \Delta_K \to \Delta_K/\Gamma$ of a doubly infinite walk in the Bruhat-Tits tree Δ_K with ends given by the pair $z^+(\gamma), z^-(\gamma)$ of fixed points of some element $\gamma \in \Gamma$.

Similarly, a bounded geodesic is an element $\omega \in \mathcal{W}(\Delta_K/\Gamma)$ which is the image, under the quotient map, of a doubly infinite walk in Δ_K with both ends on $\Lambda_\Gamma \subset \mathbb{P}^1(K)$. This implies that a bounded geodesic is a walk of the form $\omega = \pi_\Gamma(\tilde{\omega})$, for some $\tilde{\omega} \in \mathcal{W}(\Delta_\Gamma/\Gamma)$. By construction, any such walk is an axis of Δ_Γ.

Orbits of $\mathcal{W}(\Delta_\Gamma/\Gamma)$ under the action of the invertible shift T correspond bijectively to orbits of the complement of the diagonal in $\Lambda_\Gamma \times \Lambda_\Gamma$ under the action of Γ. Thus, we see that $\mathcal{W}(\Delta_\Gamma/\Gamma)_T$ gives a geometric realization of the space of "bounded geodesics" on the graph Δ_K/Γ, much as, in the case of the geometry at arithmetic infinity, we used the mapping torus of the shift T as a model of the tangle of bounded geodesics in a hyperbolic handlebody.

As in the case at infinity, we can consider the Pimsner–Voiculescu exact sequence computing the K-theory groups of the crossed product C^*-algebra $C(\mathcal{W}(\Delta_\Gamma/\Gamma)) \rtimes_T \mathbb{Z}$,

$$
\begin{aligned}
0 \to H^0(\mathcal{W}(\Delta_\Gamma/\Gamma)_T, \mathbb{Z}) \to C(\mathcal{W}(\Delta_\Gamma/\Gamma), \mathbb{Z}) &\overset{\delta=1-T}{\longrightarrow} C(\mathcal{W}(\Delta_\Gamma/\Gamma), \mathbb{Z}) \\
\to H^1(\mathcal{W}(\Delta_\Gamma/\Gamma)_T, \mathbb{Z}) \to 0. & \qquad (8.3.32)
\end{aligned}
$$

In the corresponding sequence

$$
0 \to H^0(\mathcal{W}(\Delta_\Gamma/\Gamma)_T, \kappa) \to \mathcal{P} \overset{\delta}{\longrightarrow} \mathcal{P} \to H^1(\mathcal{W}(\Delta_\Gamma/\Gamma)_T, \kappa) \to 0, \quad (8.3.33)
$$

for the cohomology for $H^*(\mathcal{W}(\Delta_\Gamma/\Gamma)_T, \kappa)$, with $\kappa = \mathbb{R}$ or $\mathbb{C}$, we can take the vector space $\mathcal{P}$ obtained, as in the case at infinity, by tensoring with κ the $\mathbb{Z}$-module $\mathcal{P} \subset C(\mathcal{W}(\Delta_\Gamma/\Gamma), \mathbb{Z})$ of functions of future coordinates where $\mathcal{P} \simeq C(\mathcal{W}^+(\Delta_\Gamma/\Gamma), \mathbb{Z})$. This has a filtration $\mathcal{P} = \cup_n \mathcal{P}_n$, where $\mathcal{P}_n$ is identified with the submodule of $C(\mathcal{W}^+(\Delta_\Gamma/\Gamma), \mathbb{Z})$ generated by characteristic functions of $\mathcal{W}^+(\Delta_\Gamma/\Gamma, \rho) \subset \mathcal{W}^+(\Delta_\Gamma/\Gamma)$, where $\rho \in \mathcal{W}^*(\Delta_\Gamma/\Gamma)$ is a finite walk $\rho = w_0 \cdots w_n$ of length $n+1$, and $\mathcal{W}^+(\Delta_\Gamma/\Gamma, \rho)$ is the set of infinite paths $\omega \in \mathcal{W}^+(\Delta_\Gamma/\Gamma)$, with $\omega_k = w_k$ for $0 \leq k \leq n+1$. This filtration defines the terms $F_n = \mathcal{P}_n/\delta\mathcal{P}_{n-1}$ in the filtration of the dynamical cohomology of the Mumford curve, as in Proposition 8.13. Again, we will use the same notation in the following for the free $\mathbb{Z}$-module $\mathcal{P}_n$ of functions of at most $n+1$ future coordinates and the vector space obtained by tensoring $\mathcal{P}_n$ by κ.

We obtain a Hilbert space completion of the space $\mathcal{P}$ of cochains in (8.3.33) by considering $\mathcal{L} = L^2(\Lambda_\Gamma, \mu)$ defined with respect to the measure on $\Lambda_\Gamma = \partial\Delta_\Gamma$ given by assigning its value on the clopen set $V(v)$, given by the ends of all paths in Δ_Γ starting at a vertex v, to be

$$
\mu(V(v)) = q^{-d(v)-1},
$$

with $q = \mathrm{card}(\mathcal{O}/\mathfrak{m})$.

In [CM] §4, it was shown how the mapping torus $\mathcal{S}_T$ of the subshift of finite type $(\mathcal{S}, T)$, associated to the limit set of the Schottky group, maps surjectively to the tangle of bounded geodesics inside the hyperbolic handlebody, through a map that resolves all the points of intersection of different geodesics. In the case of the Mumford curve, where we replace the real hyperbolic 3-space by the Bruhat–Tits building Δ_K, the analog of the surjective map from $\mathcal{S}_T$ to the tangle of bounded geodesics is a map from $\mathcal{W}(\Delta_\Gamma/\Gamma)_T$ to the dual graph Δ_Γ/Γ. Here is a description of this map.

As before, let us write elements of $\mathcal{W}(\Delta_\Gamma/\Gamma)$ as admissible doubly infinite sequences

$$\omega = \dots w_{i_{-m}} \dots w_{i_{-1}} w_{i_0} w_{i_1} \dots w_{i_n} \dots,$$

with the $w_{i_k} = \{e_{i_k}, \epsilon_{i_k}\}$ oriented edges on the graph Δ_Γ/Γ. We consider each oriented edge w of normalized length one, so that it can be parameterized as $w(t) = \{e(t), \epsilon\}$, for $0 \leq t \leq 1$, with $\bar{w}(t) = \{e(1-t), -\epsilon\}$. Since $\omega \in \mathcal{S}_A$ is an admissible sequence of oriented edges we have $w_{i_k}(1) = w_{i_{k+1}}(0) \in \Delta_\Gamma^{(0)}$.

We consider a map of the covering space $\mathcal{W}(\Delta_\Gamma/\Gamma) \times \mathbb{R}$ of $\mathcal{W}(\Delta_\Gamma/\Gamma)_T$ to $|\Delta_\Gamma|$ of the form

$$\tilde{E}(\omega, \tau) = w_{i_{[\tau]}}(\tau - [\tau]). \tag{8.3.34}$$

Here $|\Delta_\Gamma|$ denotes the geometric realization of the graph. By construction, the map $\tilde{E}$ satisfies $\tilde{E}(T\omega, \tau) = \tilde{E}(\omega, \tau + 1)$, hence it descends to a map E of the quotient

$$E : \mathcal{W}(\Delta_\Gamma)_T \to |\Delta_\Gamma|. \tag{8.3.35}$$

We then obtain a map to $|\Delta_\Gamma/\Gamma|$, by composing with the quotient map of the Γ action, $\pi_\Gamma : \Delta_\Gamma \to \Delta_\Gamma/\Gamma$, that is,

$$\bar{E} := \pi_\Gamma \circ E : \mathcal{W}(\Delta_\Gamma)_T \to |\Delta_\Gamma/\Gamma|. \tag{8.3.36}$$

Thus, we obtain the following.

Proposition 8.14. *The map $\bar{E}$ of* (8.3.36) *is a continuous surjection from the mapping torus $\mathcal{W}(\Delta_\Gamma)_T$ to the geometric realization $|\Delta_\Gamma/\Gamma|$ of the finite graph Δ_Γ/Γ.*

8.3.10 Spectral triples and Mumford curves

Let us consider the Hilbert space $\mathcal{H} = \mathcal{L} \oplus \mathcal{L}$ and the operator $\mathcal{D}$ defined as

$$\mathcal{D}|_{\mathcal{L}\oplus 0} = -\frac{2\pi}{R \log q} \sum (n+1)\, \hat{\Pi}_n \quad \mathcal{D}|_{0\oplus\mathcal{L}} = \frac{2\pi}{R \log q} \sum n\, \hat{\Pi}_n, \tag{8.3.37}$$

where $\hat{\Pi}_n = \Pi_n - \Pi_{n-1}$ are the orthogonal projections associated to the filtration $\mathcal{P}_n$, the integer R is the length of all the words representing the generators of Γ (this can be taken to be the same for all generators, possibly after blowing up a finite number of points on the special fiber, as explained in [CM03]), and $q = \mathrm{card}(\mathcal{O}/\mathfrak{m})$.

Theorem 8.15. *Consider the tree Δ_Γ of the p-adic Schottky group acting on Δ_K.*

1. *There is a representation of the algebra $C^*(\Delta_\Gamma/\Gamma)$ by bounded linear operators on the Hilbert space $\mathcal{L}$.*
2. *The data $(C^*(\Delta_\Gamma/\Gamma), \mathcal{H}, \mathcal{D})$, with the algebra acting diagonally on $\mathcal{H} = \mathcal{L} \oplus \mathcal{L}$, and the Dirac operator $\mathcal{D}$ of* (8.3.37) *form a spectral triple.*

Recall that, for a curve X over a global field $\mathbb{K}$, assuming semi-stability at all places of bad reduction, the local Euler factor at a place v has the following description ([Se70a]):

$$L_v(H^1(X), s) = \det\left(1 - \mathrm{Fr}_v^* N(v)^{-s} | H^1(\bar{X}, \mathbb{Q}_\ell)^{I_v}\right)^{-1}. \tag{8.3.38}$$

Here Fr_v^* is the geometric Frobenius acting on ℓ-adic cohomology of $\bar{X} = X \otimes \mathrm{Spec}(\bar{\mathbb{K}})$, with $\bar{\mathbb{K}}$ the algebraic closure and ℓ a prime with $(\ell, q) = 1$, where q is the cardinality of the residue field $k(v)$ at v. We denote by N the norm map. The determinant is evaluated on the inertia invariants $H^1(\bar{X}, \mathbb{Q}_\ell)^{I_v}$ at v (all of $H^1(\bar{X}, \mathbb{Q}_\ell)$ when v is a place of good reduction).

Suppose v is a place of $k(v)$-split degenerate reduction. Then the completion of X at v is a Mumford curve X_Γ. In this case, the Euler factor (8.3.38) takes the following form:

$$L_v(H^1(X_\Gamma), s) = (1 - q^{-s})^{-g}. \tag{8.3.39}$$

This is computed by the zeta regularized determinant

$$\det_{\infty, \pi(\mathcal{V}), i\mathcal{D}}(s) = L_v(H^1(X_\Gamma), s)^{-1}, \tag{8.3.40}$$

where

$$\det_{\infty, a, i\mathcal{D}}(s) := \exp\left(-\zeta'_{a, i\mathcal{D},+}(s, 0)\right) \exp\left(-\zeta'_{a, i\mathcal{D},-}(s, 0)\right), \tag{8.3.41}$$

for

$$\begin{aligned} \zeta_{a,i\mathcal{D},+}(s, z) &:= \textstyle\sum_{\lambda \in \mathrm{Spec}(i\mathcal{D}) \cap i[0,\infty)} \mathrm{Tr}(a\Pi_\lambda)(s + \lambda)^{-z} \\ \zeta_{a,i\mathcal{D},-}(s, z) &:= \textstyle\sum_{\lambda \in \mathrm{Spec}(i\mathcal{D}) \cap i(-\infty,0)} \mathrm{Tr}(a\Pi_\lambda)(s + \lambda)^{-z}. \end{aligned} \tag{8.3.42}$$

The element $a = \pi(\mathcal{V})$ is the projection onto a linear subspace $\mathcal{V}$ of $\mathcal{H}$, which is obtained via embeddings of the cohomology of the dual graph Δ_Γ/Γ into the space of cochains of the dynamical cohomology.

The projection $\pi(\mathcal{V})$ acts on the range of the spectral projections $\hat{\Pi}_n$ of D as elements Q_n in the AF algebra core of the C^*-algebra $C^*(\Delta_\Gamma/\Gamma)$.

8.4 Reduction mod ∞

In this section based on Section 7 and 8 in [CM], the "reduction mod infinity" is described in terms of the homotopy quotient associated to the noncommutative space $\mathcal{O}_A$ of the previous §8.3, and the μ-map of Baum–Connes. The geometric model of the dual graph can also be described as a homotopy quotient.

8.4.1 Homotopy quotients and "reduction mod infinity"

In the previous sections we have described the (noncommutative) geometry of the fiber at arithmetic infinity of an arithmetic surface in terms of its dual graph, which we obtained from two quotient spaces: the spaces

$$\Lambda_\Gamma/\Gamma \quad \text{and} \quad \Lambda_\Gamma \times_\Gamma \Lambda_\Gamma \simeq \mathcal{S}/\mathbb{Z}, \tag{8.4.1}$$

with $\mathbb{Z}$ acting via the invertible shift T, which we can think of as the sets of vertices and edges of the dual graph, cf. §4.2 of [CM]. Their noncommutative geometry was analyzed in terms of Connes' theory of spectral triples.

Another fundamental construction in noncommutative geometry (cf. [Co83]) is that of *homotopy quotients.* These are commutative spaces, which provide, up to homotopy, geometric models for the corresponding noncommutative spaces. The noncommutative spaces themselves, as it can be shown in our case, appear as quotient spaces of foliations on the homotopy quotients with contractible leaves.

The crucial point in our setting is that the homotopy quotient for the noncommutative space $\mathcal{S}/\mathbb{Z}$ is precisely the mapping torus which gives the geometric model of the dual graph,

$$\mathcal{S}_T = \mathcal{S} \times_{\mathbb{Z}} \mathbb{R}, \tag{8.4.2}$$

where the noncommutative space $\mathcal{S}/\mathbb{Z}$ can be identified with the quotient space of the natural foliation on (8.4.2) whose generic leaf is contractible (a copy of $\mathbb{R}$). On the other hand, the case of the noncommutative space Λ_Γ/Γ is also extremely interesting. In fact, in this case the *homotopy quotient* appears very naturally and it describes what is refered as the "reduction mod ∞" in [Man91].

Let us recall briefly how the reduction map works in the non-Archimedean setting of Mumford curves (cf. [Man91], [Mum72]).

Let K be a finite extension of $\mathbb{Q}_p$ and let $\mathcal{O}_K$ be its ring of integers. The correct analog for the Archimedean case is obtained by "passing to a limit", replacing K with its Tate closure in $\mathbb{C}_p$ (cf. [Man91] §3.1), however, for our purposes here it is sufficient to illustrate the case of a finite extension.

The role of the hyperbolic space $\mathbb{H}^3$ in the non-Archimedean case is played by the Bruhat-Tits tree $\mathcal{T}_{BT}$ with vertices

$$\mathcal{T}_{BT}^0 = \{\mathcal{O}_K - \text{lattices of rank 2 in a 2-dim } K\text{-space}\}/K^*.$$

Vertices in $\mathcal{T}_{BT}$ have valence $|\mathbb{P}^1(\mathcal{O}_K/\mathfrak{m})|$, where $\mathfrak{m}$ is the maximal ideal. Each edge in $\mathcal{T}_{BT}$ has length $\log|\mathcal{O}_K/\mathfrak{m}|$. The set of ends of $\mathcal{T}_{BT}$ is identified with $X(K) = \mathbb{P}^1(K)$. This is the analog of the conformal boundary $\mathbb{P}^1(\mathbb{C})$ of $\mathbb{H}'$. Geodesics correspond to doubly infinite paths in $\mathcal{T}_{BT}$ without backtracking. Fix a vertex v_0 on $\mathcal{T}_{BT}$. This corresponds to the closed fiber $X_{\mathfrak{a}} \otimes (\mathfrak{a}/\mathfrak{m})$ for the chosen $\mathfrak{a}$-structure $X_{\mathfrak{a}}$. Each $x \in \mathbb{P}^1(K)$ determines a unique choice of a subgraph $e(v_0, x)$ in $\mathcal{T}_{BT}$ with vertices $(v_0, v_1, v_2, \ldots)$ along the half infinite path without backtracking which has end x. The subgraphs $e(v_0, x)_k$ with vertices $(v_0, v_1, \ldots v_k)$ correspond to the reduction mod $\mathfrak{m}^k$, namely

$$\{e(v_0, x)_k : x \in X(K)\} \leftrightsquigarrow X_{\mathfrak{a}}(\mathfrak{a}/\mathfrak{m}^k).$$

Thus the finite graphs $e(v_0, x)_k$ represent $\mathfrak{a}/\mathfrak{m}^k$ points, and the infinite graph $e(v_0, x)$ represents the reduction of x.

A Schottky group Γ, in this non-Archimedean setting, is a purely loxodromic free discrete subgroup of $\mathrm{PSL}(2, K)$ in g generators. The doubly infinite paths in $\mathcal{T}_{BT}$ with ends at the pairs of fixed points $x^{\pm}(\gamma)$ of the elements $\gamma \in \Gamma$ produce a copy of the combinatorial tree $\mathcal{T}$ of the group Γ in $\mathcal{T}_{BT}$. This is the analog of regarding $\mathbb{H}^3$ as the union of the translates of a fundamental domain for the action of the Schottky group, which can be thought of as a 'tubular neighborhood' of a copy of the Cayley graph $\mathcal{T}$ of Γ embedded in $\mathbb{H}^3$. The ends of the tree $\mathcal{T} \subset \mathcal{T}_{BT}$ constitute the limit set $\Lambda_\Gamma \subset \mathbb{P}^1(K)$. The complement $\Omega_\Gamma = \mathbb{P}^1(K) \smallsetminus \Lambda_\Gamma$ gives the uniformization of the Mumford curve $X(K) \simeq \Omega_\Gamma/\Gamma$. In turn, $X(K)$ can be identified with the ends of the quotient graph $\mathcal{T}_{BT}/\Gamma$, just as in the Archimedean case the Riemann surface is the conformal boundary at infinity of the handlebody $\mathfrak{X}_\Gamma$.

The reduction map is then obtained by considering the half infinite paths $e(v, x)$ in $\mathcal{T}_{BT}/\Gamma$ that start at a vertex v of the finite graph $\mathcal{T}/\Gamma$ and whose end x is a point of $X(K)$, while the finite graphs $e(v, x)_k$ provide the $\mathfrak{a}/\mathfrak{m}^k$ points.

This suggests that the correct analog of the reduction map in the Archimedean case is obtained by considering geodesics in $\mathbb{H}^3$ with an end on Ω_Γ and the other on Λ_Γ, as described in [Man91] . Arguing as in Lemma 4.9 of [CM], we see that the set of such geodesics can be identified with the quotient $\Omega_\Gamma \times_\Gamma \Lambda_\Gamma$. The analog of the finite graphs $e(v, x)_k$ that define the reductions modulo $\mathfrak{m}^k$ is then given by the quotient $\mathbb{H}^3 \times_\Gamma \Lambda_\Gamma$.

Notice then that the quotient space

$$\Lambda_\Gamma \times_\Gamma \mathbb{H}^3 = \Lambda_\Gamma \times_\Gamma \underline{E}\Gamma, \tag{8.4.3}$$

is precisely the homotopy quotient of Λ_Γ with respect to the action of Γ, with $\underline{E}\Gamma = \mathbb{H}^3$ and the classifying space $\underline{B}\Gamma = \mathbb{H}^3/\Gamma = \mathfrak{X}_\Gamma$, (cf. [Co83]). In this case also we find that the noncommutative space Λ_Γ/Γ is the quotient space of a foliation on the homotopy quotient (8.4.3) with contractible leaves $\mathbb{H}^3$.

8.4.2 Baum-Connes map

The relation between the noncommutative spaces (8.4.1) and the homotopy quotients (8.4.2) and (8.4.3) is an instance of a very general and powerful construction, namely the μ-map (cf. [BaCo], [Co83]). In particular, in the case of the noncommutative space $C(\mathcal{S}) \rtimes_T \mathbb{Z}$, the μ-map

$$\mu : K^{*+1}(\mathcal{S}_T) \cong H^{*+1}(\mathcal{S}_T, \mathbb{Z}) \to K_*(C(\mathcal{S}) \rtimes_T \mathbb{Z}) \tag{8.4.4}$$

is the Thom isomorphism that gives the identification of (8.3.11), (8.3.12) and recovers the Pimsner–Voiculescu exact sequence (8.3.8) as in [Co81]. As explained in [CM], the map μ of (8.4.4) assigns to a K-theory class $\mathcal{E} \in K^{*+1}(\mathcal{S} \times_{\mathbb{Z}} \mathbb{R})$ the index of the longitudinal Dirac operator $\partial\!\!\!/_{\mathcal{E}}$ with coefficients $\mathcal{E}$. This index is an element of the K-theory of the crossed product algebra $C(\mathcal{S}) \rtimes_T \mathbb{Z}$ and the μ-map is an isomorphism. Similarly, in the case of the noncommutative space $C(\Lambda_\Gamma) \rtimes \Gamma$, where we have a foliation on the total space with leaves $\mathbb{H}^3$, the μ-map

$$\mu : K^{*+1}(\Lambda_\Gamma \times_\Gamma \mathbb{H}^3) \to K_*(C(\Lambda_\Gamma) \rtimes \Gamma) \tag{8.4.5}$$

is again given by the index of the longitudinal Dirac operator $\partial\!\!\!/_{\mathcal{E}}$ with coefficients $\mathcal{E} \in K^{*+1}(\Lambda_\Gamma \times_\Gamma \mathbb{H}^3)$. In this case the map is an isomorphism because the Baum–Connes conjecture with coefficients holds for the case of $G = SO_0(3,1)$, with $\mathbb{H}^3 = G/K$ and $\Gamma \subset G$ the Schottky group, cf. [Kas].

In particular, analyzing the noncommutative space $C(\Lambda_\Gamma) \rtimes \Gamma$ from the point of view of the theory of spectral triples provides cycles to pair with K-theory classes constructed geometrically via the μ-map.

To complete the analogy with the reduction map in the case of Mumford curves, one should also consider the half infinite paths $e(v,x)$ corresponding to the geodesics in $\mathfrak{X}_\Gamma$ parameterized by $\Lambda_\Gamma \times_\Gamma \Omega_\Gamma$, in addition to the finite graphs $e(v,x)_k$ that correspond to the homotopy quotient (8.4.2). This means that the space that completely describes the "reduction modulo infinity" is a compactification of the homotopy quotient

$$\Lambda_\Gamma \times_\Gamma (\mathbb{H}^3 \cup \Omega_\Gamma), \tag{8.4.6}$$

where $\underline{E\Gamma} = \mathbb{H}^3 \cup \Omega_\Gamma$ corresponds to the compactification of the classifying space $\underline{B\Gamma} = \mathbb{H}^3/\Gamma = \mathfrak{X}_\Gamma$ to $\overline{B\Gamma} = (\mathbb{H}^3 \cup \Omega_\Gamma)/\Gamma = \mathfrak{X}_\Gamma \cup X_{/\mathbb{C}}$, obtained by adding the conformal boundary at infinity of the hyperbolic handlebody.

References

[Ma-PaM] Manin Yu.I, Panchishkin A.A. (1989): Introduction to Number Theory, 1989, Edition VINITI (in Russian), 348 p.

[Ma-Pa] Manin Yu.I, Panchishkin A.A. (1995): Number Theory I: Introduction to Number Theory, Encyclopaedia of Mathematical Sciences, vol. 49, Springer-Verlag, 1995, 303 p.

[Abb97] Abbes, A.: Hauteurs et discrétude, d'après L. Szpiro, E. Ullmo et S. Zhang. Séminaire Bourbaki. Exp. No.825 (Mars 1997)

[Abr74] Abrashkin, V.A. (1974): Finding imaginary quadratic fields with an even discriminant and class–number two by the Heegner method (in Russian). Mat. Zametki, **15**, No.2, 241–246 (1974). English Translation.: Math.Notes **15**, 137–139 (1974). Zbl.292.12002

[AdHB85] Adleman, L.M., Heath–Brown, D.R. (1985): The first case of Fermat's last theorem. Inv. Math., **79**, No.2, 409-416 (1985). Zbl.557.10034

[AdHu92] Adleman, L. M., Huang, Ming-Deh A. Primality testing and abelian varieties over finite fields. Lecture Notes in Mathematics, 1512. Springer-Verlag, Berlin, 1992. viii+142 pp.

[APR83] Adleman, L.C., Pomerance, C., Rumely, R.C. (1983): On distinguishing prime numbers from composite numbers. Ann. Math., **117**,173-206 (1983). Zbl. 526.10004

[ARS78] Adleman, L.M., Rivest, R.L., Shamir, A. (1978): A method for obtaining digital signatures and public–key cryptosystems. ACM, **21** (1978), 120-126. Zbl.368.94005.

[AKS] Agrawal, M., Kayal, N., Saxena, N.: Primes is in P, Department of Computer Science & Engineering Indian Institute of Technology Kampur internet : `{manindra, Kayaln, nitinsa}@iitk.ac.in`

[SSemB87] Selected lectures of Seminar Bourbaki (1987): Algebra and number theory (with applications), (in Russian, talks translated from French and English. Mir: Moscow, 1987. Zbl.686.00005.

[AGP94] Alford W.-R., Granville, A., Pomerance, C. (1994): There are infinitely many Carmichael numbers. Ann. Math. **139**, 703–722 (1994).

[All71] Alling, N., Greenleaf, N., Foundations of the theory of Klein surfaces, Lecture Notes in Mathematics Vol. 219, Springer Verlag 1971.

[Ande82] Anderson, G.W.: Logarithmic derivatives of Dirichlet L-functions and the periods of abelian varieties, Comp. Math. **45** (1982), 315-322

[And76] Andrews, G.E. (1976): The theory of partitions. Reading, Addison–Wesley (1976). Zbl.371.10001.

[An65] Andrianov, A.N.: Representation of integers by certain quadratic forms and its relation with the theory of elliptic curves (in Russian). Izv. Akad. Nauk SSSR, Ser. mat., **29**, No.1, 227-238 (1965). English translation: Am.Math.Soc., Transl., II Ser. **66**, 191–203 (1968). Zbl.179,73

[An74] Andrianov, A.N. (1974): Siegel modular forms of genus 2. (in Russian). Uspekhi Mat. Nauk, 29, No.3 (1974), 44-109. English translation: Russ.Math.Surv. **29**, No. 3, 45–116 (1974). Zbl.304.10020

[An79a] Andrianov, A.N. (1979): The multiplicative arithmetics of the Siegel modular forms.(in Russian). Uspekhi Mat. Nauk, 34, No.1 (1979), 67-135. English translation: Russ.Math.Surv. **34**, No. 1, 75–148 (1979). Zbl. 418.10027

[An79b] Andrianov, A.N.: Modular descent and the Saito-Kurokawa conjecture. Inv. Math., **53**, No.3, 267-280 (1979). Zbl.414.10017.

[AK78] Andrianov, A.N., Kalinin V.L. (1978): On the analytical properties of the standard zeta–unctions of the Siegel modular forms. (in Russian). Mat. Sbornik, **106**, No.3, 323-339, (1978). English translation: Math. USSR , Sb. **35**, 1–17, (1979). Zbl. 389.10023.

[AP2000] Andrianov A.N., Panchishkin A.A., Singular Frobenius operators on Siegel modular forms with characters and zeta functions. Algebra i Analiz, **12**, no.2,1-34 (2000). (In Russian) English translation: St.-Petersbourg Math.J., **12**, No. 2, 233-257 (2001)

[AntChris] Antonescu, C., Christensen, E.: Spectral triples for AF C^*-algebras and metrics on the Cantor set, preprint ArXiv math.OA/0309044.

[Apo97] Apostol, T.M. (1997): Introduction to Analytic Number Theory. Springer–Verlag, 1997

[Ara74a] Arakelov, S.Yu. (1974a): The intersection theory of divisors on arithmetical surfaces. (in Russian). Izv. Akad. Nauk SSSR, Ser. mat., **38**, No.6, 1179-1192 (1974). English translation: Math. USSR , Izv. **8**, 1167–1180, (1976). Zbl. 355.14002.

[Ara74b] Arakelov, S.Yu. (1974b): Theory of intersections on the arithmetic surface. Proc. Int. Congr. Math., Vancouver 1974, Vol. **1** 405–408 (1975). Zbl.351.14003.

[AKCh87] Arkhipov, G.I., Karacuba, A.A., Chubarikov, V.M. (1987): The theory of the multiple exponential sums. (in Russian). Nauka, Moscow (1987). Zbl.638.10037

[Arne04] Arneodo, A.: L'analyse multifractale des signaux, In: "Images des mathématiques 2004", CNRS, Ed. by Et.Ghys, J.Istas. Février 2004. pp. 7-14

[Art78] Arthaud, N. (1978): On Birch and Swinnerton–Dyer's conjecture for elliptic curves with complex multiplication. Comp. Math., **37**, No.2, 209-232 (1978). Zbl.396.12001

[Ar30] Artin, E. (1930): Zur Theorie der L–Reihen mit allgemeinen Gruppencharakteren. Abh. Math. Sem. Univ. Hamburg, **8**, 292-306 (1930). Jbuch FdM.56,173

[Ar51] Artin, E. (1951): Algebraic numbers and algebraic functions. Lecture Notes J. Adamson. Princeton and New York Univ., 1951. Gordon & Breach, 1967. Zbl. 194,353

[Ar65] Artin, E. (1965): Collected Papers. Reading, Mass.: Addison–Wesley, 1967. Zbl.493.01038.

[AT51] Artin, E., Tate, J. (1951): Class field theory. Princeton Notes, 1951/1952. New–York–Amsterdam, W.Benjamin (1968) Zbl. 176,335

[AW45] Artin, E., Whaples, J. (1945): Axiomatic characterization of fields by the product formula. Bull. AMS, **51**, No.7, 469-492 (1945). Zbl.60, 83

[Arth83] Arthur, J. : The trace formula for noncompact quotients, Proceedings of International Congr of Math., Warsaw (1983), 849–859.

[ArCl89] Arthur, J., Clozel, L.: Simple algebras, base change and the advanced theory of the trace formula, in Annals of Math Studies, **120**, Princeton University Press (1989).

[AL70] Atkin A. Lehner L. (1970): Hecke operators on $\Gamma_0(m)$. Math. Ann. **185**, 134–160 (1970)

[AtMo93b] Atkin, A. O. L., Morain, F.: Elliptic curves and primality proving. Math. Comp., 61(203):29–68, July 1993.

[AB67] Atiyah, M.F., Bott, R.: A Lefchetz fixed point formula for elliptic complexes: I, Annals of Math, **86** (1967), 374-407.

[APS75] Atiyah, M.F., Patodi, V.K., Singer I.M.: Spectral asymmetry and Riemannian geometry. I, Math. Proc. Cambridge Philos. Soc. 77 (1975), 43–69.

[Ax64] Ax, J. (1964): Zeroes of polynomials over finite fields. Amer. J. Math. **86**, 255-261 (1964). Zbl.121,20.

[Ba71] Baker, A. (1971): Imaginary quadratic fields with class number 2. Ann. Math., **94**, No.1, 139-152 (1971). Zbl.219.12008.

[Ba75] Baker, A. (1975): Transcendental Number Theory. Cambridge. Univ. Press, 1975. Reissue Cambridge 1990. Zbl.297.10013

[Bak86] New advances in transcendence theory. Symp. Trans. Number Theory. Durham, July, 1986 (ed. Baker A.). Cambridge Univ. Press, 1988. Zbl.57,281.

[BaHa96] Baker, R.H., Harman G.: The Brun-Titchmarsh Theorem on average. In "Proceedings of a conference in Honor of Heini Halberstam", Vol. **1**, 39-103 (1996)

[BaRi01] Ball, K. , Rivoal, T.: Irrationalité d'une infinité de valeurs de la fonction zêta aux entiers impairs, Invent. Math. **146**, 193-207 (2001)

[BDGP96] Barré-Sirieix, K., Diaz, G., Gramain, F., Philibert, G.: Une preuve de la conjecture de Mahler-Manin. Invent. Math. **124**, no. 1-3, 1-9(1996)

[Barsky04] Barsky D.: Sur la nullité de μ-invariant d'Iwasawa des corps totalement réels. Preprint, Univ. Paris 13, 1-97, 24 mai 2004.

[Bash72] Bashmakov, M.I. (1972): Cohomology of Abelian varieties. (in Russian). Uspekhi Mat. Nauk, 27, No.6 (1972), 25-66. English translation: Russ.Math.Surv. **27**, 25-70 (1973). Zbl.256.14016

[BBBCO] Batut, C., Belabas, D., Bernardi, H., Cohen, H., Olivier, M.: The PARI/GP number theory system. `http://pari.math.u-bordeaux.fr`

[BatMan] Batyrev, V. V., Manin, Yu. I.: Sur le nombre des points rationnels de hauteur borné des variétés algébriques. Math. Ann. **286**, no. 1-3, 27–43 (1990).

[BaTsch96] Batyrev, V. V., Tschinkel, Y.: Rational points on some Fano cubic bundles. C. R. Acad. Sci. Paris Sér. I Math. **323** (1996), no. 1, 41–46.

[BaTsch98a] Batyrev, V. V., Tschinkel, Y.: Manin's conjecture for toric varieties. J. Algebraic Geom. **7** (1998), no. 1, 15–53.

[BaTsch98b] Batyrev, V. V., Tschinkel, Y.: Tamagawa numbers of polarized algebraic varieties. Nombre et répartition de points de hauteur bornée (Paris, 1996). Astérisque No. 251 (1998), 299–340.

[BKR] Baum, P., Karoubi, M., Roe, J.: On the Baum–Connes conjecture in the real case, preprint.

[BaCo] Baum P., Connes A., Geometric K-theory for Lie groups and foliations. Preprint IHES (M/82/), 1982; l'Enseignement Mathématique, **46**, 1-35, 2000.

[BelBr03] Belkale, P., Brosnan, P.: Periods and Igusa local zeta functions. Int. Math. Res. Not. 2003, no. 49, 2655–2670.

[Bea83] Beardon, A.F., The Geometry of discrete groups, Graduate Texts in Mathematics, Vol. 91, Springer Verlag 1983.

[Be79] Belyi, G.V. (1979): On the Galois extensions of the maximal cyclotomic field. (in Russian). Izv. Akad. Nauk SSSR, Ser. mat., **43**, No.2, 267-276 (1979). English translation: Math.USSR Izw. **14**, 247–256 (1980). Zbl.409.12012

[Ber03] Bernstein D.J.: Proving primality after Agrawal, Kayal and Saxena, `http://cr.yp.to/papers.html#aks, 25/01/2003.`

[Ber86] Berry, M. V.: Riemann's zeta function: a model of quantum chaos? Quantum chaos and statistical nuclear physics, Lecture Notes in Physics, **263**, 1- 17 Springer (1986). Zbl 0664.10021

[BeK] Berry, M. , Keating, J.: $H = qp$ and the Riemann zeros, 'Supersymmetry and Trace Formulae: Chaos and Disorder', edited by J.P. Keating, D.E. Khmelnitskii and I.V. Lerner (Plenum Press).

[Ber61] Bers, L.: Uniformization by Beltrami equations. Comm. Pure Appl. Math. **14**, 215–228 (1961).

[Bert92] Bertrand, D.: Groupes algébriques et équations différentielles linéaires Séminaire Bourbaki, [Exposé No 750] Février 1992

[BeDa97] Bertolini, M., Darmon, H.: A rigid analytic Gross-Zagier formula and arithmetic applications. Ann. of Math. (2) **146**, 111–147 (1997)

[BeDa04] Bertolini, M., Darmon, H.: A Birch and Swinnerton-Dyer conjecture for the Mazur-Tate circle pairing. Duke Math. J. **122**, no. 1, 181–204 (2004)

[Beu55] Beurling, A.: A closure problem related to the zeta function, Proc. Nat. Ac. Sci. **41**, 312-314 (1955).

[Bha04] Bhargava, M.: Higher composition laws. I. A new view on Gauss composition, and quadratic generalizations. Ann. of Math. (2) 159 (2004), no. 1, 217–250.

[Bi02] Bilu, Yu. F.: Catalan's conjecture, after Mihailescu. Séminaire Bourbaki. [Exposé No.909] (Novembre 2002)

[Bir61] Birch, B.J. (1961): Forms in many variables. Proc. Royal Soc. London, Sec. A, **265**, 245-263 (1961). Zbl.103,31.

[Bir63] Birch, B.J.(1963): Conjectures concerning elliptic curves. in: Theory of Numbers, Proc.Symp.Pure Math. AMS, Pasadena, 1963, **8**, 106-112 (1965). Zbl.238.14011.

[Bir75] Birch, B.J.: Heegner points of elliptic curves. in: Symp. Mat. Inst. Naz. Alta Mat. Convegni 1973 London–New York, **15**, 441–445 (1975). Zbl.317.14015.

[BD58] Birch, B.J., Davenport, H.: On a theorem of Davenport and Heilbronn. Acta Math., **100** (1958), 259-279. Zbl.82,260.

[BS83] Birch, B.J., Stephens, N.: Heegner's construction of points on the curve $y^2 = x^3 - 1728$, Progr. Math., **38** (1983), 119. Zbl.535.14012

[BSD63] Birch, B.J., Swinnerton–Dyer, H.P. (1963): Notes on elliptic curves I,II. J. reine u. angew. Math., **212** (1963), 7-25, Zbl.118,276; **218** (1965), 79-108, Zbl.147.25.

[BSD75] Birch, B.J., Swinnerton–Dyer, H.P.: The Hasse problem for rational surfaces. J. reine u. angew. Math., 274/275 (1975), 164-174. Zbl.326.14007.

[BiGS88] Bismut, J.-M. , Gillet, H., Soulé C.: Analytic torsion and holomorphic determinant bundles. I, II, III. Comm. Math.Phys. 115, no. 1, 49–78 ; no. 1, 79–126; no. 2, 301–351 (1988)

[BGS97] Bloch, S., Gillet, H., Soulé,C.: Algebraic Cycles on Degenerate Fibers, Arithmetic Geometry (Cortona 1994), 45–69, Sympos. Math. XXXVII, Cambridge Univ. Press: Cambridge, 1997.

[BK90] Bloch, S., Kato K.: L–functions and Tamagawa numbers of motives, in: The Grothendieck Festschrift, vol. 1, Prog. in Math. 86, Birkhaüser, Boston, 1990, pp. 333–400.

[Boe85] Böcherer, S.: Über die Funktionalgleichung automorpher L-Funktionen zur Siegelschen Modulgruppe. J. Reine Angew. Math. **362** (1985), 146–168.

[Bog80] Bogomolov, F.A. Torsion points on abelian varieties. (in Russian). Izv. Akad. Nauk SSSR, Ser. mat., 44, no.4 (1980), 782-803. Zbl.453.14018.

[BoTs99] Bogomolov, F. A.; Tschinkel, Yu.: On the density of rational points on elliptic fibrations. J. Reine Angew. Math. 511 (1999), 87–93.

[BoTs2000] Bogomolov, F. A.; Tschinkel, Yu.: Density of rational points on elliptic $K3$ surfaces. Asian J. Math. 4 (2000), no. 2, 351–368.

[Bom72] Bombieri, E.: Counting points on curves over finite fields (d'après S.A.Stepanov). Sém. Bourbaki, exp. 430 (1972-73), Lect. Notes in Math., **383** (1974), 234-241. Zbl.307.14011.

[BV83] Bombieri, E., Vaaler J.: On Siegel's lemma. Inv. Math., **73** (1983), 11-32. Zbl.533.10030.

[Bom90] Bombieri, E.: The Mordell conjecture revisited. Ann. Scuola Norm. Sup. Pisa Cl. Sci. (4) **17** (1990), no. 4, 615–640.

[Borch92] Borcherds, R.E.: Monstrous moonshine and monstrous Lie superalgebras, Invent. Math. 109, 405-444 (1992).

[Borch99] Borcherds, R.E.: The Gross-Kohnen-Zagier theorem in higher dimensions, Duke Math. J. **97**, No. 2, 219-233 (1999) Correction Duke math journal **105** No. 1 p.183-184 (2000)

[Bor79] Borel, A.(1979): Automorphic L–functions. Proc. Symp. Pure Math., 33 (1979), 27-61. Zbl.412.10017.

[BoCa79] Borel A., Casselman W. eds. (1979): Automorphic forms, representations and L–functions. Proc. Symp. Pure Math. Corvallis, Oregon 1977. Providence, Am. Math. Soc., Vol. **33**, Parts 1-2 (1979). Zbl.121,20.

[BS85] Borevich, Z.I., Shafarevich, I.R. (1985): Number Theory. (in Russian). 3rd ed. Nauka, Moscow (1985). English transl.: New York/London: Academic Press, 1966. Zbl.592.12001, Zbl.121,42.

[BZRKJ] Borho W., Zagier D., Rohlfs J, Kraft H.-P., Jantzen J. C. (1981): Lebendige Zahlen. Fünf Exkursionen. Boston: Birkhäuser, 1981 Zbl.539,10001. Russian translation: Moscow, Mir (1985).

[Bor03] Bornemann, F.: Primes is in P, une avancée accessible à "l'homme ordinaire", Gazette des Mathématiciens No 98, pp.14-30, Octobre 2003.

[Bo90] Bost, J.-B.: Théorie de l'intersection et théorème de Riemann-Roch arithmétiques. Séminaire Bourbaki. [Exposé 731], Novembre 1990

[Bo01] Bost, J.-B.: Algebraic leaves of algebraic foliations over number fields. Publ. Math. Inst. Hautes Études Sci. No. 93 (2001), 161–221.

[BGS94] Bost, J.-B., Gillet, H., Soulé, C.: Heights of projective varieties and positive Green forms. J. Amer. Math. Soc. **7**, 903–1027 (1994)

[Bo95] Bost, J.-B.: Périodes et isogénies des variétés abéliennes sur les corps de nombres, d'après D. Masser et G. Wüstholz. Séminaire Bourbaki. [Exposé 795], Mars 1995

[BoCo95] Bost, J.-B., Connes A.: Hecke Algebras, Type III factors and phase transitions with spontaneous symmetry breaking in number theory, Selecta Mathematica, New Series **1**, No.3 (1995), 411-457.

[Bou62] Bourbaki, N. Algèbre commutative. Paris, Hermann (1962).

[Bo] Bowen, R.: Hausdorff dimension of quasi–circles, Publ.Math. IHES 50 (1979) 11–25.

[BoHa] Boyle, M., Handelman, D.: Orbit equivalence, flow equivalence, and ordered cohomology, Israel J. Math. 95 (1996) 169–210.

[Bra50] Brauer, R. (1950): On the zeta–function of algebraic number fields.II. Amer. J. Math., 72, no. **4** (1950), 739-746. Zbl.38,176.

[Bre88] Bremner, A., Guy, R.K.: Unsolved problems. A dozen difficult Diophantine dilemmas. Amer. Math. Monthly, **95**, No.1 (1988), 31-36. Zbl.647.10017.

[Bren80] Brent, R.P. (1980): An improved Monte Carlo factorization algorithm. BIT, **29** (1980), 176-184. Zbl.439.65001.

[Bren81] Brent, R.P., Pollard, J.M. (1981): Factorization of the eighth Fermat number. Math. Comp. **36**, 627-630 (1981). Zbl.476.10007

[dlBre02] de la Bretèche, R.: Nombre de points de hauteur bornée sur les surfaces de del Pezzo de degré 5. Duke Math. J. **113**, no. 3, 421–464 (2002).

[Breu99] Breuil, Ch.: Intégration sur les variétés p-adiques, d'après Coleman, Colmez. Séminaire Bourbaki. [Exposé No.860] (1999).

[Bril81] Brillhart, J.: Fermat's factoring method and two variants. Congressus Numerantium, **32**, 29-48 (1981). Zbl.488.10006.

[BLTW83] Brillhart, J. Lehmer, D.H., Tuckerman, B., Wagstaff, S.S.: Factorization of $b^n \pm 1, b = 2, 3, 5, 6, 7, 10, 11, 12$ up to high powers. Amer. Math. Soc., 1983. Zbl.527.10001

[BrGo02] Brüdern, J., Godinho, H.: On Artin's conjecture. II. Pairs of additive forms. Proc. London Math. Soc. (3) 84 (2002), no. 3, 513–538.

[Bru61] Bruhat, F.: Distributions sur un groupe localement compact et applications á l'étude des représentations des groupes p-adiques. Bull. Soc. Math. France. **89** (1961), 43-75.

[dBr51] de Bruijn, N.G.: On the number of positive integers $< x$ and free of prime factors > y. Indag. Math., **12** (1951), $5\bar{0} - 60$. Zbl.42,42

[Bry86] Brylinski, J.–L. (1986): Transformations canoniques, dualité projective, théorie de Lefshetz, transformation de Fourier et sommes trigonométriques. Astérisque, **140/141**, 3-134 (1986). Zbl.624.32009.

[BGZ85] Buhler, J., Gross, B., Zagier., D. On the conjecture of Birch and Swinnerton Dyer for an elliptic curve of rank 3. Math.Comput., **44**, 473-481 (1985). Zbl.606.14021.

[BCSGKK3] Bump, D., Cogdell, J. W., de Shalit, E., Gaitsgory, D., Kowalski, E., Kudla, S. S.: An introduction to the Langlands program. Lectures presented at the Hebrew University of Jerusalem, Jerusalem, March 12–16, 2001. Edited by Joseph Bernstein and Stephen Gelbart. Birkhäuser Boston, Inc., Boston, MA, 2003. x+281 pp.

[Bum97] Bump, D.: Automorphic forms and representations. Cambridge University Press, Cambridge, UK, 1997

[CHM97] Caporaso, L., Harris, J., Mazur, B.: Uniformity of rational points. J. Amer. Math. Soc. **10**, no. 1, 1–35 (1997).

[Car2000] Carayol, H.: Preuve de la conjecture de Langlands locale pour GL_n: travaux de Harris-Taylor et Henniart. Séminaire Bourbaki, Vol. 1998/99. Astérisque No. 266, (2000), Exp. No. 857, 4, 191–243.

[Car93] Cartier, P.: Des nombres premiers à la géométrie algébrique (une brève histoire de la fonction zéta). Analyse diophantienne et géométrie algébrique, 51–77, Cahiers Sém. Hist. Math. Sér. 2, 3, Univ. Paris VI, Paris, 1993.

[Car95] Cartier, P.: An Introduction to zeta functions, in "From Number Theory to Physics" (Waldschmidt & al., Eds.), pp. 1–63, Springer, 1995.

[Car01] Cartier, P.: La folle journée; évolution des idées de point et de symétrie, de Grothendieck à Connes et Kontsevich, numéro spécial des Publ. Math. IHES (nov. 1998) (English translation in: Bull. Amer. Math. Soc. **38** (2001), 389–408).

[Car02] Cartier, P.: Fonctions polylogarithmes, nombres polyzêtas et groupes pro-unipotents. Séminaire Bourbaki, Vol. 2000/2001. Astérisque, No. 282, (2002), Exp. No. 885, viii, 137–173.

[CaVo90] Cartier, P., Voros, A.: Une nouvelle interprétation de la formule des traces de Selberg. The Grothendieck Festschrift, Vol. II, 1–67

[Cas59a] Cassels, J.W.S. An introduction to the geometry of numbers. Berlin, 1959. Zbl.86,262.

[Cas59b] Cassels, J.W.S. Arithmetic on curves of genus one. J. reine u. angew. Math., **202** (1959), 52-99. Zbl.90,30.

[Cas66] Cassels, J.W.S. (1966): Diophantine equations with special reference to elliptic curves. J. Lond. Math. Soc., 41 (1966), 193-291. Zbl.138.270.

[Cas78] Cassels, J.W.S. (1978): Rational quadratic forms. Acad. Press, 1978. Zbl.395.10029.

[Cas84] Cassels, J.W.S., Bremner, A., (1984): On the equation $y^2 = x(x^2 + p)$. Math. Comput., **42** (1984) 257-264. Zbl.531.10014.

[CF67] Cassels, J.W.S., Fröhlich, A., eds. (1967): Algebraic Number Theory. Proc. Int. Congr. Lond. Math. Soc. Washington DC, Thompson, 1967. Zbl.153,74.

[Ch-LT02] Chambert-Loir, A., Tschinkel, Y.: On the distribution of points of bounded height on equivariant compactifications of vector groups. Invent. Math. **148** (2002), no. 2, 421–452.

[Cha70] Chandrasekharan, K. Arithmetical functions. Berlin–Heidelberg–New York, Springer–Verlag (1970). Zbl.217,316.

[Ch-L01] Chambert-Loir, A.: Théorèmes d'algébrisation en géométrie diophantienne, d'après J.-B. Bost, Y. André et D & G. Chudnovsky. Séminaire Bourbaki. [Exposé 886] Mars 2001

[Chebo25] Chebotarev, N.G. Die Bestimmung der Dichtigheit einer Menge von Primzahlen, welcher zu einer gegebener Substitutions Klasse gehören. Math. Annalen, **95** (1925), 151-228. Jbuch FdM 51,149.

[Chebo49] Chebotarev, N.G. Introduction to the theory of algebras (in Russian). Moscow, OGIZ, 1949. Zbl.38,172.

[Cheby55] Chebyshev, P.L. Collected Works (in Russian). Moscow: Izdatelstvo Akad. Nauk SSSR, 1955. Zbl.64,1.

[Chev40] Chevalley, C. (1940): La théorie du corps de classes. Annals of Math., **41**, 394-418 (1940). Zbl.25,18.

[Chev51] Chevalley, C. (1951): Deux théorèmes d'arithmétiques. J. Math.Soc. Japan, **3**, 36-44 (1951). Zbl.44,30.

[Cho48] Chowla, S. (1949): The last entry in Gauss' diary. Proc. Nat. Acad. Sci. USA, **35**, 244-246 (1949). Zbl.32,394.

[ChSe68] Chowla, S., Selberg A.: On Epstein's zeta-function, Journ. für die reine und angew. Math. **227**, 86 - 110(1967)

[Chu47] Chudakov, N.G. (1947): Introduction to the theory of the Dirichlet $L-$ series (in Russian). Moscow, Gostekhizdat (1947). Zbl.41,399.

[Clo86] Clozel, L. Base change for $GL(n)$. Proc. Int. Congr. Math. Berkeley, Calif., Aug. 3–11, 791-797 1986, Vol.**1**. Providence Rh. I., 1987. Zbl.666.22004.

[Clo93] Clozel, L.: Nombre de points des variétés de Shimura sur un corps fini, d'après R. Kottwitz. Séminaire Bourbaki, Exp. 766, (Mars 1993)

[Coa73] Coates, J. (1983): On Iwasawa's analogue of the Jacobian for totally real number fields. Anal. Number Theory, Missouri 1972. Proc. Symp. Pure Math. (1973), 51-61. Zbl.279.12006.

[Coa84] Coates, J. (1984): The work of Gross and Zagier on Heegner points and the derivatives of $L-$series. Séminaire Bourbaki, Exp. 633, 1984. Zbl.608.14020.

[Coa89] Coates, J.: On p–adic L–functions. Sem. Bourbaki, 40eme annee, 1987-88, no 701, Asterisque (1989) 177–178.

[Coa01] Coates, J.: Iwasawa algebras and arithmetic. Séminaire Bourbaki. [Exposé No.896] (Novembre 2001)

[CFKSV] Coates, J., Fukaya, T., Kato, K., Sujata, R., Venjakob, O.: The GL_2 main conjecture for elliptic curves without complex multiplication. Preprint, 2004.

[CSS03] Coates, J., Schneider, P ., Sujatha, R.: Links between cyclotomic and GL_2 Iwasawa theory. Kazuya Kato's fiftieth birthday. Doc. Math. 2003, Extra Vol., 187–215 (electronic).

[CW77] Coates, J., Wiles, A. On the conjecture of Birch and Swinnerton–Dyer. Invent. Math., **39**, no. 3 (1977), 223-251. Zbl.279.12006.

[CoYa97] Coates, J., Yau, S.T. (Eds.): Elliptic curves, modular forms & Fermat's last theorem. Proceedings of the Conference on Elliptic Curves and Modular Forms held at the Chinese University of Hong Kong, Hong Kong, December 18–21, 1993. Second edition. International Press, Cambridge, MA, 1997. iv+340 pp.

[CKPShSh] Cogdell, J.W., Kim, H.H., Piatetski-Shapiro, I.I., Shahidi, F.: Functoriality for the classical groups. Publ. Math. Inst. Hautes Études Sci. No. 99, (2004), 163–233.

[CKM04] Cogdell, J.W., Kim, H.H., Murty, M.: Lectures on automorphic L-functions. Fields Institute Monographs, **20**. American Mathematical Society, Providence, RI, 2004. xii+283

[CoPSh94] Cogdell, J.W., Piatetski-Shapiro, I.I.: Converse theorems for GL_n, Publications Mathématiques de l'I.H.E.S n 79, 157–214 (1994)

[CoPSh02] Cogdell, J.W., Piatetski-Shapiro, I.I.: Converse theorems, functoriality, and applications to number theory. Proceedings of the International Congress of Mathematicians, Vol. II (Beijing, 2002), 119–128, Higher Ed. Press, Beijing, 2002.

[Coh96] Cohen, H.: A course in computational algebraic number theory. Graduate Texts in Mathematics, 138. Springer-Verlag, Berlin, 1996. xii+534 pp. Third printing

[Coh2000] Cohen, H.: Advanced topics in computational number theory , Graduate Text in Mathematics,193, Springer, January 2000

[CoLe83] Cohen, H., Lenstra, H. W., Jr. : Heuristics on class groups of number fields. Number theory, Noordwijkerhout 1983 (Noordwijkerhout, 1983), 33–62, Lecture Notes in Math., 1068, Springer, Berlin, 1984.

[CoLe84] Cohen, H., Lenstra, H.W., Jr. (1984): Primality testing and Jacobi sums. Math. Comput., **42**, 297-330 (1984). Zbl.578.10004.

[CM98] Coleman, R., Mazur, B.: The eigencurve. Galois representations in arithmetic algebraic geometry (Durham, 1996), 1–113, London Math. Soc. Lecture Note Ser., 254, Cambridge Univ. Press, Cambridge, 1998.

[CThS80] Colliot-Thélène J.–L., Sansuc J.–J. (1980): La descente sur les variétés rationelles. I. Journée de géometrie algébrique, Angers/France, 1979, (1980), 223-237

[CThS87] Colliot-Thélène J.–L., Sansuc J.–J. (1987): La descente sur les variétés rationelles. II. Duke Math. J. 54 (1987), 375-422.

[Colm93] Colmez P.: Périodes des variétés abeliennes à multiplication complexe, Ann. of Math. 138 (1993) 625-683

[Colm2000] Colmez, P.: Fonctions L p-adiques. Séminaire Bourbaki. Volume 1998/99. Exposés 850-864. Paris: Société Mathématique de France, Astérisque. 266, 21-58 [Exposé No.851] (2000). Zbl. 0964.11055

[Colm01] Colmez, P.: Les conjectures de monodromie p-adiques. Séminaire Bourbaki. [Exposé No.897] (Novembre 2001)

[Colm03] Colmez, P.: La conjecture de Birch et Swinnerton-Dyer p-adique. Séminaire Bourbaki. [Exposé No.919] (Juin 2003)

[Co81] Connes, A.: An analogue of the Thom isomorphism for crossed products of a C^*-algebra by an action of $\mathbb{R}$, Adv. in Math. 39 (1981), no. 1, 31–55.

[Co83] Connes, A.: Cyclic cohomology and the transverse fundamental class of a foliation. In: Geometric methods in operator algebras (Kyoto, 1983). Pitman Res. Notes in Math., 123, Longman, Harlow 1986, 52–144.

[Co89] Connes, A.: Compact metric spaces, Fredholm modules, and hyperfiniteness, Ergodic Theory and Dynamical Systems 9 (1989) 207–220.

[Co94] Connes A.: Noncommutative geometry, Academic Press, 1994.

[Co95] Connes A.: Geometry from the spectral point of view. Lett. Math. Phys. 34 (1995), no. 3, 203–238.

[Co96] Connes A., Formule de trace en Géométrie non commutative et hypothèse de Riemann, C.R. Acad. Sci. Paris Ser. A-B, (1996)

[Co99] Connes A.: Trace formula in noncommutative geometry and the zeros of the Riemann zeta function. Selecta Math. (N.S.) 5 (1999), no. 1, 29–106.

[Co2000] Connes A.: A short survey of noncommutative geometry, J. Math. Phys. 41 (2000), no. 6, 3832–3866.

[Co2000a] Connes A.: Noncommutative geometry and the Riemann zeta function. In: Mathematics: Frontiers and Perspectives, ed. by V. Arnold et al., AMS, 2000, 35–54.

[Co03] Connes, A.: Nombres de Betti L^2 et facteurs de type II_1, d'après D. Gaboriau et S. Popa. Séminaire Bourbaki. [Exposé No.920] (Juin 2003)

[CoKr01] Connes, A., Kreimer, D.: Renormalization in quantum field theory and the Riemann-Hilbert problem. II. The β-function, diffeomorphisms and the renormalization group. Comm. Math. Phys. **216** (2001), N.1, 215–241.

[CoMar04] Connes, A., Marcolli, M.: From physics to number theory via noncommutative geometry. ArXiv, math.AG/0404128, 2004

[CoMo95] Connes, A., Moscovici, H.: The local index formula in noncommutative geometry. Geom. Funct. Anal. 5 (1995), no. 2, 174–243.

[CoMo04] Connes, A., Moscovici, H.: Modular Hecke algebras and their Hopf symmetry, Moscow Mathematical Journal, **4**, 67-109 (2004)

[CoMo04a] Connes, A., Moscovici, H.: Rankin–Cohen brackets and the Hopf algebra of transverse geometry. Moscow Mathematical Journal, **4**, 111-130 (2004)

[CR01] Conrad B., Rubin K. (eds.): Arithmetic algebraic geometry (Park City, UT, 1999), 143–232, IAS/Park City Math. Ser., **9**, Amer. Math. Soc., Providence, RI, 2001.

[Cons98] Consani, C.: Double complexes and Euler L–factors, Compositio Math. 111 (1998) 323–358.

[CM] Consani, C., Marcolli, M.: Noncommutative geometry, dynamics and ∞-adic Arakelov geometry, preprint ArXiv:math.AG/0205306, to appear in Selecta Mathematica.

[CM03] Consani, C., Marcolli, M.: Spectral triples from Mumford curves, International Math. Research Notices, 36 (2003) 1945–1972.

[CM04a] Consani, C., Marcolli, M.: New perspectives in Arakelov geometry. CRM Proceedings and Lecture Notes 36 AMS (2004), 79-100 CNTA VII Meeting, Montreal May 2002.

[CM04b] Consani, C., Marcolli, M.: Archimedean Cohomology Revisited. ArXiv, math.AG/0407480, 2004

[CoNo79] Conway, J. H., Norton, S.: Monstrous moonshine, Bull. London. Math. Soc. **11**, 308-339 (1979)

[CS86] Cornell, G., Silvermann, J.H. (eds). Arithmetic Geometry. N.Y. etc; Springer–Verlag, 1986. Zbl.596.00007.

[CSS95] Cornell, G., Silverman, J. H., Stevens, G. (eds.) (1995): Modular forms and Fermat's last theorem. Papers from a conference, Boston, MA, USA, August 9–18, 1995

[Cor02] Cornut, Ch.: Mazur's conjecture on higher Heegner points. Invent. Math. 148 (2002), no. 3, 495–523.

[Cre97] Cremona, J.E. (1997): Algorithms for Modular Elliptic Curves. (2nd ed.). Cambridge Univ. Press, Cambridge (1997)

[Cu] Cuntz, J.: A class of C^*–algebras and topological Markov chains II: reducible chains and the Ext–functor for C^*–algebras, Invent. Math. 63 (1981) 25–40.

[CuEch01] Cuntz, J., Echterhoff, S.: (eds.) C^*–algebras, Springer Verlag 2001, pp.182–202.

[CuKrie] Cuntz, J., Krieger, W.: A class of C^*–algebras and topological Markov chains, Invent. Math. 56 (1980) 251–268.

[Dalbo] Dalbo, F.: Codage des géodésiques fermées des quotients de l'espace hyperbolique par un groupe libre. Influence des transformations paraboliques, Séminaire Gaston Darboux de Géométrie et Topologie Différentielle, 1994–1995 (Montpellier) **iii**, 3–19 (1995).

[Da95] Darmon, H., The Shimura-Taniyama conjecture (d'après Wiles). (In Russian). Uspekhi Mat. Nauk **50**, No. 3(303), 33–82 (1995). English translation: Russian Math. Surveys **50**, No. 3, 503–548 (1995)

[Da99] Darmon H. (1999): A proof of the full Shimura-Taniyama-Weil conjecture is announced. Notices of the AMS, December 1999, 1397-1401.

[DDT97] Darmon, H., Diamond, F., Taylor, R. (1995): Fermat's Last Theorem. In: [CoYa97], pp. 2–140.

[Dav52] Davenport, H.: The higher arithmetic. London, Hutchinson (1952). Zbl.49,309.

[Dav58] Davenport, H. The work of K.E.Roth. Proc. Int. Congr. Math., Edinburg, 1958. Cambridge Univ. Press, 1960. Zbl.119,249.

[Dav80] Davenport, H.: Multiplicative Number Theory. 2nd ed. Berlin – Heidelberg – New York, Springer–Verlag (1980). Zbl.453.10002, Zbl.159,63.

[DH34] Davenport, H., Hasse, H.: Die Nullstellen der Kongruenz Zetafunktion in gewissen zyklischen Fällen. J. reine u. angew. Math., **172**, 151-182 (1934). Zbl.10,338.

[DMR74] Davis, M., Matiyasevic, Ju., Robinson, J. (1974): Hilbert's tenth problem. Proc. Symp. Pure Math., Vol.**23**, 1974, 323-378. Zbl.346.02026.

[Ded69] Dedekind, R. Mathematische Werke. Vols. **I, II**. New York: Chelsea, 1969.

[Del68] Deligne, P. (1968): Formes modulaires et représentations l–adiques. Séminaire Bourbaki, Exp. 355, 1968/69. Zbl.206,499.

[Del72] Deligne, P. (1972): Théorie de Hodge II. Publ. Math. IHES, 40 (1972). 5-57. Zbl.219,91.

[Del73] Deligne, P.: Les constantes des equations fonctionnelles des fonctions L. Exp. 349, 1973/74. Zbl.271.14011.

[Del74] Deligne, P.: La conjecture de Weil. I. Pub. Math. IHES, **43** (1974), 273-307. Zbl.287.14001.

[Del79] Deligne, P.: Valeurs de fonctions L et périodes d'intégrales. In: [BoCa79], Part **2**, 313-346. Zbl.449.10022.

[Del80a] Deligne, P. (1980): Sommes de Gauss cubiques et revetements de $SL(2)$, d'après S. Patterson. In: Lecture Notes in Math., vol. 770, 244-277 (1980) Zbl.433.10018.

[Del80b] Deligne, P.: La conjecture de Weil.II. Publ. Math. IHES, vol. **52** (1980). Zbl.456.14014.

[Del83] Deligne, P. (1983): Preuve des conjectures de Tate et de Shafarevitch. Sém. Bourbaki, Exp. 616, 1983/84.

[De-Hu] Deligne, P., Husemöller, D.: Survey of Drinfel'd modules, Current Trends in Arithmetical Algebraic Geometry, Contemporary Mathematics, **67** (1987), 25 - 91

[DMOS] Deligne P., Milne J., Ogus A., Shih K. (1982): Hodge cycles, motives and Shimura varieties. Lect. Notes Math. 900 (1982).

[DS75] Deligne, P., Serre, J.–P.: Formes modulaires de poids 1. Ann. Sci. ENS, IV ser., **7** (1975), 507-530. Zbl.321.10026.

[DF40] Delone, B.N., Faddeev, D.K. (1940): Theory of cubic irrationalities. (in Russian). Steklov Math. Inst. Trudy, 11 (1940). Zbl.61,90.

[DLPvG] Denef, J., Lipshitz, L., Pheidas, T., van Geel, J. (eds.): Hilbert's tenth problem: relations with arithmetic and algebraic geometry, American Mathematical Society, Providence, RI, 2000, Papers from the workshop held at Ghent University, Ghent, November 2–5, 1999.

[Den91] Deninger C.: On the Γ–factors attached to motives, Invent. Math. 104 (1991) 245–261.

[Den94a] Deninger C.: Evidence for a Cohomological Approach to Analytic Number Theory. First European Math Congress, Vol. **1**, 491-510 (1992) Birkhauser, (1994).

[Den92] Deninger C.: Local L-factors of motives and regularized determinants. Invent. Math. 107 (1992), no. 1, 135–150.

[Den94] Deninger C.: Motivic L-functions and regularized determinants, in "Motives", Proceedings of Symposia in Pure Mathematics, Vol. 55 (1994) Part I, 707–743.

[Den01] Deninger, C.: Number theory and dynamical systems on foliated spaces, preprint, 2001.

[Des90] Deshouillers, J.-M.: L'étude des formes cubiques rationnelles via la méthode du cercle, d'après D. R. Heath-Brown, C. Hooley et R. C. Vaughan. Séminaire Bourbaki, [Exposé No 720], Mars 1990

[Deu68] Deuring, M.: Imaginäre quadratische Zahlkörper mit der Klassenzahl Eins. Inv. Math., 5, no.**3** (1968), 169-179.

[Dic52] Dickson, L.E. (1952): History of the Theory of Numbers. Vols. **I-III**. Chelsea, 1952.

[DH76] Diffie, W., Hellman, M.E.(1976): New directions in cryptography. IEEE Trans. Inform. Theory, **22** (1976), 644-654. Zbl.435.94018.

[Dio84] Diophantus: Arithmetics and Book on Polygonal Numbers. (in Russian). Nauka, Moscow, 1984.

[Dir68] Dirichlet, P.G.L., Dedekind, R. (1968): Vorlesungen über Zahlentheorie. New York –Chelsea (1968). (Reprint 1893)

[Dix84] Dixon, J.D. (1984): Factorization and primality tests. Amer. Math. Monthly, **91** (1984), 333-352. Zbl.548.10003.

[DGM90] Dolan, L., Goddard, P., Montague, P.: Conformal field theory, triality and the Monster group. Phys. Lett. B 236 (1990), no. 2, 165–172.

[DP68] Dolgachev, I.V., Parshin, A.N.: The different and the discriminant of regular maps. Mat. Zametki, **4** (1968), 519-523. English transl.: Math.Notes **4** (1969), 802-804. Zbl.177,242.

[Dr] Drinfeld, V.G.: Elliptic modules, I, II. Math.USSR-Sbornik, **23**,1976, 561-592; **31**, 1977, 159-170

[Dr80] Drinfeld, V.G. (1980): Langlands' conjecture for $GL(2)$ over functional fields. Proc. Int. Congr. Math. Helsinki, 15 – 23 Aug., vol. **2** (1980), 565-574. Zbl.444.12004.

[DFI97] Duke, W., Friedlander, J., Iwaniec, H.: Representations by the Determinant and Mean Values of L-Functions. In Sieve Methods, Exponential Sums and their Applications in Number Theory, Cambridge University Press, 109-115, (1997).

[Dw59] Dwork, B. (1959): On the rationality of the zeta–function. Amer. J. of Math., **82** (1959), 631-648. Zbl.173,485.

[Edx95] Edixhoven S. J., Serre's conjecture. In: [CSS95], pp. 209-241

[Edx2000] Edixhoven, Bas: Rational elliptic curves are modular, after Breuil, Conrad, Diamond and Taylor. Séminaire Bourbaki. [Exposé No.871] Mars 2000

[Edw74] Edwards, H.M. Riemann's zeta–function. New York, Academic Press, 1974. Zbl.315.10035.

[Edw77] Edwards, H.M. (1977): Fermat's last theorem. A genetic introduction to algebraic number theory. New York: Springer–Verlag, 1977. Zbl.355.120-01.

[Eich54] Eichler, M. (1954): Quaternäre quadratische Formen und die Riemannsche Vermutung für die Kongruenzzetafunktion. Arch. Math. **5**, No. 4–6, 355–366 (1954). Zbl.59,38.

[Eich73] Eichler, M. (1973): The basis problem for modular forms and the traces of the Hecke operators. Lecture Notes in Math., vol. 320, 1973, 75-151. Zbl.258.10013.

[Eis1847] Eisenstein, G., Beiträge zur Theorie der elliptischen Funktionen. J. reine u. angew. Math., **35**, 135-274 (1847).

[Elk88] Elkies, N.D.: On $A^4 + B^4 + C^4 = D^4$. Math. Comput., **51**, No. 184, 825-835 (1988). Zbl.698.10010.

[ERS91] Eskin, A., Rudnick, Z., Sarnak, P.: A proof of Siegel's weight formula. Internat. Math. Res. Notices 1991, no. 5, 65–69.

[Fal83] Faltings, G.: Endlichkeitssätze für abelschen Varietäten über Zahlkörpern. Inv. Math., **73**, No.3 (1983), 349-366. Zbl.588.14026.

[Fal84] Faltings, G.: Calculus on arithmetic surfaces. Ann. of Math. (2) **119**, no. 2, 387–424 (1984)

[Fal85] Faltings, G.: Arithmetische Kompaktifizierung des Modulraums der abelscher Varietäten. In: Lecture Notes in Math., Vol. **1111** (1985), 326-383. Zbl.597.14036.

[Fal86] Faltings, G.: Recent progress in arithmetic algebraic geometry. Proc. ICM, Berkeley, 1986. Zbl.664.14012.

[Fal91] Faltings, G. (1991): Diophantine approximation on abelian varieties Ann. Math., II.Ser. **133**, No.3 (1991), 549-576. Zbl.734.14007.

[Fal92] Faltings, G.: Lectures on the arithmetic Riemann-Roch theorem. Notes taken by Shouwu Zhang. Annals of Mathematics Studies, 127. Princeton University Press, Princeton, NJ, 1992. x+100 pp.

[Fal98] Faltings, G.: Curves and their fundamental groups, following Grothendieck, Tamagawa and Mochizuki. Séminaire Bourbaki. [Exp. No.840] Mars 1998

[Fal02] Faltings, G.: A new application of Diophantine approximations. A panorama of number theory or the view from Baker's garden (Zürich, 1999), 231–246, Cambridge Univ. Press, Cambridge, 2002.

[FW84] Faltings, G., Wüstholz, G.: Rational points. Braunschweig, Vieweg, 1984. Zbl.588.14027.

[Fel82] Feldman, N.I. (1982): The seventh Hilbert problem. (in Russian). Moscow, MGU (1982). Zbl.515.10031.

[FelNes98] Feldman, N. I.; Nesterenko, Yu. V.: Transcendental numbers. Number theory, IV, 1–345, Encyclopaedia Math. Sci., 44, Springer, Berlin, 1998.

[Fer88] Ferrero, B. (1988): Iwasawa invariants of abelian number fields. Math. Ann., **234**, No.1, 9-24 (1988). Zbl.347.12004.

[FeWa79] Ferrero, B., Washington, L.C. (1979): The Iwasawa invariant μ vanishes for Abelian number fields. Ann. Math., **109**, No.2, 377-395 (1979) Zbl.443.12001.

[Fom77] Fomenko, O.M., (1977): Applications of modular forms to number theory (in Russian). Moscow, VINITI, Itogi Nauki, **15**, 5-91 (1977). Zbl.434.10018.

[Fon81] Fontaine, J.M. (1981): Il n'y a pas des variétés abéliénnes sur $\mathbb{Z}$. Inv. Math., **81**, 515-538 (1981). Zbl.612.14043.

[Fon92] Fontaine, J.M. (1992): Valeurs spéciales des fonctions L des motifs. Séminaire Bourbaki. [Exposé No.751] Février 1992

[FP-R94] Fontaine, J.M., Perrin-Riou B.(1994): Autour des conjectures de Bloch et Kato: cohomologie galoisiennes et valeurs de fonctions L. In Motives (Seattle, WA, 1991), 599–706, Proc. Sympos. Pure Math., 55, Part 1, Amer. Math. Soc., Providence, RI, 1994]

[Fou85] Fouvry, E.: Théorème de Brun–Tithchmarsh; application au théorème de Fermat. Inv. Math., 79, no.**2** (1985), 383-407. Zbl.557.10035.

[dFr13] Franchis, M. de(1913): Un teorema sulle involuzioni irrazionali. Rend. Circ. Mat. Palermo, 36 (1913), 368. Jbuch FdM 44,657.

[FMTsch] Franke, J., Manin, Yu.I., Tschinkel, Yu. (1989): Rational points of bounded height on Fano varieties. Inv. Math., **95** (1989). Zbl.674.14012.

[Fr86a] Frey, G. (1986): Links between stable elliptic curves and certain diophantine equations. Ann. Univ. Sarav., Ser. Math. 1, No.1, 40 p. (1986)

[Fr01] Frey, G.: Applications of arithmetical geometry to cryptographic constructions. Finite fields and applications (Augsburg, 1999), 128–161, Springer, Berlin, 2001.

[Frey86] Frey, G., (1986): Some aspects of the theory of elliptic curves over number fields. Expos. Math., 4, no.**1**, 35-66 (1986). Zbl.596.14022.

[Fro75] Fröhlich, A., (1975): Galois module structure and Artin $L-$functions. Proc. Int. Congr. Math. Vancouver, vol.**1**, 351-356, (1975). Zbl.346.12006.

[Gal01] Galbraith, S.: Supersingular curves in cryptography. Advances in cryptology—ASIACRYPT 2001 (Gold Coast), 495–513, Lecture Notes in Comput. Sci., 2248, Springer, Berlin, 2001.

[GNSh84] Galochkin, A.I., Nesterenko, Yu.V., Shidlovskii, A.B. (1984): Introduction to Number Theory. (in Russian). Moscow, MGU (1984). Zbl.537.10001.

[Gal] Galois, E., (1984): Oeuvres Mathématiques. Paris, Gauthier–Villars. 1897. Jbuch FdM 28,11.

[Gan71] Gandhi, J.M., (1984): Formulae for the $n-$th prime. Proc. Washington State Univ. Conf. on Number Theory. Washington State University, Pullman, 96-106 (1971). Zbl.228.10004.

[GJ79] Garey, M.R., Johnson, D.S. (1979): Computers and intractability: A Guide to the theory of $NP-$completeness. San Francisco, Freeman, (1979). Zbl.411.68039

[Gau] Gauss, C.–F. (1966): Disquisitiones Arithmeticae. Yale University Press (1966). Zbl.136,323.

[Gel75] Gelbart, S.(1975): Automorphic forms on adèle groups. Ann. Math. Stud., **83**, (1975). Zbl.329.10018.

[Gel76] Gelbart, S. (1976): Elliptic curves and automorphic representations. Adv. Math., **21**, No.3, 235-292 (1976). Zbl.336.14003.

[Gel77] Gelbart, S. (1977): Automorphic forms and Artin's conjecture. Lecture Notes in Math., vol. **627**, 241-276 (1977). Zbl.368.10023.

[Gel95] Gelbart, S. (1995): Three lectures on the modularity of $\overline{\rho}_{E,3}$ and the Langlands reciprocity conjecture. In: [CSS95], 153–207

[GPShR87] Gelbart, S., Piatetski–Shapiro, I., Rallis, S. (1987): Explicit constructions of automorphic $L-$functions. Lecture Notes in Math., vol. **1254**, 1–152 (1987). Zbl.612.10022.

[Ge73] Gelfond, A.O. (1973): Collected Works (in Russian). Moscow, Nauka (1973). Zbl.275.01022.

[Ge83] Gelfond, A.O.: Integer solutions of equations. (in Russian). Moscow: Nauka (1983). Zbl.557.10014, Zbl.48,28.

[Gil95] Gilkey, P.B.: Invariance Theory, the Heat Equation, and the Atiyah-Singer Index Theorem. Studies in Adv. Mathematics, Second Edition. CRC Press, (1995).

[GS84] Gillet, H., Soulé, C. (1984): Intersection sur les variétés d'Arakelov. C.R. Ac. Sci. Paris, **299**, Ser. 1, no. 12, (1984), 563-566. Zbl.607.14003.

[GS85] Gillet H., Soulé C. (1985): Classes caractéristiques en théorie d'Arakelov. C.R. Acad. Sci., Paris, Ser.I 301 (1985), 439-442.

[GS91] Gillet, H., Soulé, C. (1991): Analytic torsion and arithmetic Todd genus, with an Appendix by D. Zagier, Topology **30** , 21–54 (1991)

[GS92] Gillet, H., Soulé, C. (1992): An arithmetic Riemann-Roch theorem. Invent. Math. 110 (1992), no. 3, 473–543.

[Gin85] Gindikin, S.G. (1985): Essays about mathematicians and physicists. (in Russian). Moscow: Nauka, 1985. English transl.: Boston/Basel: Birkhäuser, 1988. Zbl.566.01011.

[Gö] Gödel K. (1931): Über formal unentscheidbare Sätze der Principia mathematica und verwandter Systeme I. Monatshefte für Math. und Physik 38 (1931), 173-198.

[Gol76] Goldfeld, D.M. (1976): The class number of quadratic fields and conjectures of Birch and Swinnerton–Dyer. Ann. Scuola Norm. Sup. Pisa, 3, no.**4**, 623-663 (1976). Zbl.345.12007.

[Gol85] Goldfeld, D.M. (1985): Gauss class number problem for imaginary quadratic fields. Bull. AMS, **13** (1985), 23. Zbl.572.12004.

[Gol94] Goldfeld D., A spectral interpretation of Weil's explicit formula, Lecture Notes in Math., **1593**, Springer Verlag (1994), 135-152.

[G(ed.)86] Goldstein, C. (ed). (1986): Séminaire de la théorie de nombres. Paris, 1984-85. Boston etc., Birkhäuser (1986). Zbl.593.00007.

[GoYi03] Goldston, D., Yildirim C.Y.: Higher correlations of divisor sums related to primes, I: Triple correlations, Integers **3**, A5, 66 pp. (2003)

[Go02] Goldwasser, S.: Mathematical foundations of modern cryptography: computational complexity perspective. In: [ICM02]

[GK86] Goldwasser, S., Kilian, J. (1984): A probably correct and probably fast primality test. Preprint MIT, 1985. Proc. 18th Ann. ACM Symp. on the Theory of Comp. (STOC): Berkeley, May **28-30**, 1986.

[GK99] Goldwasser, S., Kilian, J. (1999): Primality testing using elliptic curves. J. ACM 46 (1999), no. 4, 450–472.

[GF77] Golubeva, E.P., Fomenko, O.M. (1977): On the zeta–function of a system of forms. (in Russian). Zapiski Nauchn. Sem. LOMI, 67 (1977), 156-166. English transl.: J.Sov.Math. 16 (1981), 866-870. Zbl.364.10013.

[Goo96] Goode, J. B.: H. L. M. (Hrushovski-Lang-Mordell) Séminaire Bourbaki. [Exp. No.811] Février 1996

[Gou94] Gouvêa, F.: A marvelous proof, Amer. Math. Monthly **101**, no. 3, 203–222 (1994).

[Gow01] Gowers, T.: A new proof of Szemerédi's theorem, GAFA **11**, 465-588 (2001).

[GM] Granville, A., Monagan M.B. (1988): The first case of Fermat's last theorem is true for all prime exponents up to 714 591 416 091 389. Trans. AMS, **306**, No. 1, 329-359. Zbl.645.10018.

[GR71] Grauert, H. , Remmert, R. : Analytische Stellenalgebren. Unter Mitarbeit von O. Riemenschneider. Grundlehren der mathematischen Wissenschaften 176. Springer: Berlin-Heidelberg-New York (1971).

[Gram98] Gramain, F.: Quelques résultats d'indépendance algébrique. Proceedings of the International Congress of Mathematicians, Vol. II (Berlin, 1998). Doc. Math. 1998, Extra Vol. II, 173–182 (electronic).

[Gr83] Greenberg, R. (1983): On the Birch–Swinnerton–Dyer conjecture. Inv. Math., **72**, No.2, 241-266 (1983). Zbl.546.14015.

[GrTa] Green, B.J., Tao, T.: The primes contain arbitrary long arithmetic progressions, preprint. ArXiv math.OA/0404188, 8 Apr 2004

[GZ86] Gross, B.H., Zagier, D.B. (1986): Heegner points and derivatives of $L-$series. Inv. Math., **84**, 225-320 (1986). Zbl.608.14019.

[GKZ87] Gross, B.H., Kohnen, W., Zagier, D.B. (1987): Heegner points and derivatives of $L-$series.II. Math. Ann., **278**, No. 1-4, 497-562 (1987). Zbl.641.14013.

[Gr84] Grosswald, E. (1984): Topics from the theory of numbers. 2nd ed. Birkhäuser, 1984. Zbl.532.10001.

[GrothF] The Grothendieck Festschrift: Cartier P., Illusie L., Katz N.M., Laumon G., Manin Yu., Ribet K.A. eds. (1990) The Grothendieck Festschrift. Vols I-III (1990)

[Gui77] Guillemin, V.: Lectures on spectral theory of elliptic operators, Duke Math. J., **44**, No.3 (1977), 485-517

[GuS77] Guillemin, V., Sternberg, S.: Geometric asymptotics, Math. Surveys, **14**, Amer. Math. Soc., Providence, R.I. (1977)

[GNA] Guillén, F., Navarro Aznar, V.: Sur le théorème local des cycles invariants. Duke Math. J. 61 (1990), no. 1, 133–155.

[Gu81] Guy, R.K. (1984): Unsolved problems in number theory (Problem books in Mathematics). Berlin etc., Springer–Verlag (1981). Zbl.474.10001.

[Har90] Haran S.: Riesz potentials and explicit sums in arithmetic, Invent. Math., **101** (1990), 697-703.

[HaSk02] Harari, D., Skorobogatov, A. N.: Non-abelian cohomology and rational points. Compositio Math. **130**, no. 3, 241–273 (2002).

[HaWr] Hardy, G.H., Wright, E.M. (1979): An introduction to the theory of numbers, 5th ed. Oxford Univ. Press, 1979. Zbl.58,33.

[Ha98] Harris, M.: The local Langlands conjecture for $GL(n)$ over a p- adic field, $n < p$, Invent. Math. **134** (1998), 177–210.

[HaTsch] Harris, J., Tschinkel, Y.: Rational points on quartics. Duke Math. J. **104**, no. 3, 477–500 (2000).

[Ha77] Hartshorne, R. (1977): Algebraic geometry. New York, Springer Verlag (1977). Zbl.367.14001.

[Ha50] Hasse, H. (1950): Vorlesungen über Zahlentheorie. Berlin, Springer Verlag (1950). Zbl.38,17. 2nd ed. 1964.

[Ha37] Hasse, H. (1937): The Riemann hypothesis in function fields. C. R. Congr. Int. Math. **1** (1937), 189-206 Philadelphia: Pennsylvania State University Press (1989). Zbl.18,343.

[Haz78] Hazewinkel, M. (1978): Formal groups and applications. Acad. Press, 1978. Zbl.454.14020.

[HB84] Heath–Brown, D.R. (1984): Cubic forms in 10 variables. Lecture Notes in Math., vol. **1068**, 104-108 (1984). Zbl.538.10021.

[HB86] Heath–Brown, D.R. (1986): Artin's conjecture for primitive roots. Quart. Journ. of Math., **37**, 27-38 (1986). Zbl.586.10025.

[HB88] Heath–Brown, D.R. (1988): The number of primes in a short interval. J. reine u. angew. Math., **389**, 22-63 (1988). Zbl.646.10032.

[HBP79] Heath–Brown, D.R., Patterson, S.J. (1979) The distribution of Kummer sums at prime arguments. J. reine u. angew. Math., **130**, 111-130 (1979). Zbl.412.10028.

[He29] Hecke, E. (1929) Algebraische Zahlentheorie. Leipzig. Reprint by Chelsea Publ. Comp. Inc. New York, 1929. Jbuch FdM 49,106.

[He59] Hecke, E. Mathematische Werke. Göttingen: Vandenhoeck und Ruprecht, 1959. Zbl.92,1.

[Hee52] Heegner, H. (1952): Diophantische Analysis und Modulfunktionen. Math. Zeitschrift, **56**, 227-253 (1952). Zbl.49,162.

[He97] Hellegouarch, Y. (1997): Invitation aux mathématiques de Fermat-Wiles. Enseignement des Mathématiques. Paris: Masson. vii, 397 p. (1997)

[Hel79] Hellman, M.E. The mathematics of public–key cryptography. Sci. Am., **241**, 146-157 (1979).

[Hen76] Henniart, G.: Une forme icosaèdrale de poids 1. Sém. Délange–Pisot–Poitou. Théor. de Nombres, Univ. Pierre et Marie Curie, 1976-77, **18**, No.2, 24/01-24/07. Zbl.367.10021.

[Hen86] Henniart, G.: On the local Langlands conjecture for GL(n): the cycle case, Annals of Mathematics 123, 145–203 (1986)

[Hen01] Henniart, G.: Progrès récents en fonctorialité de Langlands. Séminaire Bourbaki. [Exposé No.890] (Juin 2001)

[Hi93] Hida H. (1993): Elementary theory of L-functions and Eisenstein series, London Mathematical Society Student Texts. **26** Cambridge University Press

[Hi02] Hida H.: The Iwasawa μ-invariant of p-adic Hecke L-functions, preprint, 2002, http://www.math.ucla.edu/~hida

[Hil1900] Hilbert, D. (1900): Mathematical problems. ICM, Paris, 1900. In: Math. Developments arising from Hilbert Problems. Proc. Symp. Pure Math. AMS, **28**, 1-34 (1976).

[Hil1897] Hilbert, D. (1897): Die Theorie der algebraischer Zahlkörper. J.-ber. Deutsch. Math. Verein. **4**, 175-546 (1897) Gesammelte. Abh. New York: Chelsea (1965) 63-363. (1932) Springer–Verlag. Zbl.4,98.

[Hild86] Hildebrand, A. (1986): On the number of positive integers $> x$ and free of prime factors $> y$. J. Number Theory, **22**, 289-307 (1986). Zbl.575.10038.

[Hild88] Hildebrand, A. (1988): Gaps between prime numbers. Proc. AMS, **104**, No.1 (1988), 1-9. Zbl.663.10046.

[Hir88] Hiramatsu Toyokazu, Theory of automorphic forms of weight 1. Adv. Stud. Pure Math., Vol. **13** Invest. Number Theory. Tokyo, 1988, 32-98. Zbl.658.10031.

[Hoo] Hooley, Ch. Applications of the sieve methods in the theory of numbers. Cambridge Tracts in Mathematics, No. 70. Cambridge University Press, Cambridge-New York-Melbourne, 1976. Zbl.327.10044.

[H88] Hooley, Ch. (1988): On nonary cubic forms. J. reine u. angew. Math., **386**, 32-98 (1988). Zbl.641.10019.

[Hri85] Hriljac, P. (1985): Heights and Arakelov's intersection theory. Amer. J. Math., **107**, No. 1, 23-38 (1985). Zbl.593.14004.

[Hua59] Hua Loo–Keng (1959): Abschätzungen von Exponentialsummen und ihre Anwendungen in der Zahlentheorie. Leipzig, Teubner, 1959. Zbl.83,36.

[HW81] Hua Loo–Keng, Wang, L.: Application of number theory to numerical analysis. Berlin etc., Springer (1981). Zbl.451.10001.

[Huls94] Hulsbergen, W.W.J.: Conjectures in Arithmetic Algebraic Geometry. A Survey. Second revised ed., AMS, 1994 Zbl.0793.14011

[Hur63] Hurwitz, A. Mathematische Werke. Bd. II. Basel und Stuttgart, Birkhäuser Verlag, 1963. Zbl.122,241.

[ICM94] Proceedings of the International Congress of Mathematicians. Vol. 1, 2. Held in Zürich, August 3–11, 1994. Edited by S. D. Chatterji. Birkhäuser Verlag, Basel, 1995. Vol. 1: lxxii+717 pp.; Vol. 2: pp. i–xiii and 718–1605.

[ICM98] Proceedings of the International Congress of Mathematicians. Vol. I. Invited plenary lectures. Appendix. Held in Berlin, August 18–27, 1998. Doc. Math. 1998, Extra Vol. I. Documenta Mathematica, Bielefeld, 1998. front matter and pp. 1–662 (electronic).

[ICM02] Proceedings of the International Congress of Mathematicians. Vol. I. Plenary lectures and ceremonies. Held in Beijing, August 20–28, 2002. Edited by Tatsien Li. With 1 CD-ROM [Windows]. Higher Education Press, Beijing, 2002. x+657 pp.

[IRS(e)89] Ihara Y., K.Ribet, Serre J.-P. eds. (1989) Galois Groups over $\mathbb{Q}$. Berlin–Heidelberg–New York, Springer–Verlag (1989).

[Ing] Ingham, A.E. (1932): Distribution of Prime Numbers. Cambrige Univ.Press (1932) (Reprint (1990)) Zbl.6,397

[Ire82] Ireland, K., Rosen, M. (982): A classical introduction to modern number theory. Berlin etc.: Springer Verlag (1982). Zbl.482.10001.

[Isk70] Iskovskih, V.A. (1970): A counterexample to the Hasse principle for the system of two quadratic forms of five variables. (in Russian). Mat. Zametki, **10**, 253-257 (1970). English transl.: Math.Notes **10**, 575-577 (1972). Zbl.232.10015.

[Isk77] Iskovskih, V. A.: Fano threefolds. I. (Russian) Izv. Akad. Nauk SSSR Ser. Mat. 41 (1977), no. 3, 516–562, 717. (Reviewer: Miles Reid)

[Isk78] Iskovskih, V. A.: Fano threefolds. II. (Russian) Izv. Akad. Nauk SSSR Ser. Mat. 42 (1978), no. 3, 506–549.

[Iw87] Iwaniec, H. (1987): Spectral theory of automorphic functions and recent developments in analytic number theory. Proc. ICM Berkeley, Vol. **1**, 444-456 (1987). Zbl.663.10027.

[Iw97] Iwaniec, H. (1997): Topics in classical automorphic forms. Graduate Studies in Mathematics, 17. American Mathematical Society, Providence, RI, 1997. xii+259 pp.

[IwSa99] Iwaniec, H., Sarnak, P.: Perspectives on the analytic theory of L-functions. GAFA 2000 (Tel Aviv, 1999). Geom. Funct. Anal. 2000, Special Volume, Part II, 705–741.

[Iwa72] Iwasawa, K. (1972): Lectures on p-adic L-functions. Ann. Math. Stud. 74 (1972). Zbl.236.12001.

[Iwa75] Iwasawa, K. (1975): A note on Jacobi sums. Symp. Math., 15 (1975), 447-459. Zbl.324.12007.

[Iw86] Iwasawa K. (1986): Local Class Field Theory. Oxford University Press (translation from Japanese), 1986. Russian translation: Moscow, Mir, 1984.

[Iw01] Iwasawa K.: Collected papers. Vol. I, II. Edited and with a preface by Ichiro Satake, Genjiro Fujisaki, Kazuya Kato, Masato Kurihara and Shoichi Nakajima. With an appreciation of Iwasawa's work in algebraic number theory by John Coates. Springer-Verlag, Tokyo, 2001. Vol. I: xxii+464 pp.; Vol. II: pp. i–vi and 465–880.

[JL70] Jacquet, H. Langlands, R.P. (1970): Automorphic forms on $GL(2)$. Lecture Notes in Math., **114** (1970). Zbl.236.12010.

[JPShS] Jacquet, H., Piatetski–Shapiro I.I., Shalika, J.A. (1979): Automorphic forms on $GL(3)$. I. Ann. of Math., **109**, No.1, 169-212 (1979). Zbl.401.10037.

[JSh] Jacquet, H., Shalika, J.A. (1976): A non–vanishing theorem for zeta functions of GL_n . Inv. Math., **38**, No.1, 1-16 (1976). Zbl.349.12006.

[Sh81] Jacquet, H., Shalika, J.A. (1981): On Euler products and the classification of automorphic forms. Amer. J. Math., **103**, No. 4, 777-815 (1981). Zbl.491.10020.

[Ja90] Jannsen U. (1990): Mixed Motives and Algebraic K–Theory, Lect. Notes Math. 1400 (1990).

[JSWW76] Jones, J.P., Sato, D., Wada, H., Wiens, D. (1979): Diophantine representation of the set of prime numbers. Amer. Math. Monthly, **83**, No.6, 449-464 (1976). Zbl.336.02037.

[Jul90] Julia B.: Statistical theory of numbers, Number Theory and Physics, Springer Proceedings in Physics, **47** (1990).

[Kac59] Kac, M.: Statistical Independence in Probability, Analysis and Number Theory, Carus Math. Monographs **18** (1959)

[Kac78] Kac, V.G. (1978): Infinite–dimensional algebras, Dedekind $\eta-$function, classical Möbius function and very strange formula. Adv. Math., **30**, 85-136 (1978). Zbl.391.17010.

[Kah71] Kahn, D. (1979): The Codebreakers, the story of secret writing. Macmillan (1971).

[Kar75] Karacuba, A.A. (1975): Introduction to the analytical number theory (in Russian). Moscow, Nauka (1975). Zbl.428.10019. English transl. of the 2nd edition: Basic Analytic Number Theory. Berlin–Heidelberg–New York: Springer–Verlag, 1993. Zbl.767.11001.

[Kar85] Karacuba, A.A. (1985): On the function $G(n)$ in Waring's problem. (in Russian). Izv. Akad. Nauk SSSR, Ser. mat., **49**, No.5 (1985), 935-947. English transl.: Math.USSR, Izv.27 (1986), 239-249. Zbl.594.10041.

[Kas] Kasparov, G.G.: K–theory, group C^*–algebras, and higher signatures (Conspectus). Novikov conjectures, index theorems and rigidity, Vol. 1 (Oberwolfach, 1993), 101–146, London Math. Soc. Lecture Note Ser., 226, Cambridge Univ. Press, 1995.

[Kato99] Kato, K.: Euler systems, Iwasawa theory, and Selmer groups. Kodai Math. J. 22 (1999), no. 3, 313–372.

[Kato2000] Kato, K.: Tamagawa number conjecture for zeta values. Proceedings of the International Congress of Mathematicians, Vol. II (Beijing, 2002), 163–171, Higher Ed. Press, Beijing, 2002.

[KKS2000] Kato, K., Kurokawa, N., Saito, T.: Number Theory I: Fermat's Dream. Translations of Mathematical Monographs, Vol. 186, AMS, xv+154 pp.

[Kat76] Katz, N.M. (1976): An overview of Deligne's proof of the Riemann hypothesis for varieties over finite fields. Proc. Symp. Pure Math., **28**, 275-305 (1976). Zbl.339.14013.

[Kat88] Katz, N.M. (1988): Gauss sums, Kloostermann sums and monodromy group. Princeton Univ. Press (1988). Zbl.675.14004.

[KL85] Katz, N.M., Laumon G. (1985): Transformation de Fourier et majoration de sommes exponentielles. Publ. Math. IHES, **62**, 361-418 (1985). Zbl.603.14015.

[KS99] Katz, N., Sarnak, P.: Random matrices, Frobenius eigenvalues, and monodromy. American Mathematical Society Colloquium Publications, 45. American Mathematical Society, Providence, RI, 1999. xii+419 pp.

[KS99a] Katz, N., Sarnak, P.: Zeroes of zeta functions and symmetry. Bull. Amer. Math. Soc. (N.S.) **36**, no. 1, 1–26 (1999).

[Khi78] Khinchin, A.Ya. (1978): Continuous fractions. (in Russian). Moscow: Nauka (1978). Zbl.117,286.

[Khi79] Khinchin, A. Ya. (1979): Three gems of number theory. (in Russian). Moscow: Nauka (1979). German transl. (Reprint): Frankfurt/Main 1984. Zbl.42,40; Zbl.539.10002.

[KiSha99] Kim, H., Shahidi, F.: Symmetric cube L-functions for GL_2 are entire, Annals of Math. 150 (1999), 645–662.

[Kli62] Klingen H.: Uber die Werte Dedekindscher Zetafunktionen . Math. Ann. **145**, 265–272 (1962)

[Knu81] Knuth, D.E. (1981): The art of computer programming. Vol **2**. Seminumerical algorithms. 2nd edition. Addison–Wesley, Reading (1981). Zbl.477.65002.

[Kob77] Koblitz, N. (1977): $p-$adic numbers, $p-$adic analysis and zeta–functions. New York: Springer Verlag (1977). Zbl.364.12015.

[Kob80] Koblitz, N. (1980): $p-$adic analysis: a short course on recent work. London Math. Soc. Lecture Note Ser., London: Cambridge Univ. Press (1980).

[Kob82] Koblitz, N. (1982): Why study equations over finite fields? Math. Mag., **55**, 144-149 (1982). 1980. Zbl.439.12011

[Kob84] Koblitz, N. (1984): Introduction to elliptic curves and modular forms. New York: Springer Verlag, 1984. Zbl.553.10019.

[Kob87] Koblitz, N. (1987): A course of number theory and cryptography. New York: Springer Verlag, 1987. Zbl.648.10001.

[Kob94] Koblitz, N.: A course in number theory and cryptography. Second edition. Graduate Texts in Mathematics, 114. Springer-Verlag, New York, 1994. x+235 pp.

[Kob2000] Koblitz, N.: A survey of number theory and cryptography. Number theory, 217–239, Trends Math., Birkhäuser, Basel, 2000.

[Kob98] Koblitz, N. (1998): Algebraic aspects of cryptography, Algorithms and Computation in Math., 3, Springer-Verlag, Berlin, 1998.

[Kob01] Koblitz, N.: Cryptography. Mathematics unlimited—2001 and beyond, 749–769, Springer, Berlin, 2001.

[Kob02] Koblitz, N. (2002): Good and bad uses of elliptic curves in cryptography. Mosc. Math. J. 2 (2002), no. 4, 693–715, 805–806.

[Koch70] Koch, H.V. (1970): Galoische Theorie der $p-$Erweiterungen. Berlin VEB, Deutscher Verlag d. Wiss, 1970. Zbl.216,47.

[Koch87] Koch, H.V. (1986): Unimodular lattices and self–dual codes. Proc. ICM Berkeley, 1986, vol. **1**, 457-465, Berkeley (1987). Zbl.668.10039.

[Koch97] Koch, H.: Algebraic number theory. Translated from the 1988 Russian edition Number Theory II. Reprint of the 1992 translation. Springer-Verlag, Berlin, 1997. iv+269 pp.

[Kog71] Kogan, L.A. (1971): On the representation of integers by positively defined quadratic forms. (in Russian). Tashkent: Fan, 1971. Zbl.227.10015.

[Kol69] Kolmogorov, A.N. (1969): On the logical foundations of the information theory and probability theory. (in Russian). Probl. Teorii Peredachi Informacii, **5**, No.3 (1969), 3-7. Zbl.265.94010.

[Koly79] Kolyvagin, V.A. (1979): Formal groups and the norm residue symbol. (in Russian). Izv. Akad. Nauk SSSR, Ser. mat., **43**, no.5, 1054-1120 (1979) English transl.: Math.USSR, Izv. 15 (1980) 289-348. Zbl.429.12009.

[Koly88] Kolyvagin, V.A. (1988): Finiteness of $E(\mathbb{Q})$ and $\text{Ш}(E, \mathbb{Q})$ for a class of Weil's curves. (In Russian). Izv. Akad. Nauk SSSR, ser. mat. 52, No. 3 (1988), 522-540.English transl.: Math.USSR, Izv. 32 (1989) 523-541. Zbl.662.14017.

[Koly90] Kolyvagin V.A. (1990): Euler systems. In: The Grothendieck Festschrift (1990), Vol. II, 435-484.

[dKL89] de Koninck J.-M., Levesque Cl. (1989): Théorie de nombres. Number Theory. de Gruyter (1989).

[Kon96] Kontsevich, M. Product formulas for modular forms on O(2,n), after R. Borcherds. Séminaire Bourbaki. Exp. No.821 (Novembre 1996)

[KoZa01] Kontsevich M., Zagier, D.: Periods, Mathematics Unlimited–2001 and Beyond, Springer, Berlin, 2001, pp. 771-808.

[Kor89] Korobov, N.M. (1989): Exponential sums and their applications. (in Russian). Moscow, Nauka (1989). English transl.: Dordrecht: Kluwer 1992. Zbl.665.10026.

[Kott88] Kottwitz, R. (1988): Tamagawa numbers. Ann. of Math., **127**, 629-646 (1988) Zbl.678.22012.

[KMP74] Kozmidiadi V.A., Maslov A.N., Petri V.N. eds.(1974): Complexity of computations and algorithms. (in Russian). Kozmidiadi V.A., Maslov A.N., Petri V.N. eds. Moscow, Mir (1974). Zbl.286.00008.

[Kr26] Kraitchik, M. (1926): Théorie des Nombres. Vol. 2. Paris, Gauthier–Villars (1926). Jbuch FdM 52,137.

[Kr86] Kranakis, E. (1986): Primality and Cryptography. John Wiley and Sons (1986). Zbl.595.10001.

[Kr1863] Kronecker L. (1863): Auflösung der Pellschen Gleichung mittels elliptischer Funktionen. Monatberichte der Königlichen Preuss. Akademie der

Wissenschaften zu Berlin, (1863), 44-50 (= Kronecker L., Werke. Band 4, 219-225)

[Kub69] Kubota, T. (1969): On automorphic functions and the reciprocity law in a number field. Lect. Mat. Dept. Math. Kyoto Univ., No. 2, Kinokuhiya bookstore Co., Tokyo (1969). Zbl.231.10017.

[KuLe64] Kubota, T., Leopoldt, H.–W. (1964): Eine p–adische Theorie der Zetawerte. I. J. reine u. angew. Math., **214/215**, 328-339 (1964). Zbl.186,91.

[Ku2000] Kudla, St. S.: Derivatives of Eisenstein series and generating functions for arithmetic cycles. Séminaire Bourbaki. [Exposé No.876] (Juin 2000)

[Kum75] Kummer, E.E. (1975): Collected papers. Vol. **1**. New York: Springer Verlag (1975). Zbl.327.01019.

[Kur91] Kurokawa, N.: Multiple sine functions and Selberg zeta functions, Proc. Jap. Acad, Sci., Ser. A, 67, no. 3 (1991), 61–64

[Kuz84] Kuzmin, L.V. (1984): Fields of algebraic numbers. (In Russian). , Moscow: VINITI, Itogi Nauki, **22**, 117-204 (1984). English transl.: J.Sov.Math. 38 (1987), 1930-1988. Zbl.621.12002.

[Lab86] Labesse, J.–P. (1986): La formule des traces d'Arthur–Selberg. Astérisque, **133-134**, 73-88 (1986). Zbl.592.22011.

[Laff02] Lafforgue, L.: Chtoucas de Drinfeld et correspondance de Langlands. Invent. Math. 147 (2002), no. 1, 1–241.

[Lan97] Landi, G.: An introduction to noncommutative spaces and their geometries, Lecture Notes in Physics, Vol. m-51, Springer Verlag 1997.

[La58] Lang S. (1958): Abelian Varieties. New York: Interscience, 1958. Zbl.98,132.

[La60] Lang S. (1960): Integral points on curves. Publ. Math. Inst. Hautes Etud. Sci. 6 (1960), 27-43. Zbl.112,134.

[La62] Lang, S. (1962): Diophantine Geometry. New York, Interscience (1962). Zbl.115,387.

[La64a] Lang, S.(1964a): Les formes bilinéaires de Néron et Tate. Sém. Bourbaki, Exp. 274 (1964). Zbl.138,421.

[La64b] Lang, S. (1964b): Algebraic numbers. Reading, Mass.: Addison– Wesley (1964). Zbl.193,347.

[La65] Lang, S. (1965): Algebra. Reading, Mass.: Addison–Wesley (1965). Zbl.211, 385.

[La70] Lang S. (1970): Algebraic Number Theory. Reading, Mass.: Addison–Wesley, 1970. Zbl.211, 385.

[La72] Lang, S. (1972): Introduction to algebraic and Abelian functions. New York, Springer (1973). Zbl.255.14001. 2nd ed.: New York - Berlin - Heidelberg: Springer–Verlag, 1982.

[La73/87] Lang, S. (1973): Elliptic functions. Reading, Mass.: Addison–Wesley (1973). Zbl.316. 14001. Second Edition: New York - Berlin - Heidelberg: Springer–Verlag, 1987.

[La76] Lang S. (1976): Introduction to Modular Forms. New York - Berlin - Heidelberg: Springer–Verlag, 1976. Zbl.316.10011.

[La78a] Lang, S. (1978a): Elliptic curves. Diophantine Analysis. Berlin etc.: Springer–Verlag, 1978. Zbl.388.10001.

[La78b] Lang, S. (1978b): Cyclotomic fields. New York e.a., Springer–Verlag (1978). Zbl.395.12005.

[La83] Lang, S. Fundamentals of Diophantine Geometry. New York: Springer–Verlag (1983). Zbl.528.14013.

[La88] Lang, S. Introduction to Arakelov Theory. Springer–Verlag (1988). Zbl.667.14001.

[La90] Lang, S.: Cyclotomic fields I and II, Combined second edition, Springer, New York, 1990;

[La90a] Lang, S.: Old and new conjectured Diophantine inequalities. Bull. Amer. Math. Soc. (N.S.) 23 (1990), no. 1, 37–75.

[La91] Lang, S.: Number theory. III. Diophantine geometry. Encyclopaedia of Mathematical Sciences, 60. Springer-Verlag, Berlin, 1991. xiv+296 pp.

[LaTa] Lang, S., Tate, J. Principal homogeneous spaces over Abelian varieties. Amer. J. Math., **80**, 659-684 (1958). Zbl.97,362.

[LaTr76] Lang, S., Trotter, H.: Frobenius distributions in GL_2-extensions. Distribution of Frobenius automorphisms in GL_2-extensions of the rational numbers. Lecture Notes in Mathematics, Vol. **504**. Springer-Verlag, Berlin-New York, 1976. iii+274 pp.

[LaWe] Lang, S. Weil, A.: Number of points of varieties in finite fields. Amer. J. Math., **76** (1954), 819-827. Zbl.97,362.

[L71a] Langlands R.P. (1971a): Euler Products. Whittemore Lectures in Mathematics, 1971. Yale University Press, 1971. Zbl.231.20016

[L71b] Langlands R.P. (1971b): On Artin's L–functions. Rice Univ. Studies 56 (1971), 23-28. Zbl.245.12011.

[L76] Langlands R.P. (1976): On the functional equations satisfied by Eisenstein series. Lect. Notes Math. 544 (1976). Zbl.332.10018.

[L79] Langlands, R. P. : Automorphic representations, Shimura varieties, and motives. Ein Märchen. Automorphic forms, representations and L-functions (Proc. Sympos. Pure Math., Oregon State Univ., Corvallis, Ore., 1977), Part 2, pp. 205–246, Proc. Sympos. Pure Math., XXXIII, Amer. Math. Soc., Providence, R.I., 1979.

[L80] Langlands R.P. (1980): Base change for $GL(2)$. Ann.Math.Stud. 96, Princeton University Press (1980). Zbl.444.22007.

[L02] Langlands R.P.: Featured Review MR1875184 (2002m:11039) on [Laff02]

[LRS93] Laumon, G., Rapoport, M., Stuhler, U.: $\mathcal{D}$-elliptic sheaves and the Langlands correspondence, Invent. Math. **113**, no. 2, 217–338 (1993)

[Lau02] Laumon, G.: The work of Laurent Lafforgue. In: [ICM02], pp. 91–97

[LeV77] Le Veque, W.J. (1977): Fundamentals of number theory. Reading, Mass., Addison–Wesley (1977). Zbl.368.10001.

[Lee02] Lee, Y.: Cohen-Lenstra heuristics and the Spiegelungssatz: number fields. J. Number Theory 92 (2002), no. 1, 37–66.

[Leh32] Lehmer, D.H. (1932): Hunting big game in the theory of numbers. Scr. math., **1**, 229-235 (1932-33). Jbuch FdM 59,539.

[Leh56] Lehmer, D.H. (1956): List of prime numbers from 1 to 10006721. New york: Hafner Publishing Co., 1956. Zbl.78,31.

[Lei03] Leichtnam, E.: An invitation to Deninger's work on arithmetic zeta functions (Manuscript, November 27, 2003)

[LeA.88] Lenstra, A.K. (1988): Fast and rigorous factorization under the generalized Riemann hypothesis. Proc. Kon. Ned. Akad. Wetenskab., 91, No.4, 443-454 (1988). Zbl.669.10012.

[LeH.80] Lenstra, H.W., Jr. Primality testing algorithms (after Adleman, Rumely and Williams). Sém. Bourbaki, Exp. **576**, 1980-81. Zbl.476.10005.

[LeH.84] Lenstra, H.W., Jr. (1984): Integer programming and cryptography. Math. Intell., **6**, No.3 (1984), 14-19. Zbl.548.90050.

[LeH.87] Lenstra, H.W., Jr. (1987): Factoring integers with elliptic curves. Ann. Math., **126**, No. 3 (1987), 649-673. Zbl.629.10006.

[LeT82] Lenstra, H.W., Jr., Tijdeman, R. (ed). Computational analysis in number theory. Math. Center Tracts **154/155**, Math. Centrum, Amsterdam (1982). Zbl.497.00005.

[Leo58] Leopoldt, H. W.: Zur Strukture der 1-Klassengruppe galoischer Zahlkorper, J. Reine Angew Math. 199 (1958), 165–174.

[Li75] Li, W. New forms and functional equations. Math. Ann., **212**, 285-315 (1975). Zbl.278.10026.

[Li2000] Li, W.: Recent developments in automorphic forms and applications. Number theory for the millennium, II (Urbana, IL, 2000), 331–354, A K Peters, Natick, MA, 2002.

[LN83] Lidl, R., Niederreiter, H. (1983): Finite fields. Reading, Mass.: Addison–Wesley (1983). Zbl.554.12010.

[Lin79] Linnik, Yu.V. (1979): Collected Works.(in Russian). Leningrad: Nauka, 1979. Zbl.435.10001

[Lom78] Lomadze, G.A.(1978): Formulas for the numbers of representation of numbers by certain regular and semi−regular ternary quadratic forms belonging to genera with two classes. (in Russian). Acta Arithm., **34**, no.2, 131-162 (1978). Zbl.329.10014.

[Loe96] Loeser, F.: Exposants p-adiques et théorèmes d'indice pour les équations différentielles p-adiques, d'après G. Christol et Z. Mebkhout. Séminaire Bourbaki. Exp. No.822 (Novembre 1996)

[LRS99] Luo, W., Rudnick, Z. , Sarnak, P.: On the generalized Ramanujan conjecture for $GL(n)$, Proc. Symp. Pure Math. 66, part 2, AMS, (1999).

[LT65] Lubin, J., Tate, J. (1965): Formal complex multiplication in local fields. Ann. Math., **81** (1965), 330-387. Zbl.128.265.

[Lu37] Lutz, E. (1937): Sur l'equation $Y^2 = X^3 - AX - B$ dans les corps $p-$adiques. J. reine u. angew. Math., 177, 238-247 (1937). Zbl.17,53.

[Mac80] Macdonald, I.G.(1980): Affine Lie algebras and modular forms. Sém. Bourbaki, Exp. 577, 1980/81. Zbl.472.17006.

[Mah34] Mahler, H.(1934): Über die rationalen Punkte auf Kurven vom Geschlecht Eins. J. reine u. angew. Math., **170**, 168-178 (1934). Zbl.8,200.

[Mal62] Malyshev, A.V. (1962): On the representation of integers by positive quadratic forms. (In Russian). Steklov Math. Inst. Trudy, 65, 1962. Zbl.135,98.

[Man56] Manin, Yu.I. (1956): On the congruences of the third degree modulo a prime number. (In Russian). Izv. Akad. Nauk SSSR, Ser. mat., **20** (1956), 673–678. Zbl.72,32.

[Man58] Manin, Yu.I. (1958): Algebraic curves over fields with derivation. (In Russian). Izv. Akad Nauk SSSR, Ser. mat., **22**, No.6, 737–756 (1958).

[Man61] Manin, Yu.I. On the Hasse–Witt matrix of algebraic curves. (In Russian). Izv. Akad. Nauk SSSR, Ser. mat., **25**, 153-172 (1961). English transl.: Transl.Am.Math.Soc., II.Ser. 45 (1965), 245-269. Zbl.102,278.

[Man63a] Manin, Yu.I. (1963a): A proof of an analogue of Mordell's conjecture over functional fields. (In Russian). Doklady Akad. Nauk SSSR, ser. mat., **152**, No. 5, 1061–1063 (1963).

[Man63b] Manin, Yu.I. (1963b) Rational points of algebraic curves over functional fields. (In Russian). Izv. Akad. Nauk SSSR, ser. mat., **27**, No. 6, 1395-1440 (1963). Zbl.166,169.

[Man68] Manin, Yu.I. (1968): Correspondences, motives and monoidal transformations (in Russian). Mat. Sb., **77**, no.4 (1968), 475-507. Zbl.199,248.

[Man70a] Manin, Yu.I. (1970a): Fine structure of the Néron–Tate height. (in Russian). Mat. Sb., **83**, No.3 (1970), 331-348. English transl.: Math.USSR, Sb.12 (1971), 325-342. Zbl.214,484.

[Man70b] Manin, Yu.I. (1970b): Le groupe de Brauer–Grothendieck en géométrie Diophantienne. Actes Congr. Int. Math. Nice, 1970. Paris: Gauthier–Villars, Vol.**1**, 401-411 (1971). Zbl.239.14010.

[Man71] Manin, Yu.I. (1971): Cyclotomic fields and modular curves. (in Russian). Uspekhi, **26**, no.6 (1971), 7-78. English transl.: Russ.Math.Surv. **26**, No.6 (1972), 7-78. Zbl.241. 14014.

[Man72a] Manin Yu.I. (1972a): Cusp forms and zeta–functions of modular curves. Izv. Akad. Nauk SSSR, Ser. Mat. 36, No.1 (1972), 19-66. English transl.: Math.USSR, Izv.6 (1973), 19-64. Zbl.243.14008.

[Man72b] Manin Yu.I. (1972b): Cubic Forms. Algebra, Geometry, Arithmetic. Moscow: Nauka, 1972. Zbl.255.14002. English transl.: Amsterdam: North–Holland, 1984. Zbl.582. 14010.

[Man76] Manin, Yu.I. (1976): Non–archimedean integration and p–adic Jacquet–Langlands functions. (In Russian). Uspekhi, 31, no.1 (1976), 5-54. Zbl.336.12007.

[Man76a] Manin, Yu.I.: p-adic automorphic functions. Journ. of Soviet Math., 5 (1976) 279-333.

[Man77] Manin Yu.I. (1977): A course in mathematical logic. Graduate texts in math. 53. Springer Verlag (1977), 286

[Man78] Manin, Yu.I. (1978): Modular forms and number theory. Proc. Int. Congr. Math. Helsinki, 1978, 177-186. Zbl.421.10016.

[Man80] Manin Yu.I. (1980): Computable and non–computable. Moscow, Soviet Radio (1980) (in Russian)

[Man84] Manin, Yu.I. (1984): New dimensions in geometry. Lecture Notes n Math., vol. **1111**, Berlin e.a.: Springer, 1984. Zbl.579.14002.

[Man91] Manin, Yu.I. (1991): Three–dimensional hyperbolic geometry as ∞–adic Arakelov geometry. Invent.Math., vol. **104**, No.2, 223–243 (1991)

[Man95] Manin, Yu.I.: Lectures of zeta functions and motives (according to Deninger and Kurokawa). Columbia University Number Theory Seminar (New York, 1992). Astérisque No. 228, 4, 121–163 (1995).

[Ma99] Manin, Yu. I.: Classical computing, quantum computing, and Shor's factoring algorithm. Séminaire Bourbaki. [Exposé No.862] Juin 1999

[Man02] Manin, Yu.I.: Von Zahlen und Figuren, preprint math. AG/0201005 (2002).

[Man02a] Manin, Yu.I.: Real Multiplication and noncommutative geometry, preprint math. AG/0202109 (2002).

[ManMar2] Manin, Yu.I., Marcolli, M.: Holography principle and arithmetic of algebraic curves, Adv. Theor. Math. Phys. Vol.3 (2001) N.5.

[ManMar1] Manin, Yu.I., Marcolli, M.: Continued fractions, modular symbols, and noncommutative geometry, Selecta Mathematica, New Ser. Vol.8 N.3 (2002) 475–520.

[MaPaM] Manin Yu.I, Panchishkin A.A. (1989): Introduction to Number Theory, 1989, Edition VINITI (in Russian), 348 p.

[MaPa] Manin Yu.I, Panchishkin A.A. (1995): Number Theory I: Introduction to Number Theory, Encyclopaedia of Mathematical Sciences, vol. 49, Springer-Verlag, 1995, 303 p.

[MTs86] Manin, Yu.I., Tsfasman, M.A. (1986): Rational varieties. Algebra, Geometry, Arithmetic. (In Russian). Uspekhi, **41**, No.2, 43-94 (1986). English transl.: Russ.Math.Surv. 41, No.5 (1986), 51-116. Zbl.621.14029.

[MZ72] Manin Yu.I., Zarkhin Yu.G. (1972): Height on families of Abelian varieties. Mat. Sb., Nov.Ser. **89**, No.2, 171-181 (1972). English transl.: Math.USSR, Sb. **18** (1973), 169-179. Zbl.256.14018.

[Mar] Marcolli, M.: Limiting modular symbols and the Lyapunov spectrum, Journal of Number Theory, **98** N.2, 348–376 (2003).

[Mar04] Marcolli, M.: Lectures on Arithmetic Noncommutative Geometry, Vanderbilt University, pp. 1-100 (2004)

[Mar54] Markov A.A. (1954): The theory of algorithms. Tr.Mat.Inst.Steklova **42**, 1954. English transl.: Am.Math.Soc., Transl., II.Ser. 15 (1960), 1-14. Zbl.58,5.

[MM76] Masley, J.M., Montgomery, H.L. (1976): Cyclotomic fields with unique factorization. J. reine u. angew. Math., **286/287** (1976), 248-256. Zbl.256.14018.

[Mat72] Matiyasevich, Yu.V. (1972): Diophantine sets. (in Russian). Usp. Mat. Nauk. **27**, No.5 (1972), 185-222. English transl.: Russ.Math.Surv. **27**, No.5 (1973), 124-164. Zbl.621.14029.

[Mat04] Matiyasevich, Yu.V. (2004): Hilbert's Tenth Problem, with a foreword by Martin Davis, MIT Press, Cambridge, MA; London, England (2004), 264 pp.

[Mats70] Matsumura, H.: Commutative Algebra. Benjamin, New York, 1970.

[MaTa98] Matsuzaki, K., Taniguchi, M.: Hyperbolic manifolds and Kleinian groups, Oxford Univ. Press, 1998.

[Maz77] Mazur, B. (1977): Modular Curves and the Eisenstein Ideal. Publ. Math. I.H.E.S., 47, 33–186, (1977).

[Maz77a] Mazur, B. (1977): Rational points on modular curves. Lecture Notes in Math., vol. **601**, 1977, 107-148. Zbl.357.14005.

[Maz79] Mazur, B. On the arithmetic of special values of L–functions. Inv. Math., 55, no.3, 207-240 (1979). Zbl.426.14009

[Maz83] Mazur, B. (1983): Modular curves and arithmetic. Proc. ICM, Warszawa, 1983. Amsterdam: North–Holland, Vol.**1**, 185-211 (1984) Zbl.597.14023.

[Maz86] Mazur, B. (1986): Arithmetic on curves. Bull. AMS, **14**, No.2, 207-259 (1986) Zbl.593.14021.

[Maz87] Mazur B. (1987): On some of the mathematical contributions of Gerd Faltings. Proc. Int. Congr. Math., Berkeley/Calif. 1986. (1987) 7-12. Zbl.663.01002.

[Maz2000] Mazur B. (2000): The theme of p-adic variation. Mathematics: frontiers and perspectives, 433–459, Amer. Math. Soc., Providence, RI, 2000.

[MazRub03] Mazur, B., Rubin, K.: Studying the Growth of Mordell-Weil. Documenta Math. Extra Volume: Kazuya Kato's Fiftieth Birthday (2003) 585–607

[MazRub04] Mazur, B., Rubin, K.: Kolyvagin systems. AMS, 96 pp.

[MSD74] Mazur, B., Swinnerton–Dyer, H.P.F. (1974): Arithmetic of Weil curves. Inv. Math., **25**, 1-61 (1974). Zbl.281,14016.

[MW83] Mazur, B., Wiles, A. (1983): Analogies between function fields and number fields. Amer. J. Math., **105**, 507-521 (1983). Zbl.531.12015.

[MW84] Mazur, B., Wiles, A. (1984): Class fields of Abelian extensions of $\mathbb{Q}$. Inv. of Math., **76**, no.2 (1984), 179-330. Zbl.545.12005.

[Meh91] Mehta, M.L. : Random matrices, Academic Press,(1991).

[Men93] Menezes, A.: Elliptic curve public key cryptosystems, Kluwer Academic Publishers, Boston, MA, 1993. xiv+128 pp.

[Mer96] Merel, L.: Bornes pour la torsion des courbes elliptiques sur les corps de nombres. Invent. Math. 124 (1996), no. 1-3, 437–449.

[Me82] Mestre J.–L. (1982): Construction d'une courbe elliptique de rang ≥ 12. C. R. Acad. Sci. Paris, Ser.I 295 (1982), 643-644.

[Me84] Mestre, J.–L. (1984): Courbes de Weil et courbes supersingulières. Sém. Théorie des Nombres, Bordeaux, **23**, 1984-85. Zbl.599.14031.

[Me85] Mestre, J.–L. (1985): Courbes de Weil de conducteur 5077. CRAS Paris, **300**, No. 15 (1985). Zbl.589.14026.

[Mich98] Michel, P.: Progrès récents du crible et applications, d'après Duke, Fouvry, Friedlander, Iwaniec. Séminaire Bourbaki. Exp. No.842 (Mars 1998)

[Mich01] Michel, P.: Répartition des zéros des fonctions L et matrices aléatoires Séminaire Bourbaki. Exp. No.887 (Mars 2001)

[Mih03] Mihăilescu, P.: A class number free criterion for Catalan's conjecture. J. Number Theory 99 (2003), no. 2, 225–231.

[Mil76] Miller, G.L. (1976): Riemann's hypothesis and tests for primality. J. Comput. and Syst. Sci., **13**, 300-317 (1976). Zbl.349.68025.

[Mil80] Milne, J.S. (1980): Etale cohomology. Princeton Univ. Press, 1980. Zbl.433.14012.

[Mil68] Milnor, J.M.: Infinite cyclic coverings, Conference on the Topology of Manifolds (Michigan State Univ., E. Lansing, Mich., 1967) pp. 115–133, Prindle, Weber and Schmidt, Boston, Mass 1968.

[Miy77] Miyaoka, Y. (1977): On the Chern numbers of surfaces of general type. Inv. Math., **42**, 225-237 (1977). Zbl.374.14007.

[Mi89] Miyake, Toshitsune (1989): Modular forms. Transl. from the Japanese by Yoshitaka Maeda., Berlin etc.: Springer-Verlag. viii, 335 p. (1989).

[MF73–77] Modular functions of one variable. **I-VI**. Lecture Notes in Math., 320 (1973), 349 (1973), 416 (1975), 601 (1977), 627 (1977). 320 (1973), Zbl.255.00008; 349 (1973), Zbl.264.00002; 416 (1975), Zbl.315.14014; 601 (1977), Zbl.347.00004; 627 (1977). Zbl.355.00009.

[Mon80] Monier, L. (1980): Algorithmes de factorization d'entiers. Paris: IRIA, 1980, 313-324.

[Mont71] Montgomery, H.L. (1971): Multiplicative number theory. Berlin e.a.: Springer (1971). Zbl.216,35.

[Mont73] Montgomery H., The pair correlation of zeros of the zeta function, Analytic Number Theory, AMS (1973).

[Mor98a] Morain, F.: Primality proving using elliptic curves: an update. In J. P. Buhler, editor, Algorithmic Number Theory, volume 1423 of Lecture Notes in Comput. Sci., pages 111–127. Springer-Verlag, 1998. Third International Symposium, ANTS-III, Portland, Oregon, June 1998, Proceedings.

[Mor03] Morain, F.: La primalité en temps polynomial, d'après Adleman, Huang; Agrawal, Kayal, Saxena. Séminaire Bourbaki. [Exposé No.917] (Mars 2003).

[Mor03a] Morain, F.: Primalité théorique et primalité pratique, AKS vs. ECPP: `www.lix.polytechnique.fr/Labo/François. Morain/aks-f.pdf`

[MPTh] Moran A., Pritchard, P., Thyssen, A.: Twenty-two primes in arithmetic progression, Math. Comp. **64**, no. 211, 1337-1339 (1995).

[Mor22] Mordell, L.J. (1922): On the rational solutions of the indeterminate equations of the third and fourth degrees. Proc. Cambr. Phil. Soc., **21**, 179-192 (1922). Jbuch FdM 48,140.

[Mor69] Mordell, L.J. (1969): Diophantine equations. New York: Acad. Press, 1969. Zbl.188, 345.

[Mor93] Moree, P.: A note on Artin's conjecture. Simon Stevin **67**, no. 3-4, 255–257 (1993).

[More77] Moreno, C.J. (1977): Explicit formulas in the theory of automorphic forms. Lect. Notes in Math., vol. **626**, 1977, 73-216. Zbl.367.10023

[MoMu84] Mori, Shigefumi, Mukai, Shigeru: Classification of Fano 3-folds with $B_2 \geq 2$. Manuscripta Math. **36**, no. 2, 147–162 (1981/82).

[MoMu03] Mori, Shigefumi, Mukai, Shigeru: Erratum: "Classification of Fano 3-folds with $B_2 \geq 2$ [Manuscripta Math. **36**, no. 2, 147–162 (1981/82)]." Manuscripta Math. **110**, no. 3, p. 407

[MB75] Morrison, M.A., Brillhart, J. (1975): A method of factoring and the factorization of F_7 . Math. Comput., 29 (1975), 183-205. Zbl.302.10010.

[Mozz] Mozzochi C.J. (2000): The Fermat diary. Providence, Am. Math. Soc., xi+196 (2000)

[Mum65] Mumford, D. (1965): A remark on Mordell's conjecture. Amer. J. Math., 87, No.4, 1007-1016 (1965). Zbl.151,273.

[Mum66] Mumford, D. (1966): On the equations defining Abelian varieties.I. Inv. Math., **1**, 287-354 (1966). Zbl.219.14024.

[MSW02] Mumford, D., Series, C, Wright, D. (2002): Indra's Pearls. The visions of Felix Klein. Cambgridge University Press, New–York, 2002

[Mum72] Mumford, D. (1972): An analytic construction of degenerating curves over complete local fields. Compos. Math., **24**, 129–174 (1972).

[Mum74] Mumford, D. (1974): Abelian varieties. Oxford Univ. Press (1974). Zbl.326. 14012.

[Mum83] Mumford, D. (1983): Tata lectures on theta. I-II. Boston e.a.: Birkhäuser (1983), (1984). Zbl.549.14014.

[Mur99] Murty, R.: Review (Math. Reviews, 99k:11004) on [CSS95]

[MM81] Narasimhan, M.S., Nori, M.V. (1981): Polarizations on an Abelian variety. Proc. Ind. Ac. Sci., Math. Sci., **90**, No.2 (1981), 125-128. Zbl.509.14047.

[Nar74] Narkiewicz, W. (1974): Elementary and analytic theory of algebraic numbers. Warsaw: Polish Sci. Pub. (1974). Zbl.276.12002.

[Ner64] Néron, A. (1964): Modéles minimaux des variétés abéliennes sur les corps locaux et globaux. Publ. Math. IHES, **21**, (1964), 367-482. Zbl.132,414.

[Ner76] Néron, A. (1976): Hauteurs et fonctions theta. Rend. Sem. Mat. e Fis. Milano, **46**, 111-135 (1976). Zbl.471.14024.

[Nes99] Nesterenko, Yu. V.: Algebraic independence of π and e^π. Number theory and its applications (Ankara, 1996), 121–149, Lecture Notes in Pure and Appl. Math., 204, Dekker, New York, 1999.

[Nes02] Nesterenko, Yu. V.: On the algebraic independence of numbers. A panorama of number theory or the view from Baker's garden (Zürich, 1999), 148–167, Cambridge Univ. Press, Cambridge, 2002.

[Neuk99] Neukirch, J.: Algebraic number theory. Translated from the 1992 German original and with a note by Norbert Schappacher. With a foreword by G. Harder. Grundlehren der Mathematischen Wissenschaften [Fundamental Principles of Mathematical Sciences], 322. Springer-Verlag, Berlin, 1999. xviii+571 pp. Algebraische Zahlentheorie Berlin etc.: Springer-Verlag. xiii, 595 S. (1992). Zbl. 0747.11001,

[NSW2000] Neukirch, J., Schmidt, A., Wingberg, K.: Cohomology of number fields. Grundlehren der Mathematischen Wissenschaften [Fundamental Principles of Mathematical Sciences], 323. Grundlehren der Mathematischen Wissenschaften, Springer-Verlag, Berlin, 2000. xvi+699 pp.

[Ngo2000] Ngô, Bao Châu: Preuve d'une conjecture de Frenkel-Gaitsgory-Kazhdan-Vilonen pour les groupes linéaires généraux. Israel J. Math. **120** (2000), part A, 259–270.

[Nis54] Nisnevich, L.B. (1954): On the number of points of algebraic varieties in finite fields. (in Russian). Doklady Akad. Nauk SSSR, 99 (1954), 17-20. Zbl.459.14002.

[Nog81] Noguchi, J.(1981): A higher–dimensional analog of Mordell's conjecture over functional fields. Math. Ann., **158**, 207-221 (1981). Zbl.459.14002.

[Oe92] Oesterlé, J.: Polylogarithmes. Séminaire Bourbaki. [Exposé No.762] (Novembre 1992)

[Oe95] Oesterlé, J. (1995): Travaux de Wiles (et Taylor, ...). II. Séminaire Bourbaki, Vol. 1994/95. Astérisque No. 237, Exp. No. 804, 5, 333–355 (1996)

[Odl75] Odlyzko, A.M. (1975): Some analytic estimates of class numbers and discriminants. Inv. Math., **29**, No. 3, 275-286 (1975). Zbl.299.12010.

[Odl84] Odlyzko, A.M. (1984): Discrete logarithms in finite fields and their cryptographic significance. Adv. in Cryptology. Proc. of Eurocrypt **84**, 224-314. Springer–Verlag, 1985. Zbl.594.94016.

[Odl87] Odlyzko, A.M. (1987): New analytic algorithms in number theory. Proc. ICM Berkeley 1986, vol. I, , 446-475, Providence Rh.I. (1987). Zbl.669.10003.

[Odl87a] Odlyzko A.M.: On the distribution of spacings between zeros of zeta functions, Math. Comp. **48** (1987), 273-308.

[OR85] Odlyzko, A.M., Riele, H.J. (1985): Disproof of the Mertens conjecture. J. reine u. angew. Math., **357**, 138-160 (1985). Zbl.544.10047.

[Oe83] Oesterlé, J. (1983): Nombres de classes des corps quadratiques imaginaire. Sém. Bourbaki, Exp. **631**, 1983/84. Zbl.551. 12003.

[Oe95] Oesterlé, J. (1995) Travaux de Wiles (et Taylor,...). II. Séminaire Bourbaki. Exp. No.804 (Juin 1995)

[Ogg65] Ogg, A.P. (1965): Modular forms and Dirichlet series. Benjamin, 1965. Zbl.191.381.

[Pan81] Panchishkin, A.A.: Modular forms. (in Russian). Itogi Nauki, **19**, 135-180 (1981), Moscow, VINITI. Zbl.477.10025.

[Pan84] Panchishkin, A.A. (1984): Automorphic forms and functorality principle.(In Russian). In: Automorphic forms, representations and L–functions. Moscow, Mir, 249-286 (1984).

[Pan88] Panchishkin, A.A. (1988): Non–Archimedean automorphic zeta-functions. (In Russian). Moscow, MGU, 1988. Zbl.667.10017.

[Pan03] Panchishkin, A.A. (2003): Two variable p-adic L functions attached to eigenfamilies of positive slope, Inventiones Math., **154**, pp. 551-615 (2003)

[PaTu82] Parry W., Tuncel S., Classification problems in ergodic theory, London Math. Soc. Lecture Notes Series 67, 1982.

[Par71] Parshin, A.N. Quelques conjectures de finitude en géométrie Diophantienne. Actes Congr. Int. Math. Nice 1970, Vol. **I**, 467-471(1971). Zbl.224.14009.

[Par71] Parshin, A.N. Arithmetic of algebraic varieties, (In Russian). Itogi Nauki, **2**, 111-152 (1971). Moscow, VINITI. Zbl.284.14004.

[Par72] Parshin, A.N. Minimal models of curves of genus two and homomorphisms of abelian varieties defined over a field of finite characteristic. (in Russian). Izv. Akad. Nauk SSSR, 36, No. 1 (1972), 67-109. Zbl.249.14003.

[Par73] Parshin, A.N. Modular correspondences, heights, and isogenies of Abelian varieties. (in Russian). Steklov Math. Inst., Trudy, **122** (1973), 211-236. Zbl.305.14015.

[PSh84] Parshin, A.N., Shafarevich, I.R. Arithmetic of algebraic varieties. (in Russian). Steklov Math. Inst., Trudy, 168 (1984), 72-97. Zbl.605.14001.

[Par87] Parshin, A.N. (1987): The Bogomolov–Miyaoka–Yau inequality on the arithmetical surfaces and its applications. Sém.Théorie des Nombres. Paris, 1986/87. Boston e.a., Birkhäuser, 1987. Zbl.705.14022

[PZ88] Parshin A.N., Zarkhin Yu.G. (1988): Finiteness problems in Diophantine geometry. Appendix in Russ. ed. of: Lang S. Foundations of Diophantine geometry. Moscow: Mir, 1988, 369-438. Transl. from English. Zbl.644.14007.

[Pat88] Patterson S., An introduction to the theory of the Riemann zeta function, Cambridge Studies in advanced mathematics, **14** Cambridge University Press (1988).

[Pet85] Peterson I. (1985): Uncommon factoring. Sci. News **127**, March 30, No.13, 202-203 (1985).

[Pey95] Peyre, E.: Hauteurs et mesures de Tamagawa sur les variétés de Fano, Duke Math. J. 79 (1995), n 1, 101–218.

[Pey02] Peyre, E.: Points de hauteur bornée et géométrie des variétés (d'après Y. Manin et al.). Séminaire Bourbaki, Vol. 2000/2001. Astérisque No. 282, (2002), Exp. No. 891, ix, 323–344.

[Pey04] Peyre, E.: Obstructions au principe de Hasse et à l'approximation faible. Séminaire Bourbaki. [Exposé No.931] (Mars 2004)

[PeyTsch01] Peyre E., Tschinkel, Yu.: Rational points on algebraic varieties. Progress in Math. vol. **199**, Birkhäuser, Basel (2001)

[PSh79] Piatetski–Shapiro I.I. (1979): Classical and adelic automorphic forms. An introduction. In: Borel A., Casselman W. eds. (1979), Part 1, 185-188. Zbl.423.10017.

[PV80] Pimsner, M., Voiculescu, D.: Exact sequences for K-groups and Ext-groups of certain cross-product C^*-algebras, J. Operator Theory 4 (1980), no. 1, 93–118.

[Pl82] Platonov V.P. (1982): Arithmetic theory of algebraic groups. Usp.Mat.Nauk. 37, No. 3 (1982), 3-54. English transl.: Russ.Math.Surv. 37, No.3 (1982), 1-62. Zbl.502.20025.

[PlRa83] Platonov V.P., Rapinchuk A.S. (1983): Algebraic groups. Itogi Nauki Tekh., Ser. Algebra, Topol., Geom. 21 (1983), 80-134. English transl.: J.Sov.Math.Surv. 31, No.3 (1985), 2939-2973. Zbl.564.20023.

[Pol74] Pollard J.M. (1974): Theorems on factorization and primality testing. Proc. Camb. Philos. Soc. 76 (1974), 521-528. Zbl.294.10005.

[P74] Pólya, G.: Collected Papers, Cambridge, M.I.T. Press (1974).

[Po90] Pollicott, M.: Kleinian groups, Laplacian on forms and currents at infinity, Proc. Amer. Math. Soc. 110 (1990) 269–279.

[Pom81] Pomerance C. (1981): Recent developments in primality testing. Math. Intell. 3 (1981), 97-106. Zbl.476.10004.

[Pom82] Pomerance C. (1982): Analysis and comparison of some integer factoring algorithms. Math. Cent. Tracts 154 (1982), 89-139. Zbl.508.10004.

[Pom87] Pomerance C. (1987): Fast, rigorous factorization and discrete logarithm algorithms. In: Discrete algorithms and complexity, Proc. Semin. Kyoto 1986, Perspect. Comput. 15 (1987), 119-143. Zbl.659.10003.

[PW83] Pomerance C., Wagstaff S.S. (1983): Implementation of the continued fraction integer factoring algorithm. Proc. 12th Manitoba Conf., Winnipeg 1982, Congr. Numerantium 37 (1983) 88-118. Zbl.556.10003.

[PSW80] Pomerance C., Selfridge J.L., Wagstaff S.S. (1980): The pseudoprimes to 25.10. Math. Comput. 35 (1980), 1003-1086. Zbl.444.10007.

[Po03] Poonen, B.: Hilbert's tenth problem and Mazur's conjecture for large subrings of $\mathbb{Q}$. J. Amer. Math. Soc. 16 (2003), no. 4, 981–990

[PoTsch04] Poonen B., Tschinkel, Yu. (Eds.) Arithmetic of higher-dimensional algebraic varieties. Proceedings of the Workshop on Rational and Integral Points of Higher-Dimensional Varieties held in Palo Alto, CA, December 11–20, 2002. Progress in Mathematics, 226. Birkhäuser Boston, Inc., Boston, MA, 2004. xvi+287 pp.

[vdP79] van der Poorten A.J. (1979): A proof that Euler missed ... Apéry's proof of the irrationality of $\zeta(3)$. An informal report. Math. Intell. 1, No. 4 (1979), 195-203. Zbl.409.10028.

[vdP96] van der Poorten, A.: Notes on Fermat's last theorem. Canadian Mathematical Society Series of Monographs and Advanced Texts. A Wiley-Interscience Publication. John Wiley & Sons, Inc., New York, 1996. xviii+222 pp.

[Pos71] Postnikov A.G. (1971): Introduction to analytic number theory. Moscow: Nauka, 1971. English transl.: Providence: Transl.Math.Monogr., Vol. 68 (1988). Zbl.231. 10001. Zbl.641.10001.

[Pos78] Postnikov M.M. (1978): Fermat's Theorem. (in Russian). Moscow: Nauka, 1978.

[Pra57] Prachar K. (1957): Primzahlverteilung. Berlin–Heidelberg–New York: Springer–Verlag, 1957. Zbl.80,259.

[Pu96] Putnam, I.: C^*–algebras from Smale spaces, Can. J. Math. 48 (1996) N.1 175–195.

[PuSp-99] Putnam, I., Spielberg, J.: The structure of C*–algebras associated with hyperbolic dynamical systems, J. Funct. Anal. 163 (1999) 279–299.

[Ra80] Rabin M.O. (1980): Probabilistic algorithms for testing primality. J. Number Theory 12 (1980), 128-138. Zbl.426.10006.

[Ra77] Rademacher H. (1977): Lectures on Elementary Number Theory. Krieger Publ. Co., 1977. Zbl.363.10001.

[R99] Ramakrishna, R. (1999): Lifting Galois representations. Invent. Math. **138**, 537-595 (1999).

[Ram16] Ramanujan S. (1916): On certain arithmetical functions. Trans. Camb. Philos. Soc. **22** (1916), 159-184.

[Ran39] Rankin R. (1939): Contribution to the theory of Ramanujan's function $\tau(n)$ and similar arithmetical functions, I, II. Proc. Camb. Philos. Soc. **35**, 351-372 (1939). Zbl.21,392.

[Ray75] Raynaud M. (1975): Schémas en groupes de type $(p, \ldots, p)$. Bull. Soc. Math. Fr. **102**, 241-280 (1975). Zbl.325.14020.

[Ray83] Raynaud M. (1983): Around the Mordell conjecture for function fields and a conjecture of Serge Lang. Lect. Notes Math. **1016** (1983), 1-19. Zbl.525.14014.

[RaSi73] Ray, D.B., Singer, I.M.: Analytic torsion for complex manifolds. Ann. of Math. (2) **98** (1973), 154–177.

[Ren80] Renault, J.: A groupoid approach to C^*-algebras, Lecture Notes in Mathematics, **793**. Springer, 1980.

[Rib79] Ribenboim P. (1979): 13 Lectures on Fermat's Last Theorem. Berlin–Heidelberg–New York: Springer–Verlag, 1979. Zbl.456.10006

[Rib88] Ribenboim P. (1988): The book of prime number records. Berlin–Heidelberg–New York: Springer–Verlag, 1988. Zbl.642.10001

[Rib96] Ribenboim P. (1996): The new book of prime number records. Berlin–Heidelberg–New York: Springer–Verlag, 1996.

[Ri77] Ribet K.A. (1977): Galois representations attached to eigenforms with Nebentypus. Lect. Notes Math. **601** (1977), 17-52. Zbl.363.10015.

[Ri85] Ribet K.A. (1985): On l-adic representations attached to modular forms.II. Glasgow Math. J. **27** (1985), 185-194.

[Ri90a] Ribet K.A. (1990a): Raising the level of modular representations. Sémin. Theor. Nombres Paris, 1987-88 Progress in Math. **81** (1990), 259-271.

[Ri] Ribet, K.A. (1990): On modular representations of $\mathrm{Gal}(\overline{\mathbb{Q}}/\mathbb{Q})$ arising from modular forms. Invent. Math. 100, No.2, 431-476 (1990)

[RiSt01] Ribet, K., Stein, W.: Lectures on Serre's conjectures. In: [CR01], pp. 143–232

[Rief76] Rieffel, M.A.: Strong Morita equivalence of certain transformation group C^*–algebras. Math. Annalen, **222**, 7–23 (1976).

[Rie1858] Riemann B. (1858): Über die Anzahl der Primzahlen unter Einer Gegebenen Größe. Montasb. der Berliner Akad., **160**, 671-680 (1858).

[Rie1892] Riemann B. (1892): Gesammelte mathematische Werke. Leipzig, 1892. Jbuch.FdM. 24,21.

[Ries85] Riesel H. (1985): Prime numbers and computer methods for factorization. Prog. Math. 57, Boston: Birkhäuser, 1985. Zbl.582.10001.

[RG70] Riesel H., Göhl G. (1970): Some calculations related to Riemann's prime number formula. Math. Comput. 24 (1970), 969-983. Zbl.217,32.

[Riv01] Rivoal, T.: Propriétés diophantiennes de la fonction zêta de Riemann aux entiers impairs `http://theses-EN-ligne.in2p3.fr/documents /archives0/00/00/12/67 /index_fr.html`, PhD, T. Rivoal, Université de Caen, 2001

[Rob73] Robert G. (1973): Unités elliptiques et formules pour le nombre de classes des extensions abéliennes d'un corps quadratique imaginaire. Bull. Soc. Math. Fr. 36 (1973), 5-77. Zbl.314.12006.

[Rob01] Robertson, G.: Boundary actions for affine buildings and higher rank Cuntz–Krieger algebras, in [CuEch01]

[Rog67] Rogers H. (1967): The theory of recursive functions and effective computability. Mac-Graw Hill Publ.Co., 1967. Zbl.183,14.

[Ros84] Rosen K.H. (1984): Elementary Number Theory and its Applications. Reading, Mass.: Addison–Wesley, 1984. Zbl.546.10001.

[Roth55] Roth K.F. (1955): Rational approximations to algebraic numbers. Mathematika 2 (1955), 1-20. Zbl.64,285.

[Rub77] Rubin K. (1987): Tate–Shafarevich groups and L–functions of elliptic curves with complex multiplication. Invent. Math. 89 (1987), 527-560. Zbl.628.14018.

[Rub95] Rubin K. (1995): Modularity of mod 5 representations. In: [CSS95], pp. 463–474

[Rub98] Rubin, K.: Euler systems and modular elliptic curves. Galois representations in arithmetic algebraic geometry (Durham, 1996), 351–367, London Math. Soc. Lecture Note Ser., 254, Cambridge Univ. Press, Cambridge, 1998.

[RubSil94] Rubin K., Silverberg A. (1994): A report on Wiles' Cambridge lectures. Bull. AMS (New series) **31**, 15–38 (1994)

[Rue88] Ruelle, D.: Non–commutative algebras for hyperbolic diffeomorphisms, Invent. Math. 93 (1988) 1–13.

[Sai88] Saito, M.: Modules de Hodge Polarisable. Publ. Res. Inst. Math. Sci. 24 (1988) 849–995.

[SZ75] Samuel P., Zariski O. (1975): Commutative Algebra. Vols 1-2, 2nd ed. Berlin–Heidelberg–New York: Springer–Verlag, 1975/76. Zbl.313.13001, Zbl.322.13001.

[Sar98] Sarnak, P.: L-functions. Proceedings of the International Congress of Mathematicians, Vol. I (Berlin, 1998). Doc. Math. 1998, Extra Vol. I, 453–465 (electronic).

[SarWa] Sarnak, P., Wang, L.: Some hypersurfaces in P^4 and the Hasse-principle. C. R. Acad. Sci. Paris Sér. I Math. **321**, no. 3, 319–322 (1995).

[Scha79] Schanuel S. (1979): Heights in number fields. Bull. Soc. Math. Fr. 107 (1979), 433-449. Zbl.428.12009.

[Schm79] Schmidt W.M. (1979): Diophantine approximation. Lect. Notes Math. 785 (1976). Zbl.421.10019.

[Sch88] Schneider, P.: Introduction to the Beilinson Conjectures. In Rapoport, M. Schappacher, P. Schneider (eds) Beilinson's conjectures on special values of L-functions, 1–35, (Perspectives in Math., Vol. 4) Boston, New York: Academic Press 1988.

[Schn57] Schneider, Th. (1957): Einführung in die transzendenten Zahlen. Berlin–Heidelberg–New York: Springer–Verlag, 1957.

[Scho90] Scholl A. (1990): Motives for modular forms. Invent. Math. 100 (1990), 419–430.

[Scho98] Scholl A. (1990): An introduction to Kato's Euler systems. Galois representations in arithmetic algebraic geometry (Durham, 1996), 379–460, London Math. Soc. Lecture Note Ser., 254, Cambridge Univ. Press, Cambridge, 1998.

[Sch85] Schoof R.J. (1985): Elliptic curves over finite fields and the computation of square roots mod p. Math. Comput. 44 (1985) 483-494. Zbl.579.14025.

[Schr84] Schroeder M.R. (1984): Number theory in science and communication. Inf. Sci. ser. 7 (1984), Berlin–Heidelberg–New York: Springer–Verlag. Zbl.613.10001.

[Schr93] Schröder, H.: K–theory for real C^*–algebras and applications. Pitman Research Notes in Mathematics Series, 290. Longman Scientific Technical, Harlow 1993.

[Sei82] Seiler, E.: Gauge Theories as a problem of constructive Quantum Field Theory and Statistical Mechanics, Lecture Notes in Physics **159** Springer (1982).

[Sel51] Selberg A. (1951): An elementary proof of the prime number theorem. Ann. Math. II.Ser. 50 (1949), 305-313. Zbl.36,306.

[Sel89] Selberg, A.: Collected papers, Springer (1989).

[Selm51] Selmer E.S. (1951): The Diophantine equation $ax^3 + by^3 + cz^3 = 0$. Acta Math. 85 (1951), 203-362. Zbl.42,269.

[Selm54] Selmer E.S. (1954): Completion on the tables. Acta Math. 92 (1954), 191-197. Zbl.56,267.

[Sepp] Seppälä, M., Computation of period matrices of real algebraic curves. Discrete Comput. Geom. 11 (1994), no. 1, 65–81.

[Se56] Serre, J.– P. (1956): Géométrie algébrique et géométrie analytique. Ann. Inst. Fourier 6, 1-42 (1955/56). Zbl.0075.30401.

[Se58] Serre, J.– P. (1958): Groupes algébriques et théorie du corps de classes. Paris, Hermann, 1958. Zbl.97,356.

[Se63] Serre, J.– P. (1963): Corps locaux. Paris, Hermann, 1963. Zbl.137,26.

[Se64] Serre, J.– P. (1964): Cohomologie galoisienne. Berlin e.a.: Springer – Verlag, 1964. Zbl.128,263.

[Se65] Serre J.–P. (1965): Zeta and L–functions. In: Arithmetical Algebraic Geometry, Proc. Conf. Purdue 1963 (1965), 82-92. Zbl.171,196.

[Se68] Serre, J.– P. (1958): Zeta and L–functions, in: Arithmetical Algebraic Geometry, New York, Harper and Row, 1965, 82-92 Zbl.171,196.

[Se68a] Serre, J.–P. (1968a): Abelian l–adic representations and elliptic curves. New York: Benjamin, 1968. Zbl.186,257.

[Se68b] Serre, J.– P. (1968b): Une interpretation des congruences relatives à la fonction τ de Ramanujan. Sém. théor. de nombres Delange–Pisot–Poitou Fac. Sci. Paris, 1967-68, **9**, No. 1, 14/01-14/17. Zbl.186,369.

[Se70] Serre, J.– P. (1970): Cours d'arithmétique. Paris: Presses Univ. France, 1970. Zbl.225. 12002.

[Se70a] Serre, J.–P.: Facteurs locaux des fonctions zêta des variétés algébriques (définitions et conjectures). Sém. Delange-Pisot-Poitou, exp. 19, 1969/70.

[Se71] Serre J.–P. (1971), Cohomologie des groupes discrets, Prospects in Math., Ann. of Math. Studies **70**, Princeton Univ. Press, (1971)

[Se72] Serre, J.–P. (1972): Propriétés galoisiennes des points d'ordre fini des courbes elliptiques. Inv. Math., 15 (1972), 259-331. Zbl.235.14012.

[Se76] Serre, J.– P. (1976): Représentations l–adiques. Alg. Numb. Theory, Proc. Taniguchi Int. Symp. Div. Math., Kyoto, 1976; No.2, Kyoto, 117-1931977. Zbl.406.14015.

[Se77] Serre J.–P. (1977), Arbres, Amalgames, et SL_2 Asterisque **46** (1977); English trans. Trees, Springer-Verlag, 1980.

[Se83] Serre, J.– P. (1983): Sur le nombre des points rationnels d'une courbe algébrique sur un corps fini. C R Ac. Sci., Ser. **1**, 296, 397-402, 1983. Zbl.538.14015.

[Se86] Serre, J.– P. (1986): Oeuvres. Vols I-III. Berlin: Springer – Verlag, 1986.

[Se87] Serre, J.– P. (1987): Sur les représentations modulaires de degré 2 de Gal($\overline{\mathbb{Q}}/\mathbb{Q}$). Duke Math. Journ., 54, no.1 (1987), 179-230. Zbl.641.10026.

[Se94] Serre, J.-P. (1994) Cohomologie galoisienne : progrès et problèmes. Exp. No.783 (Mars 1994)

[Se95] Serre, J.-P. (1995): Travaux de Wiles (et Taylor,...). I. Séminaire Bourbaki. Volume 1994/95. Exposés 790-804. Paris: Societe Mathématique de France, Astérisque. 237, 319-332, Exp. No.803 (1996)

[Se97] Serre, J.-P.: Lectures on the Mordell-Weil theorem, Third ed., Friedr. Vieweg & Sohn, Braunschweig, 1997, Translated from the French and edited by Martin Brown from notes by Michel Waldschmidt, With a foreword by Brown and Serre.

[ST68] Serre, J.- P., Tate, J. Good reduction of Abelian varieties and applications, Ann. Math., **88**, No.3, 492-517 (1968). Zbl.172,461.

[Sev14] Severi, F. (1914): Sugli integrali abeliani riducibili. Rend. Acad. Lincei, Ser. V, **23**, 581-587 (1914). Jbuch FdM 45,698.

[Sey87] Seysen M.A. (1987): A probabilistic factorization algorithm with quadratic forms of negative discriminant. Math. Comput. 48 (1987), 757-780. Zbl.619.10004.

[Sha50] Shafarevich, I.R. The general reciprocity law. (in Russian). Mat. Sb., **26**, 113-146 (1950).

[Sha51] Shafarevich, I.R. (1951): A new proof of the Kronecker–Weber theorem. (In Russian). Steklov Math. Inst., Trudy, 38 (1951), 382-387. Zbl.53,355.

[Sha54] Shafarevich, I.R. Construction of algebraic number fields with a given solvable Galois group. (in Russian). Izv. Akad. Nauk SSSR, ser. Ser. mat., 18, No.6 (1954), 525-578. Zbl.57,274.

[Sha57] Shafarevich, I.R., Exponents of elliptic curves. (in Russian). Doklady Akad. Nauk SSSR, **114**, No.4 (1957), 714-716. Zbl.81,154.

[Sha59] Shafarevich, I.R. (1959): The group of principal homogeneous spaces. (in Russian). Doklady Akad. Nauk SSSR, **124**, No.1 (1959), 42-43. Zbl.115,389.

[Sha62] Shafarevich, I.R. (1962): Fields of algebraic numbers. Proc. Int. Congr. Math. Stockholm, 163-176 (1962). Zbl.126,69

[Sha65] Shafarevich, I.R., ed. (1965): Algebraic surfaces. (in Russian). Steklov Math. Inst., Trudy, 75, 1965. German transl.: Leipzig, 1968. Zbl.154,210.

[Sha66] Shafarevich, I.R. (1966): Lectures on minimal models and birational transformations of two-dimensional schemes. Bombay, Tata Inst., 1966. Zbl.164,517.

[Sha69] Shafarevich, I.R. (1969): Zeta–function. (in Russian). Moscow, MGU, 1969.

[Sha87] Shafarevich, I.R. (1987): Fundamental notions of algebra. Itogi Nauki, 11, 1987. English transl.: Encycl. Math. Sci 11. Berlin–Heidelberg–New York: Springer–Verlag, 1990. Zbl.711.16001.

[Sha88] Shafarevich, I.R. Foundations of algebraic geometry. (In Russian). 2nd ed. Vols. **1-2**. Moscow: Nauka, 1988. English transl.: Berlin–Heidelberg–New York: Springer–Verlag, 1977. Zbl.675.14001; Zbl.258.14001.

[Sh88] Shahidi F. (1988): On the Ramanujan conjecture and finiteness of poles for certain L–functions. Ann. Math., II.Ser. 127 (1988), 547-584. Zbl.654.10029.

[ShT-BT04a] Shalika, J. A., Takloo-Bighash, R., Tschinkel, Yu.: Rational points on compactifications of semi-simple groups of rank 1. Arithmetic of higher-dimensional algebraic varieties, (Palo Alto, CA, 2002), Prog.Math., vol.**226**, Birkhäuser, Boston, MA, 2004, p.205-233.

[ShT-BT04b] Shalika, J. A., Takloo-Bighash, R., Tschinkel, Yu.: Rational points and automorphic forms. Contributions to automorphic forms, geometry, and number theory, 733–742, Johns Hopkins Univ. Press, Baltimore, MD, 2004.

[ShT04] Shalika, J. A., Tschinkel, Yu.: Height zeta functions of equivariant compactifications of the Heisenberg group. Contributions to automorphic forms, geometry, and number theory, 743–771, Johns Hopkins Univ. Press, Baltimore, MD, 2004.

[Shan71] Shanks D. (1971): Class number, a theory of factorization, and genera. Proc. Symp. Pure Math. 20 (1971), 415-440. Zbl.223.12006.

[Shan85] Shanks D. (1985): Solved and Unsolved Problems in Number Theory. 3rd ed. Chelsea Publ. Co., 1985. Zbl.397.10001.

[Shid87] Shidlovski A.B. (1987): Transcendental Numbers. Moscow: Nauka, 1987. English transl.: Berlin etc.: W. de Gruyter, 1989. Zbl.629.10026.

[Shi66] Shimura G. (1966): A reciprocity law in non–solvable extensions. J. Reine Angew. Math. 221 (1966), 209-220. Zbl.144,42.

[Shi71] Shimura G. (1971): Introduction to the arithmetic theory of automorphic functions. Princeton University Press, 1971. Zbl.221.10029.

[Shi97] Shimura G. (1997): Euler products and Eisenstein series. CBMS Regional Conference Series in Mathematics, 93. Published for the Conference Board of the Mathematical Sciences, Washington, DC; by the American Mathematical Society, Providence, RI, 1997. xx+259 pp.

[Shi2000] Shimura G. (2000): Arithmeticity in the theory of automorphic forms. Mathematical Surveys and Monographs, **82**. American Mathematical Society, Providence, RI, 2000. x+302 pp.

[Shi02] Shimura G. (2002): The representation of integers as sums of squares. Amer. J. Math. **124** (2002), no. 5, 1059–1081.

[Shi04] Shimura G. (2004): Arithmetic and analytic theories of quadratic forms and Clifford groups. Mathematical Surveys and Monographs, **109**. American Mathematical Society, Providence, RI, 2004. x+275 pp.

[Shin76] Shintani, T.: On evaluation of zeta functions of totally real algebraic number fields at non–negative integers. J. Fac. Sci. Univ. Tokyo, Sec. IA, 23, No.2 (1976), 393–417

[Shl03] Shlapentokh, A.: A ring version of Mazur's conjecture on topology of rational points, Internat. Math. Res. Notices No. **7**, 411–422 (2003)

[Sho80] Shokurov V.V. (1980): Shimura integrals of cusp forms. Izv. Akad.Nauk SSSR, Ser.Mat. 44, No.3 (1980), 670-718. Zbl.444.14030.

[ShT86] Shorey T.N., Tijdeman R. (1986): Exponential Diophantine Equations. Cambridge Univ. Press, 1986. Zbl.606.10011.

[Shou] Shoup, V.: NTL: A Library for doing Number Theory, Web page: `http://shoup.net/ntl/`, 2002

[Sie29] Siegel, C.L. (1929): Über einige Anwendungen Diophantischer Approximationen. Abh. Preuss. Akad. Wiss. Phys. Math. Kl. 41-69(1929). Jbuch FdM 56,180.

[Sie35] Siegel, C.L. (1935): Über die analytische Theorie der quadratische Formen. Ann. Math., **36**, 527-606 (1935). Zbl.12,197.

[Sie39] Siegel, C.L. (1939): Einführung in die Theorie der Modulfunktionen n–ten Grades. Math. Ann, 116 (1939), 617-657. Zbl.21,203.

[Sie65] Siegel, C.L. (1965): Lectures on advanced analytic number theory. Tata Institute, Bombay, 1965. Zbl.278.10001.

[Sier64] Sierpinski W. (1964): A Selection of Problems in the Theory of Numbers. Oxford etc.: Pergamon Press, 1964. Zbl.122,44.

[Si01] Silverberg, A.: Open questions in arithmetic algebraic geometry. In: [CR01], pp.143–232.

[Silv86] Silverman J.H. (1986): The Arithmetic of Elliptic Curves. Berlin–Heidelberg–New York: Springer–Verlag, 1986. Zbl.585.14026.

[Silv88] Silverman J.H. (1988): Wieferich's criterion and the *abc*–conjecture. J. Number Theory 30, No.2 (1988), 226-237. Zbl.654.10019.

[Sim79] Simmons G.J. (1979): Cryptology: the mathematics of secure communication. Math. Intell. 4, No.1 (1979), 233-246. Zbl.411.94009.

[Sko99] Skorobogatov, A.: Beyond the Manin obstruction. Invent. Math. **135**, no. 2, 399–424 (1999).

[Sko01] Skorobogatov, A.: Torsors and rational points. Cambridge Tracts in Mathematics, 144. Cambridge University Press, Cambridge, 2001. viii+187 pp.

[SSw-D] Slater, J. B., Swinnerton-Dyer, P.: Counting points on cubic surfaces. I. Nombre et répartition de points de hauteur bornée (Paris, 1996). Astérisque No. 251, (1998), 1–12.

[Slo82] Sloane N.J.A. (1982): Recent bounds for codes, sphere packings, and related problems obtained by linear programming and other methods. Contemp. Math. 9 (1982), 153-185. Zbl.491.94024.

[SolSt77] Solovay R., Strassen V. (1977): A fast Monte–Carlo test for primality. SIAM J. Comput. 6 (1977), 84-85; erratum 7(1978), 18. Zbl.345.10002.

[Sou84] Soulé Ch. Regulateurs. Sémin. Bourb. 1984-85 Exp.No. 644, Astérisque 133/134 (1986), 237-253.

[SABK94] Soulé Ch., Abramovich, D., Burnol, J.F., Kramer J.K.: Lectures on Arakelov Geometry, CUP 1994

[Son04] Sondow, J.: Criteria for irrationality of Euler's constant `http://home.earthlink.net/~jsondow/`

[Spi91] Spielberg, J.: Free–product groups, Cuntz–Krieger algebras, and covariant maps. Internat. J. Math. 2 (1991), no. 4, 457–476.

[Spr82] Sprindzhuk V.G. (1982): Classical Diophantine Equations with two Variables. Moscow: Nauka, 1982. English transl.: Lect. Notes Math. **1559** (1993). Zbl.523.10008.

[Spr81] Springer T.A. (1981): Linear Algebraic Groups. Boston: Birkhäuser, 1981. Zbl.453. 14022.

[St67] Stark H.M. (1967): A complete determination of complex quadratic fields of class number one. Mich. Math. J. 14, No.1 (1967), 1-27. Zbl.148,278.

[St69] Stark H.M. (1969): On the "gap" in a theorem of Heegner. J. Number Theory 1, No.1 (1969), 16-27. Zbl.198,377.

[St71-80] Stark, H.: L-functions at $s = 1$, I, Adv.Math. 7 (1971), 301-343; II, 17 (1975), 60-92; III, 22 (1976), 64-84; IV, 35 (1980), 197-235

[St77a] Stark H.M. (1977a): Hilbert's twelfth problem and L–series. Bull.Am.Math.Soc. 83, No.5 (1977), 1072-1074. Zbl.378.12007.

[St77b] Stark H.M. (1977b): Class fields and modular forms of weight one. Lect. Notes Math. 601 (1977), 277-287. Zbl.363.12010.

[Stee76] Steenbrink, J.: Limits of Hodge structures. Invent. Math. 31 (1976), 229–257.

[StW.] Stein W.: An Explicit Approach to Number Theory (a forthcoming book, based on a Course Math 124 (Fall 2001), cf. http://modular.fas.harvard.edu /edu/Fall2001/124/lectures/)

[Step74] Stepanov S.A. (1974): Rational points of algebraic curves over finite fields. (in Russian). In: Aktual'nye Probl. Analit Teor. Cisel, 1974, 224-241. Zbl.347.14013.

[Step84] Stepanov S.A. (1984): Diophantine equations. Tr. Mat. Inst. Steklova 168 (1984), 31-45.

[Step94] Stepanov S.A.: Arithmetic of algebraic curves. Monographs in Contemporary Mathematics, New–York: Consultants Bureau, 1994

[Step99] Stepanov S.A.: Codes on algebraic curves. Kluwer Academic Publishers. vii, 350 p., 1999

[Ste95] Stevens G. (1995): An overview the proof of Fermat's Last Theorem. In: [CSS95], 1–16

[StTsch] Strauch, M., Tschinkel, Yu.: Height zeta functions of toric bundles over flag varieties. Selecta Math. (N.S.) **5**, no. 3, 325–396 (1999).

[Sull] Sullivan D.: On the ergodic theory at infinity of an arbitrary discrete group of hyperbolic motions. Riemann surfaces and related topics: Proceedings of the 1978 Stony Brook Conference (State Univ. New York, Stony Brook, N.Y., 1978), pp. 465–496, Ann. of Math. Stud., **97**, Princeton Univ. Press, 1981.

[SwD67] Swinnerton–Dyer H.P.F. (1967): The conjecture of Birch and Swinnerton–Dyer and of Tate. Proc. Conf. Local fields. Driebergen 1966 (1967), 132-157. Zbl.197,471.

[SwD73] Swinnerton–Dyer H.P.F. (1973): On l–adic representations and congruences for coefficients of modular forms. Lect. Notes Math. 350 (1973), 1-56. Zbl.267.10032.

[Szm75] Szemerédi, E.: On sets of integers containing no k elements in arithmetic progression, Acta Arith. **27**, 299-345 (1975).

[Sz(e)81] Szpiro L. (ed.) (1981): Séminaire sur les pinceaux de courbes de genre au moins deux. Astérisque 86, 1981. Zbl.463.00009.

[Sz83] Szpiro L. (1983): La conjecture de Mordell (d'après G. Faltings). Sémin. Bourbaki 1983/84, Exp.No. 619, Astérisque 121/122 (1985). Zbl.591.14027.

[Tan57] Taniyama Y. (1957): L–functions of number fields and zeta–functions of Abelian varieties. J. Math. Soc. Japan 9 (1957), 330-366. Zbl.213,228.

[Ta65] Tate, J. (1950): Fourier analysis in number fields and Hecke's zeta-function. Thesis. Princeton, 1950. (see also Algebraic Number Theory, Brighton 1965, 305-347 (1967). Zbl.153,74; 2nd printing Zbl.645.12001.)

[Ta65a] Tate, J. (1965a): On the conjectures of Birch and Swinnerton-Dyer and a geometric analog. Sém. Bourbaki, Exp., 1965/66. Sémin. Bourbaki 1965/66, Exp.No. 306; Adv.Stud.Pure Math. 3, 189-214 (1968). Zbl.199,556.

[Ta65b] Tate J. (1965b): Algebraic cycles and poles of zeta-functions. In: Arithmetical Algebraic Geometry, Schilling ed., Harper and Row (1965), 93-100.

[Ta66] Tate, J. (1966): Endomorphisms of Abelian varieties over finite fields. Inv. Math., **2**, 133-144 (1966). Zbl.147,203.

[Ta73] Tate, J. (1973): Algebraic formulas in arbitrary characteristic. Appendix 1 to [La73/87]

[Ta74] Tate, J. (1966): The Arithmetic of Elliptic Curves. Inv. Math., **23**, 179-206 (1974).

[Ta79] Tate J. (1979): Number–theoretic background. In: Borel A., Casselman W. (1979), Part 2, 3-26. Zbl.422.12007.

[Ta84] Tate J.: Les conjectures de Stark sur les fonctions L d'Artin en $s = 0$, Progress in Mathematics Vol.47, Birkhäuser 1984.

[Ta95] Tate J.: Finite flat group schemes. In: [CSS95], pp. 121–154

[Tay02] Taylor, R.: Galois representations. In: [ICM02], 449–474

[Ta-Wi] Taylor, R., Wiles, A. (1995): Ring-theoretic properties of certain Hecke algebras. Ann. Math., II. Ser. 141, No.3, 553-572 (1995)

[Tho79] Thompson J.G. (1979): Finite groups and modular functions. Bull. Lond. Math. Soc. 11, No. 3 (1979), 347-351. Zbl.424.20011.

[Tit51] Titchmarsh E.K. (1951): Theory of Riemann Zeta–Function. Oxford: Clarendon Press, 1951, 2nd ed., revised by D.R.Heath-Brown, 1986. Zbl.42,79; Zbl.601.10026.

[TsVl91] Tsfasman, M. A.; Vlăduţ, S. G.: Algebraic-geometric codes. Translated from the Russian by the authors. Mathematics and its Applications (Soviet Series), 58. Kluwer Academic Publishers Group, Dordrecht, 1991. xxiv+667 pp.

[Tun81] Tunnell J.B. (1983): Artin's Conjecture for representations of octahedral type. Bull. AMS (New series) **5**, 173–175 (1981).

[Tun83] Tunnell J.B. (1983): A classical Diophantine problem and modular forms of weight 3/2. Invent.Math. **72** (1983), 323-334. Zbl.515.10013.

[Vas88] Vasilenko O.N. (1988): Modern primality tests. (in Russian). Kibern. Sb., Nov.Ser. 25 (1988), 162-188. Zbl.669.10013.

[Vats03] Vatsal, V.: Special values of anticyclotomic L-functions. Duke Math. J. **116**, no. 2, 219–261 (2003).

[Vau81-97] Vaughan R.C. (1981): The Hardy–Littlewood Method. Cambridge Univ. Press, 1981. Zbl.455.10034. 2nd Edn: Cambridge Tract **125**, CUP 1997

[VaWo91] Vaughan R.C., Wooley T.D. : On Waring's problem: some refinements. Proc. London Math. Soc. (3) **63**, 35–68 (1991)

[VaWoIV] Vaughan R.C., Wooley T.D. : Further improvements in Waring's problem, IV: higher powers. Acta Arith. **94**, 203–285 (2000)

[VP] Venkov A.B., Proskurin N.V. (1982): Automorphic functions and the Kummer problem. Usp.Mat.Nauk. 37, No.3 (1982), 143-165. English transl.: Russ.Math.Surv. 37, No.3 (1982), 165-190. Zbl.494.10024.

[V81] Venkov B.A. (1981): Collected Works in Number Theory (in Russian). Leningrad: Nauka, 1981. Zbl.513.10001.

[Vi83] Viète F. (1983): The Analytic Art. Kent State University Press, 1983. Zbl.558.01041.

[Vin52] Vinogradov I.M. (1952): Collected Works. (in Russian). Moscow: Akad.Nauk SSSR, 1952. Zbl.48,31.

[Vin71] Vinogradov I.M. (1971): The Exponential Sum Method in Number Theory (in Russian). Moscow: Nauka, 1971. Zbl.229.10020.

[Vin81] Vinogradov I.M. (1981): Basic Number Theory. 9th ed. Moscow: Nauka, 1981. English transl. 1954. Zbl.547.10001, Zbl.65,270.

[VK] Vinogradov I.M., Karatsuba A.A. (1984): The trigonometric sum method in number theory. Tr. Mat. Inst. Steklova 168 (1984), 4-30. English transl.: Proc.Steklov Inst.Math. 168 (1986), 3-30. Zbl.549.10027.

[Vla91] Vlăduţ, S. G.: Kronecker's Jugendtraum and modular functions. Translated from the Russian by M. Tsfasman. Studies in the Development of Modern Mathematics, 2. Gordon and Breach Science Publishers, New York, 1991. x+411 pp.

[Voj87] Vojta P. (1987): Diophantine approximations and value distribution theory. Lect. Notes Math. 1239 (1987). Zbl.609.14011.

[Voj91] Vojta P. (1991): Arithmetic and hyperbolic geometry. Proceedings of the International Congress of Mathematicians, Vol. I, II (Kyoto, 1990), 757–765, Math. Soc. Japan, Tokyo, 1991.

[Vos77-98] Voskresenskij, V.E. (Kunyavskii, B.) (1998): Algebraic groups and their birational invariants. Transl. from the original Russsian manuscript by Boris Kunyavskii. Rev. version of "Algebraic tori", Nauka 1977. Translations of Mathematical Monographs. **179** (1998) Providence, RI: American Mathematical Society (AMS). xiii, 218 p. Zbl.499.14013, 0974.14034

[Wag86] Wagon, S. (1986): Primality testing. Math. Intell. 8, No.3 (1986), 58-61. Zbl.595. 10004.

[WaSm87] Wagstaff, S.S. Jr., Smith J.W. (1987): Methods of factoring large integers. Lect. Notes Math. 1240 (1987), 261-303. Zbl.613.10004.

[Wald96] Waldschmidt, M.: Sur la nature arithmétique des valeurs de fonctions modulaires Séminaire Bourbaki. Exp. No.824 (Novembre 1996)

[Wald2000] Waldschmidt, M.: Diophantine Approximation on Linear Algebraic Groups, Grundlehren Series 326, Springer Verlag 2000

[War36] Warning, E. (1936): Bemerkung zur vorstehenden Arbeit von Herrn Chevalley. Abh. Math. Semin. Univ. Hamburg 11 (1936), 76-83. Zbl.11,146.

[Wash78] Washington L.C. (1978): The non–p–part of the class number in a cyclotomic Γ–extension. Invent.Math. 49, No.1 (1978), 87-97. Zbl.403.12007.

[Wash82] Washington L.C. (1982): Introduction to Cyclotomic Fields. New York–Berlin–Heidelberg: Springer–Verlag, 1982. Zbl.484.12001.

[Wel80] R.O. Wells, Differential analysis on complex manifolds. Springer–Verlag, 1980.

[Wei40] Weil A. (1940): L'intégration dans les groupes topologiques et ses applications. Paris: Hermann, 1940.

[Wei48] Weil A. (1948): Variétés abéliennes et courbes algébriques. Paris: Hermann, 1948. Zbl.37,162.

[Wei49] Weil A. (1949): Number of solutions of equations in a finite field. Bull.Am.Math.Soc. 55 (1949), 497-508. Zbl.32,394.

[Wei51] Weil A. (1951): Arithmetic on algebraic varieties. Ann. Math., II.Ser. 53, No.3 (1951), 412-444. Zbl.43,270.

[Wei51] Weil A., Sur la théorie du corps de classes, J. Math. Soc. Japan, **3**, (1951).

[Wei52a] Weil A. (1952a): Jacobi sums as "Grössencharakteren". Trans. Am. Math. Soc. **73** (1952), 487-495. Zbl.48,270.

[Wei52b] Weil A. (1952b): Sur les "formules explicites" de la théorie des nombres prémièrs. Commun. Sémin. Math. Univ. de Lund (dédiée à M. Riesz) 1952, 252-265. Zbl.49,32.

[Wei57] Weil, A. (1957): Zum Beweis des Torellischen Satzes. Nachr. Akad. Wiess. Göttingen, **1**, 33-53 (1957).

[Wei64] Weil A.: Sur certains groupes d'operateurs unitaires, *Acta Math.* , **111**, (1964).

[Wei66] Weil A.: Fonctions zêta et distributions, *Séminaire Bourbaki*, **312**, (1966).

[Wei67] Weil A. (1967): Über die Bestimmung Dirichletscher Reihen durch Funktionalgleichungen. Math. Ann. 168 (1967), 149-156. Zbl.158,86.

[Wei71] Weil A. (1971): Dirichlet series and automorphic forms. Lect. Notes Math. 189 (1971). Zbl.218.10046.

[Wei72] Weil A. (1972): Sur les formules explicites de la théorie des nombres. Izv. Akad.Nauk SSSR, Ser. Mat. 36 (1972), 3-18. English transl.: Math.USSR, Izv. 6 (1972), 1-17 (1973). Zbl.245.12010, see also Zbl.49,32.

[Wei74a] Weil A. (1974a): Basic Number Theory. 3rd ed. Berlin–Heidelberg–New York: Springer–Verlag, 1974. Zbl.267.12001; Zbl.176,336.

[Wei74b] Weil A. (1974b): La cyclotomie jadis et naguère. Enseign. Math., II.Ser. 20 (1974), 247-263. Zbl.352.12006.

[Wei76] Weil A. (1976): Elliptic functions according to Eisenstein and Kronecker. Berlin–Heidelberg–New York: Springer–Verlag, 1976. Zbl.318.33004.

[Wei79] Weil A. (1979): Oeuvres scientifiques. Vol.I-III. Berlin–Heidelberg–New York: Springer–Verlag, 1979. Zbl.424.01027-29.

[Wey40] Weyl H. (1940): Algebraic Theory of Numbers. Princeton Univ. Press, 1940.

[Wi] Wiles, A. (1995): Modular elliptic curves and Fermat's Last Theorem. Ann. Math., II. Ser. 141, No.3, 443-551 (1995)

[Wi2000] Wiles, A. (2000): Twenty years of number theory. Mathematics: frontiers and perspectives, 329–342, Amer. Math. Soc., Providence, RI, 2000.

[Wil78] Williams H.C. (1978): Primality testing on a computer. Ars Comb. 5 (1978), 127-185. Zbl.406.10008.

[Wil82] Williams H.C. (1982): A $p+1$ method of factoring. Math. Comput. 39 (1982), 225-234. Zbl.492.10004.

[Wil84] Williams H.C. (1984): Factoring on a computer. Math. Intell. 6, No.3 (1984), 29-36. Zbl.548.10004.

[WD86] Williams H.C., Dubner H. (1986): The primality of $R1031$. Math. Comput. 47 (1986), 703-711. Zbl.602.10001.

[Will74] Williams, R.F.: Classification of subshifts of finite type. Ann. of Math. (2) **98** (1973), 120–153; errata, ibid. (2) **99** (1974), 380–381.

[Wun85] Wunderlich M.C. (1985): Implementing the continued fraction factoring algorithm on parallel machines. Math. Comput. **44** (1985), 251-260. Zbl.558.10001.

[Ya02] Cryptography: an introduction. Written by V. Yaschenko, N. Varnovskii, Yu. Nesterenko, G. Kabatyansky, P. Gyrdymov, A. Zubov, A. Zyazin and V. Ovchinnikov. Edited by V.Yaschenko. Translated from the 1998 Russian edition by Sergei Lando. Student Mathematical Library, 18. AMS, Providence, RI, 2002. x+229 pp.

[Yer77] Yershov Yu.A. (1977): Numeration theory. Moscow: Nauka, 1977. German transl.: Berlin: VEB dt. Verlag Wiss., Vol.I,1973; Zbl.281.02041; Vol.II,1976; Zbl.344.02031; Vol.III,1978; Zbl.374.02027.

[Yos03] Yoshida, H.: Absolute CM-Periods. Mathematical Surveys and monographs, vol. **106**, AMS, Providence, RI, 2003. x + 282 pp.

[Zag77] Zagier D. (1977): First fifty million prime numbers. Math. Intell. 0 (1977), 7-19. Zbl.392.10001.

[Zag81] Zagier D. (1981): Eisenstein series and the Riemann zeta–function. Automorphic forms, representation theory, and arithmetic. Colloq. Bombay 1979. (1981), 275-301. Zbl.484.10019.

[Zag94] Zagier D. (1994): Values of zeta functions and their applications, in "First European Congress of Mathematics", vol. II, pp. 497–512, Birkhäuser, 1994.

[Zar74] Zarhin, Yu.G. (1974): A finiteness theorem for Abelian varieties over functional fields of finite characteristic. (in Russian). Funk. Analiz, **8**, no.4, 31-34 (1974). Zbl.324.14009.

[Zar85] Zarhin, Yu. G. (1985): A finiteness theorem for unpolarized Abelian varieties over number fields with prescribed places of bad reduction. Inv. Math., **79**, 309-321 (1985). Zbl.557.14024.

[Zu] Zudilin, W.: Zeta values on the Web `http://wain.mi.ras.ru/index.html`

[Zu95] Zudilin, W.: On the estimates of the measure of linear independence for values of certain analytical functions, PhD Thesis, Moscow State University 1995

Index